One-Sample Large-Sample z Statistic for Mean (Sec. 6.B)

$$z = \frac{\bar{x} - \mu_0}{s/\sqrt{n}} \tag{6.1}$$

One-Sample Small-Sample t Statistic for Mean (Sec. 6.B)

$$t = \frac{\bar{x} - \mu_0}{s/\sqrt{n}} \quad df = n-1 \tag{6.2}$$

Two-Sample Large-Sample z Statistic for Difference of Means (Sec. 6.D)

$$z = \frac{\bar{x}_L - \bar{x}_U}{\sqrt{\dfrac{s_L^2}{n_L} + \dfrac{s_U^2}{n_U}}} \tag{6.6}$$

Two-Sample Small-Sample t Statistic for Difference of Means (Sec. 6.D)

$$t = \frac{\bar{x}_M - \bar{x}_T}{\sqrt{\dfrac{\Sigma(x_M - \bar{x}_M)^2 + \Sigma(x_T - \bar{x}_T)^2}{n_M + n_T - 2}\left(\dfrac{1}{n_M} + \dfrac{1}{n_T}\right)}} \quad df = n_M + n_T - 2 \tag{6.7}$$

Paired-Sample t Statistic for Difference of Means (Sec. 6.E)

$$t = \frac{\bar{d}\sqrt{n}}{s_d} \quad df = n-1 \tag{6.9}$$

One-Sample Chi-Square Statistic for Standard Deviation (Sec. 6.F)

$$\chi^2 = \frac{(n-1)s^2}{\sigma_0^2} = \frac{\Sigma(x - \bar{x})^2}{\sigma_0^2} \quad df = n-1 \tag{6.10}$$

Two-Sample F Statistic for Difference of Standard Deviations (Sec. 6.F)

$$F = \frac{s_N^2}{s_T^2} = \frac{\text{larger } s^2}{\text{smaller } s^2} \quad dfn = n_N - 1 \quad dfd = n_T - 1 \tag{6.12}$$

One-Sample Large-Sample z Statistic for Proportion (Sec. 6.G)

$$z = \frac{p - \pi_0}{\sqrt{\dfrac{\pi_0(1 - \pi_0)}{n}}} \tag{6.13}$$

(Continued on back endpapers)

Statistical Analysis

RESOLVING DECISION PROBLEMS
IN BUSINESS AND MANAGEMENT

Statistical Analysis

RESOLVING DECISION PROBLEMS IN BUSINESS AND MANAGEMENT

Stephen A. Book

CALIFORNIA STATE UNIVERSITY,
DOMINGUEZ HILLS

Marc J. Epstein

CALIFORNIA STATE UNIVERSITY,
LOS ANGELES

Scott, Foresman and Company

GLENVIEW, ILLINOIS

DALLAS, TEX.   OAKLAND, N.J.   PALO ALTO, CAL.   TUCKER, GA.   LONDON, ENGLAND

Library of Congress Cataloging in Publication Data

BOOK, STEPHEN A.
    Statistical analysis.

    Includes bibliographies and index.
    1. Commercial statistics. 2. Statistics.
I. Epstein, Marc J. II. Title.
HF1017.B64   519.5'024658   81–14602
ISBN 0–673–16002–5   AACR2

1 2 3 4 5 6 7 8-RRC-88 87 86 85 84 83 82 81

Cover photo: © Bonnie Freer/Peter Arnold, Inc.

To our children

Contents

Preface

In writing this book, we have made the extensive use of realistic examples and exercises (rather than mathematical theory) the major tool for illustrating the role of statistical analysis in business and management decision making. Therefore, the primary effort of our task has been to provide a large selection of examples and exercises that are especially suitable for (1) self-study by the students to determine whether or not they understand the material discussed in class, (2) classroom use by the instructor to enhance his or her lectures with real, interesting business situations, and (3) homework assignments and in-class examinations.

We intend that our book be read by students; virtually all the examples and text discussions refer to actual business problems set in a format that makes them interesting, relevant, understandable, and informative. Many of them are drawn from the authors' business and consulting experience and research and cover the areas of manufacturing, service, and not-for-profit. They are written in such a way that the beginning student (without extensive prior business and management coursework) will comprehend the situation and be able to relate it to his or her other courses and future career.

In our desire to provide problem realism, however, we have not forgotten the mathematical orientation and background of most business students. We cover all the topics typically found in the business core statistics course by fully explaining, at the student's level, any mathematical techniques he or she needs to know. Our approach to the subject matter is rooted in the philosophy that statistics cannot be taught in isolation from other business courses, but rather that all business courses, including statistics, must complement and reinforce each other.

We would like to express our appreciation to several reviewers for their assistance in shaping early versions of the manuscript into a completed text. We are especially grateful to our colleagues, Professors Paul Baum of California State University, Northridge, and Nicholas Farnum of California State University, Fullerton, for their careful reading of the manuscript through a number of drafts and for providing explicit comments, suggestions, and ideas that ultimately led to substantial revision and improvement of a large portion of the text. We are confident that they will recognize the effects of their work, and we hope that they are pleased with the result. We would also like to thank Professors Gary D. Kelley of Texas Tech University, Betty Whitten of the University of Georgia, and Jeffrey Mock of Diablo Valley College for their valuable remarks on the manuscript as a whole, and Edward Stafford of the University of South Carolina for his comments on a very early draft.

It was an enjoyable experience for us to work with our editors, Roger Holloway and Hal Humphrey, who guided the manuscript through its developmental and production phases, respectively. Their insistence on professionalism, precision, and thoroughness, from us as well as from their staffs, provided an appropriate backdrop for our writing efforts.

Finally, we would like to thank *Forbes* magazine, John Wiley & Sons, the American Statistical Association, the *Biometrika* Trustees, the Rand Corporation, and the American Society for Quality Control for their permission to reproduce certain statistical tables referred to in the course of the text. The source of each such table is specifically mentioned when the table itself appears.

Stephen A. Book
Marc J. Epstein

Statistical Analysis

RESOLVING DECISION PROBLEMS
IN BUSINESS AND MANAGEMENT

I

DESCRIPTIVE
TECHNIQUES

1

Organizing Statistical Data

Statistics exists to solve problems arising from the need to make decisions based on incomplete or uncertain information. Its goal is the organization, analysis, and explanation of facts and figures derived from studies in the social, behavioral, natural, and management sciences. It attempts to discover the general rules of behavior of economic and managerial phenomena for the dual purposes of understanding these phenomena as they are and of controlling and improving them.

Statistics as an organized system of analysis originated almost a century ago in attempts to understand the factors influencing industrial and agricultural production, in order to better control these factors and so to increase efficiency and production at reasonable cost. Today the aim of statistics is no different, although it has expanded its vista to include almost all aspects of business and management decision making.

SECTION 1.A   THE ROLE OF STATISTICS IN DECISION MAKING

Consider a business executive faced with a "make or buy" option. The executive must decide between manufacturing a product component in the company's own factory or subcontracting to another organization the job of producing the component. Naturally, he wants to choose the better procedure of the two. Unfortunately, however, he must answer several questions that are more complicated than the simple, "Which procedure is better?" A few of these questions are:

1. What do we mean by "better" and how can we measure such a quality?
2. Would the first option be "better" under some circumstances, while the second is "better" in other situations?
3. If the "better" procedure turns out to be more expensive, is it superior enough to warrant the increased cost?

In the case of the make-or-buy decision, at least two factors bear on determining which procedure is better: a quality factor and a cost factor. Will the company itself or an outside subcontractor produce a better-quality component? What is the relative cost of producing the component within the company, as compared with the cost of buying it on the outside? In the final analysis, the only true measure of which procedure is better for the company is how each would affect that portion of the company's profit that can be attributed to the component in question.

To estimate the particular component's contribution to the company's profits, a market research study could be undertaken with the objective of determining how much additional cost the market is willing to absorb in exchange for an increase in quality. As part of the study, we would conduct a sales test among prospective customers and keep detailed records regarding the level of quality they expect and the prices they are willing to pay.

Some questions will still remain unresolved, however. It may be that, had we chosen different customers for the sales test or even different ways of test-marketing our product, the "better" procedure would have come out second best. Maybe even if we conduct exactly the same market research study next year (even with the same customers), the results will be reversed, perhaps because of changing opinions and economic circumstances. How can we be sure that the procedure we have decided is "better" really is? Well, in truth, we can't be 100% sure. We can improve our degree of certainty, if we are willing to commit the time and resources necessary, by expanding the scope of our research to include several more sets of prospective customers and by carrying out the test over several time periods. How many customers and how many time periods? That depends on how certain we want to be that we have eliminated all nonessential, irrelevant, and random factors that could influence the survey results. The more information we accumulate and digest, the more certain we can be of our decision.

A different aspect of designing a statistical analysis is illustrated by the example of an art professor who has been experimenting with a new type of graphic art. To test the thought that the material is suitable for mass distribution, she conducted a survey of her 72 university students. She found that 64 of them liked the artwork and expressed a willingness to buy it. The professor was aware that national pollsters could successfully predict the outcome of an election using a very small sample of voters. She concluded that her sample of students ought to give her enough information to decide whether or not mass production of her graphics would be profitable. She was about to seek the aid of someone in the business school to help her determine cost and pricing information when a second-year business student in one of her art classes suggested that a sample survey of only her own students is not likely to produce valid information.

A number of fallacies in the professor's analysis of the problem should be readily apparent. But it is still clear that, for making even the most basic business decisions, *appropriate statistical data is essential*. Why is it that, with a small sample, national pollsters can form reasonably good estimates of the total vote count? How do we recognize whether or not the data we have is useful in making economic decisions? From a local entrepreneur trying to start a new business, to a major manufacturer involved in large-scale production, the business world has evolved to the point where businesses of all sizes and complexities must obtain and utilize statistical information in their daily decision-making processes.

The Importance of Statistical Analysis to Business

Statistical analysis is used to organize material required for informed decision making. Its importance is magnified in business, because most decisions are made under conditions of uncertainty. Uncertainty in itself exists in many aspects of our modern life. When we toss a coin to start a football or soccer game, we are uncertain of the outcome. When workers go out on strike, they are not sure they will be rehired again. When we board an airplane, bus, or car, or even cross the street, we are not really certain that we will arrive at our destination safely. While we live with uncertainties on a daily basis, we still seek the most certainty possible in these situations. For business decision makers, statistics is an organizing tool that diminishes uncertainty.

The nature of business requires that decisions about the future be made now. Typically these are monetary decisions, both short and long-term. An engineering company must decide how much office space it will need in the next five years while it renegotiates its present lease or decides to build its own office building. A life insurance company determines rates and premiums based on life expectancies sometimes 30 years in the future. The monetary considerations are vital—the wrong decisions could cost the company a lot of money.

The following examples provide additional illustrations of business situations requiring statistical analyses.

Accounting. An accounting firm doing an audit for a large national corporation in the apparel industry is involved in verifying receivables. The company has receivables all over the United States ranging in amounts from

$15 to $750. To comply with auditing standards, the accounting firm has to send out letters of verification. There are approximately 100,000 accounts receivable, and it would be too costly and time consuming to send a letter to each account. The problem is to determine how many letters to send and to whom to send them. How many responses would be required to make the survey accurate? Should letters be sent only to accounts with high balances, low balances, 30-day-old balances, over-90-day balances, or what combination of these?

Management. A small home builder in the Midwest has the opportunity to purchase 100 acres on the outskirts of a medium-sized city. The proposed construction and sales project will take two years to complete. He will either have to transfer 30 management-level people and their families to this city or plan on commuting people every week to and from the construction site 500 miles from their home office. Furthermore, the builder is not sure whether it would be more economical to use construction workers from his home city or to hire new crews at the construction site. The sales manager argues that they should employ local people for construction and sales since that would make the community more receptive to the project. The construction manager wants to use crews from the home city, as they already know the company's procedures and would therefore require less supervision. There are a lot of real issues involved here, and yet the company president senses too much individual opinion and not enough actual facts. He needs to know, among other things, whether the local real estate market can absorb the new homes and what are the prevailing wage rates for local construction workers versus wages at the home office.

Marketing. A soft-drink company is planning to market a new "anti-cola" soft drink which, they think, tastes better than those currently available. They are building an advertising campaign aimed at the young adult and young family (ages 20–35) market. Some executives feel that television advertising would be the most effective. All agree intuitively that an initial price break on the product would induce customers to try it. Later, after they've attracted a good share of the market, they would raise the price to a competitive level. One of the undecided issues is where to test market the product. (Before commiting themselves to a multimillion-dollar sales campaign, they want to be relatively sure that the product will sell.) Which city has a good-size population in the 20–35 age range, one whose income level is representative of the country at large? In order to test the price-break idea, do they need to have both a competitive-price market and a lower-price market? How will they determine whether or not the price-break idea works? Will sales decrease significantly as they bring the price up to the competition's level? If so, by how much?

Planning. An automobile manufacturer is about to start his planning for three new models to come on the market in five to seven years. The process must begin now because it takes a long time for design, testing of prototypes, retooling of assembly plants, etc. The company has suffered some business reverses in the past (for example, they came out with subcompact models in years when consumers wanted big, flashy cars). To avoid the same mistake, they want to make sure that they do adequate planning and research now. The kinds of things they need to know are as follows: How much demand for autos will there be in five years? Will consumers want big cars, little cars, sports cars, diesel cars, or what? Will gasoline be so scarce that economy cars will be in universal demand? What type of consumer will dominate the new car market—teenagers, young single adults, families, or middle-aged couples? And then there is the basic question of how much the car will cost to produce in five years. The manufacturer will need to forecast wage scales, raw material costs, advertising costs, and other production costs.

Processes of Statistical Analysis

Many of these business decisions concerning the future are based on the gathering of historical data, its statistical organization, and the particular opinions and personal experiences of the decision maker. Historical data is raw information about the past (for example, how many new cars were bought last year). Raw data in itself is not particularly useful to the above car manufacturer, but organizing this data into age groups of automobile purchasers, type of car purchased, and purchaser income level is.

While we have noted that statistics organizes information, it is also important to observe that it provides a thought pattern for training decision makers. Decision makers who know statistics can interpret and understand the data and can ask the appropriate questions whose answers are needed in making decisions.

Many people think that business decisions are now easier to make due to the increased reliance on computers to sort and analyze masses of historical data. Computers, as we know, can digest, store, and categorize information at far greater speeds and volume than can people. And with each generation of computers, this speed and volume increases and the associated costs decline. We often forget, though, that computers are merely machines. They do only what we tell them to do. So again it is the decision maker on the scene, a human being, who must understand what questions to ask and how to ask them. Only in this way will the computer be able to provide useful information for decision making.

As we remarked at the beginning of this discussion, statistics exists to solve problems. It is both a descriptive and an analytical tool, much like accounting, to be used in helping decision makers to formulate and then resolve a business or management problem in a coherent framework. In searching out and taking advantage of whatever factual information may be available, statistics provides both a logical approach to a business problem and an effective way of organizing the facts necessary to solve it.

A statistical problem begins with a question of interest in one of the various areas of management such as accounting, advertising, economics, finance, marketing, or production. However, the methods and techniques of statistical analysis enter the arena only after some data bearing on the problem are accumulated.* After the data are collected, the next step is often the organization of the data into an understandable and easily communicable format. Sometimes the data can be expressed by a pictorial or graphical structure that allows them to be perceived visually by persons not very familiar by training or experience with numerical descriptions. Visual methods, when available, are excellent vehicles for communicating statistical information quickly and clearly to those not trained in numerical thinking.

To persons accustomed to thinking in quantitative terms,† however, numerical descriptions are infinitely more useful than visual methods of data description. Numerical descriptions of data, as we will discover in future chapters, communicate a wealth of detailed and precise information that belies the relative simplicity of the methods involved.

It remains nevertheless true that visual descriptions of data, in conjunction with numerical descriptions and analysis, are very useful for purposes of communicating information. This is especially true in business organizations because of the complexity of the data, the need to understand a developing situation quickly, and the relative lack of statistical expertise on the parts of some of the executives who have to base their decisions on proper utilization of the data. We introduce in this chapter some of the important techniques of pictorially organizing a set of data so as to more easily convey the information contained therein.

SECTION 1.B.   PIE CHARTS

Segment reporting—reporting of information classified according to the various segments of the organization—has been developed in recent years to cope with the large sizes and complexities of business organizations. A reader of the annual report of a conglomerate can thereby see more easily how total sales, for example, are distributed throughout different components of the enterprise. Table 1.1 shows how annual sales of one large diversified retailing company are divided among its several operating divisions. An effective means of making the data of Table 1.1 visually more appealing and understandable is a *pie chart*.

We construct a circle (the "pie") and slice it into wedges in the usual way of dividing a pie into pieces, with each piece representing one of the

---

*The proper methods of accumulating data are studied in a distinct branch of statistics called *statistical sampling*, a brief introduction to which will be presented in Section 5.H.

†The objective of this book is to turn you into one of them!

TABLE 1.1 Annual Sales of a Diversified Retailer

Division	Annual Sales (millions of dollars)
Home products	50
Apparel	100
Commercial products	200
Food services	250
Frozen foods	150
Confections	100
Personal care	150
Total Corporate Sales	1,000

categories in the "Division" column. The size of each slice of pie is proportional to the number of millions of dollars of annual sales achieved by the division represented.

You may recall from some acquaintance with geometry that a circle can be divided into 360 degrees, denoted as 360°. This can be seen easiest in the fact that there are four 90° angles meeting at the center of the circle as in Fig. 1.1. In Fig. 1.1, the entire circle has 360°, so that each wedge (technically

FIGURE 1.1   How a Circle Has 360°

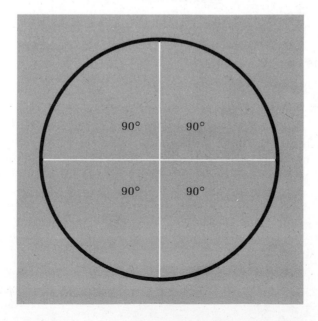

TABLE 1.2   Preliminary Computations for Constructing a Pie Chart

Division	Annual Sales (millions of dollars)	Percentage of Total Sales	Times 3.6°	Degrees Allotted
Home products	50	5%	× 3.6°	18°
Apparel	100	10%	× 3.6°	36°
Commercial products	200	20%	× 3.6°	72°
Food services	250	25%	× 3.6°	90°
Frozen foods	150	15%	× 3.6°	54°
Confections	100	10%	× 3.6°	36°
Personal care	150	15%	× 3.6°	54°
Sums	1,000	100%	× 3.6°	360°

called a *circular sector*) representing 1% of the circle has 1% of 360° or (.01)(360°) = 3.6°. Therefore, in constructing a pie chart for a set of data, we should allot 3.6° to each 1% of the data. And if a particular category contains n% of the data, it should be represented by a circular sector having (n)(3.6°). A category with 20% of the data, for example, should be allotted (20)(3.6°) = 72°.

In constructing a pie chart to illustrate the data of Table 1.1, we need to know the number of degrees that should be allotted to each category. An organized method of coming up with this information is presented in Table 1.2. The first two columns of Table 1.2 are merely the two columns in Table 1.1, and the remaining columns give the percentage of the circle and the number of degrees that should be allotted to each category.

The first two steps involved in the construction of a pie chart are illustrated in Fig. 1.2. Looking first at the category "Home Products," we see that it is necessary to allot 18° to this category. Because 18° is one-fifth of 90°, we can see that the circular section representing "Home Products" should occupy one-fifth of the first quarter (more technically, the first *quadrant*) of the circle. The second step, specifying the wedge for "Apparel," is accomplished by observing that the allotment of 36° is equivalent to two-fifths of 90°, namely twice as much as that allotted to "Home Products." If we continue to mark off in a counter-clockwise direction the sectors pertaining to each category, we come up with the completed pie chart appearing in Fig. 1.3.

The following exercises contain further examples of the construction of pie charts to illustrate business data.

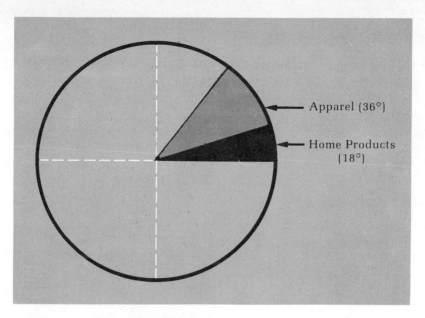

FIGURE 1.2   Starting the Pie Chart

FIGURE 1.3.   Pie Chart Illustrating Annual Sales of a Diversified Retailer

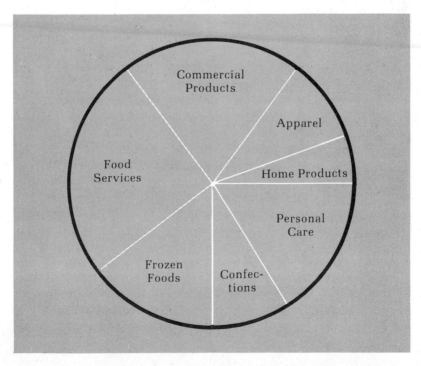

EXERCISES 1.B

1.1. A real estate agency is in the process of conducting a study of residential buildings in order to forecast future demand for family housing. The number of
buildings in each category within the city limits is listed in the following table:

High-rise apartments	200
Multiple family houses	240
Single family houses	320
Mobile homes	40

Draw a pie chart illustrating the data, carefully noting the proper number of
degrees that should be allotted to each category of residential building.

1.2. New cars sold in the United States in 1975 could be classified as to manufacturer
into the following five categories:

Manufacturer	Cars Sold (millions)
General Motors	3.0
Ford	2.0
Chrysler	1.5
American Motors	1.0
Foreign	2.5

Construct a pie chart that illustrates the information contained in the data.

1.3. The following table lists typical expenses for a family of four whose income is
$1,700 per month:

Food	$250
Housing, utilities	250
Furniture, clothing	150
Transportation	100
Entertainment	50
Savings, insurance	75
Medical	25
Taxes	300

Draw a pie chart that illustrates the family's budget.

1.4. Insurance industry analysts aim at a market for life insurance in the United
States consisting of a population of 140 million. An age breakdown of this group
is as follows:

Age Group	Millions
18–24	28
25–64	98
65 and older	14

Draw a pie chart to illustrate the age distribution of the insurance market in the U.S.

1.5. During a recent year, the seven biggest independent oil refiners in the United States had the following outputs:

Company	Barrels per Day (thousands)
Ashland	330
Clark	90
Oil Shale of L.A.	70
American Petrofina	65
Commonwealth	60
United	45
Apco	40
Others	100

Construct a pie chart showing each company's share of the production.

1.6. One diversified industrial corporation with total sales of $600 million has four major divisions whose portions of the total were as follows last year:

Division	Sales (millions)
Steel	288
Power equipment	180
Fluid control	96
Sports equipment	36

Construct a pie chart illustrating how the company's sales were divided according to division.

1.7. In late 1977, *U.S. News & World Report* asked 2000 persons attending the annual convention of the National Association of Business Economists what they felt would be the most important economic problem the country would face over the succeeding five years. The responses were as follows:

Inflation	780
Unemployment	94
Excessive government controls	566
Supply or price of energy	290
Capital shortage	146
Other	124

Construct a pie chart illustrating the range of responses.

1.8. The Newspaper Advertising Bureau surveyed 800 customers of banks and thrift institutions to find out what causes an individual to switch an account from one institution to another. The data follow:*

Reason for Switching	No. of Individuals
Change in residence or job	560
Frequent bank errors	46
Poor service	38
Rude or unhelpful employees	31
Impersonal or cold service	26
Higher interest elsewhere	29
Free checking elsewhere	19
Other varied reasons	51

Draw a pie chart illustrating the data.

1.9. For the time period of November 1–10, 1977, the number of cars sold by the four major U.S. auto manufacturers were as follows:†

Company	Number of Cars Sold
General Motors	143,927
Ford	72,925
Chrysler	29,926
American Motors	5,370
Total	252,148

Construct a pie chart illustrating how the four major U.S. auto manufacturers divided up the sales of nonimported cars in the time period indicated.

1.10. Data on the region of destination of Canada's exports in 1975 yield the following information:**

---

*Reported in *Bank Systems & Equipment*, October 1977, p. 12.

†*The Wall Street Journal*, November 16, 1977, p. 4.

**Bank of Montreal Business Review*, June 1976, p. 3.

Region	Percentage of Total Dollar Value of Canada's Exports
United Kingdom	5.4%
Six Common Market	6.9%
United States	65.6%
Japan	6.5%
All other regions	15.6%

Draw a pie chart based on the data.

SECTION 1.C   FREQUENCY DISTRIBUTIONS

Owners of small retail stores must anticipate the demand for the various items they carry in order to be reasonably sure that they will be able to meet normal demand without the risk of excessive overstocking. If asked how they know how much of a certain item to have in stock at a particular time, they often respond that they order and reorder by "gut feel." Those who are conducting a successful operation, however, usually have (in their heads) a rather sophisticated model of all the components of inventory costs—for example, the cost of stockouts, the lead times, the product interchange, etc. The development of this model typically follows a long period of gaining experience (including occasional costly mistakes) in conducting the day-to-day operations of the business.

The rest of us, however, along with managers in larger concerns, generally find it more efficient to develop a statistical model of sales levels than to wait for years of experience to pass. To get some idea of its daily demand for salad dressing, a restaurant located in a large downtown department store recorded daily usage of the item over 100 business days. It expected to use the information obtained to determine the likelihood of running out of the item at various supply and usage levels and to estimate the costs associated with proper supply maintenance. Table 1.3 contains the data collected by the restaurant.

Not much can be specified about the usage of salad dressing from a glance at the data. The problem is that we have an unorganized collection of numbers. No trends indicating possible usage levels are discernible. To make use of the mass of available data, we must first organize it into an understandable form.

To make sense out of the numbers in Table 1.3, we divide the range of data into a reasonable number of intervals. Then we count the number of data points falling into each interval. This procedure results in what is called a frequency distribution of the data, illustrated in Table 1.4. Because the minimum daily usage of salad dressing is at the level of 21 gallons and the maximum daily usage is at the 39-gallon level, a reasonable set of intervals

TABLE 1.3 Usage of Salad Dressing over 100 Days (data in gallons, rounded off)

25	32	28	26	30	21	36	27	32	32
33	23	31	29	33	31	29	33	28	31
27	29	29	33	27	28	32	23	30	25
29	32	21	24	30	31	28	32	31	34
31	26	30	30	34	26	26	29	32	22
24	31	28	32	28	37	30	35	27	33
28	30	33	30	31	30	32	28	31	28
29	26	32	24	29	32	30	22	34	26
32	27	27	34	29	31	33	39	30	33
30	28	31	30	37	31	25	32	33	28

TABLE 1.4 Frequency Distribution of Usage of Salad Dressing

Interval (number of gallons, rounded off)	Number of Days Having Usage in the Interval	Totals
20–22	IIII	4
23–25	THL III	8
26–28	THL THL THL THL III	23
29–31	THL THL THL THL THL THL IIII	34
32–34	THL THL THL THL THL I	26
35–37	IIII	4
38–40	I	1
Sum		100

to use would be 20-22, 23-25, 26-28, 29-31, 32-34, 35-37, 38-40. In choosing a set of intervals, you must watch out for the following points.

CHOOSING INTERVALS
1. Every data point must fall into *exactly* one of the intervals, no more and no less.
2. The number of intervals must not be too large, for then no substantial improvement in understanding the trend of the data would be made.
3. The number of intervals must not be too small, for then variations among the data points would be obscured.
4. In certain situations, it may not be necessary or even possible for all intervals to have the same length, but extreme fluctuations in the sizes of consecutive intervals should be avoided.

Having agreed upon the intervals, the next step is to tally the number of data points falling into each interval. Basically, this is all there is to the task of constructing a frequency distribution. The result appears in Table 1.4. As a check on the correctness of our counting, we can sum the number of days in the far right-hand column. If our counting was correct, we should get a total of 100 days, the number of days over which the restaurant's survey was taken.

Some useful information is apparent from a glance at the frequency distribution of Table 1.4, whereas almost nothing could be learned by looking at the original data of Table 1.3. For example, two things we learn from the frequency distribution are the following:

1. Approximately one-third of the time (34 days out of the 100), the usage level is near 30 gallons.
2. Usage exceeds 34 gallons only about 5% of the time (the four days corresponding to the interval 35–37 and the one day corresponding to the interval 38–40).

Histograms and Frequency Polygons

There are two common methods of pictorially communicating the information in a frequency distribution. The first of these is the *histogram* (or bar graph), and the second is the *frequency polygon* (or line graph). While the two types of graphs are closely related, there are important differences in the details of constructing them.

To construct a histogram, we modify the frequency distribution in such a way that there are no gaps between intervals. For the data in Table 1.4, we close the gap between the two intervals 20–22 and 23–25 by setting the point of division at 22.5, halfway between 22 (the end of the first interval) and 23 (the beginning of the second). To maintain the relative lengths of the intervals, we consider that the first interval starts at 19.5 and ends at 22.5, while the second starts at 22.5 and ends at 25.5. The modified frequency distribution appears in Table 1.5. Using that table, we construct the histogram of Fig. 1.4 by drawing bars one interval wide at a height corresponding to the number of days in that interval.

To construct the frequency polygon, we modify the original frequency distribution (Table 1.4) in a different manner. Here we find the midpoint of each of the intervals, for we will consider each interval to be represented by its midpoint. We also determine the midpoints of the intervals preceding 20–22 and following 38–40, as if those intervals really existed in Table 1.4. The frequency distribution modified in preparation for the frequency polygon appears in Table 1.6. The frequency polygon itself appears in Fig. 1.5. The frequency polygon is constructed by connecting dots that represent the

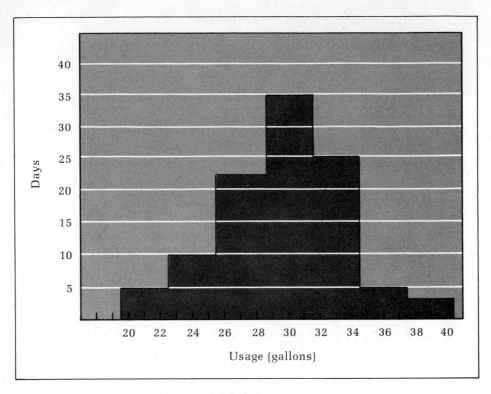

FIGURE 1.4   Histogram of Usage of Salad Dressing

TABLE 1.5   Frequency Distribution of Usage of Salad Dressing, Modified
for Construction of Histogram

Original Interval	Modified Interval Boundaries	Number of Days
20–22	19.5–22.5	4
23–25	22.5–25.5	8
26–28	25.5–28.5	23
29–31	28.5–31.5	34
32–34	31.5–34.5	26
35–37	34.5–37.5	4
38–40	37.5–40.5	1

TABLE 1.6   Frequency Distribution of Usage of Salad Dressing, Modified for Construction of Frequency Polygon

Interval	Midpoint	Number of Days
(17–19)	18	0
20–22	21	4
23–25	24	8
26–28	27	23
29–31	30	34
32–34	33	26
35–37	36	4
38–40	39	1
(41–43)	42	0

FIGURE 1.5   Frequency Polygon of Usage of Salad Dressing

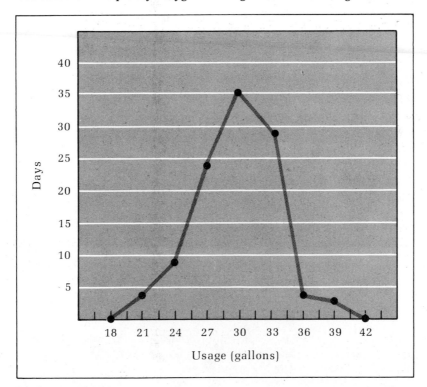

number of days that each usage level is attained, considering an interval of usage levels to be represented by its midpoint.

You should observe that the histogram and the frequency polygon both communicate the same information to a viewer. They both illustrate, for example, that usage of (and therefore demand for) salad dressing is rarely below 24 gallons per day, very often is between 26 and 34 gallons, and falls off sharply above 35 gallons. The value of a histogram or a frequency polygon as a visual aid really lies in the fact that the geometric areas set off by the graphing procedure represent the relative proportions of data points falling in the corresponding intervals.

There are some situations in which special conditions require alternative versions of the histogram or frequency polygon. In a certain type of problem, it may be sometimes appropriate to use intervals of unequal length. In other cases, there may be open-ended intervals or large gaps between intervals where no data points are located. Sometimes we may need a graph that records a running (cumulative) total of the data points. For a discussion of these and other variations of the pictorial representations we have introduced, you are invited to consult the indicated items in the bibliography at the end of this chapter.

Drawing on Additional Information

At some point in the analysis of an actual business situation, it may be useful to draw additional information out of the data; it might be helpful to further refine the data to show, say, usage on different days of the week. We know, for example, that retail stores generally do more business on Friday and Saturday than they do on Monday and Tuesday. Subject, of course, to variations depending on product and location, such information would allow a manager to arrange inventory and staffing in preparation for the anticipated large weekend volume.

Other questions may arise about seasonal fluctuations in usage. Because we have only 100 days worth of data in Table 1.3, we are unable to determine seasonal factors. If we believe that such information would be important to have, we could continue collecting data over several seasonal cycles. The additional data collection could proceed even while we are operating on the basis of information gained from the 100-day study.

EXERCISES 1.C

1.11. In December 1976, *Business Week* published some 1977 forecasts of economists and econometric models. One of the items included in the forecasts was the percentage by which prices would increase. The predictions were as follows:

7.3	5.2	5.6	5.0	5.3	5.4
5.8	5.3	5.5	5.5	6.2	5.6
6.1	5.4	5.3	5.5	5.8	5.2
5.3	5.2	5.1	5.1	5.7	5.5
6.3	6.3	5.4	5.1	5.3	5.5
5.5	6.0	4.9	4.5	5.4	

a. Use the data to construct a frequency distribution.

b. Draw a histogram that illustrates this frequency distribution.

c. Draw a frequency polygon of the data.

1.12. A popular index of common stocks traded on the New York Stock Exchange shows that current values (in dollars) of the 500 stocks involved have the following frequency distribution:

Dollar Value	No. of Stocks Having Value in Interval
0.00 to 9.99	51
10.00 to 19.99	98
20.00 to 29.99	101
30.00 to 39.99	151
40.00 to 49.99	48
50.00 to 59.99	25
60.00 to 69.99	13
70.00 to 79.99	8
80.00 to 89.99	4
90.00 to 99.99	1

Construct a frequency polygon based on the frequency distribution of stock prices.

1.13. The 50 employees of a small factory have yearly earnings distributed as follows:

Earnings Level	No. of Employees at That Level
$ 0.00– 2,999.99	3
3,000.00– 5,999.99	5
6,000.00– 8,999.99	18
9,000.00–11,999.99	15
12,000.00–14,999.99	6
15,000.00–17,999.99	0
18,000.00–20,999.99	3

Draw a frequency polygon that illustrates the data.

1.14. The U.S. Department of Commerce has released information on the 50 largest American cities ranked by median family income, based on the 1970 census data. The incomes were distributed as follows:

Income Level	No. of Cities
$ 7,000– 7,999	8
8,000– 8,999	9
9,000– 9,999	17
10,000–10,999	13
11,000–11,999	2
12,000–12,999	1

a. Draw a histogram that illustrates the data.

b. Construct a frequency polygon based on this data.

1.15. In late 1977, *U.S. News & World Report* asked 2,000 persons attending the annual convention of the National Association of Business Economists what they considered to be the lowest unemployment rate that was an appropriate goal of economic policy. The responses were classified in the following frequency distribution:

Interval	No. of Responses
2.50% – 3.49%	58
3.50% – 4.49%	192
4.50% – 5.49%	886
5.50% – 6.49%	740
6.50% – 7.49%	124

a. Construct a histogram that pictorially illustrates the frequency distribution.

b. Draw a frequency polygon based on the frequency distribution.

1.16. The United States appears to be inundated with telephones. The following set of data indicates the number of phones per 100 people in each of the fifty states:

52	62	66	54	57
46	58	65	70	58
61	71	66	56	63
52	62	48	59	63
73	63	65	64	60
68	65	59	65	62
74	51	66	63	66
75	55	79	69	48
73	57	65	60	60
64	70	74	55	65

a. Construct a frequency distribution using intervals of 5 units, beginning at 45.

b. Draw a histogram that illustrates this frequency distribution.

c. Draw a frequency polygon.

1.17. To assess the desirability of making substantial investments in rental property in a major U.S. city, an investment counseling firm selected a random sample of 100 housing units in the city and rated each according to quantity and quality of plumbing, electrical, and other physical facilities. Each unit was assigned a score from 0 to 100, and the resulting ratings had the following frequency distribution:

Score	No. of Units
0– 19	30
20– 39	15
40– 59	5
60– 79	10
80–100	40

a. Draw a histogram that illustrates the data.

b. Construct a frequency polygon based on the data.

1.18. A corporation psychologist conducted an experiment aimed at finding out how fast a newly hired employee could be taught a rather complicated quality control procedure. One hundred new hirees participated in the study. The following data records the number of attempts necessary for each to master the procedure.

1	6	17	11	4	18	24	34	2	26
4	2	7	3	12	5	10	1	15	3
16	5	3	8	9	13	1	20	2	14
6	2	8	4	9	5	14	1	2	10
1	7	7	4	5	10	5	9	15	3
6	25	3	8	8	1	6	13	19	10
20	2	7	12	4	9	2	7	9	18
1	6	11	3	19	5	8	3	8	10
21	30	2	5	4	7	17	23	4	9
10	1	4	3	6	16	22	28	32	5

a. Set up a frequency distribution of the number of attempts required to learn the lesson, using the intervals 1–5, 6–10, 11–15, 16–20, 21–25, 26–30, 31–35.

b. Construct a frequency polygon from this frequency distribution.

1.19. It is sometimes useful to construct a frequency distribution that records a running total of data points up to a certain level. For example, from the stock price data of Exercise 1.12, we may want to ask how many stocks were priced below

$50. The distribution that is constructed to answer questions like this is called a *cumulative frequency distribution* and begins as follows:

Dollar value	No. of Stocks Having Value as Indicated
below 10	51
below 20	149
below 30	250

a. Complete the cumulative frequency distribution.

b. A graph (called an *ogive*) of the cumulative frequency distribution may be constructed in a manner similar to the way a frequency polygon is constructed by appending the next lower interval as follows:

Dollar Value	No. of Stocks Having Value as Indicated
below 0	0

and then placing dots 0 units above 0, 51 units above 10, 149 units above 20, etc. We then "follow the dots." Construct the ogive of the frequency polygon of part a.

1.20. Based on the earnings data of Exercise 1.13,

a. construct the cumulative frequency distribution;

b. draw the ogive.

1.21. Using the property rating data of Exercise 1.17,

a. prepare the cumulative frequency distribution;

b. construct the ogive.

1.22. From the number-of-attempts data of Exercise 1.18 and the frequency distribution constructed there,

a. produce the cumulative frequency distribution;

b. draw the ogive.

SECTION 1.D   SCATTERGRAMS

If we obtain a set of data that represents specific measurements taken on a business situation, we can illustrate it using either pie charts or frequency distributions, as appropriate. Many questions in business, however, involve a relationship between two different characteristics, and we need some sort of pictorial representation that displays this relationship.

TABLE 1.7   Effect of Advertising on Toothpaste Sales

Region	Advertising Expenditures (millions of dollars)	Sales Volume (millions of dollars)
Northeast	1.5	8
Southeast	2.0	10
Midwest	2.5	14
Central states	1.5	10
Northwest	3.0	15
Southwest	2.0	11
Far west	2.5	12

As an example of such a situation, look at the advertising department of a major manufacturer of personal care products, which has recently been assigned the task of developing a marketing strategy and an advertising budget for a soon-to-be-introduced brand of toothpaste. In addition to the results of a marketing survey, the department has at its disposal data on the regional advertising and sales of a brand that was introduced a few years back. The department would like to use those data, which are presented in Table 1.7, to try to determine the relationship (if there is any) between the dollar amount spent on advertising and the dollar amount of sales volume. Without some organization of the data, as we have noted in previous sections, the thrust of the information is not clear.

Regional differences in the dollar amounts spent on advertising are not, of course, the only factors accounting for regional differences in sales volume. Other relevant factors might be regional differences in taste preference, attitudes toward advertising psychology, and timing. Nevertheless, as a first step toward deciding upon a budget, an analysis of the data in Table 1.7 would be useful.

The method of illustration of paired data used in Table 1.7 is called a *scattergram*. The data of Table 1.7 are displayed on the scattergram of Fig. 1.6, which is composed of a horizontal axis on which advertising expenditure levels are marked off and a vertical axis on which sales volume is indicated. Each region is represented by a single dot indicating both its advertising expenditure and its sales volume. For example, the dot for the Southeast region is 2.0 units to the right of the vertical axis because the Southeast received advertising expenditures of $2.0 million. Similarly, that dot is 10 units above the horizontal axis because the Southeast's sales volume was $10 million.

From the scattergram in Fig. 1.6, we can see that the general trend of the data is from lower left to upper right. This indicates that a large advertising expenditure can be expected to result in greater sales volume than would a

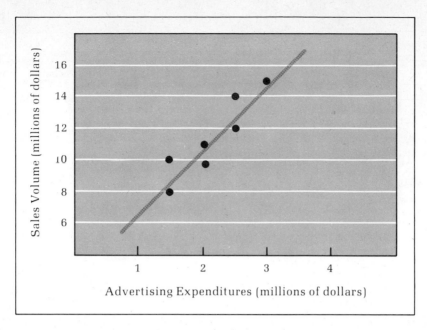

FIGURE 1.6   Scattergram of Advertising Expenditures vs. Sales Volume

small expenditure. We say in this situation that sales volume is *positively correlated* with advertising expenditure.

The scattergram in Fig. 1.7, portraying the trend of the data from upper left to lower right, illustrates a set of *negatively correlated* data. That set of

FIGURE 1.7   Scattergram of Sales Volume vs. Cost of Production

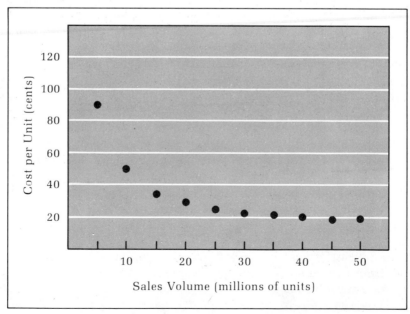

TABLE 1.8   Effect of Sales Volume on Cost of Production

Sales Volume (millions of units)	Cost of Production (per unit, cents)
5	90
10	50
15	37
20	30
25	26
30	23
35	21
40	20
45	19
50	18

data resulted from a cost accounting study undertaken preliminary to a break-even analysis of projected costs of production and projected sales volume in millions of units. The data appearing in Table 1.8, as well as the scattergram in Fig. 1.7, illustrate the fact that cost per unit to the manufacturer tends to decrease as sales volume (and so production) increases. This is a typical average cost curve, and an attuned eye can learn much from the scattergram, for example, the facts that the fixed cost is $4 million and the variable cost per unit is 10 cents. In Chapter 8, we will present a more formal and precise way of determining the components of the cost curve.

Both sets of data discussed so far have demonstrated definite directional trends. Such trends are often called *linear trends,* because they can be symbolized by a straight line, the *trend line,* following the data in the direction of the trend. Sometimes, however, no such linear trend appears on the scattergram, as illustrated in the following examples.

EXAMPLE 1.1   Sales Ability.   It is an accepted fact that ability to succeed in a sales career in heavy industrial equipment is related to general quantitative ability in an unusual way. According to the prevailing theory, persons having a very low level of quantitative ability usually cannot succeed in a field dealing with equipment of such a complex and technical nature, while those having a very high level of quantitative ability often have difficulty communicating technical information to those responsible for purchasing the equipment. One major supplier wants to check on the validity of this theory.

SOLUTION.   The sales headquarters of the supplier gave seven prospective applicants a general mathematics exam. Their exam scores were later compared with scores assigned to the individuals based on the sales record they had established with the company. The comparative scores are presented in Table 1.9, and the scattergram appears in Fig. 1.8. The scattergram of Fig. 1.8, which exhibits what is called a

TABLE 1.9   Data Relating Quantitative and Sales Ability

Salesperson	Quant Level (based on test of ability)	Sales Level (based on actual sales record)
A	0	2
B	3	9
C	1	5
D	6	2
E	2	8
F	4	8
G	5	5

FIGURE 1.8   Scattergram of Quant Level vs. Sales Level

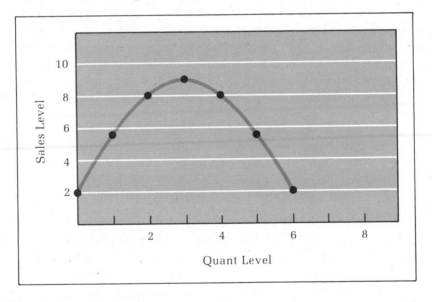

parabolic trend, shows clearly that sales ability is highest at the middle levels of quantitative ability and lowest for those having either very low or very high quantitative ability.

EXAMPLE 1.2.   Seasonal Retail Sales.   A retail firm whose merchandise sells well during the Christmas and Easter seasons, but not as well at other times of the year, might have sales totals for 12 consecutive

TABLE 1.10 Data Relating Quarter and Total Sales

Quarter	Total Sales (millions of dollars)
1	2
2	3.5
3	1
4	4
5	1.5
6	3
7	1
8	4.5
9	2
10	3
11	1.5
12	5

quarters (three-month periods), as listed in Table 1.10. The problem is to graphically illustrate the trend of the data.

SOLUTION. The scattergram of the retail sales data fluctuates up and down from quarter to quarter, indicating the presence of a *cyclical trend* in the original data of Table 1.10. The scattergram appears in Fig. 1.9.

FIGURE 1.9 Scattergram of Quarter vs. Total Sales

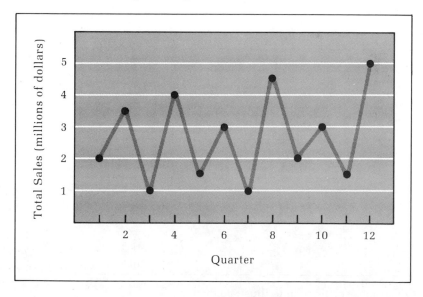

EXERCISES   1.D

1.23 The U.S. Bureau of the Census reported the following data on the median income
of men 25 years old and over by educational attainment:

Years of Education	Income Level
9.5	$ 3,500
10.4	4,500
11.3	5,500
12.0	6,500
12.1	7,500
12.3	9,000
12.6	12,250
12.9	20,000
16.1	25,000

Draw a scattergram illustrating the relationship between education and income
levels.

1.24. In a study of the efficiency of automobile engines with respect to consumption
of gasoline, one particular 3,990-pound car was operated at various speeds, and
the gasoline consumption (in miles per gallon) at each speed level was carefully
measured. The following data give the results of this experiment:

Speed (miles per hour)	Efficiency (miles per gallon)
30	20
40	18
50	17
60	14
70	11

Construct a scattergram of speed vs. efficiency, on the basis of the above data.

1.25. The Bureau of the Census gathered the following data on the income level of
households, based on the age of the head of the house. Selected data for 1974
follow:

Age of Head	Percent Having Income $10,000–$14,999
24	8.9
34	28.2
44	18.5
54	17.5
64	16.2
74	10.8

Construct a scattergram to illustrate the data.

1.26. Coronado Nautotronics of San Diego, bidding for the contract to produce the radar displays for the Navy's new fleet of patrol hydrofoils, collects the following information in an attempt to determine the cost curve for the radar displays:

Quantity Produced	Total Cost of Production Run
10	7.0
50	8.5
100	9.0
160	9.4
200	9.5
320	10.0
630	10.5
800	10.8

Draw a scattergram illustrating the cost curve of quantity produced vs. total cost of production run.

1.27. Twelve persons selected at random are asked the following two questions by a psychologist as part of a job faction study: a. "What is your hourly pay?" and b. "How would you rate your job satisfaction on a scale of 0 (low) to 10 (high)?" The results of the survey are presented below:

Respondent	A	B	C	D	E	F	G	H	I	J	K	L
Hourly pay (dollars)	8	2	6	4	4	20	10	6	6	4	11	15
Job satisfaction	6	4	5	4	3	8	7	4	1	2	9	5

Construct a scattergram of hourly pay vs. job satisfaction.

1.28. A social worker is assigned the job of finding out whether or not a particular six-month vocational training program is effective in increasing the income of its participants. The following data give the monthly incomes of 14 participants just about to begin training and also 18 months later (in hundreds of dollars):

Income at Start of Program	Income 18 Months Later
2.2	3.8
0.0	4.1
3.9	4.8
6.0	6.0
1.3	4.1
3.9	5.8
2.3	0.0
0.0	0.0
1.5	0.0
4.2	4.7
1.6	4.0
2.4	4.2
3.6	4.1
3.5	0.0

Draw a scattergram that illustrates the relationship between income at the start of the program with income 18 months later.

1.29. An agricultural research organization tested a particular chemical fertilizer to try to find out whether an increase in the amount of fertilizer used would lead to a corresponding increase in the food supply. They obtained the following data, based on seven plots of arable land:

Pounds of Fertilizer	Bushels of Beans
2	4
1	3
3	4
2	3
4	6
5	5
3	5

Draw a scattergram that illustrates the relationship between pounds of fertilizer and bushels of beans.

1.30. An 1891 study of the distribution of men and women in various occupations in England yielded the following data:*

Occupational Sector	No. of Persons Employed (thousands)	
	Men	Women
Agriculture & fishing	1,245	52
Building	833	8
Manufacture	2,609	1,530
Transport	816	10
Dealing	851	298
Industrial service	886	21
Public service & professional	563	265
Domestic service	359	1,632

Construct a scattergram to illustrate the data.

1.31. Demand for Napoleon rubies increases as the price goes down because more people are able to buy them. When the price is high, demand is also high since they are a much sought-after status symbol. At moderate prices, however, not too many are sold because the price is too high for many people yet not high enough to result in demand as status symbols. In particular, some recent data relating price and demand of these items went as follows:

*E. A. Wrigley (ed.), *Nineteenth-Century Society: Essays in the Use of Quantitative Methods for the Study of Social Data*, Cambridge University Press, 1972, p. 246.

Price per carat (hundreds of dollars)	0.5	1	3.5	5	7	8
Demand (thousands)	160	130	40	50	130	200

Draw the scattergram of data to illustrate the trend of the demand curve.

SUMMARY AND DISCUSSION

The objectives of Chapter 1 have been to introduce the role of statistical analysis in solving problems in business and management and to develop some methods of organizing and displaying statistical data.

We have shown how a mass of statistical data, originally a mere collection of numbers, can be organized in tabular or graphical form to yield a picture of the information contained within the data. The picture of a single set of data is revealed through pie charts, histograms, and frequency polygons, with each type of illustration presenting a slightly different perspective to the viewer. (Numerical, rather than pictorial, methods of describing a set of data will be dealt with in Chapters 2 and 4.) Relationships between paired sets of data may be illustrated on scattergrams, which point up the general behavior of one set of data with respect to the other (called the *correlation* between the sets of data). Such relationships will be studied in more detail in Chapters 8, 9, and 10.

CHAPTER 1   BIBLIOGRAPHY

GRAPHING TECHNIQUES

BENINGER, J. R., and D. L. ROBYN, "Quantitative Graphics in Statistics: A Brief History," *The American Statistician*, February 1978, pp. 1–11.

FIENBERG, S. E., "Graphical Methods in Statistics," *The American Statistician*, November 1979, pp. 165–178.

LUTZ, R. R., *Graphic Presentation Simplified*. New York: Funk & Wagnalls, 1949.

MAGEE, J. F., "Decision Trees for Decision Making," *Statistical Decision Series, Part III: Reprints from Harvard Business Review*. 1964, pp. 90–102.

MAGEE, J. F., "How to Use Decision Trees in Capital Investment," *Statistical Decision Series, Part III: Reprints from Harvard Business Review*. 1964, pp. 103–120.

McGILL, R., J. W. TUKEY, and W. A. LARSEN, "Variations of Box Plots," *The American Statistician*, February 1978, pp. 12–16.

SPEAR, M. E., *Practical Charting Techniques*. New York: McGraw-Hill, 1969.

READINGS IN THE APPLIED USE OF STATISTICS

HABER, A., et al. (eds.), *Readings in Statistics*. Reading, Mass.: Addison-Wesley, 1970.

MANSFIELD, E. (ed.), *Elementary Statistics for Business and Economics: Selected Readings.* New York: Norton, 1970.

TANUR, J. M., et al. (eds.), *Statistics: A Guide to Business and Economics.* San Francisco: Holden-Day, 1976.

NONTECHNICAL DISCUSSIONS OF STATISTICS

BARTHOLEMEW, D. J., and E. E. BASSETT, *Let's Look at the Figures.* Baltimore: Penguin Books, 1971.

CAMPBELL, S. K., *Flaws and Fallacies in Statistical Thinking.* Englewood Cliffs, N.J.: Prentice-Hall, 1974.

HUFF, D., *How to Lie with Statistics.* New York: Norton, 1954.

REICHARD, R. S., *The Numbers Game: Use and Abuse of Managerial Statistics.* New York: McGraw-Hill, 1972.

REICHMANN, W. J., *Use and Abuse of Statistics.* New York: Oxford University Press, 1961.

VON MISES, R., *Probability, Statistics, and Truth.* New York: Macmillan, 1939.

HOW TO READ TECHNICAL STATISTICAL ANALYSES

HUCK, S. W., W. H. CORMIER, and W. G. BOUNDS, JR., *Reading Statistics and Research.* New York: Harper & Row, 1974.

CHAPTER 1   SUPPLEMENTARY EXERCISES

1.32. The following data concern the sales volume of Europe's major automobile manufacturers during a recent one-year period.

Manufacturer	Sales Volume (billions of dollars)
Volkswagen	6.5
Renault	6.3
Daimler-Benz	5.1
Fiat	4.6
Ford-Europe	4.2
British Leyland	3.7
GM-Europe	2.6
Volvo	2.6
Chrysler-Europe	2.1

Construct a pie chart showing how Europe's major auto manufacturers divide up their portion of the sales "pie."

1.33. The following table shows the number of passengers carried during one month, according to a recent airline traffic study, by 11 of the nation's interstate carriers:

Airline	No. of Passengers (hundreds of thousands)
American	1.8
Braniff	0.7
Continental	0.6
Delta	2.2
Eastern	2.0
National	0.5
Northwest	0.7
Pan Am	0.1
TWA	1.3
United	2.9
Western	0.7

Draw a pie chart that illustrates the proportion of their combined domestic air market carried by each of the 11 major lines.

1.34. In late 1977, *U.S. News & World Report* and *Business Week* each asked samples of individuals closely connected with the behavior of securities markets how they expected the Dow Jones Industrial Average to look in 1978.

    a. *U.S. News & World Report* polled 476 members of the Securities Industry Association and asked each where the Dow would be in mid-1978. The responses were as follows:*

700–750	4.4%
750–800	11.1%
800–850	5.9%
850–900	28.9%
900–950	30.4%
950–1,000	14.1%
Over 1,000	5.2%

    Construct a pie chart illustrating the responses.

    b. *Business Week* surveyed investment advisers and big investors asking each where he or she felt that the Dow would be on December 31, 1978. The results were:†

Under 800	8.4%
800–900	18.3%
900–1,000	39.4%
1,000–1,100	12.7%
1,100–1,200	19.7%
Over 1,200	1.4%

    Draw a pie chart illustrating the data.

*U.S. News & World Report, December 12, 1977, p. 67. Do you note a defect in the way the magazine presented its data? How would a response of 850 be classified? As a point of fact, on June 30, 1978, the Dow stood at 818.95.

†Business Week, December 26, 1977, p. 67. Note the same defect in this magazine's mode of presentation. The Dow stood at 805.01 on December 31, 1978.

1.35. The following data give the price per earnings (P/E) ratios of the top 60 companies in stock market performance in 1975, as analyzed by *Forbes* magazine:*

3	7	14	25	22	17
28	15	20	4	6	4
14	32	5	14	14	7
16	4	3	4	9	8
14	14	12	14	7	8
8	5	19	35	32	12
21	11	13	8	6	16
10	5	12	12	4	5
33	7	11	5	7	14
7	7	17	27	19	5

a. Construct a frequency distribution of the P/E ratios.

b. Draw a frequency polygon that illustrates this distribution.

1.36. The Statistical Bureau of the Metropolitan Life Insurance Company has come up with the following percentages of families having children under the age of 18 for each of the 50 states:

55	62	54	62	56	58	53	57	52	56
67	57	58	53	54	55	54	54	53	56
56	55	57	56	56	47	58	56	53	52
51	56	55	57	63	57	54	55	53	56
55	54	52	56	56	48	57	55	58	57

a. Draw a histogram of the data.

b. Draw a frequency polygon using the data.

1.37. The following data record the forecasts made in June 1974, of the 1975 Consumer Price Index by 47 leading economists:

134.6	139.0	141.4	142.1	141.0	138.0
140.5	135.0	139.2	141.0	138.0	141.4
141.5	140.7	134.6	138.0	139.5	139.0
140.0	141.5	138.0	137.0	139.0	140.0
142.0	138.0	140.0	139.0	137.0	140.9
137.0	141.0	139.0	140.4	141.5	137.0
141.0	138.6	142.0	143.0	140.5	141.5
138.2	142.0	143.0	142.6	142.0	

a. Construct a frequency distribution of the data, using five intervals.

b. Draw a frequency polygon based on this frequency distribution.

1.38. An economist wants to determine the daily demand equation for rolled steel in a small industrial town. She collects the following data, relating the price with the quantity of rolled steel that can be sold at that price:

*Forbes, January 1, 1976, pp. 186–187.

Tons of Rolled Steel That Can Be Sold	Price per Ton (hundreds of dollars)
1.0	50.10
2.0	12.60
2.5	8.00
4.0	3.20
5.0	2.00
6.3	1.25

Construct a scattergram of tons of rolled steel vs. price per ton.

1.39. A recent article on the rising home prices in Southern California provided some data relating income levels to price ranges of homes. Construct a scattergram of the data below to illustrate this relationship.

Annual Household Income	Maximum Price of Affordable Home
$ 7,000	$21,489
10,000	25,781
12,000	32,225
15,000	37,596
17,500	42,967
20,000	53,709
25,000	64,450
30,000	80,225

1.40. In a study of the advertising budgets of small businesses, a consultant to the Small Business Administration collected the following data relating the size of a business' advertising budget with that business' total sales volume:

Advertising Budget (hundreds of dollars)	Total Sales Volume (thousands of dollars)
6	50
4	60
10	100
1	30
7	60
5	60
7	40

Construct a scattergram of advertising budget vs. total sales volume.

2

Numerical Descriptions of Data

As we observed from the material presented in Chapter 1, an unorganized set of data is merely a collection of numbers, seemingly listed in haphazard fashion and conveying no clear information. To discover the message inherent in the data, it is often necessary to develop an understanding of patterns and trends existing among the individual data points. Several pictorial or graphical methods of so describing the data were presented in Chapter 1. Our goal in this chapter is to develop numerical measures of a set of data that will convey an understanding of important patterns.

SECTION 2.A AVERAGES

Consider a small business having 10 persons on its payroll, whose weekly salaries are listed in Table 2.1. The salaries vary from one employee to another because of job classification, experience, and other relevant, or perhaps irrelevant, factors. One question that may be of interest is, "What is the average weekly salary of all those on the payroll?"

The usual method of computing the *average* salary is to divide the total number of dollars paid out weekly by the total number of persons on the payroll. The total number of dollars paid out is 5,000, as shown at the bottom of Table 2.1, while there are 10 persons on the payroll. Thus, this method of computing the average yields an average weekly salary of 5000/10 = $500. So far, so good.

TABLE 2.1 Weekly Salaries

Person	Weekly Salary (in dollars)
A	400
B	200
C	100
D	200
E	2,500
F	200
G	300
H	600
I	200
J	300
Total Payroll	5,000

The next questions that arise are, "What does the number 500 signify? What information does it communicate about the data? Why did we compute the average salary by such a procedure and what, in fact, are we talking about when we mention the average salary?" Consider the following three proposals for the meaning of the term *average*.

1. The average salary is the middle salary: half of those on the payroll earn at or less than the average salary, while the other half earn at or more than the average salary;
2. The average salary is the most common salary: more people earn that salary than any other single dollar amount;
3. The average salary is the dollar amount lying halfway between the highest salary and lowest salary.

While it would be difficult to think up another possible interpretation of the term average, it is a fact that the average we have computed, namely $500, satisfies none of the above three criteria. To see this, let's calculate the specific numbers that play the roles defined by the above three descriptions of the term average.

Median

First, the middle salary, which is an example of what is technically referred to as the *median*, can be determined by listing the data points from lowest to highest and then finding a number located exactly in the middle.

From Table 2.2, it is apparent that the median of this set of data is the number 250, for persons C, D, F, B, and I earn at or less than the $250 level,

TABLE 2.2   Weekly Salaries in Numerical Order

Person	Weekly Salary (in dollars)
C	100
D	200
F	200
B	200
I	200
G	300
J	300
A	400
H	600
E	2,500

while persons G, J, A, H, and E earn at or more than the $250 level. (If there is an odd number of data points, the median will be the exact middle number, while if there is an even number of data points, the median is usually taken to be the number halfway between the two middle numbers. In this case, the two middle numbers are 200 and 300, so we consider 250 to be the median.*)

The average that we computed earlier, namely 500, is certainly not the median, then, as we now know that the median salary is $250. From Table 2.2, we can see further that 8 out of the 10 persons on the payroll are earning less than the so-called average salary of $500.†

Mode

Next, the most common salary, which is an example of what is technically called the *mode*, is determined by listing each possible salary level and then noting the number of persons receiving that salary.

From Table 2.3, we see that the *modal* salary is $200, because four persons have salaries at that level, far more than the number of persons at any other salary level.

It is interesting to observe that (for this set of data, at least) the modal salary of $200, the median salary of $250, and what we have been calling the average salary of $500 are all different numbers.

---

*Occasionally when there are two middle numbers, such as 200 and 300 in our example, you will hear them jokingly referred to as "comedians."

†This observation has the somewhat entertaining sociological interpretation that "nearly everybody is below average."

TABLE 2.3   Persons Grouped by Salary Level

Salary Level (in dollars)	Persons at that Salary Level
100	1 (C)
200	4 (D, F, B, I)
300	2 (G, J)
400	1 (A)
600	1 (H)
2,500	1 (E)
Sum	10

Midrange

Finally, the dollar amount lying halfway between the highest salary ($2,500) and the lowest salary ($100) is $1,300, not the $500 we have calculated as the average.

The amount of $1,300, an example of what is technically labeled the *midrange*, differs from all the measures of average discussed previously for this particular set of data.

Mean

What, then, is the meaning of the "average," which is calculated by summing all data points and dividing that sum by the number of data points involved? Well, the precise significance of that process is not easy to explain in a few words, so we will postpone the details to the next section. We will note, however, that the technical name of the average computed in this manner is the *mean*. In particular, $500 is the mean salary of all persons on the payroll of the small business under study.

THE MEAN
The mean is the sum of all data points divided by the number of data points.

In Table 2.4, we summarize the various types of averages discussed in our attempt to determine the average salary of persons on the payroll listed in Table 2.1.

In view of the fact that the term average may mean many different things, and often different things to different people, how do we know which type of average is appropriate in any particular situation? As a rule, this decision is not based on statistical considerations, but instead on criteria existing

TABLE 2.4   Averages of Weekly Salaries

Type of Average	Numerical Value of Average
Mean	500
Median	250
Mode	200
Midrange	1,300

within the various aspects of the field of business. For example, bankers talk about the mean interest rate prevailing during the year; economists are interested in the median family income of a certain community; shoe and clothing outlets have to make sure they stock the modal sizes (for there may be relatively few individuals having exactly the mean, median, and midrange sizes, and possibly no individuals at all with these sizes, as a comparison of Tables 2.1 and 2.4 will illustrate); and investors speak about the daily midrange of stock market prices.

The moral of the story is that the term "average" may be interpreted in many different ways, and so to be sure of communicating the proper information, it is necessary to specify the particular type of average being used.

EXERCISES   2.A

2.1. The increasing emphasis on solar energy has led some homebuilders to wonder whether or not to install solar heating systems. Research on the number of years it takes to recover the cost of a solar heater through the amount of savings over conventional electric heating produced the following data:

City	Years to Recover Cost
Atlanta	14
Bismarck	14
Boston	14
Charleston	11
Columbia, Missouri	15
Dallas/Fort Worth	13
Grand Junction, Colorado	13
Los Angeles	10
Madison, Wisconsin	14
Miami	9
New York	14
Washington	15

    a. Determine the mean of the cost recovery periods.

    b. What is the mode?

    c. Locate the median.

    d. What is the midrange?

2.2. The following data are the prices, in cents per pound, of imported English cheese in 14 metropolitan areas of the United States and Canada:

250	255
260	260
280	250
270	240
260	265
240	254
254	260

    a. Calculate the mean price.

    b. Determine the median price.

    c. Find the modal price.

    d. Find the midrange of the prices.

2.3. As part of a study aimed at finding out what a typical auto body repair job costs, an insurance adjusters' organization collects the following data on repair estimates for 15 damaged cars (in dollars):

930	220	130
200	460	500
380	630	650
590	300	540
280	120	730

    a. What is the mean repair estimate?

    b. What is the median repair estimate?

    c. Is there a modal repair estimate?

    d. What is the midrange of the estimates?

2.4. Using the *Business Week* listing of forecasts of the percentage increase in prices appearing in Exercise 1.11,

    a. calculate the mean forecasted price increase;

    b. determine the median forecasted price increase;

    c. find the mode of the forecasts;

    d. locate the midrange of the forecasts.

2.5. *Forbes* magazine* conducted an analysis of 800 U.S. corporations in regard to the total remuneration paid to the chief executive of each in 1976. A random

Forbes, May 15, 1977, pp. 244–284.

selection of the corporations and the 1976 pay (rounded off) of their top executives follows.

Corporation	Total 1976 Pay of Chief Executive (dollars)
International Paper	670,000
Liggett Group	356,000
Hanna Mining	241,000
Pacific Lumber	149,000
Vulcan Materials	241,000
Southland Corporation	325,000
First Union Corporation	116,000
Burroughs	525,000
Control Data	285,000
American Broadcasting	946,000
Lubrizol	217,000
Long Island Lighting	168,000
Litton Industries	244,000
Firestone	432,000
Alabama Bancorporation	114,000

a. Calculate the mean compensation received by the sample of chief executives.

b. Determine the median 1976 total pay of the executives.

c. Find the midrange of corporate executive pay, based on the given information.

2.6. One day in October 1977, the 15 most active stocks traded on the New York Stock Exchange had the following closing prices:*

Stock	Closing Price
Am Gen Ins	20.250
Savin B Mch	26.250
Mattel Inc	10.125
Alcon Lab	28.000
U.S. Steel	29.875
Colum Pict	19.000
Chris Craft	8.000
Sambos Rst	24.625
Dow Chm	31.125
Kroger Co	25.875
Citicorp	23.625
Vetco Inc	20.875
Boeing	26.000
Ipco Hospit	6.625
Allis Chalm	23.750

*The Wall Street Journal, October 11, 1977, p. 38.

a. Calculate the mean closing price of these 15 most active stocks.

b. Calculate the median closing price.

c. Find the midrange of the closing prices.

2.7. Just as the median divides a set of data into two parts, a lower half and an upper half, we can define *quartiles* (which divide the data into four equal parts), *deciles* (ten equal parts), and *percentiles* (100 equal parts). For example, the seventh decile divides the lower 70% of the data from the upper 30%, while the third quartile divides the lower 75% from the upper 25%. Using the data of Exercise 1.18, find:

a. the median;

b. the first quartile;

c. the third decile;

d. the 57th percentile.

2.8. From the data of Exercise 1.16, determine:

a. the median;

b. the third quartile;

c. the ninth decile;

d. the 38th percentile.

2.9. If we are given only a frequency distribution of a set of data, such as in Exercise 1.12, rather than a complete list of all the data points (as in Exercise 1.11), we cannot calculate the mean, median, mode, and midrange exactly, but we can find approximations of them.

To find an approximation for the mean of a frequency distribution, using the one on usage of salad dressing in Table 1.6 of Chapter 1 for example, we would consider all 4 data points falling in the interval 20–22 to be equal to the midpoint 21 of that interval. This means that we are assuming that 4 of the 100 data points are 21s. The next 8 data points are assumed to be equal to the midpoint of the interval 23–25, namely 24. Continuing in this way, we calculate an approximation of the mean as follows:

Midpoint (m)	Frequency (f)	m × f
21	4	84
24	8	192
27	23	621
30	34	1020
33	26	858
36	4	144
39	1	39
	$\Sigma f = 100$	$\Sigma mf = 2958$

Using the symbol "≈" to denote "approximately equal to," we can use the weighted mean formula to get the approximate value of the mean:

$$\mu \approx \frac{\Sigma mf}{\Sigma f} = \frac{2958}{100} = 29.58$$

To approximate the median, we first figure out that, in a set of 100 data points, the median should be halfway between the 50th and the 51st. Looking at the frequency distribution of Table 1.6, we can see that 35 data points are 28 or below, while 69 are 31 or below. This means that the median must lie in the interval 29–31. Because 34 data points fall in the interval 29–31, we can space them equally by dividing the interval into 33 equal subintervals. We then place one point at 29 and one each at the upper end of each of the 33 subintervals. Since there are 35 data points equal to 28 or below, we need to proceed a distance of 14.5 subintervals into the interval 29–31 in order to reach the median. (Hence, we have proceeded 15.5 data points into the interval.) That means the median is 29, plus a fraction 14.5/33 of the length of the interval 29–31. Because the length of this interval is 2, we get

$$\text{median} \approx 29 + \left(\frac{14.5}{33}\right)(2) = 29 + \frac{29}{33} = 29 + .88 = 29.88.$$

For the mode, we use the midpoint of the most populous interval, namely the interval 29–31, so we have

$$\text{mode} \approx 30.$$

To approximate the midrange, we merely use the number that is halfway between the midpoint of the first interval and the midpoint of the last interval, and get

$$\text{midrange} \approx \frac{21 + 39}{2} = \frac{60}{2} = 30.$$

a. Approximate the mean dollar value of the 500 stocks of Exercise 1.12.

b. Approximate their median.

c. Approximate their mode.

d. Approximate their midrange.

2.10. Based on the earnings data of Exercise 1.13,

    a. approximate the mean yearly earnings of the employees;

    b. approximate the median;

    c. approximate the mode;

    d. approximate the midrange.

2.11. Using the unemployment rate response data of Exercise 1.15,

    a. approximate the mean response;

    b. approximate the median response;

    c. approximate the modal response;

    d. approximate the midrange.

SECTION 2.B.   THE MEAN

In the previous section we noted that the mean, although simple to calculate, is not easy to interpret, even though it is the most commonly used measure of average. Now we will explain its significance.

The Balancing Property of the Mean

Let's look again at the data on weekly salaries appearing in Table 2.1. If we imagine a seesaw in a children's playground marked off in units of measurement, and we place bowling balls at locations corresponding to the data points, we obtain a graphical description of the information presented in Table 2.3. The picture of the seesaw appears in Fig. 2.1. with the mean value of $500 clearly marked. In the picture we see one ball at location 100, four at 200, two at 300, one at 400, one at 600, and one at 2,500, in accordance with Table 2.3. What we now show is that the seesaw will balance only if its fulcrum (balancing point) is placed at the mean value, namely 500.

    One of the principles of the seesaw asserts that a 25-pound child will be able to balance a 50-pound child if the smaller child sits twice as far away from the center as does the larger child. In a similar spirit, we see that four bowling balls at distance 300 from the center will balance one bowling ball at distance 1,200 from the center. You could draw a diagram of this or could easily test it out (using bricks instead of bowling balls) at a local playground. In the seesaw of Fig. 2.1, the balls at 400 and 600 balance each other around the center position of 500, while the one ball at 2,500 balances the remaining seven balls. The equivalence of forces on the left and right sides of the 500 location causes the balancing of the seesaw at that point.

FIGURE 2.1   Balancing Property of the Mean Weekly Salary

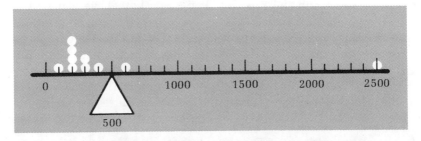

It is in the sense of the balancing point of the data that the mean can be considered a measure of average; this defines how the mean describes a set of data. When we are informed that the mean of a particular set of data is 500, we should visualize in our minds the data strung out along a seesaw that is balanced at 500.

For some types of data, nevertheless, the mean is not an appropriate measure of average because variations within the set of data are so extreme as to make the balancing point irrelevant. In the case of stock market prices exhibiting wild fluctuations, for example, the median is generally used to measure average price per share.*

Let's try to organize this balancing property into a useful technique for describing data. If we subtract the mean value of 500 from each of the data points, we get a number that tells us the distance of the data point from the mean and also tells us on which side of the mean the data point is located. Using as an example the data point 300, we calculate $300 - 500 = -200$. The negative sign indicates that 300 is to the left of the mean on the seesaw, while the 200 indicates that 300 is at a distance of 200 units away from the mean. Similarly $600 - 500 = 100$ means that the point 600 is 100 units to the right of the mean.

We call the difference between a data point and the mean the *deviation of the data point from the mean.*

DEVIATION OF THE DATA POINT FROM THE MEAN;
BALANCING PROPERTY OF THE MEAN
The deviation will be negative if the data point falls to the left of the mean and positive if the data point lies to the right of the mean. The balancing property means that the positive deviations exactly cancel the negative deviations, so that the sum of all deviations is zero.

In Table 2.5, we show how the deviations sum to zero in the case of the weekly salary data of Table 2.1. The sum of the deviations is 0 because the negative deviations and the positive deviations both have a magnitude of 2,100.

The deviations of the various data points from the mean are important in themselves for they can be used to measure whether the data points are all bunched together near the mean or whether some are dispersed far away from the mean. We will deal with this question in detail in the next section. Here let us remark only that the computational procedure carried out in Table 2.5 can be used as a check on our calculation of the mean. If the sum of

*See, for example, the article by E. F. Fama listed in the bibliography at the end of this chapter.

TABLE 2.5   Sum of Deviations of Weekly Salary Data

Person	Weekly Salary (in dollars)	Salary Minus Mean =	Deviation of Salary
A	400	400 − 500 =	−100
B	200	200 − 500 =	−300
C	100	100 − 500 =	−400
D	200	200 − 500 =	−300
E	2,500	2,500 − 500 =	2,000
F	200	200 − 500 =	−300
G	300	300 − 500 =	−200
H	600	600 − 500 =	100
I	200	200 − 500 =	−300
J	300	300 − 500 =	−200
	Sum of Deviations		0

deviations turns out to be something other than zero, we have probably made an arithmetic error in our calculation of the mean.

Labeling a Set of Data Points

In order to establish general procedures for analyzing a set of data, it is necessary to develop new techniques of labeling a set of data points. One convenient way to do this is to consider the data as a set of x's labeled as follows:

x_1 = first data point
x_2 = second data point
x_3 = third data point
.
.
.
x_k = kth data point
.
.
.
x_N = Nth data point

We will agree that x_N will be the last of our data points, and this means that we are dealing with N data points. Therefore

N = number of data points in our set.

In the case of the weekly salary data of Table 2.1, we have 10 data points so that $N = 10$. We label the points of Table 2.1 as follows:

$$
\begin{array}{ll}
x_1 = 400 & x_6 = 200 \\
x_2 = 200 & x_7 = 300 \\
x_3 = 100 & x_8 = 600 \\
x_4 = 200 & x_9 = 200 \\
x_5 = 2500 & x_{10} = 300
\end{array}
$$

To avoid writing the words "sum of the data points" over and over again, we should introduce some symbols for that expression. The symbol used in statistics to signify the word "sum" is the capital letter Σ (pronounced "sigma") of the Greek alphabet. To symbolize the sum of the data points, we write Σx, which is pronounced "sigma x" or "the sum of the x's." The meaning of that collection of symbols is therefore

$$\Sigma x = x_1 + x_2 + x_3 + \ldots + x_N.$$

For the weekly salary data,

$$
\begin{aligned}
\Sigma x &= x_1 + x_2 + x_3 + x_4 + x_5 + x_6 + x_7 + x_8 + x_9 + x_{10} \\
&= 400 + 200 + 100 + 200 + 2500 + 200 + 300 \\
&\quad + 600 + 200 + 300 \\
&= 5000.
\end{aligned}
$$

Using statistical shorthand, we can write a compact expression for the mean. In situations where the data represent a complete *population* (as do the weekly salaries of all 10 persons employed in the small business under study), rather than a statistical sample,* we use the lower case Greek letter μ (pronounced "mew") as a symbol for the mean. The expression

$$\text{the mean} = \frac{\text{sum of all data points}}{\text{number of data points}}$$

can then be translated into our new shorthand as

$$\mu = \frac{\Sigma x}{N}.$$

With the shorthand formula, we can express the mean of the weekly salary data as

$$\mu = \frac{\Sigma x}{N} = \frac{5000}{10} = 500.$$

---

*We will study statistical samples in Chapter 5.

TABLE 2.6   Sum of Deviations from the Mean

	Data Points (x)	Deviations $(x - \mu)$
	x_1	$x_1 - \mu$
	x_2	$x_2 - \mu$
	x_3	$x_3 - \mu$
	•	•
	•	•
	•	•
	x_N	$x_N - \mu$
Sums	Σx	0

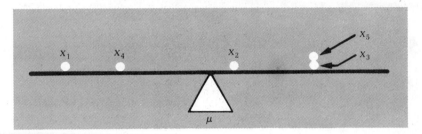

FIGURE 2.2   Balancing Property of the Mean

In terms of our system of labeling data points, we can construct a table analogous to Table 2.5, which illustrates the balancing property by pointing out that the deviations always have a sum of zero. This symbolic calculation appears in Table 2.6, while a picture of a seesaw illustrating the balancing property appears in Fig. 2.2.

It turns out that $\Sigma(x - \mu) = 0$ for any possible set of data points; that is, the sum of the deviations is zero.* This is the expression of the balancing property of the mean in the statistical shorthand language we have introduced.

EXERCISES   2.B

2.12. Draw a seesaw representation of the solar energy data of Exercise 2.1, and show that the deviations from the mean really do sum to zero.

2.13. Construct a seesaw representation of the cheese price data of Exercise 2.2, and make the calculations that verify that the deviations from the mean have a sum of zero.

* Try this on a few sets of data for yourself!

2.14. Put the auto repair estimates of Exercise 2.3 on a seesaw, and show that the deviations from the mean sum to zero.

2.15. Make up a set of any 9 numbers you can think of, and verify that the deviations from the mean really sum to zero.

2.16. Using the data of Exercise 2.1,

    a. show that the deviations from the *median do not* sum to zero;

    b. show that the deviations from the *mode do not* sum to zero;

    c. show that the deviations from the *midrange do not* sum to zero.

2.17. Using the data of Exercise 2.2,

    a. show that the deviations from the *median do not* sum to zero;

    b. show that the deviations from the *mode do not* sum to zero;

    c. show that the deviations from the *midrange do not* sum to zero.

2.18. Using the data of Exercise 2.3, show that the deviations from the median do not sum to zero.

2.19. If $x_1 = 5$, $x_2 = 8$, $x_3 = 7$, and $x_4 = 19$, calculate Σx.

2.20. If

$$
\begin{array}{ll}
x_1 = 17 & x_5 = -8 \\
x_2 = 9 & x_6 = -9 \\
x_3 = -6 & x_7 = -5 \\
x_4 = 14 & x_8 = 3
\end{array}
$$

calculate Σx.

2.21. Calculate the mean number of attempts using the learning experiment data of Exercise 1.18.

2.22. The following data record the number of working days lost through labor disputes in England during a 16-year period:*

Year	No. of Worker Days Lost (Thousands)	Year	No. of Worker Days Lost (Thousands)
1950	1389	1958	3462
1951	1694	1959	5270
1952	1792	1960	3024
1953	2184	1961	3046
1954	2457	1962	5798
1955	3781	1963	1755
1956	2083	1964	2277
1957	8412	1965	2925

*B. R. Mitchell and H. G. Jones, *Second Abstract of British Historical Statistics*, Cambridge, University Press, 1971, p. 51.

a. Calculate the mean yearly number of worker-days lost due to labor disputes during the 16-year period.

b. Construct a seesaw representation of the data.

SECTION 2.C MEASURES OF VARIATION

If we compare the sales volume of a local grocery store at two locations for a one-week period, we might come up with the data presented in the top part of Table 2.7. A check of the bottom part of Table 2.7 or an application of the methods introduced in our earlier discussion of averages shows that *both* sets of sales figures have mean, median, mode, and midrange *all* equal to $500. No matter what you consider the appropriate measure of average here, each set of data has an average of $500.

Because there is such substantial agreement on the averages of the sales data in this example, it would be important to know exactly how descriptive of a set of data the average is. In particular, when we are informed that the average (as measured by one of the methods we discussed earlier) of certain sets of data is 500, how are we to interpret that information? How are we to visualize the original set of data from the knowledge that the average is 500?

Well, as the pictorial representation shown in Fig. 2.3 demonstrates, there can be considerable differences between data sets that have the same averages. Just because two sets of data both have an average of $500, this does not mean that they exhibit the same overall characteristics.

TABLE 2.7 One-Week Comparison of Sales of Two Locations of Grocery Stores (*dollars*)

Day	Main Street	Broadway
Monday	400	100
Tuesday	500	500
Wednesday	500	200
Thursday	600	800
Friday	500	500
Saturday	500	900
Mean	500	500
Median	500	500
Mode	500	500
Midrange	500	500

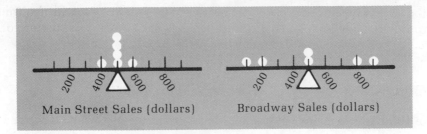

FIGURE 2.3   Seesaw Representations of Sales Data

How then do the two sets of sales data differ? As Fig. 2.3 shows, they differ primarily by how much the data points vary from the common average of $500. Daily sales at the Main Street store are heavily concentrated near their average of $500, while at the Broadway store, they vary considerably from their average of $500. In particular, Main Street sales remain between $400 and $600 per day throughout the week, while Broadway sales range from a low of $100 to a high of $900. The extent of variation from the average is an important element in the description of a set of data, but knowledge of the averages alone gives no information about the variation.

A comparison of the averages alone would lead us to believe that the two stores have essentially the same sales patterns, whereas some understanding of their internal variation may help us pinpoint serious discrepancies. In particular, we might find it useful to determine the cause of the variation and then to find out if there is anything we can (or want) to do about rectifying it. For example, perhaps Wednesday night or Thursday morning media advertisements are helping the Broadway store increase sales far beyond its normal level, but are not doing anything for sales at the Main Street store. Acting on information about the variation, then, we might decide to introduce a new advertising program oriented specifically toward customers of the Main Street store.

Knowledge of internal variation also plays a role in one prevalent theory of management, popularly dubbed *management by exception*. This theory calls for us to avoid interfering in the day-to-day operations of the business until we notice some occurrence that is different from an accepted standard. Such an occurrence may be, for example, a quality control problem, excess production costs in one phase of the operation, or lower daily sales volume. Variations from the normal routine then trigger management action, according to the management by exception theory. In accounting, we talk about "variance analysis" when we compare variations in production cost with some predetermined standard. However, in statistics, *variance* deals with deviations from the mean.

It would be useful, therefore, to have a numerical measure of variation of the individual points within a set of data, which we could use for the purpose of obtaining a more accurate description of the general pattern of the data.

Range

Consider the following proposal for a measure of variation of a set of data: *range*. The range of the data is the difference between the largest data point and the smallest.

The computation of the range is fairly simple because it involves only the two extreme data points. For the sales data appearing in Table 2.7, we can see that the daily range of Main Street sales figures is $600 - 400 = 200$ dollars, while the range of Broadway sales figures is $900 - 100 = 800$ dollars. A comparison of the ranges indicates that daily Broadway sales are more variable than daily Main Street sales, substantiating the impression given in the seesaws of Fig. 2.3.

Mean Absolute Deviation

As a second proposed measure of variation, we can look at the average deviation of points from their mean, technically referred to as the *mean absolute deviation*. This one is somewhat more difficult to compute because it involves the use of a table of all deviations, like that in Table 2.5.

The preliminary computations leading to the mean absolute deviation of each set of sales figures can be found in Table 2.8. In the portion of Table 2.8 dealing with the Main Street sales, the column headed x lists the actual sales data, while the column headed $x - \mu$ lists the deviation of each data point from the mean value of 500.

In view of the balancing property of the mean as illustrated in Table 2.6, we know that the column headed $x - \mu$ must sum to 0. But what we are interested in for the purpose of measuring variation is the *absolute magnitude* of each deviation, not whether they are positive or negative. The absolute magnitude (technically called the *absolute value*) of the deviations is listed in the column headed $|x - \mu|$.* The mean absolute deviation is then the sum of the $|x - \mu|$ column divided by the number of data points. A set of data with its data points far from their mean, on the average, will have a larger mean absolute deviation than a set having its points closer, on the average, to their mean.

---

*By the *absolute value* of a number, we mean the magnitude of the number, ignoring the + or − sign. For example, both $|5| = 5$ and $|-5| = 5$. As a result, whether a number is positive or negative (or zero), its absolute value cannot be negative.

TABLE 2.8   Computations Leading to the
Mean Absolute Deviation of Sales

Main Street Sales (x)			Broadway Sales (y)		
x	x − μ	|x − μ|	y	y − μ	|y − μ|
400	− 100	100	100	− 400	400
500	0	0	500	0	0
500	0	0	200	− 300	300
600	100	100	800	300	300
500	0	0	500	0	0
500	0	0	900	400	400
Sums 3,000	0	200	3,000	0	1,400
N = 6			N = 6		
μ = 3,000/6 = 500			μ = 3,000/6 = 500		

We abbreviate the mean absolute deviation by the symbol MAD. In the mathematical shorthand introduced earlier in this chapter, we can write that

$$\text{MAD} = \frac{\text{sum of absolute deviations}}{\text{the number of data points}} = \frac{\Sigma|x - \mu|}{N}.$$

The Main Street sales figures have MAD = 200/6 = 33.33, using the fact in Table 2.8 that the sum of the absolute deviations (the column headed by |x − μ|) is 200. On the other hand, the Broadway sales figures have MAD = 1400/6 = 233.33, again indicating strongly that daily Broadway sales are more variable than daily Main Street sales.

The Standard Deviation

Now that we have thoroughly discussed the use and computation of the range and the mean absolute deviation as possible measures of the variation of a set of data points, it is the authors' sad duty to report that neither of these measures is widely used, except in very special cases involving extremely large variations of the data where the mean absolute deviation is appropriate.* Both have been preempted in importance by a third measure of variation, technically referred to as the *standard deviation*. The standard deviation is somewhat more complicated to calculate than either of the measures already discussed, but is also more useful and efficient.

* See, for example, the article by William F. Sharpe, cited in the bibliography at the end of this chapter.

THE STANDARD DEVIATION

The standard deviation is the positive square root of the mean of the squared deviations. (In some industrial production processes, it is referred to as the *root-mean-square deviation* or the *rms deviation*.) We usually denote the standard deviation by the Greek letter σ (lower-case sigma), and in algebraic symbolism, we express it as follows,

$$\sigma = \sqrt{\frac{\Sigma(x - \mu)^2}{N}}.$$

To calculate σ in accordance with the preceding formula, we proceed as follows.

1. Calculate the deviations $x - \mu$, as in Table 2.8.
2. Square each deviation; i.e., multiply it by itself.
3. Sum the squared deviations.
4. Divide the sum by N, the number of data points.
5. Take the square root of the resulting number.*

The standard deviation, like the mean, has the unfortunate characteristic that its meaning is not immediately clear from its definition. In general, we can say that the standard deviation is the most widely used of the measures of variation because it conveys the most precise and useful information about variation of the data from the mean. The sort of information it conveys will be discussed in detail in the next section. Here we concentrate on how to calculate it.

The most direct method of calculating the standard deviation is by means of a table of preliminary computations analogous to Table 2.8. In accordance with the five-step procedure discussed above, we would record the square of each deviation rather than its absolute value. A deviation of -10 would have a square of $(-10) \times (-10) = 100$, recalling that the product of two negative numbers is a positive number. Similarly, a deviation of 40 has a square of $(40) \times (40) = 1,600$. We construct Table 2.9 with these principles in mind. From Table 2.9, we see that the sum of squared deviations of daily Main Street sales is

$$\Sigma(x - \mu)^2 = 10,000 + 0 + 0 + 10,000 + 0 + 0 = 20,000,$$

---

*The square root may be explained as follows: the square of a number, say 3, is that number multiplied by itself, namely $3 \times 3 = 9$. The square root is the reverse of this; that is, $\sqrt{9} = 3$ because $3 \times 3 = 9$. In short, for positive numbers a and b, $\sqrt{a} = b$ if and only if $b^2 = a$.

TABLE 2.9 Computations Leading to the
Standard Deviations of Sales Data

	Main Street Sales (x)			Broadway Sales (y)		
x	$x - \mu$	$(x - \mu)^2$	y	$y - \mu$	$(y - \mu)^2$
400	-100	10,000	100	-400	160,000
500	0	0	500	0	0
500	0	0	200	-300	90,000
600	100	10,000	800	300	90,000
500	0	0	500	0	0
500	0	0	900	400	160,000
Sums 3,000	0	20,000	3,000	0	500,000
N = 6			N = 6		
μ = 3000/6 = 500			μ = 3000/6 = 500		

adding up the column labeled $(x - \mu)^2$; while the sum of squared deviations
of daily Broadway sales is, by adding up the column labeled $(y - \mu)^2$,

$$\Sigma(y - \mu)^2 = 160,000 + 0 + 90,000 + 90,000 + 0 + 160,000 = 500,000.$$

It follows that, for daily Main Street sales, the standard deviation is

$$\sigma = \sqrt{\frac{(x - \mu)^2}{n}} = \sqrt{\frac{20,000}{6}} = \sqrt{3333.33} = 57.7;$$

while daily Broadway sales have standard deviation

$$\sigma = \sqrt{\frac{(y - \mu)^2}{n}} = \sqrt{\frac{500,000}{6}} = \sqrt{83,333.33} = 288.7.$$

As is clear from the pictures in Fig. 2.3, the Broadway sales figures are more
variable than the Main Street ones, and this fact is borne out by a comparison
of the standard deviations. The Broadway sales figures' standard deviation
of 288.7 exceeds by a considerable margin the Main Street figures' value of
57.7.

Square Roots

Before proceeding further in our study of the standard deviation, let's pause
a moment to talk about the square roots involved in the calculation. There
are basically three ways of finding the square root of a number: (1) using an

electronic calculator that has a square root button; (2) carrying out the paper-and-pencil calculation by hand, using the long division-type method some-times taught in elementary algebra courses; and (3) using the square root tables in the back of this textbook.

Method 1 is obviously the most efficient and most accurate method. It is the one used by those persons who handle statistics in their daily work. However, not every calculator has a square root button, so method 1 may not be practical for everyone who uses this text.

The hand calculation of method 2 is slightly more complicated and time-consuming than ordinary long division. Therefore, we also consider this method impractical. If you are interested in the details of the hand cal-culation method, you are invited to ask your local statistics expert, namely your instructor, for an explanation of it.*

This brings us to method 3. Using Table A.1 of the Appendix, we can work out the square root of any number having three significant digits. Because Table A.1 contains only the numbers 1.00 through 9.99 explicitly, it is necessary to express the number whose square root we want in terms of one of those numbers. To illustrate this procedure, we will work out the details of finding $\sqrt{3333.33}$ and $\sqrt{83,333.33}$. Because the table contains the square roots of only three-digit numbers, the best we can do using the table is to find $\sqrt{3330}$ and $\sqrt{83,300}$. We have

$$\sqrt{3330} = \sqrt{3.33 \times 1000} = \sqrt{3.33 \times 10 \times 100} = \sqrt{3.33 \times 10} \times \sqrt{100}$$

$$= 5.77 \times 10 = 57.7.$$

We arrive at this answer by looking up $n = 3.33$ and using the third column of the table, the column headed $\sqrt{10n}$, and the fact that $\sqrt{100} = 10$, which you probably already have memorized. Next, we calculate that

$$\sqrt{83,300} = \sqrt{8.33 \times 10,000} = \sqrt{8.33} \times \sqrt{10,000}$$
$$= 2.886 \times 100 = 288.6;$$

by looking up $n = 8.33$, finding $\sqrt{n} = 2.886$, and using the fact that $\sqrt{10,000} = 100$. (The round-off error gives us 288.6 instead of the more accurate $\sqrt{83,333.33} = 288.7$ obtained with a calculator.)

Additional examples throughout this text will continue to illustrate the procedure of finding square roots by using Table A.1.

Short-Cut Method

In some situations in statistics, it is more efficient to calculate the standard deviation by an alternative procedure. The alternative procedure, euphe-mistically called "the short-cut method" of computing the standard devia-

---

* See also M. Richardson, *College Algebra*, 1st ed., Prentice-Hall, 1947, pp. 45–47.

TABLE 2.10   Weekly Sales of Auto Parts

Week	Sales Volume (thousands of dollars)
#1	7
#2	5
#3	7
#4	8
#5	4
#6	9
#7	5
Sum	45

tion, does in fact simplify the computation when the mean does not come out even, namely, as a whole number.

For example, consider the data of Table 2.10, showing the weekly sales volume of a small auto parts outlet over a seven-week period. The mean sales volume is

$$\mu = \frac{\Sigma x}{N} = \frac{45}{7} = 6.43 \,,$$

which does not come out even. In Table 2.11, we illustrate the procedures

TABLE 2.11   Calculation of the Standard Deviation of Auto Parts Sales Data, Direct Method vs. Short-Cut Method

	Direct Method			Short-Cut Method	
	x	$x - \mu$	$(x - \mu)^2$	x	x^2
	7	0.57	0.3249	7	49
	5	−1.43	2.0449	5	25
	7	0.57	0.3249	7	49
	8	1.57	2.4649	8	64
	4	−2.43	5.9049	4	16
	9	2.57	6.6049	9	81
	5	−1.43	2.0449	5	25
Sums	45	−0.01	19.7143	45	309
	N = 7			N = 7	
	$\mu = 45/7 = 6.43$				

for calculating the standard deviation by both the direct method and the short-cut method. You should observe that the short-cut method does indeed save some steps in this case. The short-cut formula does not require the mean to be computed first, but rather it proceeds directly to the standard deviation. The formula involves only the sum of the data points and the sum of their squares.

SHORT-CUT METHOD OF CALCULATING THE STANDARD DEVIATION

$$\sigma = \sqrt{\frac{N\Sigma x^2 - (\Sigma x)^2}{N^2}}.$$

Both the direct and short-cut formulas give exactly the same answer in every case, except for possible differences due to rounding off decimal places. Notice that some round-off error has already occurred in the computations preparatory to the direct method. Such an error is indicated by the fact that the deviations sum to -0.01 instead of exactly 0 as the balancing property requires. The error cannot be avoided because, as a practical matter, we have to round off μ to 6.43 from its more precise value of 6.428571429. Therefore we might, in some cases, wind up with a slightly inaccurate value of σ as well.

The short-cut method, however, maintains 100% accuracy in the computation table, and is therefore the preferred method of computing σ when μ does not come out even. Furthermore, the short-cut formula is easier to use when computing the standard deviation using a calculator, for basically all that is needed are the sum of the data points and the sum of their squares. We now calculate σ both ways from the information in Table 2.11:

Direct Method:

$$\sigma = \sqrt{\frac{(x-\mu)^2}{N}} = \sqrt{\frac{19.7143}{7}} = \sqrt{2.82} = 1.68.$$

Short-Cut Method:

$$\sigma = \sqrt{\frac{N\Sigma x^2 - (\Sigma x)^2}{N^2}} = \sqrt{\frac{7(309) - (45)^2}{(7)^2}}$$

$$= \sqrt{\frac{2163 - 2025}{49}} = \sqrt{\frac{138}{49}} = \sqrt{2.82} = 1.68.$$

You should observe that both methods yield the same value of σ, namely 1.68. It is also important to observe that Σx^2 and $(\Sigma x)^2$ are not one and the same. To get Σx^2, we sum the squares, while to get $(\Sigma x)^2$, we square the sum. Here $\Sigma x^2 = 309$, and $(\Sigma x)^2 = (45)^2 = 2025$. They're quite different.

EXERCISES 2.C

2.23 From the data on the payback periods of solar energy appearing in Exercise 2.1,

    a. calculate the standard deviation of the payback periods;

    b. calculate the mean absolute deviation;

    c. calculate the range of the payback periods;

    d. calculate the standard deviation by using the short-cut formula.

2.24. Using the data points given in Exercise 2.2,

    a. find the standard deviation of the cheese prices by computations based on the direct formula;

    b. find the standard deviation by computations based on the short-cut formula;

    c. determine the mean absolute deviation of the prices;

    d. find the range of the prices.

2.25. On the basis of the data of Exercise 2.3, calculate

    a. the standard deviation of the repair estimates, using the direct formula for the standard deviation;

    b. the standard deviation, using the short-cut formula;

    c. the mean absolute deviation;

    d. the range.

2.26. Explain why the standard deviation of a set of data points will always be smaller than its range.*

2.27. Calculate the standard deviation of the 1976 executive pay from the data of Exercise 2.5.

2.28. Calculate the standard deviation of the number of attempts, using the data of Exercise 1.18.

2.29. Calculate the standard deviation of the forecasts in Exercise 1.11.

2.30. Calculate the standard deviation of the yearly number of worker-days lost, using the data of Exercise 2.22.

2.31. Using the stock market data of Exercise 2.6,

    a. calculate the standard deviation of the closing prices;

    b. calculate the mean absolute deviation of the closing prices.

2.32. If we already have an approximation for the mean of a set of data displayed in

---

*Actually, the standard deviation is always smaller than *half* the range. See the article by L. Sher, "The Range of the Standard Deviation," *The MATYC Journal*, Fall 1977, p. 197.

a frequency distribution, then we can also approximate the standard deviation. Recall the frequency distribution of Table 1.6, with which we worked in Exercise 2.9. In that exercise, we approximated the mean μ as 29.58. To approximate the standard deviation, we can write the following.

midpoint (m)	$m - \mu$	$(m - \mu)^2$	frequency (f)	$(m - \mu)^2 \cdot f$
21	-8.58	73.62	4	294.48
24	-5.58	31.14	8	249.12
27	-2.58	6.66	23	153.18
30	0.42	0.18	34	6.12
33	3.42	11.70	26	304.20
36	6.42	41.22	4	164.88
39	9.42	88.74	1	88.74
			100	1260.72

The standard deviation can then be approximated as follows:

$$\sigma \approx \sqrt{\frac{\Sigma(m - \mu)^2 \cdot f}{\Sigma f}} = \sqrt{\frac{1260.72}{100}} = \sqrt{12.61} = 3.55$$

We also can use the following analogue of the short-cut formula:

$$\sigma \approx \sqrt{\frac{(\Sigma f) \cdot (\Sigma m^2 \cdot f) - (\Sigma m f)^2}{(\Sigma f)^2}}$$

a. Approximate the standard deviation of the dollar values of the 500 stocks of Exercise 1.12, using the method illustrated above and also the calculations of Exercise 2.9.

b. Use the analogue of the short-cut formula to calculate an approximation to the standard deviation.

2.33. Based on the earnings data of Exercise 1.13 and the calculations in Exercise 2.10, approximate the standard deviation of the yearly earnings.

2.34. Approximate the standard deviation of the unemployment rate response data of Exercise 1.15, taking into consideration the calculations in Exercise 2.11.

2.35. A comparative study of prices around town shows that a bag of peanuts sells for 25 cents, on the average, with a standard deviation of 8 cents. Lincoln Continentals, on the other hand, sell for an average of $20,000, with a standard deviation of $2,000. Because bags of peanuts have a standard deviation of only

8 cents, compared with the $2,000 of the automobiles, can we say that the automobiles vary in price to greater extent than do the peanuts? Not necessarily. What we need to answer this question is a measure of relative variation that expresses the standard deviation as a fraction of the mean. This measure is called the *coefficient of variation* and is given by the following formula.

COEFFICIENT OF VARIATION

$$v = \frac{\sigma}{\mu} \times 100\%.$$

---

    a. Calculate the coefficient of variation of the bags of peanuts.

    b. Calculate the coefficient of variation for the Lincoln Continentals.

    c. For which commodity are the prices relatively more variable?

2.36. The mean attendance at college and professional football games last year in one large city was 38,000 with a standard deviation of 2,600. The mean attendance at weekly string quartet concerts was 225 with a standard deviation of 19. At which type of event was the attendance relatively more variable?

SECTION 2.D   THE IMPORTANCE OF THE MEAN AND THE STANDARD DEVIATION

The mean and the standard deviation stand out among all other possible measures of average and variation, respectively. Despite their relative complexity in comparison with the other suggested measures, they are used by people in applied work and theoretical research to a degree unmatched by any of the other measures. Why is this so? What information do the mean and standard deviation convey about the data that makes them so universally valuable? The answer to these questions is given by Chebyshev's theorem.*

    When we want to communicate information about the data, we ideally want our listener to be able to reconstruct the frequency polygon of the data after we finish talking. If we present the information, but the listener does not get a good idea of the nature of the data, then we have wasted our time making the presentation.

    Suppose you hear that Fosbert Realty takes a mean time of 14 days to sell the houses it lists, with a standard deviation of 4 days. Now you know that the waiting times in days for the houses listed with Fosbert to be sold have a mean $\mu = 14$ and standard deviation $\sigma = 4$. How well do you understand the data? Do you understand the data well enough to know, for example,

---

*The name of P. L. Chebyshev (St. Petersburg, Russia, 1821–1894) is sometimes spelled "Tschebyscheff" or other variants. These latter forms are really transliterations into German of the Russian spelling of Chebyshev's name, holdovers from the time when German was the standard language of technical communication.

how many of Fosbert's houses sell within 25 days or how many stay on the market longer than 30 days? As it turns out, Chebyshev's theorem provides useful answers to both of these questions.

Before proceeding with our work, it is therefore necessary to develop an understanding of how Chebyshev's theorem turns the mean and standard deviation into useful information about the data. Chebyshev's theorem can be stated as follows:

CHEBYSHEV'S THEOREM
If a set of data points has mean μ and standard deviation σ, then the proportion p of data lying farther from the mean than k standard deviations cannot exceed $1/k^2$.

Fig. 2.4 illustrates the meaning of Chebyshev's theorem. Forming an interval having a length of k standard deviations, namely $k\sigma$, on both sides of the mean μ, Chebyshev's theorem says that not more than a proportion $1/k^2$ of the data points can possibly fall outside this interval. This implies two equivalent assertions:

---

1. Not more than a proportion $1/k^2$ of the data points can have numerical values smaller than $\mu - k\sigma$ or larger than $\mu + k\sigma$;
2. At least a proportion $1 - (1/k^2)$ of the data points must have values between $\mu - k\sigma$ and $\mu + k\sigma$.

For example, considering an interval of 3 standard deviations about the mean, we see that $k = 3$ and we know by assertion 1 above that no more than

FIGURE 2.4   Chebyshev's Theorem

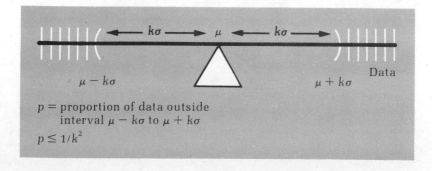

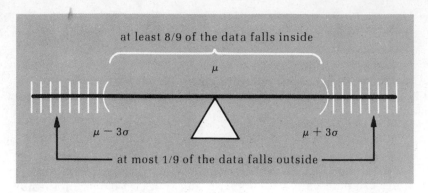

FIGURE 2.5   Chebyshev's Theorem with k = 3

$1/k^2 = 1/9$ of the data can fall outside the interval. This automatically implies assertion 2, namely, that at least $1 - (1/k^2) = 1 - (1/9) = 8/9$ must be falling inside. This situation is illustrated graphically in Fig. 2.5.

Let's take another look at the questions involving Fosbert Realty, where the relevant data have mean $\mu = 14$ and standard deviation $\sigma = 4$. The question of how many houses remain on the market longer than 30 days can be transformed into the illustration of Fig. 2.6. We are interested in the proportion of data points falling above 30. But 30 lies at a distance of $30 - 14 = 16$ from the mean $\mu = 14$, so we are interested in the proportion of data points falling farther from the mean than a distance of 16. Because $\sigma = 4$, the distance 16 represents a distance of $k = 16/\sigma = 16/4 = 4$ standard deviations. By Chebyshev's theorem, the proportion of data points farther from the mean than $k = 4$ standard deviations cannot exceed $1/k^2 = 1/4^2 = 1/16 = 0.0625 = 6.25\%$. Therefore, we know that no more than 6.25% of Fosbert Realty houses remain unsold after 30 days on the listing.

FIGURE 2.6   Fosbert Realty Data: Proportion of Houses
on Market Longer than 30 Days

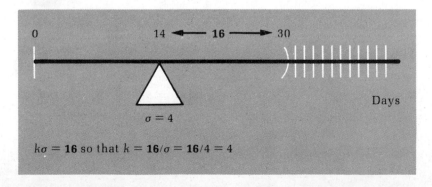

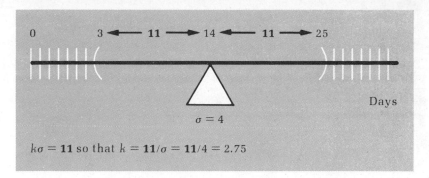

FIGURE 2.7   Fosbert Realty Data: Proportion of Houses Selling within 25 Days

The second question involving the Fosbert Realty data asks what proportion of Fosbert's houses sell within 25 days. Fig. 2.7 illustrates this question. The mean $\mu = 14$, so the relevant number 25 lies at a distance $25 - 14 = 11$ from the mean. Since $\sigma = 4$, the distance 11 can be expressed as $k = 11/4 = 2.75$ standard deviations. Chebyshev's theorem then asserts that no more than $1/k^2 = 1/2.75^2 = 1/7.5625 = 0.132 = 13.2\%$ of the data points can lie farther from the mean than 2.75 standard deviations. This means that no more than 13.2% of the data points can fall outside the interval from 3 to 25. In particular, no more than 13.2% of Fosbert Realty's houses are sold on days outside the interval 3 to 25 days after listing. Therefore, at least 86.8% ($= 100\%$ minus 13.2%) of the houses are sold between 3 and 25 days after listing. We can therefore be sure that at least 86.8% are sold within 25 days of listing.

It is possible that more than 86.8% are sold within 25 days because the 86.8% figure does not include any houses sold within 3 days. Houses in the latter category are included among the 13.2% lying outside the interval. Unfortunately, Chebyshev's theorem does not provide any breakdown of the percentages falling above 25 and below 3, but only percentages falling outside and inside various intervals.

We can summarize the sort of information available from Chebyshev's theorem in a table such as that of Table 2.12, which is based on Fig. 2.4. Using Table 2.12 together with a knowledge of only the mean and the standard deviation, we can estimate for any set of data points the proportions of data falling in and out of various intervals about the mean.

You should observe from Table 2.12 that, no matter what the set of data, no more than 1% of the data points can fall farther away from the mean than a distance of 10 standard deviations, and no more than 4% of the data points farther away than 5 standard deviations. It automatically follows that at least 96% of the data must fall within 5 standard deviations, and at least 99%

TABLE 2.12   Information Given by Chebyshev's Theorem

Distance from Mean to End of Interval (in standard deviations)	Maximum Possible Percentage of Data Outside Interval	At Least This Much Data Falls within Interval
k	$1/k^2$	$1 - (1/k^2)$
1	100.00%	0.00%
2	25.00%	75.00%
3	11.11%	88.89%
4	6.25%	93.75%
5	4.00%	96.00%
6	2.78%	97.22%
7	2.04%	97.96%
8	1.56%	98.44%
9	1.23%	98.77%
10	1.00%	99.00%

Note: k does not have to be a whole number.

within 10 standard deviations. Analogous statements can be made for any number k of standard deviations, regardless of the shape of the frequency polygon of the data.

EXAMPLE 2.1.   Job Aptitude Test.   There are 80 new jobs opening up at an airplane manufacturing plant, but 1,100 applicants show up for the 80 positions. To select the best 80 from among the applicants, the personnel department gives a combination physical and written aptitude test, which covers mechanical skill, manual dexterity, and analytical ability. The mean grade on this test turns out to be 175 and the scores have standard deviation 10. Can a person who scored 215 count on getting the job?

SOLUTION.   From the way the question is formulated, what we need to find out is the number of persons scoring above 215. This information would be useful, for if fewer than 80 applicants could have scored above 215, then a person who scored 215 would be among the top 80 and would therefore get one of the open positions.

Fig. 2.8 shows that the distance $k\sigma$ from the mean of 175 to the end 215 of the interval is $215 - 175 = 40$. But $k\sigma = 40$ means that $k = 40/\sigma = 40/10 = 4$ and $1/k^2 = 1/4^2 = 1/16 = .0625 = 6.25\%$. According to Chebyshev's theorem, this means that no more than 6.25% of the applicants could have obtained scores differing from 175 by more than

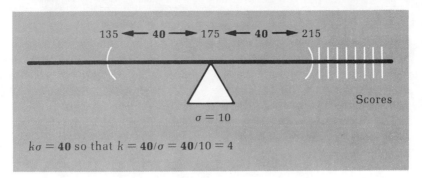

FIGURE 2.8   Aptitude Test Scores

40 points. That is, if we look at all those who scored below 135 and all those who scored above 215, then both those together will account for no more than 6.25% of all applicants. (Conversely, at least 93.75% of the applicants must have scored between 135 and 215.) In particular, the most we can say is that no more than 6.25% could have scored above 215, in view of the fact that Chebyshev's theorem does not allow us to say how much of the 6.25% falls below 135 and how much above 215.

Therefore, because $.0625 \times 1100 = 68.75$ (so that 6.25% of 1100 $= 68.75$), we can be sure that no more than 68 of the applicants could have scored higher than 215. (Even though 68.75 is rounded up to 69, there cannot be 69 applicants scoring above 215, because no more than 68.75 could have done so.) Therefore, since there are 80 positions, a person scoring 215 indeed gets one of the jobs, for no more than 68 people could have scored higher.

EXAMPLE 2.2.   Contents of Cereal Boxes.   The law in one midwestern state regulating producers of 9-ounce boxes of breakfast cereal asserts that no more than one-half of 1% of the output of any brand of cereal (in boxes labeled as containing 9 ounces) are allowed to contain less than 8.8 ounces. One company markets a 9-ounce box of Corn Puffies, which is packed by machine in such a way that the boxes have mean weight 9.0 ounces, with a standard deviation of 0.01 ounce. Is the company in violation of the law?

SOLUTION.   To test whether the law is being obeyed, it is necessary to find out what proportion of 9-ounce boxes of Corn Puffies contain less than 8.8 ounces. If this proportion turns out to be below .005 ($=$ one-half of 1%), then the company is conducting a legal operation. If not, some further investigation might be required. In Fig. 2.9, we are interested in the proportion of data falling in the shaded region, i.e., the proportion of data less than 8.8. Here the distance $k\sigma$ from the mean of

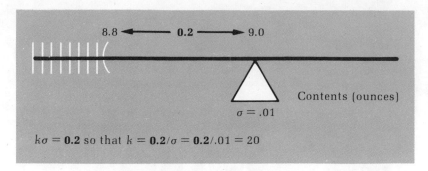

FIGURE 2.9   Cereal Box Contents (in ounces)

9.0 to the end of the interval at 8.8 is 0.2. But $k\sigma = 0.2$ means that k $= 0.2/\sigma = 0.2/.01 = 20$ and $1/k^2 = 1/(20)^2 = 1/400 = .0025$.

Therefore, by Chebyshev's theorem, no more than .0025 ($=$ one-quarter of 1%) of the Corn Puffies boxes will contain less than 8.8 ounces. Since a proportion up to .005 is allowed, the Corn Puffies company is well within legal standards, because its proportion of under-filled boxes does not exceed .0025.

Before we leave this subject, it should be emphasized that the proportion $1/k^2$ (given by Chebyshev's theorem as the maximum possible falling outside the $k\sigma$-interval surrounding μ) is often much larger than the actual proportion there. This is due to the fact that Chebyshev's theorem makes use of only the mean and standard deviation of the data and must take care of all possible distributions of data having the same mean and standard deviation. Very rarely, therefore, do we run into a set of data where the proportion of data outside the $k\sigma$-interval is as large as $1/k^2$. Most often, the actual proportion is considerably smaller than $1/k^2$. We say that estimates based on Chebyshev's theorem tend to be very conservative: although it is true that the proportion outside can be *no more than* $1/k^2$, it is very often a lot less.

EXERCISES 2.D

2.37. Even among cars produced on the same assembly line, rates of emission of environmental pollutants differ. In particular, the revolutionary new Peccarie, which gets 52 mpg at freeway speeds, emits an average of 1.2 mg. of pollutants per mile, with a standard deviation of .04 mg. New EPA regulations prohibit the sale of the entire production run of an automobile brand if more than 2% of the cars in the run emit pollutants in excess of 1.5 mg. per mile. Can the Peccarie be legally sold in the U.S.?

2.38. State guidelines on the number of units taken by college students during their undergraduate careers will soon specify, in the interests of reducing costs, that no student ought to be permitted to exceed 220 quarter units of work or equivalent: 186 units are required for graduation. If graduating students at Bolsa Chica State average 203 units, with a standard deviation of 4 units, at least what proportion of students already satisfy the new guidelines?

2.38. An electrical firm manufactures a 100-watt light bulb that, according to specifications printed on the package, has a mean life of 800 hours, with a standard deviation of 40 hours. At most, what percentage of the bulbs fail to last even 700 hours?

2.40. Since the 55 mph speed limit on freeways has been in effect, observers estimate that vehicles on the freeways travel with mean speed 55 mph and standard deviation 2 mph. If these estimates are accurate, at most what percentage of the vehicles travel at speeds in excess of 65 mph?

2.41. The A. C. Nielsen Company found that on any given night during the last three months of 1976 the average number of people watching evening television was 88 million. If the standard deviation were determined to be 3.5 million, on how many days at most did the number of viewers drop below 70 million?

2.42. Until ten years ago, the United States Smelting, Refining, and Mining Company had a mill at Midvale, Utah, which produced a mean of 26,000 tons of lead concentrate per year over the past 36 years. The standard deviation of the annual production was 3,500 tons. In how many of those 36 years at most could production have fallen below 15,500 tons?

2.43. A manufacturing company has a contract calling for several thousand units of steel pipe having a diameter between 4.56 and 4.62 inches. If a unit of pipe has diameter outside this range, it will not fit into the intended construction position and so must be wasted. Only those units having diameter within the range are acceptable to the customer. Initial output of pipe from the assembly line projects a mean diameter of 4.59 inches, with a standard deviation of .005 inches. At least what percentage of the output will be acceptable to the customer if this trend continues?

2.44. *Business Week*, in a recent article on the soft drink industry, reported that the average consumption for all age groups was 547 cans per year. If we find that the standard deviation is 21, would it be reasonable that of 400 randomly selected people, 100 drank over 610 cans per year?

2.45. The stock exchange is open 250 days each year. In a study of the fluctuation of the closing price of the stock of MGI, Inc., a computer readout indicated that the mean closing price for last year was 22.375, with a standard deviation of 5,000. At most, how many days did MGI stock close below 10?

2.46. After studying the incomes of families in a large urban area, a sociologist concludes that the mean income is $9,000, with a standard deviation of $2,000. At most, what proportion of families have incomes higher than $25,000?

2.47. An automobile battery carrying a 24-month guarantee actually has a mean lifetime of 30 months, with a standard deviation of 2 months.

a. At most, how many of these batteries fail before 24 months have passed?

b. The manufacturer is considering reducing the guarantee period in order to cut back on the expense of honoring the guarantee. At least how many of the batteries last longer than 20 months?

2.48. Auto emission standards in one eastern state allow the imposition of additional registration fees depending on the amount of vehicular pollutants emitted yearly. For example, an additional fee of $9 per year is imposed on cars emitting between 23 mg. and 29 mg. of pollutants per mile. If the average car in that state emits 26 mg. per mile with a standard deviation of 2 mg., at least what proportion of cars is assessed the additional $9 fee?

2.49. In determining the cost of a yearly service contract on a washing machine or dishwasher, the manufacturer has to take into account the likely number of repair visits required each year to the customer's home. The average customer requires 2.3 visits per year with a standard deviation of 0.4 visit. What proportion of customers, at most, require more than four visits per year?

2.50. The Suede Tortoise, a chain of fancy restaurants in the metropolitan area, obtains the bulk of its income from "three-martini" lunches sold to business executives and sales representatives. In fact, such luncheon checks average out to $23.50 per person, with a standard deviation of $0.60. At least what proportion of customers spend between $20 and $27 at the Suede Tortoise?

2.51. The mean price increase for single-family homes in the Los Angeles metropolitan area was 20% during the last calendar year, with a standard deviation of 5%. At most, what proportion of homes increased in price by 50% or more over the past calendar year?

2.52. In order to have its provisional operating license converted after one year into a permanent one, a taxi company in a large midwestern city is required to show that no more than 3% of those customers who phoned in for a taxi had to wait longer than 30 minutes for its arrival. A city-sponsored study revealed that the Purple Taxi Company's phone customers had to wait a mean of 21 minutes, with a standard deviation of 1.5 minutes. Did the company meet the city's licensing requirement?

2.53. The IRS, under current policy, sends out notices to all taxpayers who receive tax refunds in excess of $200, reminding them to have less tax withheld next year. If the mean refund is $78, with a standard deviation of $23, at most what proportion of taxpayers receive such notices?

2.54. Unemployment figures for each of the 435 congressional districts in the United States reveal that the mean unemployment rate per congressional district was at one time 9.2 percent with a standard deviation of 1.6 percent. At least what percentage of districts had unemployment rates in excess of 4 percent?

SECTION 2.E   STANDARDIZATION OF DATA

A local automobile dealer, new to the community, has decided that she needs an agressive advertising program in order to establish herself in the market. Because a limited budget precludes the possibility of advertising in both television and newspapers, she must choose between the two media by comparing the effective sizes of the markets reached.

Unfortunately for comparison purposes, local television programs are rated for advertising effectiveness on the Electronic Media Scale (EMS), whose ratings have a mean of 500 with a standard deviation of 100; while local newspapers are judged for advertising effectiveness by the Print Media Rating (PMR) system, whose scores have a mean of 60 with a standard deviation of 20. Both the EMS and the PMR provide ratings of effectiveness in reaching the automobile-buying public. The dealer would like to have a ranking of various television programs and newspapers, according to advertising effectiveness, so that she can compare costs and projected results of each. Such a ranking would allow a more logical commitment of resources to the advertising budget.

Table 2.13 presents the ratings for 3 television programs (according to the EMS) and 2 newspapers (according to the PMR). How is the dealer to compare and rank the 5 ratings, some of which are based on one numerical scale and the rest on another? For example, how do we know which is the better rating—a 700 on the EMS or a 110 on the PMR?

The way out of the problem is to adjust, if possible, all the scores so that they will all be on the same scale. Then it will be easy to see which scores are higher than others, relatively speaking. Such an adjustment procedure is called *standardization* of scores, and the resulting adjusted scores are called *standard scores* or *z scores*. (To distinguish between the z scores and the original scores, the latter are often referred to as *raw scores*.) The standardization procedure is another highly useful application involving the mean and the standard deviation, and further contributes to the desirability of these measures of average and variation.

Standard scores can be best described as follows:

STANDARD SCORES

If x is a data point of a set of data having mean μ and standard deviation σ, then the standard score (or z-score) of x is the number z, calculated by the formula

$$z = \frac{x - \mu}{\sigma}.$$

Notice that to find the z score of a data point x, we subtract the mean of its group from x and divide the resulting difference by the standard deviation of its group.

Using this formula for z scores, let's find the z score for TV program A of Table 2.13. TV program A received a rating of 700 on the EMS, a rating

TABLE 2.13   Ratings of Advertising Effectiveness

Advertising Medium	Code	Scale	Rating
After-hours movie	(TV program A)	EMS	700
Breakfast show	(TV program B)	EMS	450
Clarion Call	(Newspaper C)	PMR	70
Daytime quiz show	(TV program D)	EMS	600
Evening Standard	(Newspaper E)	PMR	110

scale that has mean 500 and standard deviation 100. Therefore the z score is

$$z_A = \frac{700 - 500}{100} = \frac{200}{100} = 2.$$

On the other hand, let's take a look at newspaper C. This advertising medium scored 70 on the PMR, a rating system on which the mean rating is 60, with a standard deviation of 20. Therefore, newspaper C has z score

$$z_C = \frac{70 - 60}{20} = \frac{10}{20} = 0.5.$$

Now the z scores of A and C are 2 and 0.5, respectively. The formula $z = (x - \mu)/\sigma$ really gives the number of standard deviations above the mean of its group that the data point x lies. In particular, TV program A scored 2 standard deviations above the mean of its group, while newspaper C scored only 0.5 (one-half) of a standard deviation above the mean of its group. Relatively speaking then, TV program A is rated somewhat higher than newspaper C.

To compute all the z scores, we set up a table of calculations like that of Table 2.14, recalling that EMS ratings have overall mean 500 and standard

TABLE 2.14   Calculation of Standard Scores (z scores)

Advertising Medium	Scale	Raw Score (x)	Mean of Group (μ)	Standard Deviation (σ)	$x - \mu$	z Score $\left(\dfrac{x - \mu}{\sigma}\right)$
A	EMS	700	500	100	200	2.0
B	EMS	450	500	100	-50	-0.5
C	PMR	70	60	20	10	0.5
D	EMS	600	500	100	100	1.0
E	PMR	110	60	20	50	2.5

TABLE 2.15   Advertising Media Ranked by z Scores

Rank	Advertising Medium	z Score
1	E	2.5
2	A	2.0
3	D	1.0
4	C	0.5
5	B	−0.5

deviation 100, while PMR ratings have overall mean 60 and standard deviation 20. Having standardized all the scores, we can now easily rank all the advertising media. Newspaper E had the highest score, 2.5 standard deviations above the mean of the PMR group, while TV program B got the lowest score, 0.5 standard deviation *below* the mean of the EMS group. The ranking of the advertising media appears in Table 2.15.

Now that we have discussed the process of standardizing sets of data that are measured according to different scales, it would be useful to know the exact way in which the data are "standardized." Standardization of each set of data does the following: it transforms the original (raw) data into a set of numbers (called z scores), which have mean 0 and standard deviation 1. In fact, no matter what the original values of μ and σ were, the z scores always have mean 0 and standard deviation 1.

Once two or more sets of data are individually standardized, they can be graphed on the same scale and compared. This graphing process is illustrated in Fig. 2.10.

FIGURE 2.10   The Process of Standardization

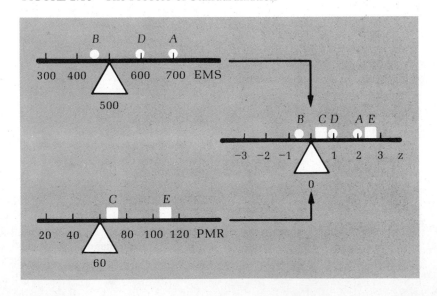

EXERCISES 2.E

2.55. Six applicants are applying for an opening as cost accountant in a small manufacturing firm that produces pipe fittings. Each applicant is given a test divided
into three parts: ability to concentrate, computer knowledge, and knowledge of
cost accounting. Their grades on each part are as follows:

Applicant	Ability to Concentrate	Computer Knowledge	Knowledge of Cost Accounting
A	10	9	5
B	9	9	6
C	5	2	0
D	5	9	10
E	1	0	6
F	0	1	9

In order to make fair comparisons, the grades are first standardized within each
competency area, and the applicant with the highest total of z scores gets the
job.

a. Standardize the test scores of each applicant on the "ability to concentrate"
section.

b. Standardize all the "computer knowledge" scores.

c. Standardize the "knowledge of cost accounting" scores.

d. Which applicant gets the job?

2.56. One business student took three exams on the same day. Her scores, as well as
the mean and standard deviation of all the scores of her group, are presented
in the following table:

Subject Matter	Student's Score	Group Mean	Group Standard Deviation
Economics	72	70.8	0.6
Mathematics	61	51.0	6.0
Psychology	50	30.0	12.0

On which exam did the student do best in comparison with the rest of her
group?

2.57. Four applicants for flight training, two from Portland, Oregon, and two from
Tampa, Florida, have submitted their scores on different flight aptitude tests.
The two from Oregon had taken the Great Open Skies Flight Aptitude Test
(GOSFAT), while the pair from Florida had taken the Gulf and Southern Flight
Aptitude Test (GASFAT). Scores on the GOSFAT have mean 50 with standard
deviation 10, while those on the GASFAT have mean 120 and standard deviation
15. The applicants and their scores are listed below:

Applicant	Test Taken	Score
#1	GOSFAT	60
#2	GOSFAT	45
#3	GASFAT	129
#4	GASFAT	115

Standardize each score and rank the applicants.

2.58. Make up a set of five data points and standardize them. Then verify by calculation that the z scores have mean 0 and standard deviation 1.

SUMMARY AND DISCUSSION

We have devoted Chapter 2 to the development of numerical characteristics of a set of data and to an analysis of their value in describing the data. We have mentioned four types of averages, the mean, the median, the mode, and the midrange, and three measures of dispersion, the standard deviation, the mean absolute deviation, and the range, and we have carefully noted the distinctions between them.

Through Chebyshev's theorem and standard scores, the mean and the standard deviation have been seen to be the most useful of the measures of average and dispersion, respectively. We have observed that when we apply Chebyshev's theorem to the mean and standard deviation of a set of data points, we can often answer very specific questions about the situation described by the set of data.

We have discussed standardization of sets of data, with the aim of being able to compare widely differing sets of data by placing all of them on the same scale. The material of Chapter 2 provides the framework for all the discussions of data to follow in the remainder of the text.

CHAPTER 2   BIBLIOGRAPHY

REVIEW OF USEFUL MATHEMATICAL BACKGROUND

AUSLANDER, L., et al., *Mathematics Through Statistics*. 1973, Baltimore: Williams & Wilkins.
BRAVERMAN, H., *Reviewing Statistics*. 1971, New York: Robin Hill.

MORE ON NUMERICAL DESCRIPTIONS

CHISSON, B. S., "Interpretation of the Kurtosis Statistic," *The American Statistician*, October 1970, pp. 19–22.
CURETON, E. C., Letter to the Editor, *The American Statistician*, December 1971, p. 61.
DARLINGTON, R. B., "Is Kurtosis Really 'Peakedness'?" *The American Statistician*, April 1970, pp. 19–22.
FAMA, E. F., "The Behavior of Stock Market Prices," *The Journal of Business*, 1965, pp. 34–105.
HAMMOOD, A. W., "On Extensions of Probabilistic Profit Budgets: A Comment," *Decision Sciences*, July 1976, pp. 567–570.

HILDEBRAND, D. K., "Kurtosis Measures Bimodality?" *The American Statistician*, February 1971, pp. 42–43.

SHARPE, W. F., "Mean-Absolute Deviation Characteristic Lines for Securities and Portfolios," *Management Science*, October 1971, pp. 1–13.

SHER, L., "The Range of the Standard Deviation," *The MATYC Journal*, Fall 1977, p. 197.

CHEBYSHEV'S THEOREM

GHOSH, M. and G. MEEDEN, "On the Non-attainability of Chebyshev's Bounds," *The American Statistician*, February 1977, pp. 35–36.

CHAPTER 2   SUPPLEMENTARY EXERCISES

2.59. A *Business Week* survey provided the following information about the cost of operating a car in various American and foreign cities. The figures, published in November 1976, included the cost of 10 gallons of gasoline and downtown parking for 5 days.

Atlanta	$13.30	Moscow	$30.00
Boston	21.20	New York	35.70
Chicago	31.60	Paris	60.00
Houston	15.00	Washington	27.00
London	29.50	San Francisco	30.00
Los Angeles	24.00	Tokyo	34.35
Mexico City	11.00		

a. Calculate the mean cost of operating a car in the cities surveyed.

b. Find the median.

c. Determine the mode.

d. Find the midrange of the costs.

e. Calculate the standard deviation.

f. Calculate the mean absolute deviation of the costs.

g. Find the range.

2.60. A local bank having 12 branch offices is considering adopting a completely computerized payroll system. As part of an analysis of the operating costs of the system, the costs of running one aspect of the system are carefully observed for the 12 branches, with the following resulting data on costs per hour of operating the system:

24	20
14	24
14	14
14	20
12	14
7	12

a. Calculate the mean operating cost for the 12-branch bank.

b. Find the median operating cost.

c. Find the modal operating cost.

d. Find the midrange of the operating costs.

e. Calculate the standard deviation of the 12 operating costs.

f. Calculate the mean absolute deviation of the costs.

g. Find the range of the operating costs.

2.61. The following data points from the files of a major life insurance company record the number of consecutive weeks that medical disability compensation was paid to 14 heart attack victims:

9	20
21	27
14	13
23	10
8	32
16	18
7	20

a. Calculate the mean number of weeks that compensation was paid.

b. Find the median number of weeks compensation was paid.

c. Find the mode.

d. Find the midrange.

e. Calculate the standard deviation of the number of weeks of paid compensation.

f. Calculate the mean absolute deviation of the number of weeks of paid compensation.

g. Find the range of the number of weeks of paid compensation.

2.62. *Business Week* regularly publishes forecasts by various economists and organizations of future unemployment rates. One recent set of predictions went as follows (percentages):

7.0	7.1	6.9	7.3	7.4	7.3
7.0	7.2	7.2	7.0	7.6	7.1
7.2	7.4	7.3	7.0	7.5	7.3
7.0	6.9	7.1	7.5	7.5	7.5
7.4	6.7	7.2	7.4	7.2	
7.2	7.0	6.9	7.4	7.3	

a. Calculate the mean.

b. Draw a seesaw diagram of the data points and indicate the location of the mean.

c. Determine the median.

d. Locate the mode.

e. Determine the midrange.

f. Calculate the standard deviation.

g. Find the range.

2.63. The A. C. Nielsen Company reported a November 1976 survey of TV viewing habits, which revealed that the average viewing time per person for one week was approximately 28.75 hours. A pilot study showed that the standard deviation was 1.1 hour. At most, what percent of the people watch TV over 36 hours per week?

2.64. Insurance claims arrive at Buzzardbait National Life & Casualty's home office at a mean rate of 140 per week, with a standard deviation of 12. At most, what proportion of weeks does the number of claims exceed 200?

2.65. An accountant commutes by automobile from his suburban home to his office in a downtown corporate headquarters. The daily trip takes, on the average, 24 minutes with a standard deviation of 4 minutes. If he were to regularly leave his home at 8:20 A.M., at least how often would he manage to make it to the office in time for the staff summary meeting at 9:00 A.M.?

2.66. A recent article described how a corporate treasurer in New England was able to lengthen the time it took to clear his checks by using a bank in California. He found that his checks to local creditors took an average of 7.5 days to clear, thus giving him that additional time to use his money. If the standard deviation were found to be .5 day:

a. At most, what percentage of the checks will take more than 9 days to clear?

b. What is the highest possible percentage of checks that could clear in under 5 days?

2.67. In an article on the soft drink industry, *Business Week* reported that those between the ages of 13 and 24 consume an average of 823 cans of soft drinks per year, with a standard deviation of 30. At most, what percentage of this group consume more than 1000 cans per year?

2.68. A multinational corporation having several offices in Europe wants to set up a program of teaching its young management trainees to read and speak German early in their careers. Preliminary language-teaching studies indicate that a trainee will need a mean of 22 months with a standard deviation of 4 months to obtain a reasonable skill level in German for business communication. At least what proportion of trainees can reach such a level within 36 months?

2.69. One local supermarket carries 20,000 different food items. With an eye toward both publicity and cost cutting, it announces that it will refuse to carry any item that has risen in price by more than 25% during the previous year. If food prices rose by an average of 12% one year, with a standard deviation of 4%, at least how many of the original 20,000 items were still on the shelves the next year?

2.70. An urban planner's data indicate that, on Los Angeles area freeways between 4 P.M. and 6 P.M. weekdays, there are 400,000 vehicles. These vehicles, furthermore, spend a mean time of 45 minutes on the freeways, with a standard deviation of 15 minutes. At most, how many vehicles spend 90 minutes or more on the freeways?

2.71. Over the past several years, the daily prime rate of interest at major New York banks averaged 10.2% with a standard deviation of 0.4%. At most, what proportion of the time has it exceeded 15.8%?

2.72. Several years ago a group of urban planners constructed a model of a "planned" ideal city, of which one goal was that no more than 2% of the residents would hav to commute longer than 20 minutes from home to work. When the actual city was completed recently, the inhabitants' mean travel time from home to work turned out to be 12 minutes, with a standard deviation of 1 minute. Was the planners' goal achieved?

2.73. Over a several-year period in the recent past the national rate of "usually employed" persons who are temporarily receiving unemployment compensation averaged 7.2% per month, with a standard deviation of 1.4%. At most, what proportion of months had a rate in excess of 12.1%?

2.74. The IRS announced in April 1977 that the average refund on the 1976 returns that had been processed up to that time was $440. If the standard deviation had turned out to be $50 and if this trend had continued, what percentage of the refunds, at most, would have exceeded $700?

2.75. Cars now on the road nationwide get 14 miles per gallon of gasoline, on the average, with a standard deviation of 3 miles per gallon. A plan now under study proposes to encourage engine efficiency by rebating all federal gasoline taxes to owners of those cars ranking among the top 10% in efficiency. If your car gets 25 miles to the gallon, would you qualify for a rebate?

2.76. The Car Leasing Division of Hertz Corporation found that the average cost of operating a 1981 intermediate two-door sedan driven 10,000 miles annually, which is kept three years (cost includes gas, insurance, oil, parts and service, interest and depreciation), is 38.1 cents per mile. Suppose that for a leasing company to find it profitable to lease this particular type of car, the cost per mile must be between 36.6 cents and 39.6 cents. If the standard deviation were determined to be .4 cent, at least what percentage of the two-door intermediates would be acceptable to the company?

2.77. A two-bedroom apartment in Tustin rents for $485 per month, on the average, with a standard deviation of $20 per month. At least what proportion of two-bedroom apartments in Tustin rent for $440 to $530 per month?

2.78. The dollars (considered as individual dollar bills) now outstanding in loans made by Oceanfront Federal Savings of Newport Beach return an average interest rate of 6.82% (some of those old loans were made at pretty low rates!), with a standard deviation of 0.58%. At most, what proportion of the dollars are returning less than 4.5%?

2.79. The Willowmart Department Stores chain appeals to consumers whose family income falls between $13,000 and $18,000 per year. In a region where the mean family income is $15,500 per year, with a standard deviation of $500, at least what proportion of families are in the target income range?

2.80. Amalgamated Marketing Enterprises, Inc., found that its products appeal mainly to those households in the $16,000-and-over income range. It turns out that radio station CLAM in Halifax attracts an audience that has a mean household income level of $21,240 with a standard deviation of $2,800. At least what percentage of CLAM's listeners are potential amalgamated customers?

2.81. A recent wage and compensation study conducted by the Ray Research Council revealed that mean employee pay increases were 9.1% over the year, with a standard deviation of 2.4%. At most what fraction of employees received pay increases in excess of the inflation rate of 18.6%?

2.82. The mean (per person) 1980 salaries and wages of all employees covered by Social Security was $17,400 with a standard deviation of $3,500. The maximum amount of salary and wages subject to Social Security taxes in 1980 was $25,900. At most, what proportion of employees had incomes in excess of the amount subject to Social Security tax?

2.83. Because of the immediate need for additional skilled technicians and assembly-line employees in the local aerospace industry (combined with an inability to recruit out-of-area employees due to excessive local housing costs), it is rather easy for persons with little experience and only the basic required skills to get a job. In fact, all applicants scoring above 50 on the aptitude test (which has mean score 120 and standard deviation 14) receive a job offer. Out of an initial pool of 1300 applicants, at least how many receive a job offer?

2.84. Since most price restrictions were removed from national and international airline fares, there has been a great variety of prices available on airline seats from Los Angeles to London. In fact, the actual prices paid per seat during a recent month was $324 with a standard deviation of $41. At most, what percentage of Los Angeles-to-London seats sold for under $200?

2.85. Wildcat oil wells in the Rocky Mountain Overthrust Belt have to be drilled through an average of 12,000 feet of hard rock with a standard deviation of 1,200 feet. What percentage of wells, at most, have to be drilled through more than 18,000 feet of hard rock?

2.86. In a Financial Accounting course at a local college, three exams are given during the semester. All three exams count equally toward the student's final grade; nothing else is taken into consideration. The following table gives the mean and standard deviation of the scores of the entire class on each of the exams:

	First Exam	Second Exam	Third Exam
Class mean	65	74	71
Class standard deviation	4	6	12

One of the students, Xerxes by name, got a 57 on the first exam, a 76 on the second exam, and a 96 on the third, while another student, Yennie, obtained a 73 on the first, an 86 on the second, and a 70 on the third. Yennie was awarded a grade of "B" for the course, while Xerxes received only a "C." Xerxes protests to the instructor, of course, because his average on the three exams is exactly the same as Yennie's, so he feels that he deserves the same grade. Unfortunately for Xerxes, however, the instructor standardized each exam score with respect to the overall class performance on that exam before computing the student's grade.

   a. Calculate each of Xerxes' standardized scores, and compute the average of the three.

   b. Standardize each of Yennie's exam scores, and average them.

   c. Explain why Yennie got a "B" and Xerxes a "C."

II

PROBABILITY
DISTRIBUTIONS

3

Probability Calculations

As we stated in the first sentence of Chapter 1, "Statistics exists to solve problems arising from the need to make decisions based on incomplete or uncertain information." The theoretical framework that has been constructed in order to measure the degree of incompleteness or uncertainty in our information is called *probability*.

Suppose, for example, that a life insurance company wants to be 99% sure that it will be able to pay all accidental-death claims out of current premium receipts, without having to sell off any assets. If the market-competitive monthly premium of a $10,000 term accidental-death policy on a 35-year-old man in a certain life-style category is $4, and there are 2,500 men in that category who have purchased policies, the company will have to calculate very carefully the probability that two or more policyholders will die by accident in the same month. If this probability is greater than .01 (= 1%), the company will not be able to meet its 99% goal. Insurance companies having problems along these lines can (1) raise the premium to cover the excess probability, (2) restrict sales to life-style categories having lower probabilities of accidental death, or (3) lower its claim-payment probability goal below 99%.

As an illustration of how probability arises in a different decision-making context, consider a national retail sales chain that test-markets a new consumer-oriented household appliance in 12 geographical areas. Before the chain acts on the decision to market the appliance in all of its outlets nation-

wide (thereby occupying limited display space, and involving sales personnel and financial resources that might be better utilized elsewhere), management wants to be reasonably sure that consumer demand for the appliance
will reach a satisfactory level. Suppose the appliance sells acceptably well
in 8 of the 12 test regions. With what probability (degree of certainty) can we
say that it will sell well in at least 400 of the chain's 600 outlets? How likely
is the appliance to sell well in at least half of the 600 outlets? To answer these
questions requires us to make probability calculations of the sort discussed
in this chapter.

SECTION 3.A   PROBABILITIES OF EVENTS

By "probability of an event" we usually mean the proportion of times the
event can be expected to occur over a long period of time. Sometimes, however, events cannot be repeated (for example, in attempting to find the probability that an earthquake will destroy Los Angeles), and then the probability
will have to mean our view of the chances of the occurrence of the event,
based on prior information in our possession.

The collection of all possible outcomes of an experiment, survey, or
other method of data collection is called the *sample space* of the experiment.
To simplify the explanation of the concepts involved and the terminology
used, we will illustrate all the basic ones in terms of a die, sometimes called
a probability cube. (The plural of die is dice.) We will conduct a detailed
analysis of the following experiment: we roll the die once only and we note
the number of dots on the side facing upward when the die comes to a stop.
If the die comes to a stop as in Fig. 3.1, with the one-dot side facing upward,
we have rolled a 1. The sample space of this experiment consists of the
numbers 1, 2, 3, 4, 5, and 6. We denote this set as

$S = \{1,2,3,4,5,6\}$.

By an *event* we mean a collection of some, but not necessarily all, of
the possible outcomes of an experiment. For example, in the probability cube
experiment, we can consider the event that we roll an even number. This
event can be symbolized by

$E = \{2,4,6\}$.

Other events that may be of interest are:

$L = \{1,2\}$ = we roll a very low number;
$H = \{5,6\}$ = we roll a very high number;
$A = \{1,2,3,4\}$ = we roll a not-high number.

FIGURE 3.1 Die (Probability Cube)

Arithmetic of Events

There is an *arithmetic of events*, a way of combining events, that plays a role somewhat analogous to the one played in ordinary numerical arithmetic by the addition and multiplication of numbers:

1. The *intersection of two events* is the set consisting of all outcomes that occur simultaneously in both of the events. For example, the intersection of the events E and L above (abbreviated $E \cap L$ where the symbol "$\cap$" is pronounced "intersect") is $E \cap L = \{2\}$, because 2 is the only outcome found in both E and L.

2. The *union of two events* is the set consisting of all outcomes that occur either in one or both of the events. For example, the union of the events E and L (abbreviated $E \cup L$ where the symbol "$\cup$" is pronounced "union") is $E \cup L = \{1,2,4,6\}$, because 1 is in L, 4 and 6 are in E, and 2 is in both E and L.

3. The *complement of an event* is the set consisting of all those outcomes in the sample space S that are not in the event itself. For example, the complement of the event A above (abbreviated A^c, which is pronounced "A-complement") is $A^c = \{5,6\}$ because 5 and 6 are the only outcomes in S that are not in A. In view of the fact that $H = \{5,6\}$ also, you should notice that $A^c = H$ because A^c and H consist of exactly the same outcomes.

One more item is also used extensively in this arithmetic, namely the empty event. The *empty event*, denoted by the symbol φ (the lower case Greek letter *phi*, pronounced "fee" or "fie") is the event that consists of no outcomes at all. It plays a role in the arithmetic of events analogous to the role played by the number zero in ordinary arithmetic of numbers. Note that (1) $L \cap H$ = φ because there are no outcomes that are both in L and in H, as $L = \{1,2\}$ and $H = \{5,6\}$; and (2) $A \cap A^c$ = φ because there are no outcomes that are in both A and A^c, as A^c consists of exactly those outcomes that are not in A. The two events L and H are said to be "mutually exclusive" or "disjoint" because $L \cap H$ = φ. In fact, if E and F are any two events such that $E \cap F$ = φ, then they are said to be *disjoint events*.

By means of pictures called *Venn diagrams*, we can illustrate the concepts of intersection, union, complement, and disjoint events. We do this in Figs. 3.2, 3.3, 3.4, and 3.5, respectively.

There is an important relationship among events that we will need to use when we get ready to solve applied problems in business and management. Suppose we have two events, labeled E and F, as in the Venn diagram of Fig. 3.6. Then

$E \cap F$ = the set of outcomes in E that are also in F;
$E \cap F^c$ = the set of outcomes in E that are not in F.

FIGURE 3.2   Intersection of Events

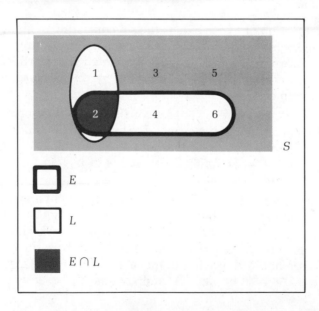

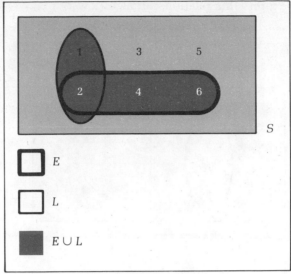

FIGURE 3.3 Union of Events

These two sets together comprise E, as an outcome in E must be either in F or not in F. (In fact, every outcome in the sample space is either in F or not in F.) Therefore, as Fig. 3.6 illustrates, E is the union of the two sets $E \cap F$ and

FIGURE 3.4 Complement of an Event

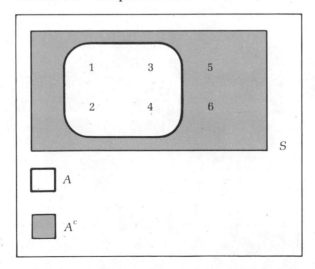

$E \cap F^c$. We can express this relationship as follows:

$$E = (E \cap F) \cup (E \cap F^c).\qquad\qquad \textbf{(3.1)}$$

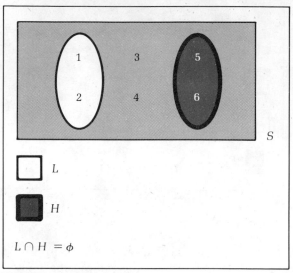

FIGURE 3.5   Disjoint Events

FIGURE 3.6 $E = (E \cap F) \cup (E \cap F^c)$

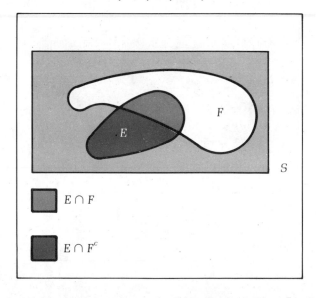

Calculating Probability

By the probability of a repeatable event, we mean the proportion of times that the event can be expected to occur relative to the number of times the experiment is repeated. Every outcome in the sample space has a probability associated with it. A die is said to be *fair* if each of the six sides has the same probability of being in the upward position after a roll. Because there are six possible outcomes in the sample space $S = \{1,2,3,4,5,6\}$, we see that each outcome has probability one-sixth (1/6).

In particular, we have recorded in Table 3.1 the facts that $P(\{1\}) = 1/6$, $P(\{2\}) = 1/6$, $P(\{3\}) = 1/6$, $P(\{4\}) = 1/6$, $P(\{5\}) = 1/6$, and $P(\{6\}) = 1/6$, where the symbols $P(\ )$ are pronounced "the probability of." These facts can be interpreted to mean that, in a large number of rolls of the die, we would expect each outcome to occur about one-sixth of the time.

Let's take another look at the following events:

$$E = \{2,4,6\} \qquad H = \{5,6\}$$
$$L = \{1,2\} \qquad A = \{1,2,3,4\}.$$

Every event has a probability associated with it. What do we mean by $P(E)$, the probability of E? Well, E is the event that we roll an even number, and since there are 3 even numbers among the 6 numbers available, we can expect to roll an even number about three sixths of the time. For this reason, we can agree that

$$P(E) = 3/6 = 1/2.$$

Let's look at $P(E)$ another way. The event E is composed of 3 distinct outcomes, each of which occurs one-sixth of the time, on the average. Therefore, the event E will occur the one-sixth of the time that $\{2\}$ occurs, the one-

TABLE 3.1 Probabilities for a Fair Die

Outcome (number of dots facing upward)	Proportion of Rolls in Which Outcome Occurs	Probability of Outcome
1	1/6	1/6
2	1/6	1/6
3	1/6	1/6
4	1/6	1/6
5	1/6	1/6
6	1/6	1/6

sixth of the time that {4} occurs, and the one-sixth of the time that {6} occurs. From this analysis, we can see that

$$P(E) = P(\{2\}) + P(\{4\}) + P(\{6\}) = 1/6 + 1/6 + 1/6 = 3/6 = 1/2.$$

Both of the analyses used above are correct methods of calculating the probability of the event E. Similar calculations give us the facts that

$$P(L) = 2/6 = 1/3, \quad P(H) = 2/6 = 1/3, \quad P(A) = 4/6 = 2/3.$$

Probabilities of Intersections

We denote by $E \cap L$ the event that the roll of the die results in a number that is both even and low. Because $E \cap L = \{2\}$, it is reasonable to agree that

$$P(E \cap L) = P(\{2\}) = 1/6.$$

In a similar manner, we find that

$$P(E \cap A) = P(\{2,4\}) = 2/6 = 1/3,$$
$$P(L \cap A) = P(\{1,2\}) = P(L) = 1/3,$$
$$P(L \cap H) = P(\phi) = 0.$$

The last assertion that $P(L \cap H) = 0$ can be considered as a statement that there is no chance of having the roll result in a number that is both low and high.

Probabilities of Unions

Now that we know that $P(L \cap H) = 0$, what about $P(L \cup H)$? The event $L \cup H$ is the event that a roll of the die results in either a low or a high (or both, if that were possible) number. As we know, a low number will occur about 1/3 of the time and a high number will occur about 1/3 of the time. It, therefore, seems reasonable to expect that the goal of obtaining an extreme number (low or high) would be attained about 2/3 of the time.

Translating that statement into probability symbols gives the assertion that

$$P(L \cup H) = P(L) + P(H) = 1/3 + 1/3 = 2/3.$$

This estimate is corroborated by the following calculations, using the direct method of counting outcomes:

$$P(L \cup H) = P(\{1,2,5,6\}) = 4/6 = 2/3.$$

Suppose we apply the same method of analysis to the problem of calculating $P(E \cup A)$. We know that an even number occurs about 1/2 of the time, while a not-too-high number occurs about 2/3 of the time. It definitely cannot be true, however, that

$$P(E \cup A) = P(E) + P(A) = 1/2 + 2/3 = 3/6 + 4/6 = 7/6,$$

because we would be saying that $E \cup A$ could be expected to occur about seven out of every six times on the average, which is more than 100% of the time. (Probabilities can never exceed one, for, if they did, the events involved would have to be occurring more than 100% of the time.)

What went wrong with the statement that $P(E \cup A) = P(E) + P(A)$? If we look at the Venn diagram in Fig. 3.7, we observe that the union $E \cup A$ consists of 5 outcomes. It, therefore, seems reasonable to believe that $P(E \cup A) = 5/6$. This belief is substantiated by the direct method of counting outcomes, namely,

$$P(E \cup A) = P(\{1,2,3,4,6\}) = 5/6.$$

How, then, did we manage to come up with the allegation that $P(E \cup A) = 7/6$? If we look closely at the statement

FIGURE 3.7   Probability of the Union of Events

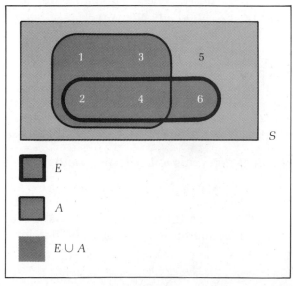

$$P(E \cup A) = P(E) + P(A) = 3/6 + 4/6 = 7/6,$$

we will see, hidden inside it, the following statement:

$$P(E \cup A) = P(E) + P(A)$$

$$= P(\{2,4,6\}) + P(\{1,2,3,4\})$$

$$= 3/6 + 4/6 = 7/6.$$

Take a look at what we have done. We have counted the outcomes 2 and 4 twice: once as part of E and once as part of A. Therefore, we have mistakenly been assuming that $E \cup A$ contains 7 outcomes, i.e., $\{1,2,2,3,4,4,6\}$, but in fact it contains only 5.

Now that we have discovered the mistake, how do we rectify it? How do we correct the erroneous statement that $P(E \cup A) = P(E) + P(A)$? When we were using $P(E) + P(A)$ to calculate $P(E \cup A)$, we were counting twice all those outcomes that are in both E and A, namely all those outcomes in $E \cap A$ = $\{2,4\}$. Because we want to count these outcomes only once, we have counted them once too often. We can correct the mistake by subtracting the overcount of 2 and 4, an amount equal to $P(E \cap A) = P(\{2,4\}) = 2/6 = 1/3$. We can, therefore, write that

$$P(E \cup A) = P(E) + P(A) - P(E \cap A)$$
$$= 3/6 + 4/6 - 2/6 = 5/6,$$

which is the correct probability. We can summarize this development in the following general rule: if E and F are any two events, then

■ $$P(E \cup F) = P(E) + P(F) - P(E \cap F). \tag{3.2}$$

Now that we have established the above general rule, let's go back a page or two and try to figure out how we managed to get away with the statement that

$$P(L \cup H) = P(L) + P(H) = 1/3 + 1/3 = 2/3.$$

If we apply the general rule to $L \cup H$, we should really be writing that

$$P(L \cup H) = P(L) + P(H) - P(L \cap H).$$

What about $P(L \cap H)$? Well, as it turns out, $L \cap H = \phi$ because none of the possible outcomes can be found in both L and H. Therefore $P(L \cap H) = P(\phi)$ = 0. It follows that

$$P(L \cup H) = P(L) + P(H) - P(L \cap H)$$
$$= 1/3 + 1/3 - 0 = 2/3,$$

and so the missing term $P(L \cap H)$ had no effect on the calculation of $P(L \cup H)$.

We should now return to a further discussion of the important relationship, formula (3.1): If E and F are any two events, we discovered that

$$E = (E \cap F) \cup (E \cap F^c).$$

What can we say about $P(E) = P([(E \cap F) \cup (E \cap F^c)])$?

Notice from Fig. 3.6 that $E \cap F$ is contained entirely inside of F, while $E \cap F^c$ is contained entirely inside of F^c. But there cannot be any outcomes that are simultaneously in both F and F^c, for F^c consists only of those outcomes that are not in F. Therefore there cannot be any outcomes that are in $(E \cap F) \cap (E \cap F^c)$. It follows that $(E \cap F) \cap (E \cap F^c) = \phi$, and so by analogy with $P(L \cup H)$,

$$P([(E \cap F) \cup (E \cap F^c)]) = P(E \cap F) + P(E \cap F^c).$$

We can therefore write that

■ $$P(E) = P(E \cap F) + P(E \cap F^c). \tag{3.3}$$

In Section 11.A, we will find formula (3.3) to be of great value in discussing Bayes' theorem.

EXERCISES 3.A

3.1. Consider the experiment of tossing a pair of dice, one of them red and the other green. The outcome of the experiment is considered to be the total number of dots facing upward on both dice together.

a. Explain why the sample space of this experiment is the set

$$S = \{2,3,4,5,6,7,8,9,10,11,12\}.$$

b. Draw a Venn diagram of the sample space and the following events:

$$W = \{7,11\} \qquad E = \{2,4,6,8,10,12\}$$
$$L = \{2,3,12\} \qquad D = \{3,5,7,9,11\}.$$

c. List the outcomes of the following events:

$W \cup L$	L^c	E^c
$W \cap L$	$W^c \cup L^c$	$E^c \cap D$
W^c	$W^c \cap L^c$	$E^c \cup D$.

3.2. a. Using the situation of Exercise 3.1, list the outcomes of each of the events $(L \cap E)^c$ and $L^c \cup E^c$.

b. Construct a Venn diagram that illustrates the fact that for any two events A and B it is always true that $(A \cap B)^c = A^c \cup B^c$.

c. Construct a Venn diagram that illustrates the fact that $(A \cup B)^c$ is always the same as $A^c \cap B^c$.

3.3. Consider the situation of Exercise 3.1 again. Because there are 6 possible ways for each of the dice to turn up (namely 1, 2, 3, 4, 5, or 6), there are 36 possible ways for the pair to turn up. For example, we can get 1 on the red and 3 on the green, or 6 on the red and 4 on the green, or 3 on the red and 1 on the green, or 5 on the red and 5 on the green, etc. If the dice are fair, each of these 36 possible ways is equally likely and so has probability 1/36 of turning up. To calculate the probability of rolling a total of 4 dots on the two dice together, we notice that 3 of the 36 possibilities have a total of 4 dots, namely, (1) 1 on the red and 3 on the green, (2) 2 on the red and 2 on the green, and (3) 3 on the red and 1 on the green. Therefore, the probability of rolling a 4 with a pair of fair dice is 3/36.

a. Find the probability of rolling a 7.

b. Find the probability of rolling an 11.

c. Find the probability of the event W, which is defined in Exercise 3.1.

d. Find the probability of the event L.

e. Find the probability of the event E.

f. Find the probability of the event D.

g. Find $P(L \cap D)$.   j. Find $P(W \cup L)$.

h. Find $P(L \cup D)$.   k. Find $P(E \cap D)$.

i. Find $P(W \cap L)$.   l. Find $P(E \cup D)$.

3.4. If A and B are two events such that $P(A) = 1/2$, $P(B) = 2/3$, and $P(A \cap B) = 1/3$, then determine the following probabilities:

a. $P(A^c)$ c. $P(A^c \cap B)$
b. $P(B^c)$ d. $P(A^c \cap B^c)$

3.5. If Q and R are two events such that $P(Q) = 1/2$, $P(R) = 3/8$, and $P(Q \cap R) = 1/4$, then determine the following probabilities:

a. $P(Q^c)$ c. $P(Q^c \cap R)$
b. $P(R^c)$ d. $P(Q^c \cap R^c)$

3.6. Explain why it is impossible to have two events U and V such that $P(U) = 2/3$, $P(V) = 4/5$, and $P(U \cap V) = 1/4$.

3.7. Explain why it is impossible to have two events T and W such that $P(T) = 1/8$, $P(W) = 1/5$, and $P(T \cup W) = 1/2$.

3.8. No matter what the events A and B are, show that $P(A \cup B)$ can never be larger than $P(A) + P(B)$.

SECTION 3.B UPDATING PROBABILITIES OF EVENTS (CONDITIONAL PROBABILITIES)

In most questions in business and management that involve uncertainty, the analyst or manager is called upon to provide an estimate of the probability of some uncertain event. Often the event of interest is not of the sort that can be repeated over and over again. Therefore, it is not possible to estimate its probability by finding the relative frequency of its occurrence. The usual procedure for finding the probability of the event would then be to gather as many as possible of the relevant facts about the situation, try to determine what effect each of the facts has upon the probability, and then come up with an estimate of the desired probability.

As an example, suppose we want to estimate the probability of the event

I = event that an individual will sustain a very serious injury within a
week.

If actuarial statistics indicate that about one out of every million individuals sustains a very serious injury every week, then

$$P(I) = 0.000001.$$

After making this estimate of $P(I)$, suppose that new information comes in about the particular individual under discussion. Say, for example, that we have the information

P = individual is a licensed pilot of small airplanes.

Since the injury rates for pilots of small planes are somewhat higher than those of the general public, perhaps

$$P(I|P) = 0.0001,$$

namely, one out of ten thousand.

The symbol $P(I|P)$ is pronounced "the probability of I, given P" and it means the probability of I, updated so as to reflect the new information contained in statement P. In particular, $P(I|P)$ is the probability that an individual will sustain a very serious injury within the week, if that individual is a licensed pilot of small airplanes. Updated probabilities such as $P(I|P)$ are technically referred to as *conditional probabilities*, for they measure the probabilities of events updated to take changing conditions into account.

In most cases of interest, new information about a changing situation continues to flow in, and the probabilities must be continually updated. Suppose the following fact about our current problem becomes known:

C = individual regularly pilots crop-dusting plane.

Because piloting a crop-dusting plane is somewhat more hazardous than piloting other small planes (because crop-dusting planes commonly fly about 10 to 20 feet above the ground, with trees, wires, etc., in their paths), it would be reasonable to update the probability of injury to

$$P(I|C) = 0.001,$$

or one chance out of a thousand. More information flows in, with each piece of information having an effect on the probability of injury. The bits of information and their effects on the probability of injury are listed in Table 3.2, which shows that the conditional probability fluctuates up and down as conditions change. The events P, C, F, B, and MH tend to exert upward pressure on the probability of I, while the events H and MB tend to exert downward pressure.

To illustrate the calculation of more concrete conditional probabilities, let's return to the probability cube example of the last two sections, and the following four events:

E = {2,4,6} $P(E)$ = 1/2
L = {1,2} $P(L)$ = 1/3
H = {5,6} $P(H)$ = 1/3
A = {1,2,3,4} $P(A)$ = 2/3.

We know that the probability of L is 1/3. Suppose that we have received new information to the effect that the roll of the die resulted in the occurrence of the event A. What is the updated probability of L, in view of the new information that A has occurred?

We come up with the numerical value of $P(L|A)$ by reasoning as follows: the fact that A has occurred means that the number of dots on the upward facing side of the die was either 1, 2, 3, or 4. The possibility that 5 or 6 might have been rolled was excluded by the occurrence of event A. We have no

---

TABLE 3.2   Effect of New Information on the Probability of Injury

Time of Receipt of Information	Gist of New Information	Symbol	Updated Probability
1:00 P.M.	He is licensed to pilot small planes.	P	$P(I|P)$ = .0001
2:00 P.M.	He regularly pilots crop-dusters.	C	$P(I|C)$ = .001
3:00 P.M.	He fell out of his plane while crop-dusting.	F	$P(I|F)$ = .99
3:30 P.M.	There was an extremely large haystack below the plane.	H	$P(I|H)$ = .50
3:45 P.M.	There was a bull sleeping in the haystack.	B	$P(I|B)$ = .75
3:47 P.M.	He missed the bull.	MB	$P(I|MB)$ = .05
3:48 P.M.	He missed the haystack.	MH	$P(I|MH)$ = .99

I = event that individual will sustain very serious injury within week
$P(I)$ = .000001 (at noon)

information asserting that any of the numbers 1, 2, 3, and 4 was more likely
to occur than any of the others, so the nature of the probability cube requires
that the chances of each of these numbers be updated equally to 1/4. This is
due to the fact that there are now only four possible outcomes, taking the
new information into account that the outcomes 5 and 6 definitely did not
occur. Since the only possible outcomes were 1, 2, 3, and 4, the updated
probability of $L = \{1,2\}$ is then

$$P(L|A) = 2/4 = 1/2,$$

because L includes two of the four possible outcomes.

While the original probability of L was 1/3, we see that the conditional
probability of L, given A, is 1/2. Therefore, the occurrence of A has *increased*
the probability of L, because $P(L|A) > P(L)$.

If, instead of the occurrence of A, we were given the information that
the event H had occurred, we would be interested in calculating $P(L|H)$, the
updated probability of L. The information that H has occurred means that the
number rolled on the die was definitely either a 5 or a 6. There is no longer
any possibility that a 1, a 2, a 3, or a 4 was rolled. But $L = \{1,2\}$, so that there
is no chance that event L could have occurred. It follows that:

$$P(L|H) = 0.$$

Because the original probability of L was 1/3 and the conditional probability
of L, given H, is 0, we can see that $P(L|H) < P(L)$. The occurrence of H has
decreased, (quite substantially in this case) the probability of L.

Suppose, finally, that the information we were given was that the event
E had occurred. The occurrence of E would mean that the actual number
rolled was either 2, 4, or 6. The only possible outcome that would therefore
result in the occurrence of L would be the 2. There is no chance that L's
outcome 1 would have occurred, for the occurrence of E specifically excludes
the occurrence of 1, 3, and 5. Therefore, the numbers 2, 4, and 6 would each
be assigned an updated probability of 1/3, since those are now the only three
possible outcomes. The occurrence of 2 would imply the occurrence of L,
while the occurrence of 4 and 6 would not. Therefore the updated probability
of L, in view of the information that E has occurred, is

$$P(L|E) = 1/3,$$

because L includes one of the three possible outcomes, 2, 4, and 6.

We note something unusual here: the original probability of L was 1/3,
while the conditional probability of L, given E, also turned out to be 1/3.
Therefore, it seems that the occurrence of E *did not affect* in any way the
chances of L, because $P(L|E) = P(L)$. We describe this situation by saying that
the event L is *independent* of the event E.

We summarize the types of relationships between the original and the
updated probabilities in Table 3.3.

TABLE 3.3   Conditional Probability Relationships

Verbal Expression	Probability Expression
C increases likelihood of D	$P(D|C) > P(D)$
C decreases likelihood of D	$P(D|C) < P(D)$
D is independent of C	$P(D|C) = P(D)$

Formulas for Computing Conditional Probability

In many cases of applied interest, it will be useful to have a formula for computing conditional probabilities in terms of original probabilities. In order to develop such a formula, let's sit down and try to figure out what the conditional probability of C, given D, really is.

Consider the Venn diagram on the left side of Fig. 3.8. The original probability of C can be viewed as the proportion of the sample space S that is occupied by C. If C is a large part of S, the probability of C will be high, while if C is a small part of S, the probability of C will be small.

What happens to C after D occurs? This situation is illustrated on the right side of Fig. 3.8. The information that D has occurred has the effect of reducing the sample space to D, because the outcomes not in D can no longer be considered as possible outcomes. The probability of C in this situation is then the proportion of the *new sample space* D that is occupied by C. The part of D that is occupied by C is the set $C \cap D$. Therefore, if $C \cap D$ is a large part of D, the conditional probability of C, given D, will be large. If $C \cap D$ is

FIGURE 3.8   The Conditional Probability of C, Given D

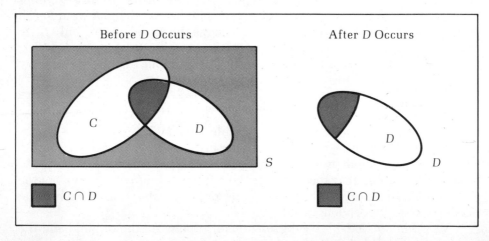

a small part of D, the conditional probability of C, given D, will be small. To express the conditional probability as the proportion of D occupied by $C \cap D$, we have the following formula.

Formula for Conditional Probability:

$$P(C|D) = \frac{P(C \cap D)}{P(D)} \, . \tag{3.4}$$

Before illustrating the use of this formula in applied work, let's use it to recalculate the conditional probabilities that we found directly for the roll of a die:

$$P(L|A) = \frac{P(L \cap A)}{P(A)} = \frac{P(\{1,2\})}{P(\{1,2,3,4\})} = \frac{1/3}{2/3} = 1/2,$$

$$P(L|H) = \frac{P(L \cap H)}{P(H)} = \frac{P(\phi)}{P(\{5,6\})} = 0,$$

$$P(L|E) = \frac{P(L \cap E)}{P(E)} = \frac{P(\{2\})}{P(\{2,4,6\})} = 2/6 = 1/3.$$

Naturally, these results were exactly the same as those we obtained earlier.

If, in formula (3.4) we multiply both sides of the equation through by $P(D)$, we see that

$$P(C|D) \, P(D) = \frac{P(C \cap D)}{P(D)} \cdot P(D) = P(C \cap D)$$

as the two $P(D)$'s cancel out. The resulting expression,

$$P(C \cap D) = P(C|D) \, P(D) \tag{3.5}$$

is very useful in applications, perhaps just as useful as formula (3.4) itself.

EXAMPLE 3.1. Tax Returns. Based on a preliminary reading of all tax returns prepared professionally by Ax-Your-Tax Associates, the IRS selects for formal audit 90% of those returns that contain serious errors and 2% of those that don't. On the average, about 5% of all Ax-Your-Tax returns contain serious errors. What percentage of returns by Ax-Your-Tax are audited by the IRS?

SOLUTION. We label the following events:

E = event that return contains serious *errors*
A = event that return is *audited*

In the language of these events, we know that $P(E) = .05$ and so $P(E^c) = .95$ because E^c is the event that a return does not contain serious errors. Given that a return contains serious errors, the probability is 90% that the IRS will audit it. This means that $P(A|E) = .90$. On the other hand, $P(A|E^c) = .02$. We want to calculate $P(A)$. According to formula (3.3).

$$P(A) = P(A \cap E) + P(A \cap E)^c).$$

Furthermore, by formula (3.5),

$$P(A \cap E) = P(A|E)P(E) = (.90)(.05) = .045$$

$$P(A \cap E^c) = P(A|E^c)P(E^c) = (.02)(.95) = .019.$$

Therefore $P(A) = .045 + .019 = .064 = 6.4\%$ of all Ax-Your-Tax returns are audited by the IRS.

In addition to the technical details of calculating the exact answers to questions like the above, we will discuss the communication of the sense of the data by means of pictorial descriptions. The appropriate vehicle for organizing data of this sort is called a *tree diagram*.

Fig. 3.9 illustrates the first steps to be taken in the construction of a tree diagram based on the Ax-Your-Tax data. The total percentage of tax returns, namely 100%, is shown as divided into two categories: those that contain serious errors (5%) and those that do not (95%). The completed tree diagram, which appears in Fig. 3.10, exhibits several points of information. First, of that 5% of Ax-Your-Tax returns containing serious errors, 90% of them are audited by the IRS. Taking 90% of 5% (by multiplying .90 times .05 to get .045 = 4.5%), we see that 4.5% of the returns both contain serious errors and are audited. Secondly, of that same 5%, 10% are not audited; so multiplying

FIGURE 3.9   First Steps in Constructing a Tree Diagram   *(Ax-Your-Tax data)*

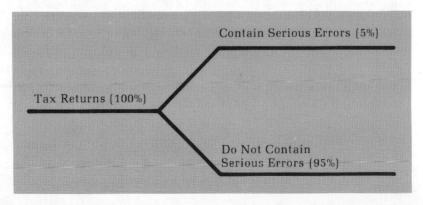

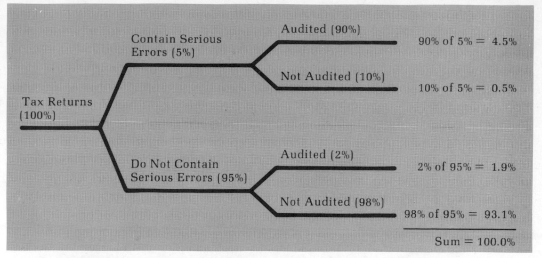

FIGURE 3.10   Tree Diagram of Audits vs. Errors   *(Ax-Your-Tax data)*

yields that (.05)(.10) = .005 = 0.5% (one-half of 1%) of the original group of
returns have serious errors but are not audited. Finally, we use similar multiplication techniques to conclude that 1.9% of the returns are audited but
do not contain serious errors, while the remaining 93.1% have no serious
errors and are not audited.

From the information conveyed by the tree diagram, we can actually
determine the answers to three questions based on the situation of Example
3.1, as follows.

1. What percentage of returns prepared by Ax-Your-Tax are audited by the
   IRS?

   Answer. We can see from Fig. 3.10 that 4.5% of returns contain serious
   errors and are audited, while 1.9% do not contain serious errors but are
   audited. Therefore 4.5% + 1.9% = 6.4% of the returns are audited.
   (This specific situation is illustrated in the tree diagram of Fig. 3.11.)

2. Of those Ax-Your-Tax returns that are audited, what percentage are free
   of serious errors?

   Answer. Of the 6.4% that are audited (according to the previous
   answer), only 1.9% are free of serious errors, while the remaining 4.5%
   are not. It follows that, of those returns which are audited, a fraction
   1.9/6.4 = .297 = 29.7% are free of serious errors.

3. Of those Ax-Your-Tax returns that are audited, what percentage contain
   serious errors?

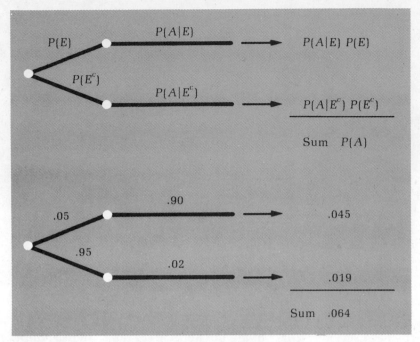

FIGURE 3.11   Tree Diagram of Tax Returns Example

Answer. Using the previous answer, this answer is 100% − 29.7% = 70.3%. Reasoning also directly from Fig. 3.10, we can see that, of the 6.4% that are audited, only 4.5% contain serious errors. Percentagewise, this means that returns containing serious errors comprise 4.5/6.4 = .703 = 70.3% of audited returns.

Once we construct a tree diagram, it is not difficult to formulate answers to other questions that arise in the course of analyzing a business situation. For example, suppose we want to know what percentage of nonaudited Ax-Your-Tax returns do in fact contain serious errors. Well, we know from Fig. 3.10 that 0.5% + 93.1% = 93.6% of all returns are not audited. Of these, the 0.5% come from the group that contains serious errors. Therefore, the percentage of nonaudited returns containing serious errors is 0.5/93.6 = .0053 = 0.53% = 53/100 of 1%, a very small percentage indeed.

EXAMPLE 3.2.   Credit Risks.   Eighty percent (80%) of major appliance purchasers are good credit risks. Seventy percent (70%) of good credit risks have a charge account with at least one department store, while only 40% of bad credit risks have a charge account. What percentage of those with a charge account are bad credit risks?

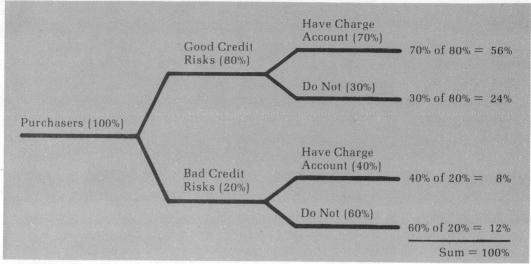

FIGURE 3.12   Tree Diagram of Credit Risks vs. Charge Accounts

SOLUTION.   Beginning with 100% of major appliance purchasers, we construct the tree diagram of Fig. 3.12. Following the branches of the tree from left to right, we see that 56% of purchasers have a charge account and are good credit risks, while 8% have a charge account and are bad credit risks. Therefore 56% + 8% = 64% have a charge account. It follows that a fraction 8/64 = .125 = 12.5% of those with a charge account are bad credit risks.

What we have really used in answering the question asked in Example 3.1 was the expression:

■ $$P(A) = P(A|E) P(E) + P(A|E^c) P(E^c). \qquad \textbf{(3.6)}$$

This very useful formula is a combination of formulas (3.3) and (3.5).
Suppose the sample space is divided into three disjoint events F, G, and H as in Fig. 3.13, where A is an event that intersects all of them. Then, by analogy with Fig. 3.6 and formula (3.1), we can see that

$$A = (A \cap F) \cup (A \cap G) \cup (A \cap H).$$

We then have the following analogue of formula (3.3),

$$P(A) = P(A \cap F) + P(A \cap G) + P(A \cap H),$$

which we can combine with formula (3.5) to get

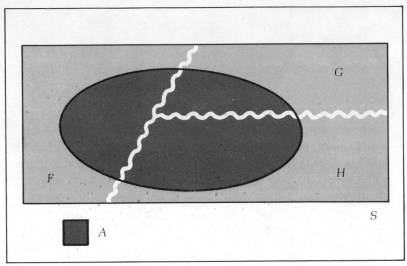

FIGURE 3.13 $A = (A \cap F) \cup (A \cap G) \cup (A \cap H)$

$$P(A) = P(A|F)\,P(F) + P(A|G)\,P(G) + P(A|H)\,P(H). \qquad (3.7)$$

This formula works only if $F \cup G \cup H = S$. Formula (3.7) is needed in the following example.

EXAMPLE 3.3.   Automobile Insurance.   An insurance company issues three types of automobile policies: Type G for good risks, Type M for moderate risks, and Type B for bad risks. The company's clients are classified 20% as Type G, 40% as Type M, and 40% as Type B. Accident statistics reveal that a Type G driver has probability .01 of causing an accident in a 12-month period, a Type M driver has probability .02, and Type B driver has probability .08. What proportion of the company's clients will cause an accident in the next 12 months?

SOLUTION.   The pertinent events are as follows:

A = event that client will cause accident in next 12 months
G = event that client is Type G driver
M = event that client is Type M driver
B = event that client is Type B driver.

The answer to the question is $P(A)$, the proportion of clients who will cause an accident in the next 12 months. From formula (3.7), we know that

$$P(A) = P(A|G)\,P(G) + P(A|M)\,P(M) + P(A|B)\,P(B).$$

The information given in the example translates into the following mathematical symbolism:

$P(G) = .20$ $P(A|G) = .01$
$P(M) = .40$ $P(A|M) = .02$
$P(B) = .40$ $P(A|B) = .08.$

Therefore

$$P(A) = (.01)(.20) + (.02)(.40) + (.08)(.40)$$

$$= .002 + .008 + .032$$

$$= .042$$

so that 4.2% of the company's clients can be expected to cause an accident in the next 12 months. The calculation is illustrated by a tree diagram in Fig. 3.14.

FIGURE 3.14   Tree Diagram of the Accident Insurance Example

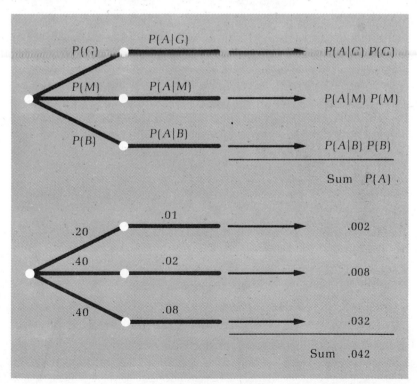

An expanded analysis of the conditional probability relationships among dependent events appears in Chapter 11, where Bayes' theorem is introduced. If desired, you can proceed directly to Section 11.A at this point and continue the study of conditional probability. The material in Section 11.A is not dependent on anything in Chapters 4 through 10.

EXERCISES 3.B

3.9. Consider the experiment of tossing a pair of fair dice, which was discussed previously in Exercises 3.1 and 3.3. Calculate the following conditional probabilities:

a. $P(W|D)$ d. $P(E|L)$
b. $P(L|E)$ e. $P(W|E)$
c. $P(D|W)$ f. $P(L|D)$

3.10. A Census Bureau survey disclosed that 25.3% of households headed by persons under 25 years of age contain a color television set, while 41.5% of other households do. Furthermore, 7.3% of all households are headed by persons under 25. What percentage of households contain a color television set?

3.11. A watthour meter is inspected by two testers before leaving the factory where it was manufactured. If a defective meter comes down the assembly line, the probability is 1/10 that it will get by the first inspector. Of those defective articles that get by the first inspector, 3 out of 10 also get by the second inspector. What proportion of defective meters escape both testers?

3.12. An automobile dealer sells the two models of the high-priced, low-mileage Peccarie that experience shows to be in equal demand. Half the customers want Model P and the other half want Model PH. The dealer decides to stock six of each model.

a. What is the probability that the first six purchasers, acting independently of each other, all want Model PH?

b. What is the probability that the first seven customers all want Model PH, and therefore one must be turned away?

3.13. Marketing often targets a certain segment of the society as its prime market and exerts its efforts toward that one particular segment. An advertising executive is considering aiming a new campaign for color TV sets at people under 25. She found (from the Census Bureau) that 25.3% of the households headed by persons under 25 contain a color TV while 41.5% of other households do. Also, 7.3% of all households are headed by persons under 25.

a. Construct a tree diagram that indicates the relationship between the age of the head of the household and color TV ownership.

b. Use the data from the tree diagram to calculate what percentage of the population are color TV owners and are under 25.

    c. Use the data from the tree diagram to determine what percentage of house-holds contain a color TV.

3.14. A national real estate marketing survey indicates that 15% of all adults are college graduates, while 85% are not. Furthermore, 80% of all college graduates own their own homes, while only 30% of nongraduates are homeowners.

    a. Construct a tree diagram illustrating the relationship between college grad-uates and homeowners.

    b. Determine from the tree diagram what percentage of adults are homeowners.

    c. Determine from the tree diagram what percentage of homeowners are college graduates.

    d. Use conditional probability analysis to find out what percentage of adults are homeowners.

3.15. A local freeway bond referendum is defeated at the polls in a region where 30% of the voters travel long distances to work or shop, and 70% do not. It is esti-mated on the basis of residential areas that 90% of those who travel long dis-tances voted for the bond issue, while only 20% of the others did so.

    a. Construct a tree diagram classifying the voters as to how they voted and whether or not they often traveled long distances to work or shop.

    b. Use the tree diagram of part a to determine what percentage of voters voted against the bond issue.

    c. Use the tree diagram of part a to find out what percentage of those who voted against the bond issue did not travel long distances.

    d. Determine from a conditional probability analysis the percentage of voters that voted against the bond issue.

3.16. A psychological test is designed to separate new employees into good prospects and not-so-good prospects. Among those who later performed satisfactorily during the year, 80% had passed the test. Among the employees who did unsat-isfactory work in their first year, only 40% had passed the test. On the whole, 70% of the employees tested did satisfactory work that year.

    a. Draw a tree diagram of satisfactory work vs. the psychological test.

    b. Use the tree diagram to find out what percentage of new employees passed the psychological test.

    c. Determine from the tree diagram the percentage of those passing the test who went on to do satisfactory work.

    d. On the basis of a conditional probability analysis, find the proportion of new employees who passed the test.

3.17. A life insurance underwriting survey has revealed that 60% of all applicants for life insurance are in excellent health, 25% have mild preexisting diseases, which would ordinarily allow them to buy life insurance anyway, while the remaining 15% have serious preexisting diseases that would disqualify them from buying insurance. Of those in excellent health, 95% have their applications for insurance approved, as do 50% of those with mild diseases and 10% of those with (evidently undetected) serious diseases.

a. Draw a tree diagram illustrating the situation.

b. Determine what percentage of applicants for insurance have their applications approved.

c. Of those individuals who have their applications approved, what percentage actually have serious preexisting diseases?

d. Use a conditional probability analysis to determine the percentage of applicants who have their applications approved.

SECTION 3.C   PERMUTATIONS, COMBINATIONS, AND INDEPENDENT EVENTS

Independent Events

In Section B, we noticed that $P(C|D)$, the conditional probability of C, given D, could be either less than, equal to, or greater than the original probability of C. In those cases where $P(C|D)$ is different from $P(C)$, the knowledge of the occurrence of D has had an influence on our view of the likelihood of the occurrence of C. We can describe this situation by saying that the event C "depends on" the event D, in the sense that the occurrence of D affects the probability of C.

Suppose, however, that $P(C|D)$ is exactly the same, numerically speaking, as $P(C)$. This means that the occurrence of D has had no influence at all on the probability of C. In the situation that $P(C|D) = P(C)$, we therefore say that the event C is *independent* of the event D.

The concept of independent events is often mistaken for the concept of disjoint events. In fact, however, independence and disjointness are extreme opposites of each other. If the events C and D are disjoint, then $C \cap D = \phi$ so that, by formula (3.4),

$$P(C|D) = \frac{P(C \cap D)}{P(D)} = \frac{P(\phi)}{P(D)} = \frac{0}{P(D)} = 0,$$

no matter what the original probability of C was. What this means is that, even if $P(C)$ is very high, the occurrence of D eliminates all chances of C's occurrence, due to the fact that C and D are mutually exclusive. Therefore, C depends very heavily on D, as the conditional probability of C given D, is very small compared with the original probability of C.

A useful fact about independent events is that, if C is independent of D, then D is independent of C. Specifically, if the occurrence of C does not affect the probability of D, then the occurrence of D would not affect the probability of C. This fact can be expressed as follows:

If $P(C|D) = P(C)$, then $P(D|C) = P(D)$.

Its justification involves the use of the formulas for conditional probability, (3.4) and (3.5). If $P(C|D) = P(C)$, then

$$P(D|C) = \frac{P(D \cap C)}{P(C)} = \frac{P(C \cap D)}{P(C)} = \frac{P(C|D)\,P(D)}{P(C)}$$

$$= \frac{P(C)\,P(D)}{P(C)} = P(D),$$

where we have replaced $P(C|D)$ by its equal, $P(C)$. Because of this symmetry of meaning of the word "independence," we can refer to C and D as a pair of independent events.

There is an algebraic expression of the meaning of independence that reflects this symmetry. The independence of C and D means that $P(C) = P(C|D)$, so that

$$P(C) = \frac{P(C \cap D)}{P(D)},$$

which, by cross multiplication, becomes

$$P(C)\,P(D) = P(C \cap D).$$

It then follows that the events C and D are independent if and only if

$$P(C \cap D) = P(C)\,P(D). \tag{3.8}$$

In dealing with more than two independent events, we need an expanded version of formula (3.8). We will say that n events $C_1, C_2, \ldots, C_n$ are independent if every event is independent not only of every other event, but also of every possible combination of all other events. It turns out that

$$P(C_1 \cap C_2 \cap \ldots \cap C_n) = P(C_1)\,P(C_2) \ldots P(C_n), \tag{3.9}$$
if the events $C_1, C_2 \ldots, C_n$ are independent.

This means that the probability that all the events will occur is given by the product of all their original probabilities.

Sequences (Permutations)

Suppose now that we have a jar containing n billiard balls, each having a number 1 to n, inclusive, painted on its surface. (For an ordinary game of pool, n would be 15.) Out of this set of n balls in the jar, suppose we have to draw k of them in sequence, where k is a number less than or equal to n. (For example, if $n = 15$, we could take $k = 8$ in order to select 8 out of the 15 balls in the jar.)

The question we want to ask first is, "How many possible different 'sequences' of k balls can be selected from the n balls in the jar?" As a clarification of what a sequence is, we should point out that, if you draw ball 3 first, ball 8 second, and ball 6 third, and I draw ball 6 first, ball 3 second, and ball 8 third, our sequences are different although we both have drawn the same three balls. A sequence depends not only on the balls that are drawn, but also on the order in which they are drawn.

We turn to the tree diagrams introduced in Section 3.B to help us answer the question of how many different sequences of size k can be drawn from a set of n elements. In Fig. 3.15, we illustrate the various sequences of size 3 that can be drawn from a jar containing 5 balls.

As Fig. 3.15 shows, there are 60 possible sequences of size 3 that can be drawn from a set of 5 elements. As you can imagine, the number of sequences rapidly becomes astronomical as the number of elements involved grows. Clearly we need a formula for the number of sequences so that we don't have to draw a tree diagram every time.

Fortunately, the tree diagram of Fig. 3.15 contains within it the seeds of the formula we need. Notice that we have 5 options on the first draw, 4 on the second (for we are already holding the one element we had drawn first), and 3 on the third (for we are holding two elements drawn previously). Therefore, for each of the 5 options on the first draw, we have 4 options on the second, yielding $5 \times 4 = 20$ different sequences resulting from the first two draws. Having in hand these 20 two-element sequences, we then have 3 options for the third draw. So, each of the 20 two-element sequences can be augmented by one of 3 remaining elements, resulting in 60 possible three-element sequences.

In summary, we can say that the total number of possible different sequences of size 3 from a set of 5 elements is

$$5 \times 4 \times 3 = 60.$$

Now, what about drawing k elements in sequence from a set of n elements? Well, on the first draw we have n options; on the second, we have $n - 1$; on the third, we have $n - 2; \ldots$; and on the kth draw, we have $n - (k - 1) = n - k + 1$ options. Therefore, the total number of possible different sequences of size k from a set of elements is given by the following formula.

■ $$P(n,k) = n \times (n - 1) \times \ldots \times (n - k + 1). \tag{3.10}$$

Here $P(n,k)$ is read, "the number of permutations of n elements taken k at a time." The word *permutation* is the technical term for sequences in this context. If we insert $n = 5$ and $k = 3$ into the formula for $P(n,k)$ we get that

$$n - k + 1 = 5 - 3 + 1 = 3,$$

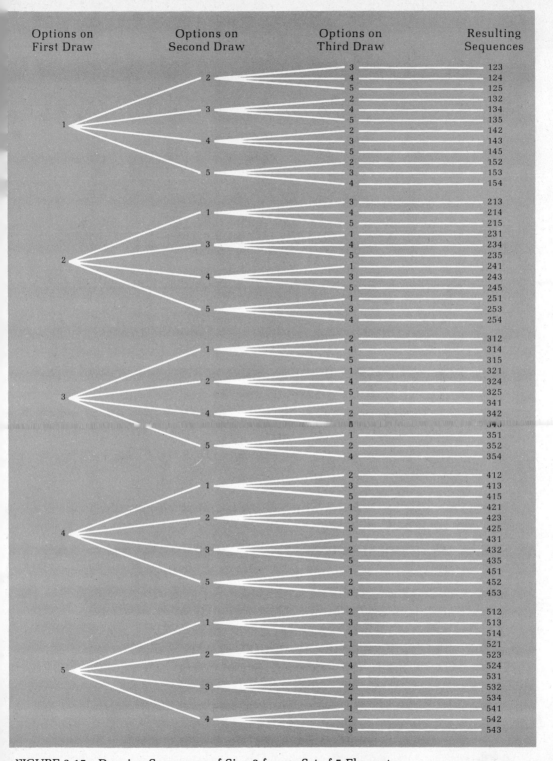

FIGURE 3.15 Drawing Sequences of Size 3 from a Set of 5 Elements

so that

$$P(5,3) = n \times (n-1) \times \ldots \times (n-k+1)$$

$$= 5 \times 4 \times 3$$

$$= 60,$$

which is, of course, the same result we obtained from the tree diagram.

At this point, we introduce some new mathematical symbolism so we can simplify the formulas to come. We define n! (pronounced, "n-factorial") to be

$$n! = n \times (n-1) \times (n-2) \times \ldots \times 1.^*$$

For example,

$$3! = 3 \times 2 \times 1 = 6$$
$$5! = 5 \times 4 \times 3 \times 2 \times 1 = 120$$
$$6! = 6 \times 5 \times 4 \times 3 \times 2 \times 1 = 720.$$

Furthermore, 6! = 6 × (5!) = 6 × 5 × 4 × (3!). Using this notation, we can see that

$$P(n,k) = n \times (n-1) \times \ldots \times (n-k+1)$$

$$= \frac{n \times (n-1) \times \ldots \times (n-k+1) \times (n-k) \times (n-k-1) \times \ldots \times 1}{(n-k) \times (n-k-1) \times \ldots \times 1}$$

$$= \frac{n!}{(n-k)!},$$

because the number $(n-k) \times (n-k-1) \times \ldots \times 1$ cancels out of both the numerator and the denominator. If we try out the permutation formula

■
$$P(n,k) = \frac{n!}{(n-k)!} \qquad (3.11)$$

on n = 5 and k = 3, we get

$$P(5,3) = \frac{5!}{(5-3)!} = \frac{5!}{2!} = \frac{120}{2} = 60,$$

exactly the same result we obtained earlier, of course. If we are drawing k = 8 balls at random out of a set of n = 15, there would be

$$P(15,8) = \frac{15!}{(15-8)!} = \frac{15!}{7!} = \frac{1,307,674,368,000}{5040} = 259,459,200$$

---

*Due to certain technicalities, 0! turns out to be 1. This is due to the fact that 0! = P(0,0), and there is only one sequence of 0 elements in a set of 0 elements.

different sequences obtainable, justifying our earlier description of the number of possible sequences as "astronomical."

Subsets (Combinations)

Now we are ready to ask a second question: "How many possible different *subsets* of k balls can be selected from the n balls in the jar?" By a subset here, we mean simply the collection of the k balls drawn, without regard to the order in which they are drawn. Now, are there more sequences of size k or more subsets of size k? Clearly, there are more sequences than subsets, because one subset can be drawn in several different orders, each order counting as a separate sequence. In particular, each subset of size k can be drawn as one of

$$P(k,k) = k \times (k - 1) \times \ldots \times 1 = k!$$

different sequences, using formula (3.10) with $n = k$. Therefore, because

$$\begin{pmatrix} \text{number of subsets} \\ \text{of size } k \end{pmatrix} \times \begin{pmatrix} \text{number of sequences} \\ \text{per subset} \end{pmatrix} = \begin{pmatrix} \text{number of sequences} \\ \text{of size } k \end{pmatrix}$$

we see that

$$\begin{aligned} \frac{\text{number of subsets}}{\text{of size } k} &= \frac{\text{number of sequences of size } k}{\text{number of sequences per subset}} \\[2mm] &= \frac{P(n,k)}{P(k,k)} = \frac{\dfrac{n!}{(n-k)!}}{k!} \\[2mm] &= \frac{n!}{(n-k)!} \cdot \frac{1}{k!} = \frac{n!}{k!(n-k)!} \end{aligned}$$

The technical term for the word subset in this context is *combination*, so that we can express "the number of combinations of n elements taken k at a time" by the following formula.

Combination Formula:

$$C(n,k) = \frac{n!}{k!(n - k)!} \tag{3.12}$$

Using the combination formula, let's calculate the number of subsets of size 3 of a set of 5 elements. Here $n = 5$ and $k = 3$ so that there are

$$C(5,3) = \frac{5!}{3!(5-3)!} = \frac{5!}{(3!)(2!)} = \frac{120}{(6)(2)} = 10$$

possible subsets. An examination of the tree diagram in Fig. 3.12 confirms this, the 10 different subsets being

$$\{1,2,3\}, \quad \{1,2,4\}, \quad \{1,2,5\}, \quad \{1,3,4\}, \quad \{1,3,5\},$$
$$\{1,4,5\}, \quad \{2,3,4\}, \quad \{2,3,5\}, \quad \{2,4,5\}, \quad \{3,4,5\}.$$

Each of the 60 sequences is merely a permutation of one of these 10 combinations.

EXAMPLE 3.4. Automobile Loan Approvals. Suppose one bank's records indicate that 60% of all automobile loan applications submitted to it are eventually approved. If we were to select a random sample of 7 applications from a certain section of a major city, what is the probability that a majority of those applications would not be approved?

SOLUTION. We denote by A the number of applications in the sample that are approved. Then the probability

$$P(\text{a majority of the sample is not approved}) = P(A \leq 3).$$

We now proceed to calculate $P(A \leq 3)$. First of all, we see that

$$P(A \leq 3) = P(A = 0) + P(A = 1) + P(A = 2) + P(A = 3).$$

Let's look at $P(A = 3)$. For A to equal 3, the random sample of 7 applications must contain 3 that are approved and 4 that are not approved. One way to have $A = 3$ would be to have $A_1 \cap A_2 \cap A_3 \cap A_4^c \cap A_5^c \cap A_6^c \cap A_7^c$ where A_k is the event that the kth application is approved, and A_k^c is the event that the kth application is not approved. Because the 7 randomly selected applications are presumably acted upon independently, these 7 events may be considered as independent so that

$$P(A_1 \cap A_2 \cap A_3 \cap A_4^c \cap A_5^c \cap A_6^c \cap A_7^c) = P(A_1)P(A_2)P(A_3)P(A_4^c)P(A_5^c)P(A_6^c)P(A_7^c)$$
$$= (.60)(.60)(.60)(.40)(.40)(.40)(.40)$$
$$= (.60)^3(.40)^4$$
$$= (.216)(.0256) = .0055296.$$

(Because the probability of an application's being approved is .60, the probability of its not being approved is .40.) Now this is only one way to fill the 7 positions in the random sample with three A_k's and four A_k^c's so as to have $A = 3$. There are several other ways to have $A = 3$; in fact, there is one way corresponding to each subset of three positions in which to put the A_k's. The number of ways of having $A = 3$ is therefore equal to the number of subsets of size 3 of a set of 7 elements. By formula (3.12), this number is

$$C(7,3) = \frac{7!}{3!4!} = \frac{5040}{(6)(24)} = 35.$$

So there are 35 ways to get $A = 3$ and each of these has probability .0055296. It follows that

$$P(A = 3) = (35)(.0055296) = .19354.$$

The quantity A, the number of applications in the random sample that are approved, is technically called a *random variable*. A random variable is a quantity whose value can be exactly determined only after a complete investigation of the random sample of data resulting from an experiment or survey. If we use the symbols $n =$ number of applications in the random sample and $p =$ overall proportion of applications that are approved, then the following general rule can be extracted from the above discussion.

Binomial Probability Formula:

$$P(A = j) = C(n,j)(p)^j(1-p)^{n-j} \qquad (3.13)$$

Because its probabilities can be determined by this formula, A is said to have the "binomial distribution with parameters n and p." Binomial probability distributions will be discussed in greater detail in Section 4.B.

In our situation, we want to calculate $P(A = 3)$, and so $j = 3, n = 7$, and $p = .60$. Using formula (3.13), we therefore see that

$$P(A = 3) = C(7,3)(.60)^3(.40)^{7-3}$$

$$= (35)(.60)^3(.40)^4 = .19354.$$

Using formula (3.13) again, we can complete the calculation:

$$P(A = 2) = C(7,2)(.60)^2(.40)^{7-2}$$

$$= \frac{7!}{2!\,5!}(.60)^2(.40)^5$$

$$= (21)(.36)(.01024) = .07741.$$

$$P(A = 1) = C(7,1)(.60)^1(.40)^6$$

$$= (7)(.60)(.004096) = .01720.$$

$$P(A = 0) = C(7,0)(.60)^0(.40)^7$$

$$= (1)(1)(.0016384) = .00164.$$

From the results of these calculations, we find that

$$P(A \leqslant 3) = P(A = 0) + P(A = 1) + P(A = 2) + P(A = 3)$$

$$= .00164 + .01720 + .07741 + .19354$$

$$= .2898.$$

EXERCISES 3.C

3.18. A corporation has offices in 10 major cities in South America, and the chief operating officer would like to visit 5 of these on her tour of the area next month.

   a. How many different groups of 5 cities can she arrange to visit? Is this a problem of permutations or combinations?

   b. If she is willing to visit any 5 of the 10 cities, how many different routes can her travel agent consider in his planning of the trip? Is this a problem of permutations or combinations?

3.19. The United States Senate has 100 members. A lobbyist attempting to predict the outcome of the voting on a tax reform bill wants to choose a sample of 20 senators on which to base his prediction.

   a. How many different samples of 20 senators can the lobbyist select? Is this a problem of permutations or combinations?

   b. If the lobbyist chooses the 20 members of his sample by walking from one senator's office to another until he has covered 20 offices, how many different routes can he follow in knocking on 20 of the 100 office doors? Is this a problem of permutations or combinations?

3.20. Four-letter "words" in which no letter appears more than once can be formed by choosing in order 4 out of the 26 letters of the English alphabet.

   a. How many four-letter "words," with no letters appearing more than once, are there in the English language?*

   b. Is this a problem of permutations or combinations?

   c. If we allow letters to appear more than once, how many four-letter "words" are there in the English language?

3.21. In an attempt to find out if any of several self-proclaimed stock-market clairvoyants actually have ESP ("extra-sensory perception"), a psychologist conducts an experiment, selecting 20 stocks traded on the New York Stock Exchange and asking each candidate for clairvoyance to pick those 5 stocks that will perform best next year.

   a. How many different groups of 5 stocks is it possible for the candidate to select?

   b. Of these, how many contain the 5 stocks that will perform best over the next year?

   c. If the candidate does not really have ESP, what is the probability that he will pick exactly the 5 best-performing stocks of the year?

3.22. In a study of the purchasing habits of 1000 families, a marketing researcher needs a sample of 10 families. How many different samples can she choose?

3.23. An accountant is assigned the task of verifying a group of 400 transactions. How many samples of 15 transactions each are available for his detailed analysis?

---

*Of course, not all of these are real words. Consider "lorf," for example.

3.24. In 1976, the Bureau of Labor Statistics reported that 74% of the nation's labor force were unhappy with their jobs.* If this figure is correct, what would be the probability that more than half of 61 randomly selected employees are satisfied with their jobs?

SECTION 3.D   ACCEPTANCE SAMPLING AND QUALITY CONTROL

A well-known dealer in used and rebuilt appliances has the opportunity to purchase a lot of 30 refrigerators at an exceptionally good price. There is a chance, however, that one or more of the refrigerators might require extensive repairs, and this would tend to reduce the attractiveness of the opportunity. Because of the expense required for a complete inspection of all 30 refrigerators, the dealer decides to pick 3 at random (with each of the 30 equally likely to be picked) and to subject them to a thorough inspection. If those 3 turn out to be acceptable, he agrees to buy the entire lot of 30.

What the dealer is doing here is conducting a procedure called *acceptance sampling*. Due to practical difficulties and costs in time and money that prevent the testing of an entire lot of items, the prospective purchaser usually tests a random sample of 10% or less of the lot of items. Then, if a reasonable number of items in the sample are acceptable, the dealer will consider the entire lot to be acceptable.

Now for the bad news. Unfortunately, just because all the items in a 10% random sample may be acceptable, it does not necessarily follow that all the items in the entire lot are in acceptable condition. Statistics enters the discussion when we try to determine the relationship between the proportion of acceptable items in the sample and that in the entire lot.

Suppose, for example, that (unknown to the dealer) 6 of the 30 refrigerators are really in unacceptable condition. What is the probability that none of these 6 will make it into the dealer's random sample? To rephrase the question: what are the chances that the 3 refrigerators in the dealer's random sample will all be in acceptable condition, even though 20% (6 out of 30) of the entire lot is unacceptable?

To compute the probability that all of the three refrigerators in the random sample are acceptable, we must first determine how many samples of size 3 are available to the dealer, and then we must calculate how many of these consist of acceptable refrigerators only. If we denote the event

A = event that dealer accepts lot

* As quoted in *Orange County Business*, November/December 1977, p. 54.

then

P(A) = proportion of possible samples that contain acceptable refrigerators only

$$= \frac{\text{number of samples consisting of acceptable refrigerators only}}{\text{total number of possible samples of size 3}}$$

First, it is relatively easy to calculate the total number of samples of size 3 that the dealer can select from the 30 refrigerators. This is $C(30,3)$, the number of combinations of 30 elements taken 3 at a time. The denominator of $P(A)$ is therefore

$$C(30,3) = \frac{30!}{(30-3)!3!} = \frac{(30)(29)(28)}{(3)(2)(1)} = 4,060.$$

It is only slightly more difficult to calculate the number of samples consisting of acceptable refrigerators only. In order to select a sample of 3 refrigerators out of the 30 in such a way that all 3 are acceptable, we divide the lot of 30 refrigerators in two groups: group A^* of acceptable ones and group U^* of unacceptable ones. Group A^* contains 24 refrigerators, while group U^* contains 6. We must select 3 from group A^* and then 0 from group U^*. Since these two selection operations are carried out independently of each other, we can combine each of the $C(24,3)$ sets of acceptable refrigerators with each of the $C(6,0)$ sets of unacceptable ones. Therefore, the total number of samples that consist of acceptable refrigerators only is

$$C(24,3)C(6,0) = \frac{24!}{(24-3)!3!} \times \frac{6!}{(6-0)!0!} = \frac{(24!)(6!)}{(21!)(3!)(6!)(0!)}$$

$$= \frac{(24)(23)(22)}{(3)(2)(1)} = 2,024.$$

It follows that the probability that the dealer's random sample contains acceptable refrigerators only is

$$P(A) = \frac{C(24,3)C(6,0)}{C(30,3)} = \frac{2,024}{4,060} = .4985 = 49.85\%.$$

Our conclusion may be stated as follows: if 20% of the 30 refrigerators are *not* in acceptable condition, the probability is about 1/2 that the dealer's acceptance sampling plan *will* recommend purchase of the entire lot anyway.

The dealer may consider an acceptance probability of 49.85% to be unreasonably high, when 6 out of the 30 refrigerators will probably require extensive repairs. If he does, the only option he has open is to choose a more stringent acceptance sampling plan, perhaps by increasing the sample examined. Suppose he were to thoroughly check out a random sample of 4, instead of only 3, refrigerators. Then the number of samples of size 4 that contain only acceptable refrigerators will be those that have 4 selected from group A^* and 0 from group U^*. Their number will be

$$C(24,4)C(6,0) = \frac{24!}{20!4!} \times \frac{6!}{6!0!} = \frac{(24)(23)(22)(21)}{(4)(3)(2)(1)}$$

$$= 10{,}626,$$

while the total number of possible samples of size 4 selected from the entire lot of 30 will be

$$C(30,4) = \frac{30!}{26!4!} = \frac{(30)(29)(28)(27)}{(4)(3)(2)(1)} = 27{,}405.$$

Based on this acceptance sampling plan, then, the dealer will have a probability of

$$\frac{C(24,4)C(6,0)}{C(30,4)} = \frac{10{,}626}{27{,}405} = .3877 = 38.77\%,$$

of deciding to purchase the lot of 30 refrigerators if 6 of them are really in unacceptable condition.

Naturally, the more refrigerators the dealer uses in his acceptance sampling plan, the less likely he is to make an incorrect assessment of the true number of unacceptable ones in the lot. Table 3.4 compares the size of the sample used with the probability of acceptance of the lot.

Let's now develop a general formula for use in acceptance sampling situations. We need the following symbols:

n = number of items in the entire lot
r = number of items in the randomly selected sample
m = the true number of acceptable items in the lot
N = the number of acceptable items in the random sample.

What we have to calculate is $P(N = k)$, the probability that a sample of size r contains k acceptable items. First we should observe that, for a sample of size r to contain k acceptable items, it must be composed of k items from group A^* and $r-k$ items from group U^*. Since group A^* contains m items, there are $C(m,k)$ different sets of k acceptable items that may be selected for the sample. Since group U^* contains $n-m$ items (= number of items in entire lot, minus number of acceptable items in lot), there are $C(n-m, r-k)$ different sets of $r-k$ unacceptable items that may be selected for the sample. Combining each of the $C(m,k)$ sets of acceptable items with each of the $C(n-m,r-k)$ sets of unacceptable items, we find that there are

TABLE 3.4   Acceptance Probabilities for Various Acceptance Sampling
Plans   *(Lot size 30, of which 6 are unacceptable; accept the lot if sample
consists of acceptable items only.)*

Sample Size k	$C(24,k)C(6,0)$	$C(30,k)$	Acceptance Probability $\dfrac{C(24,k)C(6,0)}{C(30,k)}$
1	24	30	.8000
2	276	435	.6345
3	2,024	4,060	.4985
4	10,626	27,405	.3877
5	42,504	142,506	.2983
6	134,596	593,775	.2267
7	346,104	2,035,800	.1700
8	735,471	5,852,925	.1257
9	1,307,504	14,307,150	.0914
10	1,961,256	30,045,015	.0653

$$C(m,k)C(n-m,r-k)$$

samples of r items that contain exactly k acceptable items.

Therefore, the probability that a sample of size r contains k acceptable
items is calculated by the following formula.

Acceptance Sampling Formula:

■ $$P(N=k) = \frac{\text{number of samples that contain } k \text{ acceptable items}}{\text{number of samples of size } r}$$

$$= \frac{C(m,k)C(n-m,r-k)}{C(n,r)}. \tag{3.14}$$

By analogy with the binomial distribution mentioned in Section 3.C, we say
that N, the number of acceptable items in the sample, is a random variable
that has the *hypergeometric distribution*. We will study the hypergeometric
distribution in more detail in Section 4.F.

EXAMPLE 3.5.   Quality Control.   A machine that puts out plastic
pipe is, according to contractual specifications, supposed to be producing pipe of 4 inches in diameter. Pipe that is either too large or too small
will not meet the construction requirements. In order to insure that the
machine is operating properly (so that the pipe it is producing falls
within the contractual specifications), its output must be carefully mon-

itored. Every hour 50 pieces of pipe are produced, and a random sample of 8 pieces is selected from the lot of 50. If 7 or more of the 8 pieces sampled are acceptable according to the specifications, the entire lot of 50 is packed up and shipped to the customer. If 10 out of the 50 pieces are really defective, what is the probability that the lot will be shipped?

SOLUTION. In formula (3.14), we have $n = 50$, $r = 8$, and $m = 40$. (Since 10 are defective, the remaining 40 are acceptable.) If we set

N = number of acceptable items in the random sample,

then the probability that the lot will be shipped is, according to formula (3.14),

$$P(N \geqslant 7) = P(N = 7) + P(N = 8)$$

$$= \frac{C(40,7)C(10,1)}{C(50,8)} + \frac{C(40,8)C(10,0)}{C(50,8)}$$

$$= \frac{\dfrac{40!}{33!7!} \times \dfrac{10!}{9!1!}}{\dfrac{50!}{42!8!}} + \frac{\dfrac{40!}{32!8!} \times \dfrac{10!}{10!0!}}{\dfrac{50!}{42!8!}}$$

$$= \frac{\dfrac{(40)(39)(38)(37)(36)(35)(34)}{(7)(6)(5)(4)(3)(2)(1)} \times \dfrac{10}{1}}{\dfrac{(50)(49)(48)(47)(46)(45)(44)(43)}{(8)(7)(6)(5)(4)(3)(2)(1)}}$$

$$+ \frac{\dfrac{(40)(39)(38)(37)(36)(35)(34)(33)}{(8)(7)(6)(5)(4)(3)(2)(1)} \times \dfrac{1}{1}}{\dfrac{(50)(49)(48)(47)(46)(45)(44)(43)}{(8)(7)(6)(5)(4)(3)(2)(1)}}$$

$$= \frac{(40)(39)(38)(37)(36)(35)(34)(10)(8)}{(50)(49)(48)(47)(46)(45)(44)(43)}$$

$$+ \frac{(40)(39)(38)(37)(36)(35)(34)(33)}{(50)(49)(48)(47)(46)(45)(44)(43)}$$

$$= \frac{12,429,040}{35,791,910} + \frac{5,126,979}{35,791,910}$$

$$= .3473 + .1432 = .4905 = 49.05\%.$$

Our conclusion, therefore, is that our acceptance sampling procedure will give us a probability of approximately 49% of shipping a lot of 50 pieces of pipe if 10 are actually unacceptable.

If a probability of 49% seems too high, we can either choose a larger random sample (more than 8 pieces) or we can make an acceptance rule more stringent (ship the lot only if *all* pieces in the sample are acceptable.) We could construct a table analogous to Table 3.4 in order to get an idea of the acceptance probabilities based on various acceptance sampling plans.

Operating Characteristic (OC) Curve

All calculations we have made so far in this section have been carried out under the assumption of a specific number of unacceptable items in the lot. For example, in the pipe manufacturing situation, we assumed that 10 of the 50 pieces in the lot were defective. In reality, however, we do not know how many of the items in the lot are actually unacceptable. It would, therefore, be useful to have a way of measuring how a proposed acceptance sampling plan operates in the presence of various possible numbers of unacceptable items in the lot. The *operating characteristic* (or OC) curve provides just such a way of measuring the effectiveness of an acceptance sampling plan.

Let's reconsider the situation of Example 3.5. When 10 of the 50 pieces of pipe are unacceptable, our sampling plan gives us a probability of 49.05% of shipping the lot. What would the acceptance probability be if 20 of the 50 pieces were unacceptable? Well, in formula (3.14), we would take $n = 50$, $r = 8$, and $m = 30$, so the probability of shipping the lot would be

$$P(N \geqslant 7) = P(N = 7) + P(N = 8)$$

$$= \frac{C(30,7)C(20,1)}{C(50,8)} + \frac{C(30,8)C(20,0)}{C(50,8)}$$

$$= \frac{(30!)(20!)(42!)(8!)}{(23!)(7!)(19!)(1!)(50!)} + \frac{(30!)(20!)(42!)(8!)}{(22!)(8!)(20!)(0!)(50!)}$$

$$= \frac{(30)(29)(28)(27)(26)(25)(24)(20)(8)}{(50)(49)(48)(47)(46)(45)(44)(43)}$$

$$+ \frac{(30)(29)(28)(27)(26)(25)(24)(23)}{(50)(49)(48)(47)(46)(45)(44)(43)}$$

$$= \frac{271,440}{3,579,191} + \frac{3,393}{311,234}$$

$$= .0758 + .0109 = .0867 = 8.67\%.$$

TABLE 3.5   Acceptance Probabilities for One Sampling Plan   *(Lot size 50;*
select 8 and ship lot if at least 7 are acceptable.)

Actual Number of Unacceptable Pieces $(50-m)$	$P(N = 7)$	$P(N=8)$	Acceptance Probability
2	.2743	.7029	.9772
4	.3988	.4860	.8848
6	.4282	.3301	.7583
8	.4019	.2198	.6217
10	.3473	.1432	.4905
12	.2821	.0911	.3732
14	.2178	.0564	.2742
16	.1602	.0338	.1940
18	.1129	.0196	.1325
20	.0758	.0109	.0867
22	.0486	.0058	.0544
24	.0293	.0029	.0322
26	.0171	.0014	.0185
28	.0090	.0006	.0096
30	.0037	.0002	.0039

N = number of acceptable pieces in sample of 8

The calculation we have just made shows that our acceptance sampling procedure (namely, choosing 8 at random and shipping the lot if at least 7 of
them are acceptable) gives us a probability of only 8.67% of shipping a lot of
50 pieces when 20 of the 50 are actually unacceptable.

   Table 3.5 lists the probabilities of shipping the lot of 50 pieces of pipe,
based on the results of our acceptance sampling plan, corresponding to various amounts of actually unacceptable pieces in the lot. The OC curve constructed from Table 3.5 appears in Fig. 3.16 and graphically compares the
actual number of unacceptable pieces with its corresponding acceptance
probability.

EXERCISES   3.D

3.25. A local electronics supply store uses the following acceptance sampling plan
      on lots of 19 CB radio crystals imported from Hong Kong: select 4 of the 19 at
      random and buy the lot only if all 4 are in perfect condition.

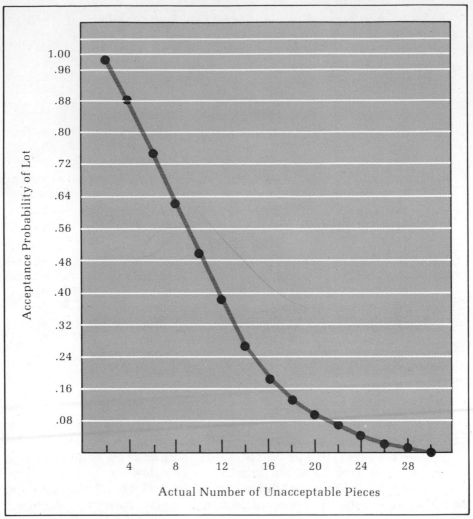

FIGURE 3.16   Operating Characteristic (OC) Curve for One Sampling Plan   *(Lot Size 50; select 8 and ship lot if at least 7 are acceptable.)*

a. Find the probability that all 4 in the sample will be in perfect condition, even if 5 of the 19 crystals are defective.

b. Construct a table of acceptance probabilities for this type of sampling plan for samples of size 1, 4, 7, 10. Use Table 3.4 of the text as a model.

c. Construct a table of acceptance probabilities for the sampling plan based on a sample of size 4, when the actual number of defective crystals is 0, 1, 3, 5, 7, 9, 11. Use Table 3.5 of the text as a model.

d. Draw the OC curve corresponding to the table constructed in preceding step c.

3.26. An automobile assembly plant agrees to buy a lot of 16 cartons of bolts only if 2 randomly selected cartons both pass a detailed inspection.

   a. What is the probability that both cartons in the random sample will pass inspection if 3 of the 16 cartons are actually in unacceptable condition?

   b. Construct a table of acceptance probabilities, analogous to Table 3.4 of the text, for samples of size 1, 2, 3, 4, 5, 6, 7, 8.

   c. Construct a table of acceptance probabilities, analogous to Table 3.5 of the text, for the actual number of unacceptable cartons being 0, 1, 2, 3, 4, 5, 6, 7, 8.

   d. Draw the OC curve corresponding to the table in part c.

3.27. The consumer credit office of a department store decided to mail a credit card to each of 60 households in its immediate neighborhood if, in a random sample of 5 households, at least 4 are found to have good credit ratings.

   a. If 6 of the 60 households are bad credit risks, what are the chances that at least 4 out of the sample of 5 would turn out to be good risks?

   b. Construct a table of acceptance probabilities for this sampling plan, analogous to Table 3.5 of the text, in case the actual number of bad credit risks were 0, 2, 4, 6, 8, 10, 12, 14, 16, 18, 20, 22, 24, 26, 28, 30.

   c. Draw the OC curve based on the table of part b above.

3.28. An advertising manager has to select a sample of 6 papers in a chain of 35 in which to place her advertisements.

   a. If 30 of the 35 papers reach the desired market, what is the probability that at least 5 of the 6 papers she selects will do so?

   b. Construct a table, analogous to Table 3.5 of the text, listing the probabilities that 5 of the 6 papers she selects actually reach the desired market, if 0, 1, 2, 3, 4, 5, 6, 7, 8, 9, 10, respectively, of the 35 papers in the chain fail to reach the market.

   c. Draw the OC curve based on the table of part b.

SUMMARY AND DISCUSSION

Our focus in Chapter 3 has been probability theory. Our direct concern has been the updating of probabilities of events, taking account of newly available information. We have seen that the presence of new information represents a change in circumstances that can increase or decrease or have no effect on probabilities of particular events. In the special case in which occurrence of one event has no effect on the probability of the other, the two events are said to be independent.

We introduced the concepts and calculational formulas of permutations and combinations, and showed how they are used to analyze the behavior of collections of independent events.

Finally we touched briefly on the binomial and hypergeometric probability formulas, in anticipation of further discussion of these topics in the next chapter. Conditional probability will be discussed in Section 11.A in terms of Bayes' theorem. Section 11.A can be studied now, directly after Chapter 3, because it does not require any knowledge of the intervening chapters.

CHAPTER 3   BIBLIOGRAPHY

GENERAL DISCUSSIONS OF PROBABILITY

FINE, T. L.: *Theories of Probability: An Examination of Foundations*, 1973, New York: Academic Press.
MAISTROV, L. E. (S. KOTZ, tr.): *Probability Theory: An Historical Sketch*, 1974, New York: Academic Press.
WEAVER, W.: *Lady Luck*, 1963, New York: Doubleday.

INDEPENDENT EVENTS

HUFF, B. W.: "Another Definition of Independence," *Mathematics Magazine*, September 1971, pp. 196–197.

CHAPTER 3   SUPPLEMENTARY EXERCISES

3.29 Decide which of the following statements are always true and which are sometimes false:

a. $P(A) + P(A^c) = 1$

b. $P(A|B) + P(A^c|B) = 1$

c. $P(A|B) + P(A|B^c) = 1$

d. $P(A|B) + P(A^c|B^c) = 1$

3.30. If we roll a fair tetrahedron (a four-sided rather than a six-sided die) once, the sample space for the experiment is $S = \{1,2,3,4\}$, and all four outcomes are equally likely. Consider the events

$$A = \{1,2\} \quad B = \{1,3\} \quad C = \{1,4\}$$

a. Show that A, B, and C are pairwise independent; namely, that A is independent of B, B is independent of C, and C is independent of A.

b. Show, however, that A is not independent of the event $B \cap C$.

3.31. If A and B are events such that $P(A) = 1/2$, $P(B) = 2/3$, and $P(A|B) = 1/4$, then calculate

a. $P(A^c|B)$ c. $P(B|A)$

b. $P(A|B^c)$ d. $P(A^c|B^c)$

3.32. Data compiled by the U.S. National Center for Education Statistics on college degrees conferred in 1974 revealed that 14% of the bachelor's degrees were awarded in the fields of business and management. In addition, the researchers found that 87% of the business and management degrees were awarded to males, even though males accounted for only 51% of the degrees in other fields.

a. Draw a tree diagram illustrating the relationship between sex and degrees.

b. Use the tree diagram to find out what percentage of 1974 college graduates were male students receiving their degrees in business and management.

c. Determine what percentage of male college graduates received their degrees in business and management.

d. What percentage of female graduates received business and management degrees?

3.33. A "stop-smoking" clinic advertised that, in a test region last August, 80% of those who tried and were able to stop smoking had participated in its program, while only 30% of those who tried and failed to stop smoking had participated. Research by the Consumer Fraud Division supported these assertions, but also revealed that, of all smokers in the region who tried to "kick the habit" last August, 90% had failed, while only 10% succeeded.

a. Draw a tree diagram illustrating the relationship between the ability to stop smoking and the clinic's program.

b. Determine, from the information contained in the tree diagram, what percentage of those who tried to kick the habit had participated in the clinic's program.

c. Determine from the tree diagram what percentage of the clinic's customers actually were able to stop smoking.

3.34. The noxious oxides of nitrogen comprise 20% of all pollutants in the air, by weight, in a certain metropolitan area. Automobile exhaust accounts for 70% of those noxious oxides, but only 10% of all other pollutants in the air.

a. Construct a tree diagram of the relationship between pollutants in the air and automobile exhaust in the metropolitan area under study.

b. From the tree diagram, find out what percentage of pollutants in the air are accounted for by automobile exhaust.

c. Use the tree diagram to find out how much of the pollution contributed by automobile exhaust can be classified as noxious oxides of nitrogen.

3.35. A Georgeson-Lind survey of the opinions of individual shareholders, taken in late 1977, showed that only 6% of the respondents felt that corporations provided shareholders with more useful financial information than did stockbrokers.*

a. If the survey results are valid, what is the probability that, of 80 randomly selected shareholders, three or fewer feel that way?

b. The *Poisson distribution* may be used to approximate the binomial distribution for "large" values of n and "small" values of π. (The general rule of thumb says that the approximation works as long as $n\pi < 8$).

*Reported in *The Wall Street Journal*, December 28, 1977, p. 10.

THE POISSON APPROXIMATION

The *Poisson approximation* asserts that

$$C(n,k)(\pi)^k(1-\pi)^{n-k} \approx \frac{(n\pi)^k}{k!} e^{-n\pi}$$

where the symbol "$\approx$" means "is approximately equal to." (The number $e^{-n\pi}$ can be found in Table A.6 of the Appendix.)

Use the Poisson approximation to calculate* the probability asked for in part a.

3.36. Of all major spectator sports, major-league, minor-league, and college hockey showed the greatest increase in paid attendance between 1966 and 1976[†]. In 1966, 216.6 million paying spectators watched major sports events, and 6.8 million of these (or 3.14%) saw hockey. In 1976, there were 314.0 million paying spectators in all, and 22.7 million (or 7.23%) saw hockey.

   a. Of 100 randomly selected spectators at games held in 1966, what are the chances that 5 or more were watching a hockey game?

   b. Compute the probability in part a, using the Poisson approximation.

   c. Of 100 randomly selected spectators at games held in 1976, what is the probability that 5 or more were watching a hockey game?

   d. Compute the probability in part c, using the Poisson approximation.

3.37. The Abdec Oil Corporation has to decide quickly whether or not to bid on leases for 18 new offshore tracts just opened by the Interior Department. The company's management decides to bid for the leases if 3 randomly selected tracts subjected to detailed paleontological analysis all showed definite indications of the presence of oil.

   a. If 5 of the 18 tracts would really be considered unacceptable by the company, what are the chances that the company's acceptance sampling plan would lead them to bid on the 18 tracts anyway?

   b. Construct a table of acceptance probabilities, analogous to Table 3.4 of the text, for samples of size 1, 2, 3, 4, 5, 6, 7, 8, 9, 10.

   c. Construct a table of acceptance probabilities, analogous to Table 3.5 of the text, for actual numbers of unacceptable tracts 0, 1, 2, 3, 4, 5, 6, 7, 8, 9, 10.

   d. Draw the OC curve based on the table of part c above.

---

*This method of approximation dates back to the 1800s, but note that it simplifies the calculation even now in the computer age.

†*U.S. News & World Report,* May 23, 1977, p. 63.

3.38. A quality control supervisor for a chain of discount clothing stores checks 6 suits out of every lot of 40, and rejects the lot if 4 or more out of the sample of size 6 are defective.

    a. If 10 of the 40 suits are really defective, what is the probability that the supervisor will accept the lot anyway?

    b. Construct a table of acceptance probabilities, analogous to Table 3.5 of the text, when the actual number of unacceptable suits ranges from 0 to 30.

    c. Draw the OC curve based on the table of part b.

4

The

Basic Probability

Distributions

In the first three chapters, several techniques for describing a set of data were introduced to organize the data into an understandable format. Our work so far, however, has dealt only with the individual organization of a single set of data. All the data sets that have been discussed in the text and the exercises generally have had very different pictorial representations and numerical descriptions. It seemed that there was no way of telling in advance how the frequency polygon will look or what the mean and standard deviation will be. Yet as it turns out there are often discernible patterns common to sets of data derived from many different sources.

In this chapter, our objective will be to further our understanding of statistical data by studying these general patterns of behavior of sets of data. If we can fit a particular data set into one of these general patterns, then we will be able to say that it has all the properties of the general pattern to which it belongs, in addition to those special properties that can be discovered by using the methods of the first two chapters. These general patterns of data behavior are called *probability distributions*.

SECTION 4.A PROBABILITIES AND PROPORTIONS

By the probability of an event, we usually mean the proportion of time that the event can be expected to occur in the long run. For a simple example, consider the tossing of a coin and the observation of which of the two sides, heads or tails, is facing upward when the coin lands. If the coin is tossed 10,000 times and falls heads on 3,000 of these tosses, we are probably justified in asserting that the probability of heads for this particular coin is 3/10. Remember that by a *fair* coin, we mean a coin whose probability of heads is 1/2.

To give an example of how probabilities are determined from a set of data, let's take a look at the subject matter of Table 1.4, which we reproduce in a briefer format as Table 4.1. We can see from Table 4.1 that, on 23 of the 100 days involved in the study, the usage of salad dressing was between 26 and 28 gallons, inclusive. We can rephrase this fact in probability language as follows: the study indicates that the probability of daily usage of salad dressing between 26 and 28 gallons is 23/100 or 0.23. Using symbols, we can write that

$$P(26 \leqslant D \leqslant 28) = 0.23,$$

where $P(\ )$ is pronounced "probability that" and D stands for "daily usage of salad dressing." Technically speaking, D is called a *random variable*. Random variables are variable quantities that take one of several possible values, depending on which outcome of the sample space has actually occurred. Recall what we said at the beginning of Section 3.A: "the collection of all possible outcomes . . . is called the *sample space*." Now these outcomes, and therefore the possible values of the random variable, occur with various probabilities.

In the salad dressing example there is a specific probability based on the collected data that $D = 26$, a specific probability that $D = 27$, and a specific probability that $D = 28$. The sample space here is the collection of all possible numbers that might be the daily usage in gallons. The random variable D takes on a numerical value that is determined by the particular outcome that occurred. The probability distribution of D is the list of all

TABLE 4.1   Frequency Distribution of Usage of Salad Dressing

Usage (number of gallons)	Days
20–22	4
23–25	8
26–28	23
29–31	34
32–34	26
35–37	4
38–40	1
Sum	100

possible values (or ranges of values) of D, together with the probabilities that
D equals each of those values (or that D falls in those ranges).

Looking at the frequency distribution of Table 4.1, we can see that the
random variable D is at least as large as 26 (namely, $26 \leq D$), but no larger
than 28 (namely, $D \leq 28$), 23% of the time. This is the origin of the statement
that $P(26 \leq D \leq 28) = 0.23$. Similarly, we can write down the following
remaining probabilities:

$P(20 \leq D \leq 22) = 0.04$
$P(23 \leq D \leq 25) = 0.08$
$P(29 \leq D \leq 31) = 0.34$
$P(32 \leq D \leq 34) = 0.26$
$P(35 \leq D \leq 37) = 0.04$
$P(38 \leq D \leq 40) = 0.01$

The seven probabilities calculated above comprise (insofar as we can determine from the frequency distribution in Table 4.1) what is called the *probability distribution* of the daily usage of salad dressing. More simply, we can
refer to the "probability distribution of D." In Table 4.2, we formally display
the probability distribution of D.

As we have already shown in Chapter 1, the probability distribution of
D can be illustrated graphically by the histogram of Fig. 1.4 or the frequency
polygon of Fig. 1.5. All that is necessary is to divide the number of days on
the vertical scale by 100 (because a total of 100 days are represented by the
data) to obtain the proportions.

In Table 4.3, we present the probability distribution for the toss of the
coin discussed earlier in this section. You should recall that the coin in
question fell heads 3,000 times in 10,000 tosses. The coin fell tails, of course,

TABLE 4.2   Probability Distribution of Daily Usage of Salad Dressing (D)

a (number of gallons)	b (number of gallons)	$P(a \leq D \leq b)$
20	22	0.04
23	25	0.08
26	28	0.23
29	31	0.34
32	34	0.26
35	37	0.04
38	40	0.01
	Sum	1.00

the remaining 7,000 times. We use the random variable H, the number of heads (either 0 or 1) obtained on an individual toss, to represent the coin-tossing process. From the data we can say that $P(H = 1) = 3/10$ and $P(H = 0) = 7/10$.

A probability distribution is said to be *uniform* if all of the possible outcomes have the same probability. A simple example of a uniform distribution would be the probability distribution of one toss of a fair coin. Such a coin would fall heads about half the time and tails the other half. We could then write

$$P(H = 1) = 1/2 \qquad \text{and} \qquad P(H = 0) = 1/2$$

for fair coin. In any particular coin-tossing experiment with a fair coin, it is unlikely that we will get exactly half heads and half tails. What we actually do is to use the results obtained from a long series of experiments to help us determine the underlying probability distribution of the coin.

If a quantity is uniformly distributed and has n possible outcomes, it is automatic that each of the n possible outcomes has probability $1/n$. For example, if 16 evenly matched golfers play in a tournament, then each has probability 1/16 of winning. This estimate of probabilities comes from the observation that, if the tournament were repeated many, many times, each golfer would win a proportion of one out of every 16 times, on the average. Of course, this is true only for evenly matched golfers. Often in practical situations, we talk as though the uniform distribution is valid in general, even when things are not evenly matched. By proper understanding of probability distributions, however, we can avoid committing this error.

In the remaining sections of this chapter, we will study in detail what are perhaps the five most useful probability distributions in the statistical study of elementary business and management situations: namely, the binomial, normal, Poisson, hypergeometric, and exponential distributions.

TABLE 4.3   Probability Distribution of Random Variable H (for a coin which fell heads 3,000 times in 10,000 tosses)

Outcome	H	Probability
heads	1	3/10
tails	0	7/10
Sum		1

EXERCISES 4.A

4.1. From the data of Exercise 1.11, construct the probability distribution of the price forecasts.

4.2. Using the data of Exercise 1.18, set up the probability distribution of the number of attempts necessary.

4.3. Construct a probability distribution of the daily output of oil by the 7 biggest independent oil refiners in the United States. Use the data of Exercise 1.5.

4.4. Based on the information presented in Exercise 1.17, set up a probability distribution of the housing ratings.

4.5. Use the data of Exercise 1.7 to construct a probability distribution of a business economist's opinion of the most important economic problem the country would face over the succeeding five-year period.

SECTION 4.B   THE BINOMIAL DISTRIBUTION

The *repeated* tossing of a coin provides perhaps the simplest model of a process that generates binomially distributed data. There are only two possible outcomes on each run of the experiment, heads and tails, with P(heads) = 1/2 and P(tails) = 1/2 if the coin is fair. Furthermore, each toss of the coin is independent of all previous tosses.* These are the characteristic properties of what is called a *binomial experiment:* "binomial," from Greek and Latin roots, means "having two names." In a binomial experiment there are only two possible outcomes; their probabilities remain constant throughout the experiment although they do not necessarily have to equal 1/2; and each trial of the experiment is independent of all previous trials. Many examples of interest in business and management can be organized as binomial experi-

---

*This statement about coin-tossing has generated substantial philosophical controversy over the years, even though it can be proved by simple experiment. The allegation is that, if a fair coin falls heads ten times in a row, it is very likely to fall tails on the eleventh toss. Much money has been lost by gambling on this untrue assertion. Even though the law of averages implies that a coin will fall heads about half the time and tails the other half, this does not place a moral obligation on the individual coin to pay back those 10 tails it owes. In fact, during the eleventh toss, the coin itself does not even remember that it has fallen heads on the last 10 consecutive tosses. All it knows while it is spinning in the air is that, because of its structure, it is equally likely to fall heads or tails on any given toss, including the eleventh.

You are invited to check this out by conducting a coin-tossing experiment. Take a handful of, say, 500 pennies and toss them in the air in a closed room having no holes in the floor. About half the coins will fall tails. Remove the ones that fell tails, and toss the remaining ones a second time. Again, about half will have fallen tails. Remove them and proceed. Each time only half the coins will fall tails, even though every coin tossed has fallen heads on all previous tosses.

ments, and therefore any resulting data taken in such a situation will have the binomial distribution. Consider the following few cases:

1. *A Marketing Survey.* Several shoppers near the dairy case of a supermarket are asked the question, "Would you try a carton of chocolate-flavored buttermilk if such a product were developed?" The two possible outcomes are "yes" and "no." Presumably each shopper's opinion would not depend on that of any other shopper if the questioning were properly conducted. The probability of primary concern involved here is $\pi = P$ (yes), the proportion of shoppers who would be willing to try the new product. (This proportion, which represents "shoppers" in general, may be subject to change if we know we are dealing with only male shoppers, or only female shoppers, or only teenage shoppers, or only shoppers of a particular ethnic group or income level. In detailed marketing surveys, all these factors must be taken into account.)

2. *An Auditing Procedure.* In auditing the records of a large company, the public accounting firm mails letters to a randomly selected number of the company's customers asking for verification of their accounts. The customers will respond that the records of their accounts are either "correct" or "not correct." Each customer acts independently, and the relevant probability is $\pi = P$ (records are correct), the proportion of accounts that are correctly reported in the company's records.

3. *An Automobile Loan Approval.* Banks continually receive automobile loan applications from individuals. After an application is submitted, the two possible outcomes are "approved" and "denied." Each person's application is presumably judged on its own merits and therefore does not depend on the situation of any other application. The probability $\pi = P$ (application is approved) represents the proportion of loan applications that are approved.

4. *An Insurance Problem.* Several men, each age 45, apply for a 10-year term life insurance policy. The two possible outcomes are "died before age 55" and "lived to age 55." Whether or not one particular man dies before he reaches the age of 55 does not seem to be dependent on what happens to any other policyholder. The relevant probability is $\pi = P$ (death before age 55), the proportion of 45-year-old men who die before they reach age 55.

All binomial experiments can be completely characterized by two numbers. The first is π, the probability of one of the outcomes; the other outcome must then have probability $1 - \pi$, for if P(heads) = 1/3, then P(tails) = 2/3, and if P(yes) = 4/5, then P(no) = 1/5. The second is n, the number of times the experiment is repeated. All properties and mathematical formulas involved in the study of binomial experiments can be expressed in terms of the numbers n and π. Numbers that are characteristic of a situation are called the *parameters* of the problem. For this reason, much of statistics is con-cerned with the study of parameters, because if we can figure out what the parameters of a situation are, then we can determine everything else from them.*

EXAMPLE 4.1.   Coin Tossing.   Suppose we take three fair coins, toss them, and count the number of heads we have obtained. Here the param-eters are $n = 3$ and $\pi = 1/2$, where $\pi = P(\text{heads})$. We can denote by H the total number of heads obtained with the three coins. The quantity H is a random variable taking the possible values 0, 1, 2, and 3, because with three coins, we can get none, one, two, or three heads. (There is no way to get four or more heads with three coins.) Compute the following probabilities:

$P(H = 0)$ = probability of getting no heads with three coins;
$P(H = 1)$ = probability of getting one head with three coins;
$P(H = 2)$ = probability of getting two heads with three coins;
$P(H = 3)$ = probability of getting three heads with three coins.

SOLUTION.   As described in Section 4.A, the probabilities above com-prise the probability distribution of the number of heads obtained in the toss of three coins. In this case, it should not be too difficult to recognize the random variable H as having the binomial distribution mentioned briefly in Section 3.C. It has the parameters

$$n = 3 \text{ and } \pi = 1/2.$$

Now, how do we compute these probabilities? In Table 4.4, we list all possible outcomes of the toss of the three coins. As can be seen from Table 4.4, there are eight possible ways the three coins, as a group, can turn up. Because $\pi = 1/2$, we know that, for each of the three coins, heads and tails are equally likely. Therefore outcome No. 1 is just as likely to occur as outcome No. 8, and, in fact, all eight outcomes are equally likely to occur. This means that we can expect each outcome to occur about 1/8 or 12.5% of the time.

We are not directly interested in these outcomes, however, but in the number of heads obtained in each case. From Table 4.4, we can see that we obtain three heads from outcome No. 1, two heads from outcomes No. 2, No. 3, and No. 4; one head from outcomes No. 5, No. 6, and No. 7; and no heads from outcome No. 8. Therefore, we can expect to get

*However, some types of questions do not easily lend themselves to the analysis of parameters. To answer these questions, we need an entirely different class of procedures called *nonpara-metric* statistics, a brief introduction to which appears in Chapters 12 and 13.

TABLE 4.4 Outcomes of the Toss of Three Coins

Outcome	First Coin	Second Coin	Third Coin
No. 1	heads	heads	heads
No. 2	heads	heads	tails
No. 3	heads	tails	heads
No. 4	tails	heads	heads
No. 5	tails	heads	tails
No. 6	tails	tails	heads
No. 7	heads	tails	tails
No. 8	tails	tails	tails

three heads whenever outcome No. 1 occurs, namely 12.5% of the time. We express this in mathematical symbols by writing $P(H = 3) = .125$. We can expect to get two heads the 12.5% of the time when outcome No. 2 occurs, the 12.5% of the time when outcome No. 3 occurs, and also the 12.5% of the time when outcome No. 4 occurs. Therefore, we would expect to obtain two heads a total of 12.5% + 12.5% + 12.5% = 37.5% of the time. In mathematical symbols, we write $P(H = 2) = .375$. By similar reasoning, it can be shown that $P(H = 1) = .375$ and $P(H = 0) = .125$. By analogy with Table 4.2, we collect all these results and display them in the probability distribution of Table 4.5. In Fig. 4.1, we present the histogram of the probability distribution of Table 4.5.

FIGURE 4.1 Histogram of the Binomial Distribution with Parameters $n = 3, \pi = 1/2$

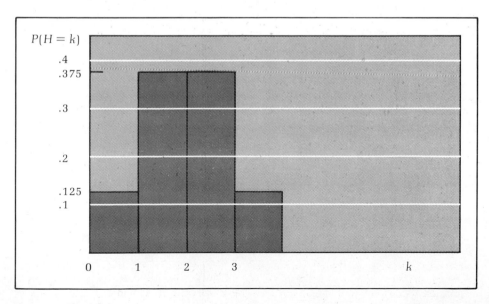

139

TABLE 4.5   Binomial Distribution with
Parameters $n = 3$, $p = 1/2$ (number of heads, H)

k	$P(H=k)$
0	.125
1	.375
2	.375
3	.125
Sum	1.000

In our calculation of the probability distribution of H, we relied very heavily on the fact that all eight outcomes of the coin tossing were equally likely to occur. This situation was reflected in value of the parameter π, which was equal to 1/2. In case $\pi \neq 1/2$, we no longer will have equally likely outcomes. When $\pi > 1/2$, we are more likely to get a head than a tail, and when $\pi < 1/2$, we are more likely to get a tail. In such cases, the calculation of the probability distribution is somewhat more difficult. In addition to the formula that was developed in Section 3.C for calculating the various probabilities involved, we have collected in Table A.2 of the Appendix the results of those calculations.

How to Use the Table of the Binomial Distribution

Let's recompute the material in Table 4.5 using Table A.2 instead of the facts about coin tossing. The first fact we must note about Table A.2 is that the numbers there are values of $P(H \leq k)$ instead of $P(H = k)$. With this in mind, we reproduce in Table 4.6 the portion of Table A.2 dealing with $n = 3$ and $\pi = 1/2$ ($\pi = .50$ in the language of Table A.2). We then proceed to derive

TABLE 4.6   Table A.2 for $n = 3$, $\pi = .50$

n	k	$\pi = .50$
3	0	.1250
	1	.5000
	2	.8750
	3	1.0000

Table 4.5 from the information of Table 4.6. As has been just mentioned, the numbers in the right-hand column of Table 4.6 are probabilities of the type $P(H \leq k)$. This means that:

$$.1250 = P(H \leq 0) = P(H = 0)$$

$$.5000 = P(H \leq 1) = P(H = 0) + P(H = 1)$$

$$.8750 = P(H \leq 2) = P(H = 0) + P(H = 1) + P(H = 2)$$

$$1.0000 = P(H \leq 3) = P(H = 0) + P(H = 1) + P(H = 2) + P(H = 3).$$

From the above information, we would like to compute the individual probabilities $P(H = 0)$, $P(H = 1)$, $P(H = 2)$, and $P(H = 3)$. First of all, since H can never be negative, $P(H \leq 0)$ is the same as $P(H = 0)$, so that

$$P(H = 0) = .1250 = .125.$$

Then, because $P(H = 0) + P(H = 1) = .5000$, it follows that

$$P(H = 1) = .5000 - P(H = 0) = P(H \leq 1) - P(H \leq 0)$$

$$= .5000 - .1250 = .3750 = .375.$$

Furthermore, because $P(H = 0) + P(H = 1) + P(H = 2) = .8750$, we have that

$$P(H = 2) = .8750 - [P(H = 0) + P(H = 1)]$$

$$= P(H \leq 2) - P(H \leq 1)$$

$$= .8750 - .5000 = .3750 = .375.$$

And, as $P(H = 0) + P(H = 1) + P(H = 2) + P(H = 3) = 1.0000$, we get, finally,

$$P(H = 3) = 1.0000 - [P(H = 0) + P(H = 1) + P(H = 2)]$$

$$= P(H \leq 3) - P(H \leq 2)$$

$$= 1.0000 - .8750 = .1250 = .125.$$

The Formula for the Binomial Distribution

If H is a binomial random variable having parameters n and π, we have seen in Section 3.C that we can calculate $P(H = k)$ as follows:

$$P(H = k) = C(n, k)\pi^k(1 - \pi)^{n - k} \tag{4.1}$$

where

■ $$C(n, k) = \frac{n!}{k! \, (n-k)!} \qquad (4.2)$$

As you should recall, $C(n, k)$ is called "the number of combinations of n elements taken k at a time" and the symbol m! (pronounced "m-factorial") means $1 \times 2 \times \ldots \times m$, the product of all whole numbers between 1 and m, inclusive.* The symbol π^k means $\pi \times \pi \times \ldots \times \pi$ k-times, the product of k π's. For example,

$$\left(\frac{1}{2}\right)^4 = \frac{1}{2} \times \frac{1}{2} \times \frac{1}{2} \times \frac{1}{2} = \frac{1}{16} = .0625$$

and $\left(\frac{2}{3}\right)^5 = \frac{2}{3} \times \frac{2}{3} \times \frac{2}{3} \times \frac{2}{3} \times \frac{2}{3} = \frac{32}{243} = .132.$

In Example 4.1, we had $n = 3$ and $\pi = 1/2$. Suppose we want to calculate $P(H = 1)$, using the formula. Then we would set $k = 1$, $n = 3$, and $\pi = 1/2$. We get

$$C(n, k) = C(3, 1) = \frac{3!}{1!(3-1)!} = \frac{3!}{(1!)(2!)} = \frac{6}{(1)(2)} = 3$$

so that

$$P(H = 1) = C(n, k)\pi^k(1 - \pi)^{n-k} = 3\left(\frac{1}{2}\right)^1\left(1 - \frac{1}{2}\right)^{3-1}$$

$$= 3\left(\frac{1}{2}\right)\left(\frac{1}{2}\right)^2 = 3\left(\frac{1}{2}\right)\left(\frac{1}{4}\right) = \frac{3}{8} = .375.$$

Of course, this is exactly the same answer we got using Table A.2. We can simplify the computation somewhat by looking up the numerical value of the binomial coefficient $C(n,k)$ in Table A.3 of the Appendix, rather than calculating it from formula (4.2).

We observe, in summary, that the entry in Table A.2 corresponding to parameters n and π is the number

■ $$P(X \leq k) = \sum_{j=0}^{k} P(X = j) = \sum_{j=0}^{k} C(n,j)\,\pi^j(1 - \pi)^{n-j} \qquad (4.3)$$

where X is a binomial random variable with parameters n and π.

As a further illustration of the techniques by which Table A.2 was developed, let's calculate the entry in Table A.2 corresponding to $n = 9$, k

*Due to certain technicalities, 0! turns out to be 1, as noted in Section 3.C.

= 4, and π = .55. According to formula (4.3), this entry should be

$$P(X \leq 4) = \sum_{j=0}^{4} C(9,j)(.55)^j(1 - .55)^{9-j}$$

$$= C(9,0)(.55)^0(.45)^9 + C(9,1)(.55)^1(.45)^8 + C(9,2)(.55)^2(.45)^7$$

$$+ C(9,3)(.55)^3(.45)^6 + C(9,4)(.55)^4(.45)^5$$

$$= (1)(1)(.000757) + (9)(.55)(.001682) + (36)(.3025)(.003737)$$

$$+ (84)(.166375)(.008304) + (126)(.091506)(.018453)$$

$$= .00076 + .00833 + .04070 + .11605 + .21276$$

$$= .3786.$$

A glance at Table A.2 reveals that the entry for n = 9, k = 4, and π = .55 is indeed .3786. In the above computation, we have used the facts (available from Table A.3) that

$$C(9,0) = \frac{9!}{0!9!} = 1$$

$$C(9,1) = \frac{9!}{1!8!} = 9$$

$$C(9,2) = \frac{9!}{2!7!} = 36$$

$$C(9,3) = \frac{9!}{3!6!} = 84$$

$$C(9,4) = \frac{9!}{4!5!} = 126.$$

As we continue to work with binomially distributed data in future portions of this text, we will refer to Table A.2 whenever numerical calculations are required.*

EXAMPLE 4.2   Automobile Loan Approvals.   Suppose one bank's records indicate that 60% of all automobile loan applications submitted to it are eventually approved. If we were to select a random sample of 7 applications from a certain section of a major city, what is the probability that a majority of those applications would not be approved?

SOLUTION.   This is the same situation as the one in Example 3.3. Here we are dealing with a binomial experiment with n = 7. This can be viewed as analogous to tossing 7 coins (representing the 7 applications, each marked "approved" on one side and "not approved" on the other).

---

*However, you are welcome to use the formula if you prefer.

TABLE 4.7   Table A.2 for $n = 7$, $\pi = .60$

n	k	$\pi = .60$
7	0	.0016
	1	.0188
	2	.0963
	3	.2898
	4	.5801
	5	.8414
	6	.9720
	7	1.0000

The fact that 60% of the applications are eventually approved can be translated into coin-tossing language as the statement that $\pi = P$ ("approved") $= .60$. Therefore, if we abbreviate by A the number of applications that are *approved*, we see that A has the binomial distribution with parameters $n = 7$ and $\pi = .60$. In Table 4.7, we reproduce the relevant portion of the table of binomial distribution, Table A.2 of the Appendix.

The question asked was what the probability is that a majority of applications in the sample is not approved. To say that a majority of the 7 applications is not approved means that 4 or more are not approved. This is equivalent to saying that 3 or fewer are approved, or that $A \leq 3$. It follows that

P(a majority of the sample is not approved)
$= P(3$ or fewer are approved$) = P(A \leq 3)$,

in view of the fact that A represents the number of loan applications that are approved. Now, as we have pointed out earlier, Table A.2 (and also Table 4.7) contain exactly probabilities of the form $P(A \leq k)$. Therefore, in Table 4.7, to the right of 3 (of the k column), we find $P(A \leq 3)$ $= .2898$. Therefore, the probability is 28.98%, or almost 29%, that the majority of our sample will be not approved, even though 60% of all loan applications are eventually approved. This is, of course, the same result we obtained in Example 3.3.

In Table 4.8 and Fig. 4.2, respectively, we record for illustrative purposes the probability distribution and the histogram of A. The calculations used to derive the information for Table 4.7 are also shown.

As the preceding example shows, there is more than one chance in four that a survey of seven applications would mislead a government investigator

TABLE 4.8   Binomial Distribution with Parameters $n = 7$,
$\pi = .60$ (number of loan applications approved, A)

k	$P(A \leq k) - P(A \leq k-1) =$	$P(A = k)$
0	.0016 − .0000 =	.0016
1	.0188 − .0016 =	.0172
2	.0963 − .0188 =	.0775
3	.2898 − .0963 =	.1935
4	.5801 − .2898 =	.2903
5	.8414 − .5801 =	.2613
6	.9720 − .8414 =	.1306
7	1.0000 − .9720 =	.0280
	Sum =	1.0000

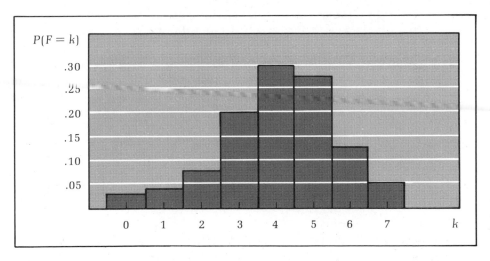

FIGURE 4.2   Histogram of the Binomial Distribution
with Parameters $n = 7$, $\pi = .60$

into believing that the bank was discriminating against residents of the section of the city under discussion, even if it were not doing so. The problem
here is that 7 applications do not constitute a sample large enough to inspire
real confidence in the results.

Suppose, for purposes of comparison, we had taken instead a sample
of 19 loan applications instead of 7. Then, to assert that a majority is not
approved means that 9 or fewer are approved, namely $A \leq 9$. In Table A.2,
for $n = 19$ and $\pi = .60$, we read that $P(A \leq 9) = .1861 = 18.61\%$. Therefore,

working with 19 applications instead of 7, we would be able to reduce our chances of being seriously misled from 28.98% down to 18.61%. Our chances of error would naturally be even smaller were we to take a sample of 50 or 100 or more applications. While Table A.2 contains the binomial probabilities only for values of n up to n = 20, we will learn in the next two sections how to use Table A.4 to calculate binomial probabilities having parameter n larger than 20.

> **EXAMPLE 4.3.   An Auditing Procedure.**   Suppose that it is really true that only 80% of the records of individual accounts on one company's books are correct. If the public accounting firm, conducting a formal audit, checks on the records of 15 randomly selected accounts, what are the chances that 12 or more of them turn out to be correct?

> **SOLUTION.**   What we have here is a set of binomial data with parameters $n = 15$ and $\pi = .80$. If we denote by C the number of records that turn out to be correct, we are interested in $P(C \geq 12)$, which is the same as $P(C > 11)$. Table A.2, as it is constructed, works better with $P(C > 11)$ than it does with $P(C \geq 12)$. In fact, the table contains neither of these items, but it does show that $P(C \leq 11) = .3518$ for $n = 15$, $\pi = .80$, and $k = 11$. This means that the chances are 35.18% that 11 or fewer records are correct. It follows automatically that the chances are 100% − 35.18% = 64.82% that more than 11 records are correct. We can condense this calculation by writing

$$P(C > 11) = 1 - P(C \leq 11) = 1 - .3518$$

$$= .6482 = 64.82\%.$$

Therefore, if 80% of all records are correct, the probability is 64.82% that 12 or more of 15 randomly selected records will turn out to be correct.

In Table 4.9 and Fig. 4.3, respectively, we derive the probability distribution and graph the histogram of a binomial random variable having parameters $n = 15$ and $\pi = .80$.

Mean and Standard Deviation of the Binomial Distribution

Before going on to the next section, we should call attention to a few more facts about the binomial distribution that we will find useful in our future work with statistics. These facts involve the mean and standard deviation of a binomial distribution that, as the material of Chapter 2 has shown, are important in understanding the behavior of random variables having that distribution.

TABLE 4.9 Binomial Distribution with Parameters $n = 15$, $\pi = .80$ (number of records that are correct, C)

n	k	$P(C \leq k)$	$P(C \leq k) - P(C \leq k - 1)$	=	$P(C = k)$
15	0	.0000*	.0000 − .0000	=	.0000*
	1	.0000*	.0000 − .0000	=	.0000*
	2	.0000*	.0000 − .0000	=	.0000*
	3	.0000*	.0000 − .0000	=	.0000*
	4	.0000*	.0000 − .0000	=	.0000*
	5	.0001	.0001 − .0000	=	.0001
	6	.0008	.0008 − .0001	=	.0007
	7	.0042	.0042 − .0008	=	.0034
	8	.0181	.0181 − .0042	=	.0139
	9	.0611	.0611 − .0181	=	.0430
	10	.1642	.1642 − .0611	=	.1031
	11	.3518	.3518 − .1642	=	.1876
	12	.6020	.6020 − .3518	=	.2502
	13	.8329	.8329 − .6020	=	.2309
	14	.9648	.9648 − .8329	=	.1319
	15	1.0000	1.0000 − .9648	=	.0352
			Sum	=	1.0000

*A probability of .0000 signifies less than 1 chance in 20,000.

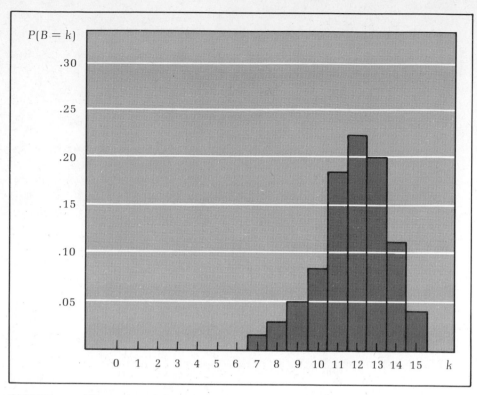

FIGURE 4.3　Histogram of the Binomial Distribution
with Parameters $n = 15, \pi = .80$

To introduce the concept of the mean of a probability distribution, let's take a look at some coin-tossing examples. Suppose we take a fair coin, having $\pi = P(\text{heads}) = .50$, and toss it 10 times. We would expect to get 5 heads in the 10 tosses. In fact, if several people got together, and each tossed a fair coin 10 times, the number of heads tossed by each would average out to something near 5. It would therefore be reasonable to agree that the mean number of heads obtained in 10 tosses of a fair coin is 5.

Now, what would the situation be if we were to toss another coin that was not fair but instead had probability of heads $\pi = .80$? Then we would expect to obtain 8 heads in our 10 tosses, and so it would be reasonable to say that mean number of heads in 10 tosses of the other coin is 8. For larger values of π, therefore, the mean number of heads would also be larger.

Suppose, on the other hand, we were to toss each coin 50 times instead of 10? Well, with a fair coin we would expect to get 25 heads, while with the coin having $\pi = .80$ we would expect 80% heads, or 40 heads out of the 50 tosses.

In general, suppose we take a coin having probability π of heads, and suppose we toss it n times. How many heads can we expect to get? We would expect that a proportion π of the n tosses would turn up heads. Therefore, the number of heads expected would be $\pi \times n = n\pi$ heads. Notice that this is exactly the procedure we followed in the specific cases above. In 10 tosses ($n = 10$) of a fair ($\pi = .50$) coin, we would expect $n\pi = 10 \times (.50) = 5$ heads, while in 50 tosses ($n = 50$) of a coin having $\pi = .80$, we would expect $n\pi = 50 \times (.80) = 40$ heads. We summarize these calculations and some more like it in Table 4.10.

As Examples 4.1, 4.2, and 4.3 made clear, coin tossing is merely one special case of a large class of situations in which the data follow binomial distributions. Furthermore, as we have seen, whenever a random variable has a binomial distribution, all relevant probabilities can be determined from Table A.2 of the Appendix, which is classified according to the parameters n and π of the random variable under study.

Therefore, all binomially distributed random variables having the same numerical values of n and π as parameters are statistically the same because they take the same numerical values with the same probabilities. By similar reasoning, we can see that all these random variables having the same n and π will also have the same mean, namely $n\pi$, regardless of whether they are the mean number of heads tossed, the mean number of loan applications approved, the mean number of records that are correct, or the mean of any other binomially distributed random variable having parameters n and π.

We are therefore justified in using the following formula to calculate the mean of any set of binomially distributed data having parameters n and π.

Formula for Mean of the Binomial Distribution:

$$\mu = n\pi \qquad\qquad (4.4)$$

TABLE 4.10   The Mean Number of Heads in Coin Tossing

Number of Tosses (n)	Probability of Heads (π)	Expected Number of Heads ($n\pi$)
10	.30	3
10	.50	5
10	.80	8
50	.30	15
50	.50	25
50	.80	40
200	.30	60
200	.50	100
200	.80	160

TABLE 4.11   Calculation of μ and σ for Some Binomial Distributions

n	π	$\mu = n\pi$	$1 - \pi$	$n\pi(1 - \pi)$	$\sigma = \sqrt{n\pi(1 - \pi)}$
10	.30	3	.70	2.1	1.45
10	.50	5	.50	2.5	1.58
10	.80	8	.20	1.6	1.26
50	.30	15	.70	10.5	3.24
50	.50	25	.50	12.5	3.54
50	.80	40	.20	8.0	2.83

An analogous argument, although a somewhat more complicated one, would show that we should use the following formula to calculate the standard deviation of a set of binomial data having parameters n and π.

Formula for Standard Deviation of the Binomial Distribution:

$$\sigma = \sqrt{n\pi(1 - \pi)} \tag{4.5}$$

We will have occasion to use these formulas for μ and σ in applied contexts from time to time as we proceed through this course. In Table 4.11, we present some sample calculations of μ and σ.

EXERCISES 4.B

4.6. Suppose we take four fair coins, toss them, and count the number of heads obtained, denoting the resulting number of heads by the random variable H.

a. What are the parameters of the binomial random variable H?

b. What are the possible values of H?

c. Construct the probability distribution of H.

4.7. An executive in charge of new product development at a local dairy feels that about 10% of the population would be willing to try a carton of chocolate-flavored buttermilk, while the other 90% would not be at all interested. To support her feeling with some observable facts, she conducts a survey at a local supermarket, asking each of 15 persons whether or not they would be willing to try a carton. She denotes by Y the random variable indicating the number of persons responding "yes" to the question.

a. What are the parameters of the random variable Y?

b. What are the possible values of Y?

c. Calculate the mean of Y.

d. Calculate the standard deviation of Y.

e. Construct the probability distribution of Y.

4.8. The *Los Angeles Times* recently reported that a power company in Vermont has decided to offer customers lower rates if they agree to use energy mainly at offpeak hours. According to the *Times*, the company expects only 20% of their customers to take advantage of their offer, but it wants to survey the customers to verify this expectation. Twenty people are asked whether or not they would be willing to subscribe to this service, and Y denotes the random variable indicating the number of persons responding "yes."

a. What are the parameters of the random variable Y?

b. What are the possible values of Y?

c. Calculate the mean of Y.

d. Calculate the standard deviation of Y.

e. Construct the probability distribution of Y.

4.9. After reading the results of a Gallup Poll that indicated that approximately 70% of all Americans over 18 drink alcoholic beverages, a local soft drink distributor decided to check on these disturbing (to him) findings. He contacts 10 randomly selected adults, and denotes by A the random variable indicating the number of adults that drink alcoholic beverages.

a. State the parameters of the random variable A.

b. List the possible values of A.

c. Calculate the mean of the random variable A

d. Calculate the standard deviation of the random variable A.

e. Set up the probability distribution of A.

4.10. The manufacturer of a new brand of soap guarantees that, with the aid of her unique marketing strategies, exactly 80% of the bars will be sold quickly. To check on her assertion, we market 17 bars as directed and denote by S the number that are sold quickly.

a. What should the parameters of S be?

b. What are the possible values of S?

c. What is the probability that fewer than 10 of our 17 bars will be sold?

d. Construct the probability distribution of S.

4.11. Twenty men aged 45, who are members of the same occupational grouping, apply for a 10-year term life insurance policy. In order to know how much to charge for the policy, the issuing company needs to be able to estimate how many of the applicants will die before age 55, in which case the company will have to pay off on the policy. Actuarial studies indicate that, of all 45-year-old men in the applicants' occupational grouping, 5% will die before they reach age 55.

a. Find the parameters of the random variable D which denotes the number of applicants who die before they reach age 55.

b. What is the probability that none of the 20 applicants will die before age 55, making it a very profitable deal for the insurance company?

c. Find the probability that exactly one of the 20 will die during the 10-year term.

d. What are the chances that more than 5 of the 20 applicants will die before age 55?

e. What is the probability that fewer than 10 of the applicants will die during the specified period?

f. Calculate the probability that more than 5, but fewer than 10, applicants will not make it to 55.

4.12. After reading a news reporter's allegation that 65% of the members of the various state legislatures favor a certain controversial change in the banking laws, a staff assistant to the major lobbying group opposing the change conducts a quick straw poll of 14 randomly selected House members.

a. If the reporter's story is correct, what are the chances that fewer than half of the 14 members surveyed would favor the change?

b. What is the probability that exactly half of those surveyed favor the change?

c. What would be the probability that more than 12 of the 14 members polled would favor the change?

d. What are the chances that more than half, but fewer than three-fourths, of those questioned would favor the change?

4.13. A consumer confidence survey conducted by the University of Michigan's Survey Research Center for the first quarter of 1976 revealed that 56% of the respondents expected prices to increase in the next twelve months.

a. If you were to randomly select 12 people, what are the chances that fewer than half believe that prices will increase (assuming that the survey results are valid)?

b. What is the probability that exactly one-half believe that prices will rise?

c. What is the probability that more than 10 of those in your survey expect the increases?

d. What are the chances that more than half, but fewer than three-fourths, of those polled anticipate the increase?

4.14. A 1977 survey by the A. C. Nielsen Company found that 45% of all homes in the United States have more than one television set.* What is the probability that, of 20 randomly selected homes, only 5 have more than one television set?

4.15. It was reported that 61% of all lawyers opposed the 1977 Supreme Court decision that lawyers may advertise their services.† If this figure is correct, what is the probability that, of 23 attorneys surveyed, more than 15 opposed the Court decision?

*Quoted in *U.S. News & World Report*, September 12, 1977, p. 23.

†*Orange County Business*, November/December 1977, p. 39.

4.16. Calculate the entry in Table A.2 corresponding to $n = 5$, $k = 3$, and $\pi = .85$.

4.17. Calculate the entry in Table A.2 corresponding to $n = 11$, $k = 4$, and $\pi = .35$.

4.18. Calculate the entry in Table A.2 corresponding to $n = 16$, $k = 13$, and $\pi = .70$.

SECTION 4.C THE NORMAL DISTRIBUTION

There are two major sources from which normally distributed sets of data arise in business and management situations: repetition of certain kinds of industrial or commercial processes over and over again (such as machine filling of 8 oz. cans of tomato sauce), and artificial construction of "standardized" tests in such a way that the test scores or indexes (such as scores on the Graduate Management Admissions Test) automatically turn out to be normally distributed. A set of data can often (but not always!) be recognized as having the normal* distribution if its frequency polygon is a *bell-shaped* curve, as illustrated in Figure 4.4.

A random variable that has the normal distribution is a particular kind of continuous variable, that is to say, one which can take on as a possible value any number at all within some specified range. (The binomial distribution, on the other hand, can assign only whole numbers to its random variables. Random variables that can take on only particular isolated values, such as the binomial that takes whole numbers only, rather than all numbers

FIGURE 4.4   A Bell-Shaped Curve

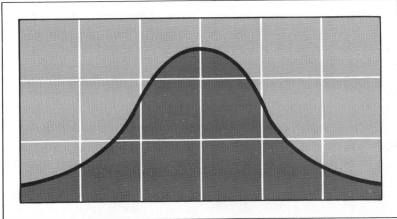

*The word "normal" comes from historical usage and is not meant to imply that other distributions are abnormal in any sense.

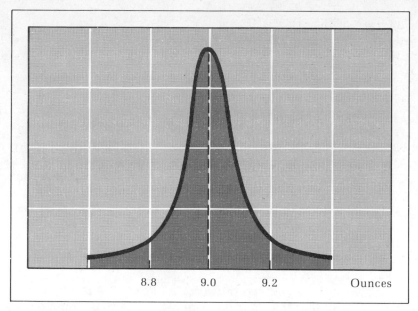

FIGURE 4.5   Frequency Polygon of Normally Distributed Contents
of Corn Puffies Boxes ($\mu = 9$, $\sigma = .01$)

in an entire interval, are called *discrete*.) As examples of the occurrence of
normally distributed sets of data, consider the following.

1. *Contents of cereal boxes*. It is a well-known fact that Corn Puffies are
packed by machine into 9-ounce boxes. The great majority of boxes contain
amounts very close to 9 ounces, a little more or less, but every once in a while
a box having contents quite a bit more or less shows up. Contents in ounces
of boxes of Corn Puffies are normally distributed with mean $\mu = 9$ and
standard deviation $\sigma = .01$. The fact that the standard deviation is so small,
relatively speaking, means that virtually all the boxes have almost exactly 9
ounces of cereal, while very, very, very few differ appreciably from it. The
intense concentration of data points near 9 results in the tall and thin bell-
shaped frequency polygon of Figure 4.5.

2. *Foot sizes (for shoe purchase) of individuals*. People living in a certain area, whether the specified area is a political subdivision such as a city,
state, or nation or a geographical subdivision such as a valley, plain, delta
region, etc., have normally distributed heights, weights, and other descriptive
measures such as chest or foot sizes. (Foot sizes are to be contrasted with
commercially available shoe sizes, which come in whole numbers and whole-
plus-a-half numbers only and are therefore discrete.) A particular group of
men may have mean foot size $\mu = 9.53$ with a standard deviation of $\sigma = 0.6$.
The degree of concentration near the mean would therefore be not as extreme

as in the case of the Corn Puffies data of the previous example. It follows logically that the lessened degree of concentration near the mean would result in a flatter bell-shaped curve. The normal curve of men's foot sizes appears in Figure 4.6.

This latter example points up the fact that it is very important in marketing to understand the nature and shape of a probability distribution. Of particular interest to managers of shoe departments and shoe stores would be the shape of the curve of foot sizes, along with any changes in the shape as we cross different geographic areas of the country. Do men in the East have bigger feet than men in the West? If they do, it will surely force us to change our marketing and distribution policies. Even if there is very little difference in foot sizes across the country, there certainly are differences in income and spending habits, and this induces differences in shoe-purchasing patterns across regional and urban-rural lines. Furthermore, many businesses mistakenly believe that knowledge of the mean or median is satisfactory. However, as this example illustrates, for marketing analyses a knowledge of the entire distribution is necessary if we are to avoid overstocks of certain sizes and understocks of others.

3. *Scores on standardized tests.* Scores of all students taking the Graduate Management Admissions Test (GMAT) are artificially constructed so as to have a normal distribution with mean $\mu = 500$ and standard deviation $\sigma = 100$. The purpose of such artificial labeling of scores is to allow persons who use the results of such tests to be better able to interpret the results in terms of the relative performances of several candidates. The frequency polygon of the GMAT scores is graphed in Fig. 4.7.

Each set of normally distributed data can be completely characterized by two numbers, its mean μ and its standard deviation σ. As explained in

FIGURE 4.6   Normal Curve of Men's Foot Sizes ($\mu = 9.53$, $\sigma = .6$)

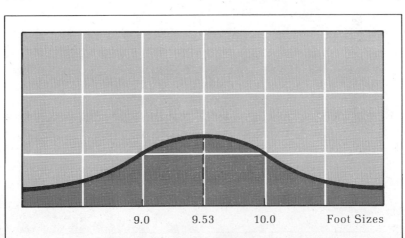

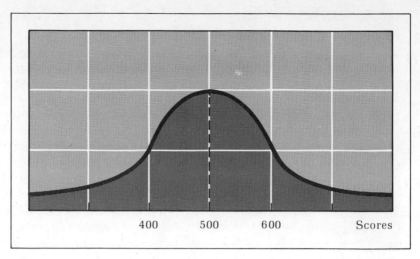

FIGURE 4.7   Frequency Polygon of GMAT Scores ($\mu = 500$, $\sigma = 100$)

Chapter 2, μ specifies the location of the mean, while σ measures the extent to which the data points are concentrated near the mean. The numbers μ and σ, since they completely describe the data, are the parameters of the normal distribution.

EXAMPLE 4.4.   The Graduate Management Admissions Test.   As we know, verbal scores on the GMAT are normally distributed with mean $\mu = 500$ and standard deviation $\sigma = 100$. If a student scores 643, what percentage of those who took the test ranked higher than she did?

SOLUTION.   The graph of the normal distribution involved in this example is shown in Fig. 4.8. The problem is to calculate the proportion

FIGURE 4.8   Proportion of GMAT Scores above 643 ($\mu = 500$, $\sigma = 100$)

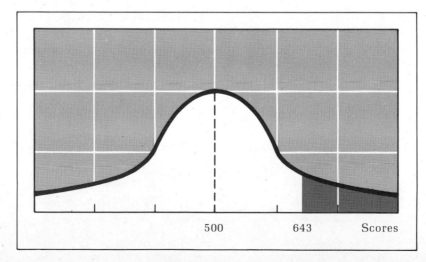

of data falling in the shaded region to the right of the number 643. There is a formula for calculating the proportions of a set of normally distributed data that fall in various intervals. Unfortunately, however, to be able to read this formula out loud, let alone to know how to use it, requires a considerable knowledge of calculus. (Details will appear in Section 4.G.) Fortunately, we have in Table A.4 of the Appendix a table of the various probabilities connected with a set of normally distributed data.

It should also be pointed out here that there are several types of probability distributions whose frequency polygons appear bell-shaped.* Although these distributions seem, at first glance, to be normal, their statistical characteristics, such as means and standard deviations, will be very different from those of normal distributions unless their probabilities are consistent with those of Table A.4.

How to Use the Table of the Normal Distribution

First of all, as the commentary in Table A.4 indicates, it is a table of the *standard* normal distribution. What does this mean? If you recall the discussion of standard scores, or z scores, in Section 2.E, you should note that we can calculate the z scores of any set of data and that these z scores have a probability distribution with mean $\mu = 0$ and standard deviation $\sigma = 1$. If we calculate the z scores of a normally distributed set of data (namely, if we "standardize" a set of normally distributed data), we will obtain a set of normally distributed data having mean $\mu = 0$ and standard deviation $\sigma = 1$. The resulting normal distribution having the parameters $\mu = 0$ and $\sigma = 1$ is called the standard normal distribution.

It would be grossly impractical to print tables of normal distributions for *every possible combination* of the numbers μ and σ, for this would require an extremely large number of pages. Fortunately, however, in problems dealing with *any* normal distribution, it is sufficient to use the table of the standard normal distribution, merely by working with the z scores instead of the original data points. To see how this is done, let's return to the GMAT example.

What we want to know is the proportion of scores falling above 643. This proportion will be exactly the same as the proportion of z scores falling above the z score corresponding to 643. Now, as you may recall from Section 2.E, the z score corresponding to the number x is given by the formula

$$z = \frac{x - \mu}{\sigma}$$

---

* See the article by Fama and Roll listed in the bibliography at the end of this chapter.

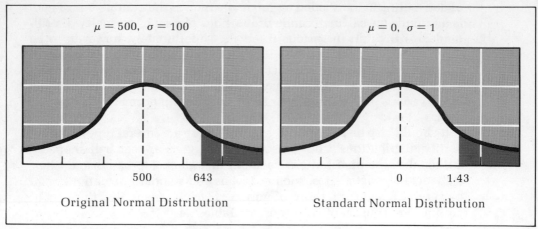

FIGURE 4.9   Original and Standardized GMAT Scores

and represents the number of standard deviations above the mean that x is located. (All normal distributions have exactly the same proportion of scores lying beyond z standard deviations above the mean.) Therefore the z score corresponding to 643 is

$$z = \frac{643 - 500}{100} = \frac{143}{100} = 1.43,$$

because here $\mu = 500$ and $\sigma = 100$.

The original and standardized normal distribution involved are illustrated in Fig. 4.9. As the pictures indicate, we can determine the proportion of candidates scoring above 643 on the verbal portion of the GMAT merely by calculating the proportion of a normal distribution that falls *above* the z score 1.43. Table A.4 is basically a list of z scores, together with the proportion of a set of standard normal data falling *below* the z score.

Looking for the z score $z = 1.43$ in Table A.4, we find next to it the proportion .9236, which indicates that 92.36% of a standard normal distribution lies *below* the z score $z = 1.43$. We reproduce the relevant portion of Table A.4 in Table 4.12. It follows, then, that $100\% - 92.36\% = 7.64\%$ of

TABLE 4.12   A Small Portion of Table A.4

z	Proportion
1.43	.9236

the distribution must lie *above* z = 1.43. Returning to the GMAT data, we can conclude that 7.64% of all candidates taking the GMAT score above 643. A student scoring 643 can therefore be sure of easily qualifying for the top 10% of all students taking the GMAT, since only 7.64% could have scored higher.

EXAMPLE 4.5.   The Graduate Management Admissions Test (Another View).   Now that we know that only 7.64% of students score above 643, it would be useful to know what score will qualify a student for the top 5% of all candidates.

SOLUTION.   As it turns out, we also can answer this question by using Table A.4. A graph illustrating the situation appears in Fig. 4.10, with the letter x denoting the (unknown) test score that separates the lower 95% of candidates from the upper 5%. Because $\mu = 500$ and $\sigma = 100$, the z score of the unknown x can be written as

$$z = \frac{x - 500}{100}.$$

Because of this algebraic relationship between x and z, we will be able to compute x if we know the numerical value of z. As the picture of the standard normal distribution of Fig. 4.10 shows, 5% or .0500 of the z scores will be above z, since 5% of the original scores are above x. This

FIGURE 4.10   Original and Standardized GMAT Scores

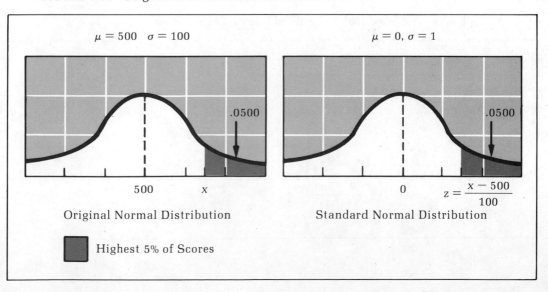

$\mu = 500 \quad \sigma = 100$ $\mu = 0, \sigma = 1$

.0500   .0500

500   x   0 $z = \dfrac{x - 500}{100}$

Original Normal Distribution   Standard Normal Distribution

Highest 5% of Scores

implies that 95% or .9500 of the z scores will be below z. From Table A.4, we can find the particular z that has 95% of the z scores below it. All we have to do is to find the *proportion* in Table A.4 that is closest to .9500, and then we read off the corresponding value of z.

The relevant portion of Table A.4 is reproduced in Table 4.13. Unfortunately, our table is not sufficiently detailed so as to include the number .9500 exactly. As Table 4.13 shows, there are two proportions, .9495 and .9505, which are equally close to .9500. In a case like this, it is customary to choose the z score that is closer to 0, so let's say that the z score in question is $z = 1.64$. For purposes of comparison, the graph of the standard normal distribution illustrating the position of $z = 1.64$ appears in Fig. 4.11 together with the standard normal distribution of Fig. 4.10. From the pictures in Fig. 4.11, we note the following two facts:

1. 5% of a standard normal distribution lies to the right of $z = (x - 500)/100$.

2. 5% of a standard normal distribution lies to the right of $z = 1.64$.

TABLE 4.13   A Small Portion of Table A.4

z	Proportion
1.64	.9495
1.65	.9505

FIGURE 4.11   A Comparison of Standard Normal Distributions

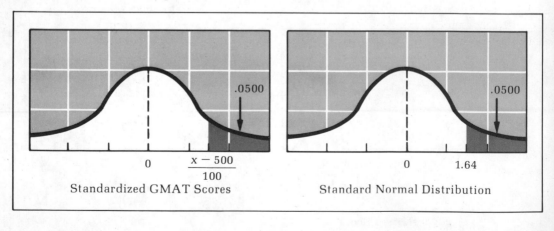

A comparison of the two pictures in Fig. 4.11 then shows that

$$\frac{x - 500}{100} = 1.64.$$

Successive algebraic operations then yield that

$$x - 500 = (1.64)(100) = 164;$$

and so

$$x = 164 + 500 = 664.$$

We have therefore found the numerical value (664) of x, the test score that separates the lower 95% of candidates from the upper 5%. It follows that a score above 664 on the verbal portion of the GMAT will put a student in the top 5% of all persons taking the test.

EXAMPLE 4.6.   Suit Sizes.   Men living in a certain urban area have mean suit size 38 with a standard deviation of 2.5. Furthermore, their suit sizes are normally distributed. If a small ready-to-wear clothing store plans to begin business with an initial stock of 500 suits, how many of these should be of sizes between 35 and 40?

SOLUTION.   First of all, we should remark that men's actual suit sizes are *continuous*, while ready-to-wear suits are manufactured only in *discrete* sizes. For example, an individual man may have size 35.8430067, but he will have to buy a size 36 suit even though it will not fit him exactly. To compute the proportion of men who are to be fitted with size 36 suits, a reasonable way would be to find the proportion of men having actual physical sizes between 35.5 and 36.5. We would in fact be computing the probability that a normal random variable with mean 38 and standard deviation 2.5 will have a value between 35.5 and 36.5.

To return to the question asked, we need to compute the proportion of men having *actual* suit sizes between 34.5 and 40.5. If we do this, we will know what proportion of men will need suits labeled between 35 and 40. The diagrams appear in Fig. 4.12, where the z score of 34.5 is

$$z(34.5) = \frac{34.5 - \mu}{\sigma} = \frac{34.5 - 38}{2.5} = \frac{-3.5}{2.5} = -1.40$$

and the z score of 40.5 is

$$z(40.5) = \frac{40.5 - \mu}{\sigma} = \frac{40.5 - 38}{2.5} = \frac{2.5}{2.5} = 1.00.$$

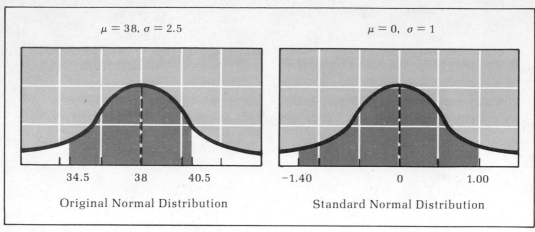

$\mu = 38, \sigma = 2.5$

$\mu = 0, \sigma = 1$

| 34.5 | 38 | 40.5 |

| −1.40 | 0 | 1.00 |

Original Normal Distribution

Standard Normal Distribution

FIGURE 4.12   Original and Standardized Suit Sizes

TABLE 4.14   A Small Portion of Table A.4

z	Proportion
−1.40	.0808
1.00	.8413

From Table A.4, or from the portion of it appearing in Table 4.14, we see that 8.08% of the z scores lie below −1.40, while 84.13% lie below 1.00. Of the 84.13% that lie below 1.00, then, 8.08% also lie below −1.40, and so the difference 84.13% − 8.08% = 76.05% falls between −1.40 and 1.00. Returning to the original normal distribution of suit sizes, we conclude that 76.05% of men have sizes between 34.5 and 40.5 and would therefore be fitted with suits labeled 35 to 40. Of the initial stock of 500 suits, then, the store should order 76.05% or 380 suits of sizes between 35 and 40.

EXAMPLE 4.7.   Approval of Mortgage Loans.   The dollar amounts requested of the First Phreemuny State Bank in home loan applications are normally distributed with mean $70,000 and standard deviation $8,000. Bank policy requires that applications involving the lowest 2% of amounts requested be submitted to its special low-income housing committee for approval. What size home loan requests are submitted to the low-income housing committee?

SOLUTION.   If we denote by x the (unknown) largest dollar amount that must be submitted to the committee, then bank policy implies that 2% of the normally distributed loan amounts fall below (i.e., to the left of) x. This means that 2% of a standard normal distribution lies to the left of

162

TABLE 4.15   A Small Portion of Table A.4

z	Proportion
− 2.06	.0197
− 2.05	.0202

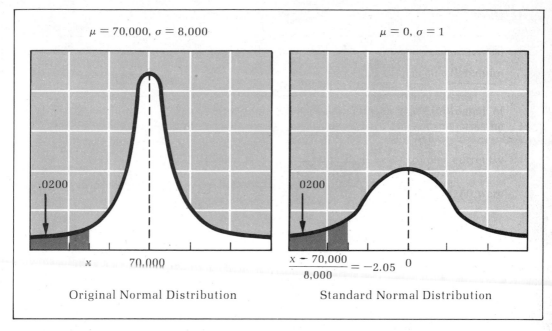

FIGURE 4.13   Original and Standardized Sizes of Home Loan Applications

$$z = \frac{x - \mu}{\sigma} = \frac{x - 70,000}{8,000}.$$

From Table A.4, or the remnant of it appearing in Table 4.15, we see that the z score of x must be − 2.05, because .0202 is closer to .0200 than is .0197. Therefore

$$\frac{x - 70,000}{8,000} = -2.05,$$

and some algebraic transformations yield that

$$x - 70,000 = (-2.05)(8,000) = -16,400,$$

$$x = -16,400 + 70,000 = 53,600.$$

It follows that all loan applications requesting $53,600 or less must be submitted to the special committee. The diagrams of the relevant normal distributions appear in Fig. 4.13.

163

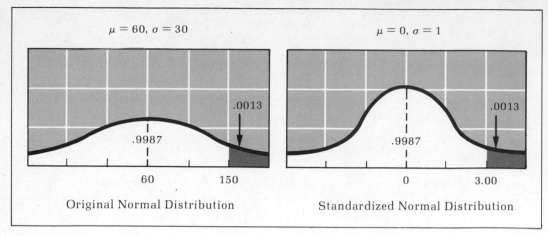

$\mu = 60, \sigma = 30$ $\mu = 0, \sigma = 1$

.0013

.9987

60   150

Original Normal Distribution

.0013

.9987

0   3.00

Standardized Normal Distribution

FIGURE 4.14   Original and Standardized Collection Times (Days)

EXAMPLE 4.8.   Aging of Accounts Receivable.   Burns & Antelope, a leading department store chain, gets a large amount of credit sales. Detailed analysis of experience collecting on its outstanding credit indicates the collection time of such a debt is a normally distributed random variable with mean 60 days and standard deviation 30 days. Classical credit studies have shown that, after 150 days, such debts are generally uncollectable and should be written off. What percentage of the chain's total credit sales will eventually be written off as bad debts?

SOLUTION.   The question asks for the probability that a normally distributed random variable with mean $\mu = 60$ and standard deviation $\sigma = 30$ exceeds $x = 150$. A diagram of the situation is presented in Fig. 4.14. A transformation to z scores gives the z score of $x = 150$ as

$$z = \frac{x - \mu}{\sigma} = \frac{150 - 60}{30} = \frac{90}{30} = 3.00.$$

From the portion of Table A.4 that appears in Table 4.16, we see that a standard normal random variable is less than or equal to 3.00 about 99.87% of the time. It follows that such a variable exceeds 3.00 about .13% = 13/100 of 1% of the time, and we can therefore conclude that 13/100 of 1% of the chain's credit sales remain uncollected after 150 days and are therefore written off as bad debts.

TABLE 4.16   A Small Portion of Table A.4

z	Proportion
3.00	.99865

TABLE 4.17   A Small Portion of Table A.4

z	Proportion
− 1.29	.0985
− 1.28	.1003
− 1.27	.1020

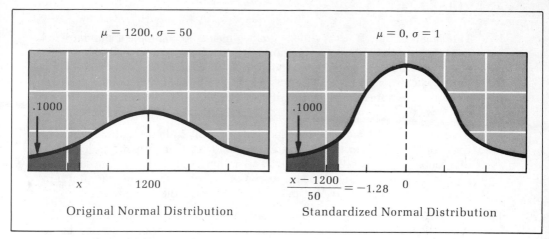

FIGURE 4.15   Original and Standardized Battery Lifetimes

EXAMPLE 4.9   Automobile Batteries.   A small auto parts company markets a 12-volt battery that, according to statistical studies, lasts an average of 1,200 days, with a standard deviation of 50 days. The lifetimes of the batteries were also shown to be normally distributed. The company would like to put a guarantee on its product so that no more than 10% of the batteries will fail before the guarantee runs out. For how many days should the battery be guaranteed?

SOLUTION.   We denote by x the (unknown) number of days for which the battery should be guaranteed. The requirement that no more than 10% of the batteries fail before x days translates into the assertion that 10% of the normally distributed battery lifetimes fall below x. It follows that 10% of the z scores will fall below the z score of x, which is

$$z = \frac{x - \mu}{\sigma} = \frac{x - 1200}{50}.$$

A glance at Table 4.17, which reproduces the relevant section of Table A.4, reveals that the z score of x must be −1.28. A diagram appears in Fig. 4.15. Solving the equation $(x - 1200)/50 = -1.28$ for the unknown x, we get

$$x = (-1.28)(50) + 1200 = -64 + 1200 = 1136.$$

Therefore, if the company guarantees its battery for 1136 days (a little more than 3 years), it can be reasonably sure that no more than 10% of the batteries sold will fail before the guarantee runs out.

EXERCISES 4.C

4.19. The length of time it takes college finance majors to complete a standardized financial analyst aptitude test is a normally distributed random variable having mean 58 minutes and standard deviation 9.5 minutes. The professional committee who designed the exam wants it to be completed by 90% of the students who take it, within a specified time allowed for the test. How much time should be allowed?

4.20. Records of the Wetspot Washing Machine Company indicate that the length of time their washing machines operate without requiring repairs is normally distributed with mean 4.3 years and standard deviation 1.6 years. The company repairs free any machine that fails to work properly within one year after purchase. What percentage of Wetspot machines require these free repairs?

4.21. The mathematics portion of the nationally administered Graduate Management Admissions Test is graded in such a way that the scores are normally distributed with mean 500 and standard deviation 100.

   a. What proportion of applicants score 682 or higher?

   b. What proportion score between 340 and 682?

4.22. Men's shoe sizes nationwide are normally distributed with mean 10.5 and standard deviation 1.2. What proportion of men have shoe sizes between 8.25 and 12.25?

4.23. Persons who take the Standard Tire and Hubcap Retail Inventory Forecasting Test (THRIFT) score a mean of 150 points with a standard deviation of 10. Furthermore, the scores on this standardized test are normally distributed. What score do you need in order to rank among the top 2% of all those who take the test?

4.24. Lamps used in residential area street lighting are constructed to have a mean lifetime of 400 days, with a standard deviation of 30 days. Furthermore, their lifetimes are normally distributed.

   a. What percentage of such lamps last longer than one year (365 days)?

   b. What percentage last between 375 and 425 days?

   c. What percentage last longer than 480 days?

4.25. The price of gasoline (in cents) in 1978 in California was normally distributed with mean 65.7 and standard deviation of 8.9. What proportion of gasoline prices lay between 59.9 and 70.3 cents?

4.26. A cafeteria vending machine dispenses 6 ounces of coffee per cup, on the average, in such a way that the amount dispensed per cup is a normally distributed random variable. How fine should the machine be "tuned"; namely, to what level should the standard deviation be set, so that 99% of the cups are filled with at least 5.9 ounces of liquid?

4.27. One accounting instructor makes sure that every set of examination scores in her course is normally distributed. Then she assigns A grades to the top 15% of the scores, while the bottom 15% are awarded F's. A particular set of exam scores turns out to have a mean of 60 and a standard deviation of 16. What score is the dividing line between A and B, and what score between D and F?

4.28. A well-known law asserts that no more than one-half of one percent of a company's cereal boxes, labeled as containing 9 ounces of cereal, may contain less then 8.8 ounces. A company planning to produce a 9-ounce box of "Rice Puffies" purchases a newly-devised box-filling machine that fills cereal boxes in such a way that the contents of an individual box is a normally distributed random variable having *standard deviation* 0.05 ounce. The machine has a dial with which the company can set the *mean* contents to any desired level. What is the lowest value to which the company can set the mean and still remain within the law?

4.29. The New York Stock Exchange is open 250 days each year. Closing prices of shares of Brownline Copper Co. (BCC) averaged 21½ in the past year, with a standard deviation of 4.3. Furthermore, BCC stock closed above 32¼ on 28 different days during the year. Were closing prices of BCC shares normally distributed?

4.30. Among those who audit accounts of large public corporations, it is known that the dollar value of errors present in accounts of a certain type is a normally distributed random variable. The accounting firm of Kutt, Payst, and Tehppe discovered that United Motor Products' accounts of this type had a mean error of $22.40 and that 8% of the accounts had errors in excess of $40.00.

a. Determine the standard deviation of the errors on such accounts.

b. What proportion of accounts had errors below $10.00 in value?

4.31. The tax season lasts from January 1 to April 15, with an overwhelming majority of the work being done after April 1. The completion times of returns are approximately normally distributed with a mean of 100 days and a standard deviation of 10 days. What percentage of the tax returns can be expected to be completed by April 1, namely within 90 days?

4.32. The A. C. Nielsen Co. published the results of a 1976 survey that indicated that men between the ages of 18 and 24 watched television an average of 21 hours and 20 minutes a week, with a standard deviation of 80 minutes, and that the watching times were normally distributed.

a. What percentage of such young men watch more than 18 hours?

b. What percentage watch between 20 hours and 22 hours?

c. What percentage watch more than 25 hours of television per week?

4.33. Scores on the Colorado Operations Management Skills Aptitude Test (COMSAT) are standardized so as to be normally distributed with mean 80 and standard deviation 10. Persons scoring in the top 3% are automatically assigned to a management training program. What score is required to rank among the top 3% of all applicants?

4.34. Imperial Oil, Ltd. of Canada operates an oil refinery on the outskirts of Edmonton, Alberta, which produces an average of 140,000 barrels of oil per day, with a standard deviation of 8,000 barrels. The daily output of the refinery is also normally distributed. On what proportion of days is the output in excess of 160,000 barrels?

4.35. A survey of several thousand shoppers was conducted by the California Beefeaters Council to find out the highest price consumers would be willing to pay for a pound of steak. The responses were normally distributed with a mean of $1.80 and a standard deviation of $0.15. What price level would be so high that 10% of consumers would stop buying steak?

4.36. In order to maintain the quality of the carbonated beverages it markets, an interstate phosphate pop company demotes the lowest 8% of its employees each year, according to their performance on the Phosphate Industry Zero Water Emission Level (PHIZWEL) Test. Scores on the PHIZWEL test are normally distributed with mean 150 and standard deviation 14. What minimum score is required on the test to avoid being demoted?

4.37. Amalgamated Junque Foods, Inc., manufactures a candy bar that contains 1.2 ounces net weight. Of this, an average of 0.15 ounces per bar consists of peanuts, while the rest is chocolate. The peanut content per candy bar has standard deviation 0.01 ounces, and is a normally distributed random variable. What percentage of such candy bars contain between 0.14 and 0.17 ounces of peanuts?

4.38. A county-wide survey conducted by Quickie Gas, Inc., a chain of discount gasoline stations in Los Angeles County, showed that the prices of unleaded gasoline throughout the county are normally distributed with mean 131.1 cents per gallon and standard deviation 3.2 cents per gallon. If Quickie wants to be certain of having its prices among the lowest 6% in the county, how much should it charge per gallon?

4.39. Federated Engine Hoses (FEH) has a contract to produce 11,000 Type CR hoses of diameter 1.3 cm. The customer does not have to pay for any hoses that vary from the contracted diameter by more than 0.05 cm. (i.e., ones whose diameter is between 1.25 and 1.35 cm.). FEH's assembly line produces hoses with diameters that are normally distributed with mean 1.31 cm. and standard deviation 0.025 cm. What proportion of hoses delivered does the customer have to pay for?

4.40. An auto dealers' business organization collected data that indicated that the actual sticker prices of imported small cars are normally distributed with a standard deviation of $720.00. Furthermore, 20% of the cars in the class studied have sticker prices below $5,200.00. What is the mean sticker price of all imported small cars in the class surveyed?

4.41. Purchasers of adult-size basketball shoes have mean shoe size 11.3 (although the shoes themselves come in only full and half sizes) and standard deviation 1.4. Furthermore, these shoes sizes are normally distributed. "The Court Jogger," a new chain of athletic shoe stores, has to know how many shoes to order and stock in each size range. Assuming that all persons whose measured shoe sizes fall between 10.75 and 11.25 will purchase size 11 shoes, what proportion of the store's basketball shoes should be in size 11?

SECTION 4.D   THE NORMAL APPROXIMATION TO THE BINOMIAL DISTRIBUTION

Table A.2, the table of probabilities of the binomial distribution, contains information only for situations when the parameter n has a value of 20 or less. We have been unable up to now to calculate, for example, the probability of tossing 40 or fewer heads in 100 tosses of a fair coin because here $n = 100$. Fortunately, there is a result from advanced statistical theory, called the *central limit theorem* (to be discussed in greater detail in Section 5.B), which gives the surprising information that the normal distribution can be used to find that probability.

What the central limit theorem says in regard to this problem is that when $n\pi$ and $n(1-\pi)$ are both greater than 5 in magnitude, a binomial random variable with parameters n and π can be considered as a normally distributed one with parameters $\mu = n\pi$ and $\sigma = \sqrt{n\pi(1-\pi)}$.*

A glance at Figures 4.1, 4.2, and 4.3 will illustrate how the binomial histograms tend to get to look bell-shaped like the normal distribution for large values of n. If n is sufficiently large, so that $n\pi$ and $n(1-\pi)$ both exceed 5, the difference between the binomial and normal histograms turns out to be insignificant. In this form, the central limit theorem is referred to as "the normal approximation to the binomial distribution."

In the case of the 100 tosses of a fair coin, the parameters of the binomial random variable H (the number of heads obtained in the 100 tosses) are n $= 100$ and $\pi = .50$. Therefore $n\pi = 100(.50) = 50 > 5$ and $n(1-\pi) = 100(.50) = 50 > 5$, so that the normal approximation to the binomial distribution applies to this problem. H can then be considered as a normal random variable with parameters $\mu = n\pi = 100(.50) = 50$ and $\sigma = \sqrt{n\pi(1-\pi)} = \sqrt{100(.50)(1-.50)} = \sqrt{25} = 5$. Furthermore, since H is a discrete random variable, as discussed in Example 4.6 above, we actually should calculate the probability that $H < 40.5$, namely that we toss fewer than 40.5 heads, if we are looking for the probability of tossing 40 or fewer heads in 100 tosses of the coin.† The bell-shaped curves for the coin-tossing situation are presented in Fig. 4.16. The z score of 40.5 is

$$z = \frac{40.5 - \mu}{\sigma} = \frac{40.5 - 50}{5} = \frac{-9.5}{5} = -1.90$$

and Table A.4 indicates that the proportion of a standard normal distribution falling below -1.90 is .0287. Therefore the probability of tossing 40 or fewer heads in 100 tosses of a fair coin is only 2.87% or 287 chances out of 10,000.

---

*These formulas come from formulas (4.4) and (4.5) of Section 4.B.

†If we don't put the dividing line between 40 heads and 41 heads at 40.5, then the probability that a (continuous!) normal random variable lies between 40 and 41 will be lost. The act of putting the dividing line halfway between the two values of the discrete variable is called the *continuity correction.*

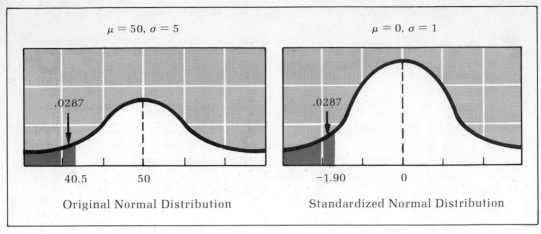

$\mu = 50, \sigma = 5$ $\mu = 0, \sigma = 1$

.0287   .0287

40.5   50   −1.90   0

Original Normal Distribution   Standardized Normal Distribution

FIGURE 4.16   Original and Standardized Number of Heads Tossed

EXAMPLE 4.10. **Automobile Loan Approvals.** (This example is a modified version of Example 4.2.) Suppose one bank's records indicate that 60% of all automobile loan applications submitted to it are eventually approved. If we were to select a random sample of 70 applications from a certain section of a major city, what is the probability that a majority of those applications would not be approved?

SOLUTION. Here we are dealing with a binomial random variable A, the number of applications that are *approved*, having parameters $n = 70$ and $\pi = .60$. The normal approximation is valid because $n\pi = 70(.60) = 42$ and $n(1 - \pi) = 70(.40) = 28$ are both greater than 5. The probability that a majority of the 70 applications are not approved is the same as the probability that A is fewer than 34.5, and this latter probability is the same as the proportion of a standard normal distribution that falls below

$$z = \frac{34.5 - \mu}{\sigma}$$

where

$$\mu = n\pi = 70(.60) = 42,$$

and

$$\sigma = \sqrt{n\pi(1 - \pi)} = \sqrt{16.8} = 4.1.$$

We then have that

$$z = \frac{34.5 - 42}{4.1} = \frac{-7.5}{4.1} = -1.83.$$

From Table A.4, we find out that the probability that a standard normal random variable falls below -1.83 is .0336 or 3.36%. Therefore, if we take a sample of 70 applications, our chances of being misled into believing that a majority are *not* approved (and that the bank may be guilty of discrimination) are only 3.36%.

Referring back to Example 4.2 of Section B, we note that our chances of being so misled were 28.98% when we used a sample of 7 applications, and 18.61% when we used a sample of 19 applications. As we have just shown, the probability of being misled drops considerably to 3.36%, when a sample of 70 applications is used. This demonstrates convincingly that, by increasing the sample size, we can obtain a substantial decrease in our probability of error.

EXERCISES   4.D

4.42. As part of a coin-tossing experiment to check on the laws of probability (presumably to find out if they should be repealed), a fair coin is tossed several times; then the number of heads obtained is carefully counted and denoted by H.

   a. If the coin is tossed 4 times, find the probability that H, the number of heads, will be at least 2, and no larger than 3.

   b. If the coin is tossed 40 times, find the probability that H will be at least 20, and no larger than 30.

   c. If the coin is tossed 400 times, find the probability that H will be at least 200, and no larger than 300.

   d. If the coin is tossed 4,000 times, find the probability that H will be at least 2,000, and no larger than 3,000.

4.43. The manufacturer of a new brand of soap guarantees that, with the aid of her unique marketing strategies, exactly 80% of the soap will be sold quickly. If the manufacturer's assertion is correct:

   a. With what probability can we expect more than 13 of 17 bars of soap marketed will be sold quickly?

   b. With what probability can we expect that more than 130 of 170 bars of soap marketed will be sold quickly?

   c. With what probability can we expect that more than 1,300 of 1,700 bars of soap marketed will be sold quickly?

   d. With what probability can we expect that more than 13,000 of 17,000 bars of soap marketed will be sold quickly?

4.44. Actuarial studies indicate that, of all 45-year old men in a certain occupational grouping, 5% will die before they reach age 55. As part of the process of deciding whether or not to offer a 10-year term group life insurance policy to all 45-year-old men in that occupational grouping, an insurance company has to know the answers to the following questions:

a. If 20 men apply for the policy, what are the chances that more than 2 of them will die before reaching age 55?

b. If 200 men apply, what are the chances that more than 20 of them will die before age 55?

c. If 2,000 men apply, what are the chances that more than 200 of them will die before age 55?

d. If 20,000 men apply, what are the chances that more than 2,000 of them will die before age 55?

4.45. A news reporter alleges that 65% of the members of the various state legislatures favor a certain controversial change in the banking laws. If the reporter's story is correct,

a. find the probability that no more than half of 10 randomly selected members support the change;

b. find the probability that no more than half of 100 randomly selected members actually support the change;

c. find the probability that no more than half of 400 randomly selected members actually support the change.

4.46. A new credit policy aimed at reducing bad debts seems to be 80% effective, according to an intensive pretesting program.

a. If the new credit policy is used for 15 customers, with what probability can we expect fewer than 11 to pay their accounts?

b. If the credit policy is used for 150 customers, how likely is it that fewer than 110 will pay their accounts?

c. Of 1,500 new customers, what are the chances that fewer than 1,100 will pay their accounts under this new credit policy?

d. With 15,000 new customers, what is the probability that fewer than 11,000 will pay their accounts under this new credit policy?

SECTION 4.E   THE POISSON DISTRIBUTION

Criminology data for one major West Coast metropolitan area indicate that attempted bank robberies occur at the rate of 2.4 per day, on the average. The security* division of the Bank of Bel Air, which has over a hundred branches throughout the area, would like to be able to answer the following questions:

1. What is the probability that more than 6 bank robberies will be attempted tomorrow?

2. What are the chances that no robberies will be attempted next Tuesday?

*Not "securities"!

TABLE 4.18   A Small Portion of Table A.5

	N	
λ	0	6
2.40	.0907	.9884

If we denote by R the number of attempted bank robberies that occur on any particular day, then R is a random variable that has the Poisson distribution with parameter λ = 2.4. The parameter λ (pronounced "lambda") represents the *average* number of attempted robberies per day, while the random variable R represents the *actual* number of attempts on any particular day. To answer question 1, we have to find $P(R > 6)$, the probability that more than 6 bank robberies will be attempted; to answer question 2, we need $P(R = 0)$, the probability that none will be attempted.

Table A.5 of the Appendix contains Poisson probabilities of the form $P(R \leq k)$. Since R has parameter λ = 2.4, we look in the row of Table A.5 corresponding to λ = 2.4. Because $P(R > 6) = 1 - P(R \leq 6)$, $P(R = 0) = P(R \leq 0)$, we use the columns of Table A.5 headed by k = 6 and k = 0. The specific portion of Table A.5 is reproduced in Table 4.18.

From Table 4.18, we see that the probability that more than 6 bank robberies will be attempted tomorrow is

$$P(R > 6) = 1 - P(R \leq 6) = 1 - .9884 = .0116 = 1.16\%.$$

The chances that no robberies will be attempted next Tuesday are

$$P(R = 0) = .0907 = 9.07\%.$$

There is a formula for Poisson probabilities that can be used for values of λ and k not found in Table A.5 (or even for those that can). This formula asserts that

■
$$P(R = k) = \frac{\lambda^k}{k!}e^{-\lambda} \qquad (4.6)$$

where e is a number arising from calculus that is approximately equal to 2.7183. Table A.6 of the Appendix contains the numbers $e^{-\lambda}$ for various values of λ.

Let's see how to calculate $P(R > 6)$ and $P(R = 0)$ using formula (4.6). First of all, we need to know $e^{-\lambda}$ for λ = 2.4, and we find this in Table A.6, an abbreviated version of which appears in Table 4.19. Since $e^{-2.4} = .09072$, as we see in Table 4.19, and since $(2.4)^0 = 1$ and $0! = 1$, we find easily from formula (4.6) that

$$P(R = 0) = \frac{(2.4)^0}{0!}e^{-2.4} = \frac{1}{1}(.0907)$$

$$= .0907 = 9.07\%.$$

TABLE 4.19   A Small Portion of Table A.6

x	e^{-x}
2.40	.0907

TABLE 4.20   Calculation of $P(R \leqslant 6)$

k	$(2.4)^k$	k!	$\dfrac{(2.4)^k}{k!} e^{-2.4}$	=	$P(R = k)$
0	1	1	$\dfrac{1}{1}(.0907)$	=	.0907
1	2.4	1	$\dfrac{2.4}{1}(.0907)$	=	.2177
2	5.76	2	$\dfrac{5.76}{2}(.0907)$	=	.2612
3	13.824	6	$\dfrac{13.824}{6}(.0907)$	=	.2090
4	33.178	24	$\dfrac{33.178}{24}(.0907)$	=	.1254
5	79.626	120	$\dfrac{79.626}{120}(.0907)$	=	.0602
6	191.103	720	$\dfrac{191.103}{720}(.0907)$	=	.0241
			Sum $P(R \leqslant 6)$	=	.9883

To calculate $P(R > 6) = 1 - P(R \leqslant 6)$, we first have to compute $P(R = k)$ for $k = 0, 1, 2, 3, 4, 5,$ and 6 and then sum all those quantities. We do this in Table 4.20, which shows that

$$P(R \leqslant 6) = .9883.$$

It follows that

$$P(R > 6) = 1 - P(R \leqslant 6) = 1 - .9883 = .0117 = 1.17\%$$

as before (except for rounding discrepancies).

EXAMPLE 4.11. Customer Checkout Queues. Customers arrive at the checkout area of a large discount drug and variety store at the average rate of 3.8 per minute during peak hours. In order to appropriately assign checkout personnel, the assistant manager needs to know how likely it is that between 3 and 7 customers, inclusive, will arrive at the checkout area in the next minute.

TABLE 4.21 A Small Portion of Table A.5

	N	
	2	7
3.80	.2689	.9599

SOLUTION. The number of customers, C, arriving at the checkout area in the next minute is a Poisson random variable with parameter $\lambda = 3.8$. We want to compute the probability $P(3 \leq C \leq 7)$.

First of all, we write this probability as follows:

$$P(3 \leq C \leq 7) = P(C \leq 7) - P(C \leq 2)$$

to conform with the requirements of Table A.5.

We next isolate the relevant portion of Table A.5 in Table 4.21, where we see that $P(C \leq 7) = .9599$ and $P(C \leq 2) = .2689$.

It follows that the probability that between 3 and 7 customers, inclusive, will arrive in the next minute is

$$P(3 \leq C \leq 7) = P(C \leq 7) - P(C \leq 2)$$

$$= .9599 - .2689 = .6910 = 69.1\%.$$

When is it appropriate to use the Poisson distribution as a model for calculating probabilities? The basic situation is one in which certain events (for example, bank robberies or customer arrivals) occur at an average rate of λ per period of time or unit of space. If these events occur on a random basis and independently of each other, then the total number of them that occur in any particular length of time or space is a random variable having the Poisson distribution.

The Poisson Approximation to Binomial Distribution

Recall from Section 4.D that the binomial distribution may be approximated by the normal distribution if $n\pi$ and $n(1 - \pi)$ are both *greater* than 5. It turns out that, if n is large but $n\pi$ is small (*less* than 8, say), then Poisson probabilities based on the parameter $\lambda = n\pi$ can be used to approximate those binomial probabilities. Specifically, this means that when X is a random variable that has the binomial distribution with parameters n and π, and n is large but $n\pi$ is small, then

$$P(X = k) \approx \frac{(n\pi)^k}{k!} e^{-n\pi}$$

where the symbol "$\approx$" means "is approximately equal to."

To illustrate the use of the Poisson approximation, suppose the discount store of Example 4.11 now has 500 customers inside, all behaving independently, and each has probability .007 of arriving at the checkout area in the next minute. What is the probability that exactly 5 customers will arrive in the next minute? Well, $n = 500$ is large and $n\pi = 500(.007) = 3.5$ is small (< 8), so that the Poisson approximation is usable. If we denote by X the actual number of customers that arrive in the next minute, then the above formula implies that

$$P(X = 5) \approx \frac{(n\pi)^5}{5!}e^{-n\pi} = \frac{(3.5)^5}{5!}e^{-3.5}$$

$$\approx \frac{525.22}{120}(.0302) = 0.132,$$

using Table A.6 of the Appendix.

This calculation contains a broad hint as to why binomial probabilities may be approximated by Poisson probabilities: if there are 500 customers and each has independently a probability .007 of arriving at the checkout counter in the next minute, then an average of $n\pi = (500)(.007) = 3.5$ customers will arrive in the next minute. Setting $\lambda = 3.5$, we can see that the actual number of customer arrivals fits rather neatly into the Poisson model.

EXERCISES 4.E

4.47. Circle Imports, Inc., an independent local auto dealer, sells three cars per day, on the average, on weekdays and five cars per day, on the average, on weekends.

a. What are the chances that Circle Imports sells five or more cars next Wednesday?

b. Find the probability that the dealer sells three or fewer cars next Saturday.

4.48. One accounting major with a particularly spectacular academic record, including especially fantastic grades in statistics, is currently receiving job offers in the mail at the rate of approximately 0.3 per day.

a. Find the probability that she receives no job offers tomorrow.

b. What is the probability that she gets at least one job offer tomorrow?

c. What are the chances that she gets three or more job offers tomorrow?

d. The rate of 0.3 per day is equivalent to a rate of 3.0 job offers in a 10-day period. Find the probability that the student gets 2 or more job offers in the next 10 days.

e. The rate of 0.3 per day is equivalent to what rate for a five-day period?

f. What are the chances that she gets at least one job offer in the next five days?

4.49. An actuary working for a small insurance company serving five southeastern states observed that, on the average, four deaths per month occur among the company's policyholders aged between 45 and 48. Find the probability that none of the company's policyholders of this age group dies next December.

4.50. The auto loan desk at one branch of the Bank of Bel Air receives, on the average, 3.4 applications for new car loans per week. What is the probability that it receives six or more applications this week?

4.51. On the average, there are 1.8 breakdowns every five working days in major factory equipment at the main plant of General Fabricators, according to the plant's maintenance department.

    a. Find the probability that there will be fewer than 2 breakdowns during the next 5 working days.

    b. The rate of breakdowns, namely 1.8 per 5 working days, is equivalent to a rate of 3.6 per 10 working days. Find the probability that fewer than 4 breakdowns occur in the next 10 working days.

    c. The rate of breakdowns is also equivalent to a rate of 0.36 breakdown per working day. What are the chances that there will be one or more breakdowns next Tuesday?

    d. What is the probability that there will be one or more breakdowns during the next three-working-day period?

    e. Find the probability that fewer than two breakdowns will occur during the next six working days

4.52. Northern National Overnight Express (NONOX, Inc.), a major air carrier of fragile merchandise, has an average of 1.2 planeloads of air freight arrive at its transfer center in Memphis every five minutes.

    a. Find the probability that five or more planeloads arrive during the next 15 minutes.

    b. What are the chances that two or fewer planeloads arrive during the next 25 minutes?

4.53. A study of its customers' usage of credit cards showed Stokey's Department Store that, on the average, its own card was being handed to a salesperson 45 times every hour. What is the probability that at least one customer will hand a Stokey's credit card to a salesperson during the next two minutes?

4.54. Plateau Telephone Company of Texas observes that 2.6 calls per minute come in on a certain line, on the average. What is the probability that 10 or more calls will come in on that line during the next two minutes?

4.55. An aerospace subcontracting firm deposits cash reserves into its sick leave fund on the assumption that its employees will require sick leave pay for 7.4 days, on the average, each year. What proportion of employees will call in sick 11 or more days this year? (The fund assumes that employees call in sick independently of each other—that no epidemics will spread through the company.)

4.56. A major computer manufacturing company receives silicon chips in standard batches from a subcontractor. There is an average of 6 defective chips per batch shipped by the subcontractor. What proportion of batches contain 3 or more defective chips?

4.57. It has been reported that well-traveled New York City streets contain an average of 5.4 "potholes" per mile.

    a. Find the probability that a particular half-mile stretch contains no potholes.

    b. What is the probability that a particular half-mile stretch contains 6 potholes?

4.58. An automatic bolt-producing machine turns out defective bolts at the average rate of 0.8 per lot of 100. (The other 99.2 are satisfactory.)

    a. What is the probability that there are no defective bolts in the next lot produced?

    b. Find the probability that there is exactly one defective bolt among the next three lots.

4.59. United Stereo Outlets, Inc., a statewide chain, sells an average of 3.7 GATT-091 stereo units per outlet per month. What is the probability that the outlet you manage in Augusta will sell 5 or more of the units?

4.60. Save-Mart, a suburban discount department store, bought at auction several hundred cases of a discontinued line of personal deodorant. To move the stock as fast as possible, store management has decided to display the product prominently at each checkout counter. Preliminary sales analysis shows that an average of 1.8 cases is sold per checkout counter each day. What is the probability that checkout counter No. 6 (staffed by a newly hired employee) would sell 3 or more cases tomorrow?

SECTION 4.F   THE HYPERGEOMETRIC DISTRIBUTION

Attempting to verify the inventory of a company that markets 20 different products, an auditor feels he has staff and resources sufficient for a complete check of only 8 of the products. Based on his experience with the company's inventory, however, he suspects strongly that 3 of the company's 20 products will show an inaccurate inventory count, although he doesn't know which three products will be involved. Since only 8 of the 20 products can be checked, the auditor would like to know the answers to the following questions.

    1. What are the chances that the 8 randomly selected products to be checked will include all 3 of the products that show an inaccurate inventory count?

    2. What are the chances that 2 of the inaccurately counted products will be among the 8 products selected for detailed checking?

3. What are the chances that only 1 of the inaccurately counted products will show up among the 8 products checked?

4. What is the probability that none of the inaccurately counted products will be uncovered during the auditor's check of 8 randomly selected products?

The auditor is concerned with the random variable

N = number of inaccurately counted products appearing among the 8 products randomly selected for a complete check.

Random variables of N's type are said to have the *hypergeometric distribution* with parameters n, r, and m, where

n = total number of products from which selection will be made
r = number of products that will be selected for verification
m = number of products that are really inaccurately counted

In this situation, $n = 20$, $r = 8$, and $m = 3$. The probability

$$P(N = k) = \frac{C(m, k)C(n - m, r - k)}{C(n, r)} \tag{4.7}$$

where the combination formula (4.2) is referred to, gives the probability that among the 8 completely checked products, exactly k of the inaccurately counted ones will be found.*

To carry out the computation of $P(N = k)$ for the possible values of k = 0, 1, 2, 3, we can either use formula (4.2) or refer to Table A.3 of the Appendix. Let's first calculate $P(N = 3)$, so that $k = 3$, $n = 20$, $r = 8$, and $m = 3$ in formula (4.7). Then

$$P(N = 3) = \frac{C(3, 3)C(20 - 3, 8 - 3)}{C(20, 8)} = \frac{C(3, 3)C(17, 5)}{C(20, 8)}.$$

The relevant portions of Table A.3 are reproduced in Table 4.22, where we see that

$C(3, 3) = 1$
$C(17, 5) = 6,188$
$C(20, 8) = 125,970.$

---

*A complete discussion of the logical origin of the formula for the hypergeometric distribution was presented in Section 3.D.

TABLE 4.22   A Small Portion of Table A.3

n	C(n, 3)	C(n, 5)	C(n, 8)
3	1	–	–
17	–	6,188	–
20	–	–	125,970

Therefore

$$P(N = 3) = \frac{(1)(6,188)}{125,970} = .049 = 4.9\%.$$

This means that the auditor, in selecting 8 of the 20 products at random, has a probability of 4.9% (roughly one chance out of 20) of successfully discovering the 3 products that have erroneous inventory counts. Furthermore, the chances of discovering 2 of the 3 inaccurately counted products are given by (here $k = 2$)

$$P(N = 2) = \frac{C(3,2)C(20 - 3, 8 - 2)}{C(20,8)} = \frac{C(3,2)C(17,6)}{C(20,8)}$$

$$= \frac{(3)(12,376)}{125,970)} = \frac{37,128}{125,970} = .295 = 29.5\%,$$

or slightly less than 1 chance in 3. Similarly, the probability of detecting only 1 of the inaccurately counted products is

$$P(N = 1) = \frac{C(3,1)C(17,7)}{C(20,8)} = \frac{(3)(19,448)}{125,970}$$

$$= .463 = 46.3\%,$$

or slightly less than 1 chance in 2.

Finally, the probability that *all* 3 inaccurately counted products will successfully *evade* the auditor's net is

$$P(N = 0) = \frac{C(3,0)C(17,8)}{C(20,8)} = \frac{(1)(24,310)}{125,970}$$

$$= .193 = 19.3\%,$$

or roughly 1 chance in 5.

To summarize the results of the calculations, we can say that the probability of the auditor's discovering all, 2, 1, or none of the 3 inaccurately

---

TABLE 4.23   Probability Distribution of N,
the Number of Inaccurately Inventoried
Products Successfully Discovered by Auditor

k	P(N = k)
0	.193
1	.463
2	.295
3	.049
Sum =	1.000

---

counted products is, respectively, 4.9%, 29.5%, 46.3%, and 19.3%. The probability distribution of the hypergeometric random variable N appears in Table 4.23. Notice that, as required, the column of probabilities sums to 1 or 100%.

EXAMPLE 4.12.   Major Defects in New Cars.   Every three years, Beach City Realty purchases a fleet of 5 new cars to place at the disposal of its small sales staff. It chooses the 5 cars from among 40 of the desired make right on the lot of a nearby dealer. Generally 10% of such new cars need a major engine adjustment before they are able to function properly. What is the probability that, of the 5 new cars purchased, 2 will require a major engine adjustment?

SOLUTION.   We want to know $P(N = 2)$ where N is the number of cars among the 5 purchased that require a major engine adjustment. Because the total number of cars from which a selection will be made is 40, the number of cars that really require a major adjustment is 10% of them or 4 cars. Therefore N is a hypergeometric random variable with parameters $n = 40$, $r = 5$, and $m = 4$. Since $k = 2$, we see from formula (4.7) that

$$P(N = 2) = \frac{C(4,2)C(40 - 4, 5 - 2)}{C(40,5)} = \frac{C(4,2)C(36,3)}{C(40,5)}.$$

Since Table A.3 contains $C(n, k)$ for $n \leq 20$, we will have to compute $C(36, 3)$ and $C(40, 5)$ by using formula (4.2). Starting with $C(4, 2)$, we have

$$C(4,2) = \frac{4!}{2!(4-2)!} = \frac{(4)(3)(2!)}{(2)(1)(2!)} = \frac{(4)(3)}{(2)(1)}$$

$$= \frac{12}{2} = 6$$

$$C(36,3) = \frac{36!}{3!(36-3)!} = \frac{(36)(35)(34)(33!)}{(3)(2)(1)(33!)}$$

$$= \frac{(36)(35)(34)}{6} = \frac{42,840}{6} = 7,140$$

$$C(40,5) = \frac{40!}{5!(40-5)!} = \frac{(40)(39)(38)(37)(36)(35!)}{(5)(4)(3)(2)(1)(35!)}$$

$$= \frac{(40)(39)(38)(37)(36)}{(5)(4)(3)(2)(1)} = \frac{(39)(38)(37)(36)}{(3)(1)}$$

$$= \frac{1,974,024}{3} = 658,008.$$

Therefore the probability that 2 of the 5 new cars purchased will require a major engine adjustment is

$$P(N = 2) = \frac{C(4,2)C(36,3)}{C(40,5)} = \frac{(6)(7,140)}{658,008} = \frac{42,840}{658,008}$$

$$= .065 = 6.5\%,$$

or, roughly, 1 chance in 16.

Just as the Poisson distribution turned out to be related to the binomial distribution, there is a relationship between the hypergeometric distribution and the binomial. Suppose the question in Example 4.12 were reworded as follows: "If 10% of new cars need a major engine adjustment before they are able to function properly, what is the probability that, of 5 new cars purchased, 2 will require a major engine adjustment?" The answer to this question, as reworded (ignoring the number of cars on the lot), would be the probability that a binomial random variable N with parameters $n = 5$ and $\pi = .10$ is equal to 2. Using the formula (4.1) for binomial probabilities, we calculate that

$$P(N = 2) = C(5,2)(.10)^2(1-.10)^{5-2} = \frac{5!}{2!\,3!}(.01)(.90)^3$$

$$= \frac{120}{(2)(6)}(.01)(.729) = .0729 = 7.29\%.$$

This is somewhat different from the answer $P(N = 2) = 6.5\%$ to the question worded exactly as in Example 4.12.

What is the difference between the two questions? The difference is that, in Example 4.12 as originally worded, the cars are *not selected independently* of each other. There are 40 cars on the lot, of which 4 need a major engine adjustment. The company purchases 5 new cars. The probability that the first car picked needs an adjustment is 4/40 = .10. The conditional probability that the second car picked needs an adjustment *depends* on whether or not the first car picked needed one:

P(second needs adjustment | first did) = 3/39

because, of the 39 cars left on the lot, 3 need an adjustment; and

P(second needs adjustment | first didn't) = 4/39

because, of the 39 cars left on the lot, 4 need an adjustment. This particular type of dependence is what characterizes the hypergeometric distribution.

The binomial distribution, on the other hand, is characterized by the properties mentioned at the beginning of Section 4.B: the probabilities remain constant (.10 in the reworded situation) from pick to pick, and each car is selected independently of the others.

EXERCISES   4.F

4.61. A local electronics supply store has an option to buy a lot of 19 citizens' band radio crystals recently shipped from a manufacturer in Hong Kong. Experience has shown the store's manager that a good way to judge a lot of size 19 for defective crystals is to select 4 of the 19 at random and then to buy the lot only if all 4 are in perfect condition. What is the probability that all 4 in the sample will be in perfect condition, even if 5 of the 19 crystals are defective?

4.62. An automobile assembly plant producing the sleek and sophisticated Peccarie purchases every month a truckload of 16 cartons of engine bolts. Since these are precision items, they must fit exactly or they will be no good to the factory. Because the expense of checking through 16 cartons, each consisting of 1,728 bolts, is prohibitive, the factory's procedure is to carefully examine 2 of the 16 cartons each month, and to accept the lot of 16 only if the 2 randomly selected cartons pass inspection. If one lot of 16 cartons actually contains 3 cartons that are unacceptable, what are the chances that the factory would be led to buy the lot anyway because its random sample of 2 cartons failed to detect any of the 3 unacceptable ones?

4.63. When The Dallas House, a department store chain specializing in top-of-the-line items, recently opened a new store near a high-income residential area, its first impulse was to mail a credit card to all 60 households in the immediate neighborhood in order to generate some quick business. Upon further thought, though, they decided to run a careful credit check on 5 randomly selected households. (By judicious use of free merchandise, they were able to obtain the consent of the individuals involved in the check.) If 6 of the 60 households were actually bad credit risks, what are the chances that at least 4 out of the 5 selected for the sample would turn out to be good credit risks?

4.64. The advertising manager of a plant that manufactures agricultural accessories conducted a survey of the market reached by a chain of 35 local newspapers serving rural communities. It appears to her that, in any given week, only 30 of the 35 papers effectively reach the desired market. If she chooses 6 papers at random in which to place her advertisements next week, what are the chances that all 6 will reach the desired market effectively?

4.65. In order to evaluate the financial statements of a national department store chain in accordance with generally accepted accounting principles, a CPA firm carefully checks a random sample of customers' statements of outstanding credit. Suppose 10% of a group of 30 statements about to be mailed contain inaccuracies. If an accountant selects a sample of 4 statements from the group, what is the probability that all 4 will be fully accurate, thereby fooling the accountant into believing that the entire group contains no inaccuracies?

SECTION 4.G   THE EXPONENTIAL AND OTHER CONTINUOUS DISTRIBUTIONS

Recall the criminology data mentioned in Section 4.E, which indicated that attempted bank robberies occur at the rate of 2.4 per day, on the average. It turned out that the actual number R of robberies attempted on any particular day is a Poisson random variable with parameter $\lambda = 2.4$, a random variable to which formula (4.6) applies. In a time period of length t days (where t may be a positive whole number or a positive fraction), the average number of attempted robberies is 2.4t, namely 2.4 attempts per full day. Therefore, if we set

R_t = number of robberies attempted in a period of t days,

then R_t is a Poisson random variable having parameter $\lambda = 2.4t$. It follows from formula (4.6) that, for $t \geqslant 0$,

$$P(R_t = k) = \frac{(\lambda t)^k}{k!}e^{-\lambda t} \tag{4.8}$$

What we now want to investigate is how long a time (from right now) we will have to wait for the next attempted robbery to occur. We will denote the random variable involved by

W = waiting time until next attempted robbery occurs.

While the number of attempts in any particular length of time t must be a whole number (none, one, two, etc.), the waiting time until the next attempt can be any positive number at all, because the attempt can occur at any time. Therefore, W is a continuous random variable, and it makes no sense at all to ask for the probability $P(W = t)$.

We describe the continuous random variable W not by a probability distribution of the sort discussed in Section 4.A, but by a probability distribution *function*. The probability distribution function is based on probabilities of intervals, as we found to be appropriate in our study of the normal distribution. Recall that Table A.4, the table of the standard normal distribution, lists probabilities of the form $P(Z \leq z)$, where Z is a standard normal random variable. The waiting time W is, as Z is also, a continuous random variable, and we therefore will study it in terms of the probabilities $P(W \leq t)$ or $P(W > t)$.

The event that $W > t$ can be expressed in words by saying that we have to wait longer than t days for the next attempted robbery to occur or, equivalently, that *no* robberies occur in the time interval up to time t. Using formula (4.8), we can therefore write that

$$P(W > t) = P(R_t = 0) = e^{-\lambda t}, \tag{4.9}$$

setting $k = 0$. Because the event $W \leq t$ is the complement of the event $W > t$, we can see that

$$P(W \leq t) = 1 - P(W > t) = 1 - e^{-\lambda t} \tag{4.10}$$

The function $F(t) = P(W \leq t)$ is called the *probability distribution function* of the continuous random variable W. A random variable with distribution function (4.10) is said to have the "exponential distribution with parameter λ."

Using formulas (4.9) and (4.10), we can calculate the probabilities that we will have to wait various amounts of time for the next bank robbery to occur. For example, the probability that the next robbery occurs within half a day can be calculated as follows:

$$P(W \leq 0.5) = 1 - e^{-\lambda(0.5)} = 1 - e^{-(2.4)(0.5)}$$

$$= 1 - e^{-1.2} = 1 - .3012 = .6988$$

(using Table A.6 of the Appendix), while the probability that we will have to wait longer than 2 days for the next robbery is

$$P(W > 2) = e^{-\lambda(2)} = e^{-(2.4)(2)} = e^{-4.8} = .0082.$$

The probability that we will have to wait between 1 and 3 days is

$$P(1 < W \leq 3) = P(W > 1) - P(W > 3) = e^{-(2.4)(1)} - e^{-(2.4)(3)}$$

$$= e^{-2.4} - e^{-7.2} = .0907 - .0007 = .0900.$$

In general, the probability that we will have to wait between a and b days is

$$P(a < W \leqslant b) = P(W > a) - P(W > b) = e^{-\lambda a} - e^{-\lambda b} \tag{4.11}$$

The Probability Density Function

How do we calculate the mean (or expected) waiting time until the occurrence of the next bank robbery? To answer this question, we will briefly return to a discussion of the general concept of a continuous probability distribution.

A continuous probability distribution is defined by its distribution function $F(x)$. In most, if not all, situations of practical interest, this distribution function is differentiable: that is, we can calculate its derivative. It turns out that the derivative $f(x) = F'(x)$ is much more usable in making calculations about and describing the characteristics of the distribution than is the distribution function itself. For this reason, the derivative of a probability distribution function has been given its own special name; it is called a *probability density function*. A continuous distribution that can be described by a probability density function is said to be "absolutely continuous."

The exponential distribution is absolutely continuous, and its probability density function is

$$f(t) = F'(t) = \frac{d}{dt}(1 - e^{-\lambda t}) = \lambda e^{-\lambda t} \tag{4.12}$$

for $t \geqslant 0$. Since W can never be negative, $F(t)$ and therefore $f(t)$ are always equal to 0 when $t < 0$. A graph of the probability density function (4.12) for $\lambda = 2.4$ appears in Fig. 4.17. This graph is the theoretical version of the frequency polygon that we discussed in Chapter 1.

Suppose we calculate the area under the probability density function from $t = a$ to $t = b$. This area can be expressed as the integral

$$\int_a^b f(t)dt = \int_a^b \lambda e^{-\lambda t}dt = -e^{-\lambda t}\Big|_a^b = e^{-\lambda a} - e^{-\lambda b},$$

which is the probability $P(a < W \leqslant b)$, as shown in formula (4.11). It turns out, then, that probabilities of random variables falling in intervals can be calculated as areas under the probability density function over those intervals. This concept is superimposed on Fig. 4.17.

Recall the bell-shaped curve of the standard normal distribution that we discussed in Section 4.C. This curve is actually the graph of the standard normal probability density function, which has the formula

$$g(x) = \frac{1}{\sqrt{2\pi}}e^{-x^2/2} \tag{4.13}$$

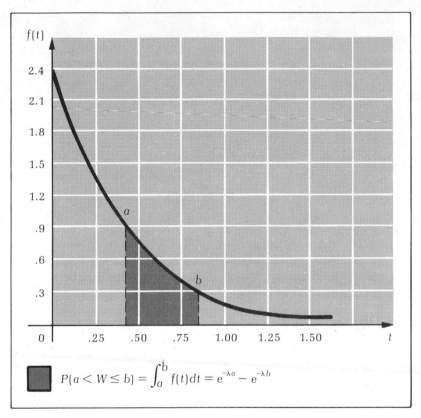

FIGURE 4.17   The Exponential Probability Density Function
$f(t) = \lambda e^{-\lambda t}$ $(\lambda = 2.4)$

Here $\pi \approx 3.1416$ is the number appearing in the geometrical formula for the area of a circle. Calculating the first and second derivatives of the density function and setting them equal to 0, we can see that the curve has a horizontal slope at $x = 0$ and points of inflection at $x = 1$ and $x = -1$. These characteristics are crucial in establishing the bell-shaped nature of the normal probability density function. If Z is a standard normal random variable, then using the area under the curve,

$$P(Z \leq z) = \frac{1}{\sqrt{2\pi}} \int_{-\infty}^{z} e^{-x^2/2} dx. \qquad (4.14)$$

This integral cannot actually be worked out using usual calculus methods (because there is no standard function whose derivative is $e^{-x^2/2}$), but approximations can be calculated to any desired degree of accuracy. Numerical values of the probability (4.4) have been calculated and collected in Table A.4 of the Appendix.

The fact that probabilities can be envisioned (and calculated) as areas under the probability density curve is only one of a long list of valuable properties of the probability density function. Another important property is that the mean μ and the standard deviation σ of an absolutely continuous probability distribution can be expressed, respectively, as the following integrals:

$$\blacksquare \qquad \qquad \mu = \int_{-\infty}^{\infty} xf(x)dx \qquad \qquad (4.15)$$

and

$$\blacksquare \qquad \qquad \sigma = \sqrt{\int_{-\infty}^{\infty} (x-\mu)^2 f(x)dx} \qquad \qquad (4.16)$$

where the distribution has density function $f(x)$.

Let's use formula (4.15) to calculate the mean waiting time based on the random variable W. Recall that W has an exponential distribution with parameter λ, whose density function $f(x)$ is given by formula (4.12) for $x \geq 0$, and $f(x) = 0$ for $x < 0$. Therefore the mean waiting time is

$$\mu = \int_{-\infty}^{\infty} xf(x)dx = \int_{-\infty}^{0} xf(x)dx + \int_{0}^{\infty} xf(x)dx = 0 + \int_{0}^{\infty} x\lambda e^{-\lambda x}dx$$

$$= -xe^{-\lambda x}\bigg|_{0}^{\infty} + \int_{0}^{\infty} e^{-\lambda x}dx = 0 + \left(-\frac{1}{\lambda}e^{-\lambda x}\right)\bigg|_{0}^{\infty} = \frac{1}{\lambda},$$

where we have used integration by parts with $u = -x$, $dv = -\lambda e^{-\lambda x}dx$. This value of μ makes logical sense because, if there were 2 bank robberies per day on the average, we could expect to wait half a day for the next one, while if there were 100 per day, we could expect to wait 1/100 of a day. If $\lambda = 2.4$, as it is in our example, the mean waiting time is $1/\lambda = 1/2.4 = .417$ of a day.

The standard deviation of an exponential random variable with parameter λ can be calculated using formula (4.16), and also turns out to be $\sigma = 1/\lambda$. If $\lambda = 2.4$, then the standard deviation must also be .417.

EXAMPLE 4.13. Computer Downtime. Wide River National Bank of Council Bluffs has a computerized system of recording and retrieving all information concerning its transactions. Unfortunately, the computer that supports the system "goes down" once every three weeks (i.e., once every 120 business hours) on the average. Furthermore, when it is down, the waiting time ("downtime") until it comes "up" is an exponentially distributed random variable with mean 0.75 hour. Find the following probabilities: (a) the probability that the system will go down during the next week (40 hours); (b) if it goes down, the probability that the downtime will exceed 3 hours; and (c) the probability that the down period will last between 4 and 8 hours.

SOLUTION.   a. The probability that the system will go down during the next 40 hours is the probability that $W \leqslant 40$, where

W = waiting time until system goes down.

Because W is an exponential random variable with parameter $\lambda = 1/120 = .0083$ (where the time unit is one hour), we can apply formula (4.10) to find that

$$P(W \leqslant 40) = 1 - e^{-(\lambda)(40)} = 1 - e^{-(1/120)(40)}$$

$$= 1 - e^{-1/3} = 1 - .717 = .283.$$

Therefore the probability is .283 that the system will go down during the next 40 hours.

b. The probability that the downtime will exceed 3 hours is the probability that $D > 3$, where

D = waiting time until system comes up.

The random variable D has an exponential distribution with mean $\mu = 0.75$ hour, and because $\mu = 1/\lambda$, it follows that $\lambda = 1/\mu = 1/0.75 = 4/3$. Using formula (4.9), we can see that

$$P(D > 3) = e^{-(\lambda)(3)} = e^{(4/3)(3)} = e^{-4} = .0183.$$

Therefore the probability is .0183 that the downtime will exceed 3 hours.

c. The probability that the downtime will last between 4 and 8 hours can be expressed by formula (4.11) or the integral of the exponential density function (4.12) with parameter $\lambda = 4/3$. In either case, it comes down to

$$P(4 < D < 8) = e^{-(4/3)(4)} - e^{-(4/3)(8)} = e^{-16/3} - e^{-32/3}$$

$$= .0048 - .0000 = .0048.$$

So the probability that the system will be down for 4 to 8 hours is .0048.

There are a number of other important continuous distributions, and each one is completely characterized by its probability density function. As a final example, consider the uniform probability distribution over the interval a to b. It has density function

$$u(x) = \frac{1}{b - a} \qquad \text{when } a \leqslant x \leqslant b. \tag{4.17}$$

and $u(x) = 0$ for all other values of x. The uniform density function is illustrated graphically in Fig. 4.18.

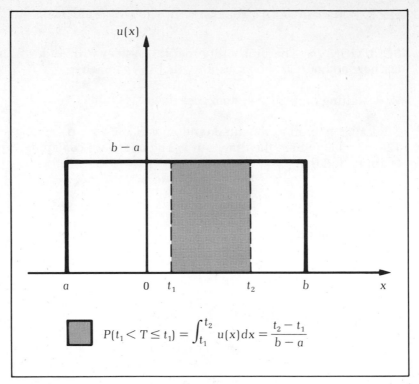

FIGURE 4.18   The Uniform Probability Density Function

$u(x) = \dfrac{1}{b - a}$ when $a \leqslant x \leqslant b$

EXAMPLE 4.14.   Scheduled Deliveries.   Jackrabbit Starts Delivery Service (JSDS) of Boston guarantees that its scheduled 11 A.M. delivery to your company will be uniformly distributed over the interval 10:30 A.M. to 11:30 A.M., as long as you are located in the metropolitan area. Calculate (a) the probability that the 11 A.M. delivery is made before 10:50 A.M.; (b) the probability that the 11 A.M. delivery is made after 11:25 A.M.; (c) the mean (expected) delivery time; and (d) the standard deviation of the delivery time.

SOLUTION.   The earliest possible time that delivery can occur is 10:30 A.M. Let's denote by T the random variable representing the time (in minutes) after 10:30 A.M. that the delivery is made. According to JSDS, T is uniformly distributed over the interval $a = 0$ (minutes after 10:30 A.M.) to $b = 60$ (minutes after 10:30 A.M.). In view of formula (4.17), this means that T has the probability density function

$$u(x) = \frac{1}{60} \qquad \text{for } 0 \leqslant x \leqslant 60.$$

190

and $u(x) = 0$ for other values of x. We calculate each of the desired answers in turn.

a. The probability that the delivery occurs before 10:50 A.M. is

$$P(T \leqslant 20) = \int_{-\infty}^{20} u(x)dx = \int_{0}^{20} (1/60)dx = (x/60) \Big|_{0}^{20} = 20/60 = .3333.$$

b. The probability that delivery occurs after 11:25 A.M. is

$$P(T > 55) = \int_{55}^{\infty} u(x)dx = \int_{55}^{60} (1/60)dx = (60 - 55)/60 = 5/60 = .0833.$$

c. The expected value of T is, according to formula (4.15),

$$\mu = \int_{-\infty}^{\infty} xu(x)dx = \int_{0}^{60} (x/60)dx = (x^2/120) \Big|_{0}^{60} = (60)^2/120 = 30$$

minutes, so that the expected delivery time is 10:30 A.M. plus 30 minutes, or 11 A.M..

d. According to formula (4.16), the standard deviation of T is the square root of

$$\sigma^2 = \int_{-\infty}^{\infty} (x-30)^2 u(x)dx = \int_{0}^{60} [(x-30)^2/60]dx = \int_{0}^{60} [(x^2 - 60x + 900)/60]dx$$

$$= \frac{1}{60}(x^3/3 - 30x^2 + 900) \Big|_{0}^{60} = \frac{1}{60}(216,000/3 - 108,000 + 54,000)$$

$$= 18,000/60 = 300$$

Therefore the standard deviation of T is $\sigma = \sqrt{300} = 17.32$ minutes.

EXERCISES   4.G

4.66. Amalgamated Glue's company stationery store receives an average of 6.2 requests per week (five working days) for a box of 100 Type FP file folders.

a. How soon can we expect the next request to come in?

b. What is the probability that the time until the next request comes in will exceed 3 working days (0.6 week)?

c. What is the probability that the next request will come in within one working day (0.2 week)?

4.67. Recall the accounting major of Exercise 4.48 who receives job offers in the mail at the rate of 0.3 per day. What is the probability that she will have to wait longer than 6 days for the next offer to arrive?

4.68. Mountain Lakes Life & Casualty Insurance Company of New Mexico writes specialized automobile policies for residents of 10 counties in the northern portion of the state. Claims are received at company headquarters in Santa Fe at the rate of 0.8 per day.

a. What is the mean waiting time in days for the next claim to be received?

b. Find the probability that the waiting time for the next claim will exceed four days.

4.69. According to a report referred to in Exercise 4.57, well-traveled New York City streets contain an average of 5.4 potholes per mile.

a. If a city repair crew proceeds along a major street, looking for potholes, how far should they expect to travel before encountering one?

b. What is the probability that they can proceed more than one-half mile without reaching a pothole?

4.70. Dexterglass Industries has a Stagg automatic fiberglass stamping machine whose reliability can be characterized by saying that the time until it fails to operate satisfactorily is an exponential random variable with mean 10 weeks. Maintenance personnel from Stagg visit the plant on a regular basis for one whole week every eight weeks. Dexterglass can save a considerable amount in repair costs if equipment failure occurs during these visits, because no special trips to the plant are required.

a. The Stagg people have just arrived. What is the probability that the next failure occurs within one week?

b. The Stagg people have just left. What is the probability that the next failure occurs during their next visit between 7 and 8 weeks from now?

4.71. The Citibus Schedule indicates that Bus No. 30 will arrive at the intersection closest to your place of work at 8:28 A.M. In fact, however, the arrival time is uniformly distributed over the interval 8:25 A.M. to 8:37 A.M.

a. What is the mean arrival time?

b. What is the standard deviation of the arrival time?

c. Find the probability that the bus arrives later than 8:35 A.M.

4.72. Another absolutely continuous distribution of some importance is Pareto's distribution, which has probability density function as shown below.

PARETO'S DISTRIBUTION

$$p(x) = \frac{\nu\theta^\nu}{x^{\nu+1}}$$

when $x \geq \theta$,

and $p(x) = 0$ for $x < \theta$. Here ν (Greek letter *nu*) and θ (Greek letter *theta*) are the parameters of the distribution. The values of the parameters depend on the special situation involved.

---

a. The random variable I that describes the annual family income in dollars of residents of Myersville may be considered to be a Pareto random variable with parameters $\nu = 2$ and $\theta = 10,000$. Calculate the mean annual family income of this town.

   b. What percentage of Myersville families have annual incomes in excess of
      $100,000?

   c. What percentage of Myersville families have annual incomes below $20,000?

   d. The median family income m can be calculated using the fact that $P(I \leqslant m)$
      $= 0.50$ if m is the median. (This is valid for any continuous distribution.)
      But $P(I \leqslant m) = \int_0^m p(x)dx$, so that we can find the median m if we can solve
      the calculus equation $\int_0^m p(x)dx = 0.50$. Find the median annual income of
      Myersville families.

   e. Draw a graph of the probability density function of the random variable I.

4.73. The number E of employees of a real estate firm in Ohio may be considered to
      have the Pareto distribution with parameters $\nu = 3$ and $\theta = 40$.

   a. Draw a graph of the density function of E.

   b. Find the median number of employees of real estate firms in Ohio.

   c. Calculate the mean number of employees of Ohio real estate firms.

   d. What percentage of Ohio real estate firms have 200 or more employees?

   e. What percentage of such firms have fewer than 45 employees?

4.74. The population distribution of cities, towns, and villages in New York can be
      modeled as a Pareto distribution with parameters $\nu = 1.6$ and $\theta = 3000$.

   a. Draw a graph of the population density function of such municipal units.

   b. Find the mean population of such units.

   c. Find the median population of such units.

   d. What percentage of New York's municipal units have populations in excess
      of 100,000?

   e. What percentage have populations below 5,000?

SUMMARY AND DISCUSSION

In Chapter 4, we have continued our development of probability theory by
studying general patterns of data called probability distributions. We have
concentrated on the binomial, normal, Poisson, hypergeometric, and exponential distributions, the most common types of distributions that describe
the behavior of data arising out of applications in business and management.

   The techniques developed in this chapter will be utilized throughout
the remainder of the book as we enter the field of *statistical inference*, the
drawing of conclusions (inferences) from a study of patterns inherent in a set
of data. Here we have seen how to recognize some kinds of data that have the
various probability distributions.

   Future chapters will be devoted to special methods of analyzing binomial, normal, and other patterns of data. In Chapter 13, we will introduce a
formal procedure for determining whether or not a particular set of data is
binomial, normal, Poisson, or none of these. The notion of a probability
distribution is a fundamental idea in the organized study of statistics, for we
learn from it that sets of data are not merely haphazard collections of numbers, but are, in fact, based upon general rules of behavior.

CHAPTER 4   BIBLIOGRAPHY

ADDITIONAL TYPES OF PROBABILITY DISTRIBUTIONS AND THEIR PROPERTIES

DERMAN, C., L. GLESER, and I. OLKIN. *A Guide to Probability Theory and Its Application.* 1973, New York: Holt, Rinehart & Winston, pp. 247–518.

MORRISON, D. G. "A Probability Model for Forced Binary Choices," *The American Statistician*, February 1978, pp. 23–25.

OFFICER, R. R., "The Distribution of Stock Returns," *Journal of the American Statistical Association*, December 1972, pp. 807–812.

TABLES OF PROBABILITY DISTRIBUTIONS

FAMA, E., and R. ROLL, "Some Properties of Symmetric Stable Distributions," *Journal of the American Statistical Association*, September 1968, pp. 817–836.

HARTER, H. L., and D. B. OWEN, *Selected Tables in Mathematical Statistics*, vol. I (1970), vol. II (1974), vol. III (1976). Providence, R.I.: American Mathematical Society.

RANDOM VARIABLES AS A FOUNDATION FOR STATISTICS

CHAPMAN, D. G., and R. A. SCHAUFELE, *Elementary Probability Models and Statistical Inference*, 1970, Waltham, Mass.: Xerox, pp. 10–207.

THE POISSON DISTRIBUTION

MULLET, G. M., "Simeon Poisson and the National Hockey League," *The American Statistician*, February 1977, pp. 8–12.

CHAPTER 4   SUPPLEMENTARY EXERCISES

4.75. The 1976 income tax forms caused the IRS as many headaches as they did the taxpayers. As of April 1977 the Internal Revenue Service reported that approximately 14% of the short forms contained an error. If this pattern were to hold true:

    a. What is the probability that more than 26 of the next 200 tax returns examined will contain an error?

    b. What is the probability that more than 260 out of the next 2,000 returns will contain an error?

4.76. A recent Gallup Poll revealed that about 70% of Americans over 18 years of age consumed alcoholic beverages in 1976. If these findings are correct:

    a. What are the chances that at least 12 out of 17 adults subsequently interviewed drank alcoholic beverages that year?

    b. What are the chances that at least 120 out of 170 of the adults contacted in another poll drank alcoholic beverages in 1976?

4.77. The latest figures indicate that 15% of all licensed drivers in one major metropolitan area have a drinking problem. If, on one particular day, 170 drivers apply for the normal three-year renewal of their licenses, what is the probability that more than 25% of them are problem drinkers?

4.78. One bank analyst feels that 55% of the adult residents of a particular major city have had bank loans at one time or another.

a. If his figures are correct, find the probability that, of a random sample of 18 residents, fewer than 5 have ever had a bank loan.

b. Under the banker's assumptions, what would be the probability that fewer than 500 of 1,800 residents have had a bank loan?

4.79. An accountant reports to her superiors that 35% of all accounts that are currently overdue will eventually require legal action to force payment. If her estimate is correct,

a. What is the probability that, of 12 currently overdue accounts, fewer than 3 will actually require legal action?

b. Find the probability that, of 120 overdue accounts, fewer than 30 will require legal action.

4.80. A psychologist specializing in marriage counseling claims that his program of "conscientious communication" can prevent divorce in 80% of the cases he works on.

a. If his claims are true, what is the probability that, of 20 couples currently involved in his program, more than 18 will stay together?

b. What is the probability that, of 200 cases in his files, fewer than 150 were able to avoid divorce?

4.81. The time it takes a job applicant to complete a standard psychological test is a normally distributed random variable with mean 80 minutes and standard deviation 10 minutes.

a. How much time should be allowed for the test so that 90% of the applicants will be able to finish it?

b. How much time should be allowed for the test so that 95% of the applicants will be able to finish it?

c. How much time should be allowed for the test so that 99% of the applicants will be able to finish it?

4.82. The A. C. Nielsen Co. conducted its annual TV report of the viewing habits of its cross section of American TV households in November 1976. They concluded that 77% of the nation's 71.2 million TV households have color TV sets.

a. If these figures are correct, find the probability that, of a random sample of 15 households, fewer than 10 have color TV sets.

b. What would be the probability that fewer than 1,000 of a sample of 1,500 households have color TV's?

4.83. Diebschau Realty, Inc., selects its sales staff from among the top 7.5% of all applicants taking the West End Real Estate Novice Test (WERENT). Scores on the WERENT are normally distributed with mean 120 and standard deviation 15.

a. What proportion of applicants score above 100?

b. What proportion score above 120?

c. What proportion score above 140?

d. What score is required to rank among the top 7.5% of all applicants?

4.84. A particular job requires very relaxed individuals. There is available a psychological test of aptitude for this job on which more relaxed persons score higher than less relaxed ones. The scores on this test are standardized so as to be normally distributed with mean 70 and standard deviation 10. What score is required to rank among the top 5% of all those taking the test?

4.85. The main plant of General Fabricators is lighted by several thousand light bulbs whose length of life is normally distributed with mean 2,000 hours and standard deviation 100 hours. Instead of replacing each bulb as it burns out (individual attention requires greater expense overall), the plant replaces all bulbs at the same time on a regular schedule. The management would like to set up the replacement schedule in such a way that no more than 1% of the bulbs will have burned out before being replaced. After how many hours of use should all the bulbs be replaced in order to achieve this goal?

4.86. A cafeteria vending machine dispenses coffee into 6-ounce cups in such a way that the actual amount dispensed into any particular cup is a normally distributed random variable with a standard deviation of 0.07 ounce. The vending machine has a dial in its interior with which the company can set the mean amount dispensed to any desired level. At what level should the mean be set so that a 6-ounce cup will overflow only 2% of the time?

4.87. A major nut company markets a bag of peanuts labeled as containing 10 ounces of nuts. The bags are packed by machine in such a way that the bags can have any desired mean weight, but their standard deviation will always be 0.1 ounce. Furthermore the weights will always be normally distributed.

    a. To what mean weight should the machine be set so that not more than one-fourth of one percent of the bags contain fewer than 9.8 ounces?

    b. If the mean is set at the level decided upon in part a, what percentage of the bags will contain more than 10.3 ounces?

4.88. An economist studying the fluctuation of vegetable prices in small grocery stores in rural areas concludes that the weekly average of prices per pound of 10 basic vegetables has mean 60 cents and standard deviation 7.8 cents.

    a. If the economist does not know the probability distribution of the vegetable prices, what should his answer be to the question, "At most what proportion of the time does the weekly average exceed 78 cents per pound"?

    b. If the economist tests the data (using methods discussed in Section 13.C) and finds the data to be normally distributed, how should he answer the question?

    c. Explain why the two answers are different, and explain the relationship between them.

4.89. The Diebschau Property Management Company knows that the maximum rent level apartment renters in the exclusive section of Lago Seco are willing to pay averages $320 per month for a two-bedroom apartment, with a standard deviation of $30. Furthermore, the maximum rent level that renters are willing to pay is a normally distributed random variable. What is the highest level at which the rent can be set and still have 98% of the renters willing to pay it?

4.90. An analyst working for the National Trucking Association is assigned the task of measuring the reaction times of experienced interstate truck drivers when they are confronted with an animal crossing the highway in front of them at night. She discovers that the reaction times are approximately normally distributed with mean 2.1 seconds and standard deviation 0.7 seconds. Earlier studies of animals crossing highways have found that, if the driver does not react before 3.6 seconds have passed, the animal will be hit. What proportion of the time is the animal hit?

4.91. One marketing instructor at a large university always gives A's to 10% of the students in his large introductory course, B's to 20%, C's to 40%, D's to 20%, and F's to the remaining 10%. On one final exam, the scores were normally distributed with mean 70 and standard deviation 10. What ranges of scores would qualify students for A's, B's, C's, D's, and F's?

4.92. A team of accountants checking on the financial situation of the Amalgamated Glue Corporation hypothesizes that there are, on the average, 3.6 erroneous statements out of every 100 in one particular file.

   a. If their hypothesis is correct, what are the chances of finding 2 or fewer errors in the next 100 statements checked?

   b. What is the probability that 100 randomly selected statements will contain 7 or fewer errors?

   c. What are the chances that 100 randomly selected statements will contain more than 9 erroneous ones?

   d. To expect 3.6 erroneous statements out of every 100 is equivalent to expecting 1.8 out of every 50. What is the probability that the next 50 statements checked contain 2 or fewer erroneous ones?

4.93. Fire insurance claims come in to her office at the rate of 4 per hour, on the average, according to an insurance adjuster in a heavily populated region of one southern state.

   a. Find the probability that, in the next hour, fewer than 2 claims will come in.

   b. What is the probability that, during the next hour, 8 or more claims will come in?

   c. Determine the mean waiting time for the next claim to arrive.

4.94. A supermarket manager notes that 1.6 customers, on the average, arrive at the checkout stands each minute. In order to arrange the hiring and scheduling of checkers, he needs to know the answers to questions like the following:

   a. What is the probability that more than 5 customers will arrive at the stands during the next minute?

   b. What are the chances that the next minute will pass with no customers having arrived at the checkstands?

   c. What is the probability that 2 or more customers will arrive during the next minute?

   d. To say that 1.6 customers arrive each minute is equivalent to asserting that

3.2 arrive in each two-minute period. What is the probability that fewer than 4 customers arrive at the checkstands during the next two minutes?

e. What are the chances that more than 9 customers arrive at the checkstands during the next two minutes?

f. Find the mean waiting time for the next customer to arrive at the checkstands.

4.95. The Abdec Oil Corporation has to decide quickly whether or not to bid on leases for 18 new offshore tracts just opened by the Interior Department. Based on prior experience, the company feels that only 13 of the 18 tracts will eventually prove to be worthwhile investments. If the company's feeling is correct, what would be the probability that, of 3 tracts subjected to detailed paleontological analysis, all showed definite indications of the presence of oil?

4.96. A quality control supervisor at a truck assembly plant is in charge of certifying the dimensions and hardness of the wheel bearings before the trucks are shipped. A random sampling scheme is used, with a sample of 6 of every lot of 40 trucks being thoroughly inspected. If a particular lot of 40 trucks contains 3 with defective bearings, what is the probability that at least one of the 3 will be detected by the sampling process?

4.97. Foot (i.e., shoe) sizes of adult males are normally distributed with mean 10.5 and standard deviation 1.2. Uncle Al's Footwear, Inc., advertises that it carries only those sizes required by the 10% of men having the longest feet. What is the smallest foot size that Uncle Al can fit?

4.98. The office of player personnel of the Dayton Rollers baseball club has determined that major-league players have spent an average of 3.7 seasons, with a standard deviation of 0.8 seasons, in the minor leagues. Furthermore, the time a player spends in the minors is a normally distributed random variable. What proportion of major-leaguers have spent between 2.5 and 4.5 seasons in the minors?

4.99. Child psychologists are exhibiting a considerable amount of interest in so-called "Kid Vid" commercials, i.e, television advertisements aimed at children. A recent university study of 233 Los Angeles schoolchildren investigated the possible effect of educational sessions intended to reduce children's belief levels in such TV ads. On a scale of 0 (no belief) to 100 (complete belief), the children participating in the study averaged 55, with a standard deviation of 20, after attending the sessions. The belief scores also turned out to be normally distributed. What percentage of children still had belief scores above 90?

4.100. Because of the immediate need for additional skilled technicians and assembly-line employees in the local aerospace industry (combined with an inability to recruit out-of-area employees, due to excessive local housing costs), it is rather easy for persons with little experience and only the basic required skills to get a job. In fact, all applicants scoring in the top 85% on the aptitude test (the scores on which are normally distributed with mean 120 and standard deviation 14) receive a job offer. What is the minimum score that an applicant can get and still qualify for a job offer?

4.101. The Pferd Motor Company, manufacturers of the Pinto, the Mustang, and the Bronco, has expressed its goal for the model year 1984: no more than three-fourths of one percent of its line of cars will have actual gasoline efficiency below 34 miles per gallon (MPG). Current specifications call for a mean MPG of 38 miles per gallon, and the per-car MPG levels are also normally distributed. What must the per-car MPG standard deviation be in order for the company to meet its goal?

4.102. The National Organization of Distributors, Entrepreneurs, Buyers, Traders, and Sellers (NODEBTS) completed a research analysis that showed that the time it takes a customer to pay a bill is a normally distributed random variable with mean 58.7 days and standard deviation 22.8 days. The report further recommends that the latest 2% of all bills be turned over to professional collection agencies. If that recommendation is adopted, how many days would a customer be given to pay a bill before it is turned over for collection?

4.103. The price of a gallon of unleaded gasoline in the county was once $1.66, on the average, with a standard deviation of $0.028. Furthermore, prices at the time were normally distributed. You strongly suspected that the friendly station on the corner near your residence was among the 5% most expensive stations in the county. If that were true, what would the price (at least) have been at that station?

III

STATISTICAL

INFERENCE

5

Estimation

What we have been discussing so far in this course is generally referred to as *descriptive statistics*. Descriptive statistics is that aspect of statistics whose primary objective is to describe, graphically and numerically, the characteristic behavior of sets of data. However, in most cases of applied interest, the data we have available will be only a small fraction of the totality of existing information on the subject, so we cannot be certain that the data we have will be a true representation of the existing situation.

For example, suppose a distributor of citrus fruits wants to know the mean weight of California grapefruits for the purpose of estimating shipping costs. The way to calculate it would, of course, be to weigh each California grapefruit, add all the weights, and divide by the total number of grapefruits. To do this would be very difficult, if it were at all even possible. The best that can be done is to select a *random sample* of a relatively few grapefruits (a few compared to the total number of California grapefruits in existence) and calculate the mean weight of all the grapefruits in the sample. One of the most important questions in statistics then surfaces: "What is the relationship between this *sample mean* and the item we are looking for, the *true mean* weight of all California grapefruits?" To answer this question requires the development of a second aspect of statistics, the aspect called *statistical inference*.

In statistical inference, our goal is to determine what conclusions about a situation may be validly and logically drawn ("inferred") on the basis of a random sample of statistical data. The remainder of this book is primarily concerned with statistical inference.

SECTION 5.A THE PROBLEM OF ESTIMATING THE TRUE MEAN

Let's denote the "true" mean weight of all California grapefruits by the Greek letter μ. How can we calculate μ? Well, we need to weigh all California grapefruits and then calculate the mean of all the weights so obtained.

As a practical matter, however, it is virtually impossible to do this. Some of the reasons why the exact value of μ cannot be directly calculated are: (1) it would involve too much work and expense to remove every single mature California grapefruit from its tree and weigh it; (2) it would create an instantaneous glut on the grapefruit market exceeding the capacity of all cold storage facilities, thus wiping out a substantial portion of the citrus industry; (3) by the time the results were announced, they would be invalid because many of the grapefruits used in the study would have been eaten while new ones (of possibly different weights) would have matured on the trees; and (4) the results, even if obtainable, would be valid for only one season anyway, and it would be ridiculous to repeat this procedure every year. What we therefore need is an indirect way of getting a reasonably accurate measurement of μ. It is the objective of statistical inference to come up with a practical method of determining the numerical value of μ.

The Sample Mean and the Population Mean

Because it is not possible to weigh every single grapefruit, how many of them should we weigh? Well, it seems reasonable that the more grapefruits we weigh, the closer our average of those weighed will be to the "true" mean weight (if such a thing indeed exists) of all California grapefruits. Suppose, for example, we take a sample of n randomly selected California grapefruits. By "randomly selected" we mean chosen in such a way that every possible combination of n grapefruits in the state has an equal chance of being our sample. From our random sample of n grapefruits, we compute the following numbers:

x_1 = weight of the 1st grapefruit of the sample,
x_2 = weight of the 2nd grapefruit of the sample,

.
.
.

x_k = weight of the kth grapefruit of the sample,
.
.
.

x_n = weight of the nth grapefruit of the sample.

Using these n numbers, which are the weights of the n grapefruits in our random sample, we can compute the *sample mean*,

$$\bar{x} = \frac{\sum x}{n} = \frac{x_1 + x_2 + \ldots + x_n}{n}.$$ (5.1)

Here the sample mean, the mean of the numbers in the sample, is denoted by the symbol $\bar{x}$ (pronounced "x-bar"). By way of contrast, μ is often referred to as the *population mean*, to signify that it is the mean of an entire population, not merely of a sample.

   Let's describe, if we can, the relationship between the sample mean $\bar{x}$ and the true mean μ. Does $\bar{x} = \mu$? If we choose n grapefruits at random and weigh them, will their mean weight be the same as the mean weight of all California grapefruits? In answer to this question, we'd have to say "probably not," for it would be highly fortuitous if we were to pick n grapefruits (10; 500; 2,000; or whatever number n may be) whose mean turned out to be exactly the same as the overall mean of all California grapefruits. In fact, suppose I pick n grapefruits, you pick n grapefruits, and six other people we know each pick n grapefruits. Then each of us will have his or her own personal $\bar{x}$. More likely than not, these eight $\bar{x}$'s not only will differ from μ, but they will also differ from each other. All eight of us will each come up with a different value for $\bar{x}$.

   Now that we have agreed that $\bar{x}$ is probably not the same as μ, it would be useful to know how close it actually is to μ. The surprising answer to this question is that we really have no idea how close $\bar{x}$ is to μ. Our difficulty here is rooted in the fact that we don't really know the numerical value of μ; therefore, no matter what the value of $\bar{x}$ is, we can't tell how close it is to μ.

   Well, the chances for an accurate estimate of μ look sort of bad, don't they? It looks as though we have gone just about as far as we can go. At this point, fortunately, we have available some major results from statistical theory. It is the job of theoretical research statisticians to figure out ways of getting us out of situations like this, and their methods will be presented in the next section.

EXERCISES   5.A

5.1. Give at least one reason why each of the following true means cannot be calculated exactly:

   a. The true mean protein content of eggs produced in Denmark

   b. The true mean height of American adults

   c. The true mean cost of weekly groceries for a family of four

   d. The true mean number of persons per household in the U.S.

   e. The true mean number of pupils in second-grade classrooms in the Chicago public school system

   f. The true mean ozone level in a cubic meter of air over Los Angeles County today

   g. The true mean rainfall in the state of Louisiana last week

5.2. Explain why each of the following true means can be calculated exactly:

   a. The true mean paid attendance per game at the home games of the New York Mets last season

   b. The true mean grade-point average of all students attending the University of Iowa last term

   c. The true mean number of votes cast per congressional district during the 1980 election

   d. The true mean number of empty seats on all nonstop commercial flights between Boston and Chicago last month

SECTION 5.B   CONFIDENCE INTERVALS BASED ON LARGE SAMPLES

Our objective in this section is to develop procedures for estimating μ from a random sample consisting of a large number of data points. We refer to such random samples, which as a rule of thumb contain more than 30 data points, as *large samples*. By *small samples* we mean sets of data consisting of 30 or fewer data points.

The Law of Averages

We noted earlier that when we choose a sample of n data points and calculate their sample mean $\bar{x}$, we really don't know how close $\bar{x}$ is to the true mean μ. It seems reasonable, however, that the more data points we have in our sample, the closer $\bar{x}$ will be to μ.

     For example, if I tried to estimate the true mean weight of California grapefruits by randomly selecting 25 grapefruits and averaging their weights, while you tried to do the job with a sample of 200 grapefruits, we'd have to agree that your $\bar{x}$ would probably be closer to μ than would my $\bar{x}$. The formal statement of this principle is a statistical theorem called the *strong law of large numbers* or, more popularly, the *law of averages*. A fairly simple wording of the strong law of large numbers reads as follows:

THE STRONG LAW OF LARGE NUMBERS (THE LAW OF AVERAGES)
If a population has the true mean μ, and $\bar{x}$ is the sample mean of a random sample of n members of the population, then, as n grows larger and larger (as we include more and more data points in our sample), our sample mean $\bar{x}$ becomes closer and closer to the true mean μ.

Because of the strong law of large numbers, we know that we can improve our estimate of μ by using random samples of larger and larger amounts of data points. However, we unfortunately still do not know exactly how far we can expect our $\bar{x}$ to be from μ.

The Central Limit Theorem

A second statistical theorem, called the *central limit theorem*, goes a long way toward breaking this impasse. Roughly speaking, it asserts that, for large values of n, the set of all possible sample means of random samples consisting of n data points is a set of normally distributed data. More precisely, we can state the central limit theorem as follows:

THE CENTRAL LIMIT THEOREM
If a population has the true mean μ and the true standard deviation σ, then the probability distribution of the set of sample means of all possible samples consisting of n members of the population becomes closer and closer to the normal distribution with mean μ and standard deviation $\sigma/\sqrt{n}$, as n grows larger and larger.

For all practical purposes, if n is larger than 30, namely if we are working with samples of more than 30 data points, then the distribution of all possible sample means is usually close enough to the normal distribution for us to consider it to be the normal distribution. For large samples, then, we consider the numbers $\bar{x}$ to be normally distributed with mean μ and standard deviation $\sigma/\sqrt{n}$. This situation is illustrated in Fig. 5.1.

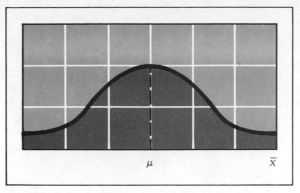

FIGURE 5.1 Distribution of Sample Means
for Large Samples

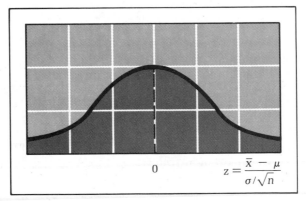

FIGURE 5.2 Distribution of z Scores
of Sample Means for Large Samples

In any particular large-sample applied problem under study, it is therefore appropriate to use Table A.4 of the Appendix, the table of the normal
distribution. Now Table A.4, as you will recall, is actually a table of the z
scores of the normal distribution, and it is therefore more useful to work with
z scores of the possible sample means than with the original sample means
themselves. A graph of the distribution of the z scores of the possible sample
means, a standard normal distribution, appears in Fig. 5.2.

Percentage Points of the Normal Distribution

To be better able to specify the location on the normal distribution of these
z scores, it is useful to introduce here a system of labeling the distribution.
Using the Greek letter α (pronounced "alpha") to represent a proportion of

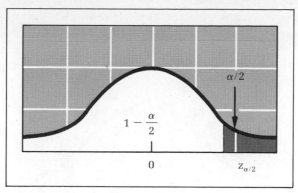

FIGURE 5.3   The Meaning of $z_{\alpha/2}$

TABLE 5.1   A Small Portion of Table A.4

z	Proportion
1.96	.9750

data, we will use the symbol $z_{\alpha/2}$ to signify the z value of a standard normal distribution that has, to its right, a fraction $\alpha/2$ of the data. The meaning of $z_{\alpha/2}$ is illustrated in Fig. 5.3. Because $z_{\alpha/2}$ has a proportion $\alpha/2$ of the data to its right, it must have the remainder, a proportion $1 - \alpha/2$, to its left. For example, taking $\alpha = .05$, we have $\alpha/2 = .025$. Here $z_{\alpha/2}$ is $z_{.025}$, the z score that has .025 or 2.5% of the data to its right, and therefore .975 or 97.5% to its left. We can find the numerical value of $z_{.025}$ by using Table A.4 of the Appendix because $z_{.025}$ is the z score corresponding to the proportion .9750. The portion of Table A.4, the part reproduced in Table 5.1, shows that $z_{.025}$ = 1.96.

Suppose for $\alpha = .05$, we needed to know z_{α}, the z score that has a proportion α of the data to its right. As illustrated in Fig. 5.4, this means that a fraction $1 - \alpha$ of a set of standard normal data lies to the left of z_{α}. When $\alpha = .05$, z_{α} is $z_{.05}$, which has 5% of the data to its right and $1 - \alpha = 1 - .05$ = .95 = 95% to its left. From the portion of Table A.4 that is reproduced in Table 5.2, we see that the exact value of $z_{.05}$ lies somewhere between 1.64 and 1.65. As we have agreed to choose the z score that is closer to 0 in cases like this, we shall set $z_{.05} = 1.64$.

In Table 5.3, we list z_{α} and $z_{\alpha/2}$ for several selected values of α. The number z_{α} corresponds to the numbers $1 - \alpha$ of Table A.4, while the $z_{\alpha/2}$'s correspond to the $1 - \alpha/2$'s. In Fig. 5.5, we illustrate the position of various percentage points of the normal distribution. We will make extensive use of percentage points throughout the rest of the book, beginning now.

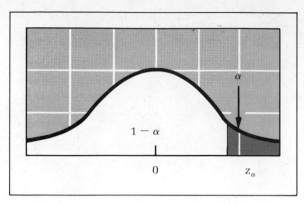

FIGURE 5.4   The Meaning of z_α

TABLE 5.2   A Small Portion of Table A.4

z	Proportion
1.64	.9495
1.65	.9505

TABLE 5.3   Some Selected Percentage Points of the Normal Distribution

α	$1 - \alpha$	z_α	$\alpha/2$	$1 - \dfrac{\alpha}{2}$	$z_{\alpha/2}$
.20	.8000	0.84	.10	.9000	1.28
.10	.9000	1.28	.05	.9500	1.64
.05	.9500	1.64	.025	.9750	1.96
.025	.9750	1.96	.0125	.9875	2.24
.02	.9800	2.05	.01	.9900	2.33
.01	.9900	2.33	.005	.9950	2.57
.005	.9950	2.57	.0025	.9975	2.81

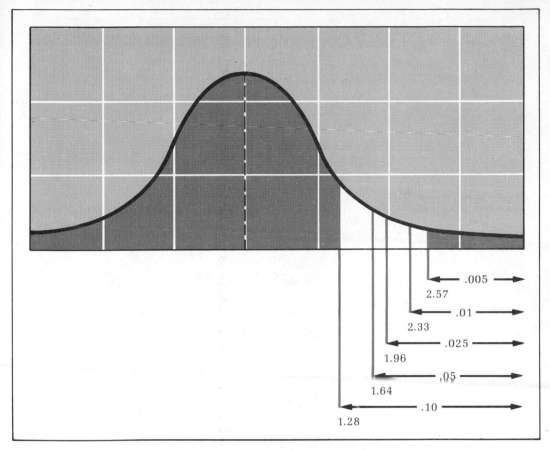

FIGURE 5.5   Some Percentage Points of the Normal Distribution

In terms of the percentage points of the normal distribution, this means that a fraction $1 - \alpha$ or a percentage $(1 - \alpha) \times 100\%$ of the z scores fall between the numbers $-z_{\alpha/2}$ and $z_{\alpha/2}$. For example, taking $\alpha = .05$, so that $1 - \alpha = 1 - .05 = .95$ and $(1 - \alpha) \times 100\% = (.95) \times 100\% = 95\%$, we see that 95% of the z scores fall between -1.96 and 1.96. This situation is illustrated in Fig. 5.6. In probability language, we can express this fact by saying that the probability is 95% that the z score

$$z = \frac{\overline{x} - \mu}{\sigma/\sqrt{n}}$$

of whatever sample mean $\overline{x}$ we obtain will fall between -1.96 and 1.96.

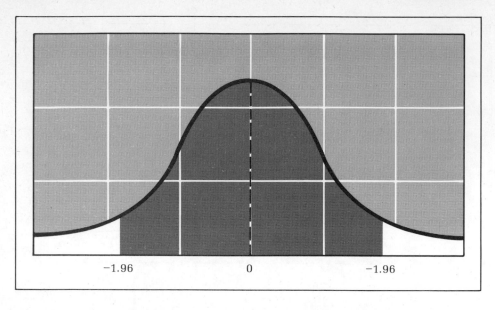

FIGURE 5.6   The Middle 95% of z Scores of Sample Means (Large Samples)

Fig. 5.7 provides another view of how the central limit theorem works. That graph shows that 10% of the z scores fall above 1.28. We can therefore say that 10% of the sample means result in a z score above 1.28, so that there is a 10% chance that your z score (or ours or anybody else's, for that matter) will turn out to be larger than 1.28.

FIGURE 5.7   The Upper 10% of z Scores of Sample Means (Large Samples)

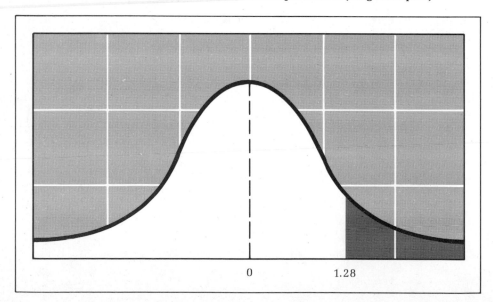

Unfortunately, although we can be fairly sure that 10% of the z scores will be larger than 1.28, we can never tell whether a particular z score is among that upper 10%. Why is this so? Well, the reason goes back to the fact that we do not know what μ is, and therefore we can never calculate the numerical value of our z score

$$z = \frac{\overline{x} - \mu}{\sigma/\sqrt{n}}.$$

In fact, not only do we not know what μ is, but we also do not generally know what σ is. Therefore we cannot substitute enough numbers into the formula for z in order to calculate it. However, even though we cannot tell whether our particular z score is, for example, between -1.96 and 1.96, we do know that 95% of them are. And we can use this information to help us in estimating the numerical value of the true mean μ.

We first express in mathematical symbolism the statement that 95% of the z scores fall between -1.96 and 1.96. In algebraic language, we can be 95% sure that $-1.96 < z < 1.96$*

Replacing z by its formula, we see that we can be 95% sure that, using our value $\overline{x}$ of the sample mean,

$$-1.96 < \frac{\overline{x} - \mu}{\sigma/\sqrt{n}} < 1.96 \, .$$

Multiplying across the above inequality (an equation with $<$ or $>$ instead of $=$) by the denominator $\sigma/\sqrt{n}$ gives us 95% certainty that

$$-1.96 \frac{\sigma}{\sqrt{n}} < \overline{x} - \mu < 1.96 \frac{\sigma}{\sqrt{n}} \, .$$

Applying a final bit of algebra, we see that we can be 95% sure that

$$\overline{x} - 1.96 \frac{\sigma}{\sqrt{n}} < \mu < \overline{x} + 1.96 \frac{\sigma}{\sqrt{n}} \, .$$

Constructed as it is from the data points of a random sample, the interval $\overline{x} - 1.96(\sigma/\sqrt{n})$ to $\overline{x} + 1.96(\sigma/\sqrt{n})$ is called a *random interval*. Now is a good time to go back to the English language and reflect upon what the algebraic statement just above really means. A direct translation would read as follows: We can be 95% sure that the random interval $\overline{x} - 1.96(\sigma/\sqrt{n})$ to $\overline{x} + 1.96(\sigma/\sqrt{n})$ contains the true mean μ.

---

*The notation $-1.96 < z < 1.96$ represents a combination of the two statements $-1.96 < z$ (z is greater than -1.96) and $z < 1.96$ (z is less than 1.96).

Let's return to the problem introduced in Section 5.A, the problem of finding the true mean weight of California grapefruits. Suppose we try to estimate the true mean μ by selecting a sample of $n = 64$ grapefruits. We weigh each of these grapefruits, and they turn out to have mean weight $\bar{x} =$ 11.36 ounces. What we now want to know is how close we can expect our sample mean 11.36 to be to the true mean μ. Well, as we have discussed above, we can assert (taking note of the facts that $\sqrt{64} = 8$ and $1.96/8 = .25$) that the interval $11.36 - .25\sigma$ to $11.36 + .25\sigma$ is one of those intervals that contain the true mean weight μ. Because 95% of the intervals we obtain in this way will actually contain μ, the probability is $0.95 = 95\%$ that our assertion will be correct.

Now σ is the true standard deviation of the weights of all California grapefruits. Until we find out the numerical value of σ, we will not know how good our estimate of μ is. Unfortunately, the same difficulties that prevented us from directly measuring μ also prevent us from directly measuring σ.

If we already know the numerical value of σ (from previous studies, etc.), then there is no problem. However, if we don't, it turns out that for large values of n (say, $n > 30$) there is a way of estimating σ to a degree of accuracy sufficient for completing the problem at hand. What we do is to replace σ in the formula we have been using by a number s, called the *sample standard deviation*, which is an estimate of the true standard deviation calculated from the n data points of the random sample. The formula for the sample standard deviation s is a slight variation of the formula for σ introduced in Section 2.C:

$$s = \sqrt{\frac{\Sigma(x-\bar{x})^2}{n-1}} \qquad\qquad (5.2)$$

Comparing this latter formula with the one for σ in Section 2.C, we note two changes: (1) μ is replaced by $\bar{x}$; and (2) in the denominator N is replaced by $n-1$. The change from μ to $\bar{x}$ is necessary because we are dealing with a random sample of which we know (or can calculate) the sample mean, but we have no access to the true mean μ. The replacement of N by $n-1$ is less easy to understand. Basically the idea here is that, if we calculate s according to formula (5.2), the mean of all possible values of s^2 (called the *sample variance*) obtainable from samples of size n will be σ^2 (called the *variance*), in the same way that the mean of all values of $\bar{x}$ is μ. If we use n, however, instead of $n-1$ in the denominator of s, the possible values of s^2 will average out to something slightly smaller than σ^2.

As an equivalent short-cut formula for the sample standard deviation, we have

$$s = \sqrt{\frac{n\,\Sigma x^2 - (\Sigma x)^2}{n(n-1)}} \qquad\qquad (5.3)$$

As we mentioned earlier in connection with the short-cut formula for the population standard deviation, this formula is the one more adaptable for use with a hand calculator.

Suppose our random sample of 64 grapefruits turned out to have sample standard deviation $s = 1.76$ ounces. Replacing σ in the expression

$$11.36 - 25\sigma < \mu < 11.36 + .25\sigma$$

by $s = 1.76$, we could assert that

$$11.36 - (.25)(1.76) < \mu < 11.36 + (.25)(1.76)$$

$$11.36 - .44 < \mu < 11.36 + .44$$

$$10.92 < \mu < 11.80 .$$

Our conclusion about the mean weight of California grapefruits can now be expressed as follows: On the basis of our random sampling procedure, we have a 95% probability of coming up with an interval that contains the true mean. Therefore we can be 95% sure that the interval 10.92 to 11.80 ounces is one of those that contain the true mean weight of all California grapefruits.

It is important to remember, in making a statement of our being 95% sure that μ is between 10.92 and 11.80, that the true mean μ is not the variable. The interval 10.92 to 11.80 is the value of the variable that is determined on the basis of the random sample of data. A more precise statement of our conclusion would be that since 95% of all intervals we obtain in this way will contain μ, we can be 95% sure that the interval 10.92 to 11.80 is one that does. Because it is obtained from a random sample of data, the interval $\bar{x} - 1.96(\sigma/\sqrt{n})$ to $\bar{x} + 1.96(\sigma/\sqrt{n})$ is sometimes called a *random interval*. The way in which random intervals fall around or away from the true mean is illustrated in Fig. 5.8.

The interval 10.92 to 11.80 is said to be a 95% *confidence interval* for μ. It allows us to come up with not only an estimate of μ, but also a measure of how accurate we can expect our estimate to be. A picture of this confidence interval appears in Fig. 5.9.

The confidence interval we have just found is only one of many, many confidence intervals that may be used to estimate μ. In fact, it is not *the* confidence interval for μ, but it is *a* 95% confidence interval for μ based on *a* random sample of size 64. If, for example, we wanted to be 98% sure (based on the same random sample) of having come up with an interval in which μ fell, we would have to modify our present interval by replacing the 1.96 in the formula (and in Fig. 5.6) by $z_{.01} = 2.33$. Based on the same sample of 64 data points, our 98% confidence interval would be

$$11.36 - (2.33)\left(\frac{1.76}{\sqrt{64}}\right) < \mu < 11.36 + (2.33)\left(\frac{1.76}{\sqrt{64}}\right),$$

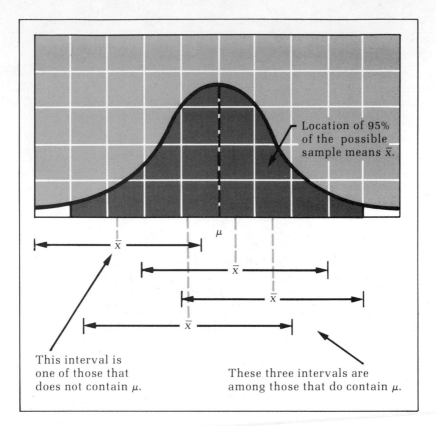

FIGURE 5.8   The True Mean and Some Random Intervals

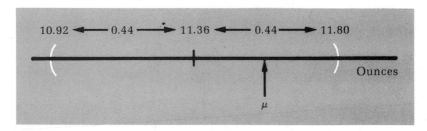

FIGURE 5.9   A 95% Confidence Interval for the Mean Weight of
California Grapefruits

namely, $10.85 < \mu < 11.87$. This says that there is a 98% chance that the
interval 10.85 to 11.87 is one of those that contain μ.

Notice that the 98% confidence interval for μ is wider than the 95%
interval based on the same data. This is logical, because if we want to be
more certain of having an interval that contains μ, we are going to have to

TABLE 5.4   Various Confidence Intervals for the True Mean Weight of California Grapefruits   *(Based on a sample of n = 64 grapefruits having sample mean x̄ = 11.36 and sample standard deviation s = 1.76.)*

Degree of Confidence $(1-\alpha)100\%$	α	$\alpha/2$	$z_{\alpha/2}$	$z_{\alpha/2}\dfrac{s}{\sqrt{n}}$	Confidence Interval $\bar{x}-z_{\alpha/2}\dfrac{s}{\sqrt{n}}<\mu<\bar{x}+z_{\alpha/2}\dfrac{s}{\sqrt{n}}$
80%	.20	.10	1.28	.28	$11.08 < \mu < 11.64$
90%	.10	.05	1.64	.36	$11.00 < \mu < 11.72$
95%	.05	.025	1.96	.44	$10.92 < \mu < 11.80$
98%	.02	.01	2.33	.51	$10.85 < \mu < 11.87$
99%	.01	.005	2.57	.57	$10.79 < \mu < 11.93$
99.9%	.001	.0005	3.3	.73	$10.63 < \mu < 12.09$

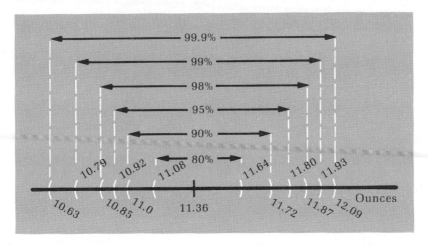

FIGURE 5.10   Graphical Representations of Various Confidence Intervals for the True Mean Weight of California Grapefruits

give ourselves a little more room by widening our interval. Similarly, if we were to narrow our interval, there might be only an 80% chance of coming up with one that contains μ. We summarize this discussion in Table 5.4, which demonstrates the procedure of obtaining confidence intervals of various degrees of confidence. For purposes of comparison, the various intervals are graphed in Fig. 5.10. The calculations there are based on the fact that the middle $(1-\alpha)100\%$ of the standard normal distribution lies between the numbers $-z_{\alpha/2}$ and $z_{\alpha/2}$. This fact results in the replacement of the 1.96 in our discussion by $z_{\alpha/2}$. We can therefore be $(1-\alpha)100\%$ sure that

$$\bar{x} - z_{\alpha/2}\frac{s}{\sqrt{n}} < \mu < \bar{x} + z_{\alpha/2}\frac{s}{\sqrt{n}} \qquad (5.4)$$

based on a random sample of n data points having sample mean $\bar{x}$ and sample standard deviation s. Formula (5.4) is the general formula for computing confidence intervals for means, based on large samples.

EXAMPLE 5.1.   Automobile Usage   A random sample of 100 private automobiles registered in the province of Ontario put a mean yearly mileage of 6,500 on their odometers, with a standard deviation of 1,400 miles. These facts were brought out as part of a statistical study of automobile usage commissioned by a local insurance agency. Find a 99% confidence interval for the true mean yearly mileage of all private cars in the province.

SOLUTION.   For a 99% confidence interval, we have that $(1-\alpha)100\%$ = 99%. Dividing through by 100%, we get that $1-\alpha = .99$ so that $-\alpha$ = $.99-1 = -.01$, and finally $\alpha = .01$. It follows that $\alpha/2 = .005$, from which we conclude that

$$z_{\alpha/2} = z_{.005} = 2.57.$$

The data give us the information that

$$n = 100 \qquad \bar{x} = 6,500 \qquad s = 1,400.$$

Applying formula (5.4) above, where μ is the true mean yearly mileage of all private cars in the province, we insert the actual numerical values of the quantities involved:

$$6,500 - (2.57)\frac{1,400}{\sqrt{100}} < \mu < 6,500 + (2.57)\frac{1,400}{\sqrt{100}}$$

$$6,500 - \frac{(2.57)(1,400)}{10} < \mu < 6,500 + \frac{(2.57)(1,400)}{10}$$

$$6,500 - 359.8 < \mu < 6,500 + 359.8$$

$$6,140.2 < \mu < 6,859.8.$$

Therefore, we can assert that the interval 6,140.2 to 6,859.8 miles is a 99% confidence interval for the true mean yearly mileage of all private cars in the province of Ontario.

Bounds on Error of Estimation

When we say, in the grapefruit weight example, that we are 95% sure of obtaining an interval in which the true mean μ falls, we are saying that we are 95% sure that μ is within a distance $1.96(\sigma/\sqrt{n})$ of $\bar{x}$. This way of express-

TABLE 5.5   Probable Bounds on Error of Estimation of the True Mean
Weight of California Grapefruits   (*Based on a sample of n = 64 grapefruits
having sample mean $\overline{x}$ = 11.36 and sample standard deviation s = 1.76*)

Degree of Confidence $(1 - \alpha)100\%$	α	$\alpha/2$	$z_{\alpha/2}$	Probable Bound on Error of Estimation $E = z_{\alpha/2}s/\sqrt{n}$
80%	.20	.10	1.28	.28
90%	.10	.05	1.64	.36
95%	.05	.025	1.96	.44
98%	.02	.01	2.33	.51
99%	.01	.005	2.57	.57
99.9%	.001	.0005	3.3	.73

ing the situation can be seen in Fig. 5.9, where attention is drawn to the fact
that 10.92 is located a distance .44 to the left of 11.36, while 11.80 is .44 to
the right. Therefore, if we were to use the sample mean of 11.36 ounces as an
estimate of the true mean μ, we could be 95% sure of being in error by no
more than .44 ounces.

   If we look at the general situation this way, we can view the quantity

$$E = z_{\alpha/2} \frac{s}{\sqrt{n}} \tag{5.5}$$

as follows: we can be $(1-\alpha)100\%$ sure that the maximum possible error in
using $\overline{x}$ as an estimate of μ cannot exceed E. In terms of error of estimation,
then, we can recast the subject matter of Table 5.4 into the form used in Table
5.5. These two procedures, confidence interval and error of estimation, are
merely ways of viewing the same information from different vantage points.

   The salient fact that is especially apparent from Table 5.5 is the parallel
behavior of the degree of confidence and the probable bound on the error of
estimation. As the degree of confidence increases, so does the probable bound
on the error of estimation.

   As we have pointed out earlier in the discussion, this is entirely logical:
the larger our allowable error of estimation, the more confident we can be of
having included the true mean μ within the induced bounds (and the less
precise we are in estimating μ). There is a definite trade-off between confidence and precision.

   EXAMPLE 5.2.   Reading Scores.   It has recently been suggested that
   new job applicants do not have the same level of basic skills as do
   present employees. The personnel director of a large corporation would
   like to estimate, for comparison purposes, the mean reading level, as
   measured by the standard Zeta-Epsilon Basic Educational Skills Test
   (ZEBEST) of all present employees. She chooses a random sample of 60

employees and gives the ZEBEST to each. They obtained a mean score of 550 points with a standard deviation of 90. How sure can the personnel director be that she has the true mean for the entire corporation estimated correctly to within 20 points?

SOLUTION. Here α is the unknown item, for that is related to the degree of confidence, $(1-\alpha)100\%$. From the results of the study, we know that

$$n = 60 \qquad \bar{x} = 550 \qquad s = 90$$

($\bar{x}$ turns out to be unneeded for the solution of this problem). The desired probable bound on the error is $E = 20$. Substituting into formula (5.5), we obtain the relationship

$$20 = z_{\alpha/2} \frac{90}{\sqrt{60}} .$$

A little algebra yields that

$$z_{\alpha/2} = \frac{(20)\sqrt{60}}{90} = 1.72 ,$$

because $\sqrt{60} = \sqrt{6 \times 10} = 7.746$. (Algebra applied directly to formula (5.5) would give us the general formula $z_{\alpha/2} = (E\sqrt{n})/s$.) From Fig. 5.11 and Table A.4 of the Appendix, we see that

$$\alpha/2 = 1.0000 - .9573 = .0427$$

FIGURE 5.11   Finding α when $z_{\alpha/2} = 1.72$

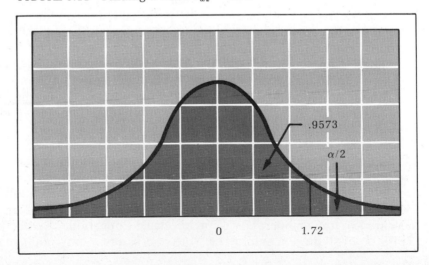

so that

$$\alpha = 2(.0427) = .0854$$

and then $\qquad 1 - \alpha = 1 - .0854 = .9146 \ .$

Therefore the personnel director can be $(1-\alpha)100\% = (.9146)100\% = 91.46\%$ sure that she has the true employee mean score estimated correctly to within 20 points.

EXERCISES   5.B

5.3. For each of the following numerical values of α, find z_α:

a. $\alpha = .30$ e. $\alpha = .015$

b. $\alpha = .15$ f. $\alpha = .008$

c. $\alpha = .075$ g. $\alpha = .002$

d. $\alpha = .03$ h. $\alpha = .001.$

5.4. For each of the following numerical values of α, find $z_{\alpha/2}$:

a. $\alpha = .30$ e. $\alpha = .015$

b. $\alpha = .15$ f. $\alpha = .008$

c. $\alpha = .075$ g. $\alpha = .002$

d. $\alpha = .03$ h. $\alpha = .001.$

5.5. If $z_\alpha = 1.41$, find α.

5.6. If $z_\alpha = 2.93$, find α.

5.7. If $z_{\alpha/2} = 2.19$, find α.

5.8. If $z_{\alpha/2} = 1.20$, find α.

5.9. For 64 randomly selected months in the past, the Oryx Veterinary Supply Company (OVS) has sold an average of $16,224.72 worth of goods each month, with a standard deviation of $205.00. Find a 90% confidence interval for the true mean monthly sales level of OVS.

5.10. The car leasing division of Hertz Corporation conducted a survey on the cost of operating a car in 1976. An analysis of 100 cars indicated that the average cost of operating a car (a 1976 intermediate two-door sedan driven 10,000 miles annually and kept for three years), including gas, insurance, oil, parts and service, interest, and depreciation, was 28.1¢ per mile with a standard deviation of 4¢ a mile. Find a 95% confidence interval for the true mean cost per mile of running this type of automobile.

5.11. Initial flights by British Airways' supersonic Concorde from London to New York and back to London averaged 2 hours and 14.3 minutes (134.3 minutes) turnaround time at New York, with a standard deviation of 4.8 minutes, in 36 trials during the fall of 1977. Find a 98% confidence interval for the Concorde's true mean turnaround time at New York.

5.12. A statewide real estate brokerage company would like to be able to forecast the true mean price of a house with an accuracy within $1,500. Such information will allow the company to develop accurate estimates of its projected earnings. A survey of 100 recently sold houses shows a mean price of $83,400 with a standard deviation of $7,100. How certain can the company be that the true mean is within $1,500 of its estimate of $83,400?

5.13. An auto parts dealer plans on advertising the average lifetime of the auto batteries he sells. To find out what that average lifetime is, he gives away 100 randomly chosen batteries to 100 randomly selected customers, with the proviso that the customer use the battery continuously with no maintenance, except for adding water as needed and cleaning the terminals, until it fails. The lifetimes, in terms of the 100 batteries turn out to have mean 2.0 years with a standard deviation of 0.156 years.

a. Determine a 90% confidence interval for the true mean lifetime of the dealer's batteries.

b. Determine a 95% confidence interval for the true mean lifetime.

c. Determine a 99% confidence interval for the true mean lifetime.

5.14. Find a 90% confidence interval for the true mean price forecast, based on the data of Exercise 1.11.

5.15. Find a 98% confidence interval for the true mean number of attempts required by a new employee to learn the quality control procedure discussed in Exercise 1.18.

5.16. A sample of 36 congressional districts in the United States had a mean 1981 inflation rate of 13.3 percent, with a standard deviation of 0.7 percent. Find a 90% confidence interval for the true mean inflation rate per congressional district for the entire nation.

5.17. A *U.S. News & World Report* survey reported the following unemployment rates in a random sample of cities across the United States in October 1977.*

City	Rate (%)	City	Rate (%)	City	Rate (%)
Boston	6.1	Buffalo	8.7	San Antonio	8.3
Jacksonville	5.6	Miami	7.2	Boise	3.0
Pittsburgh	6.8	Savannah	8.3	Honolulu	7.3
Wilmington	7.4	Indianapolis	5.5	Sacramento	7.7
Des Moines	3.3	New Orleans	8.3	Seattle	6.7
Knoxville	4.2	Charlotte	4.7	Hartford	7.2
Mobile	6.7	Newark	8.1	Norfolk	6.3
Tulsa	4.4	Rochester	7.0	Fargo	2.9
Fresno	6.6	Cleveland	5.5	Memphis	5.7
Phoenix	6.5	Grand Rapids	5.5	Topeka	4.0
San Diego	10.0	Milwaukee	5.2		

Find a 97.5% confidence interval for the true mean unemployment rate in cities across the United States.

*U.S. News & World Report, October 24, 1977, pp. 73–75.

5.18. An analysis of municipal bond markets around the country showed that tax-exempt bonds issued by 81 randomly selected small cities and rural counties were yielding an average return of 5.45%, with a standard deviation of 0.36%. Find a 98% confidence interval for the true mean yield of all tax-exempt bonds issued by small cities and rural counties.

5.19. A random sample of 40 corporations whose stock is listed on the New York Stock Exchange (NYSE) had mean earnings per share of 4.09 dollars during the past 12 months with a standard deviation of 1.74 dollars.* Find a 96% confidence interval for the true mean earnings per share of all corporations whose stock is listed on the exchange.

SECTION 5.C CONFIDENCE INTERVALS BASED ON SMALL SAMPLES

In the previous section we have shown how to estimate the true mean of a population if we have a random sample of more than 30 data points. The procedures discussed there also work for smaller samples (containing 30 or fewer data points) if the underlying population has the normal distribution and we know the value of the true standard deviation σ. Sometimes, however, it is not possible or economically feasible to use a large sample and, more often than not, σ is not known.

For example, a steel mill may want to estimate temperatures, pressures, etc., in various parts of its furnaces, and time or other constraints may not allow as many as 30 readings to be taken. A marketing manager may want to study the purchasing patterns of a community with regard to a particular type of product, but there may not be data available for 30 similar products that were marketed in that community. In situations such as these, where sufficient data are unobtainable, the central limit theorem of Section 5.B simply does not apply. We therefore have no guarantee that the z scores of the sample means are normally distributed.

We can handle the case of small samples (30 or fewer data points, according to standard usage) when σ is not known, if we have reason to believe that the data points came from a normally distributed population. This will be the case, for example, in situations satisfying the criteria for normally distributed data presented at the beginning of Section 4.C. Just as the central limit theorem revealed (in the large-sample case) that the numbers

$$z = \frac{\overline{x} - \mu}{s/\sqrt{n}}$$

are approximately normally distributed, theoretical calculations show that, if the overall population from which a random sample is selected has the

* Selected from *Forbes*, January 9, 1978, pp. 201–218.

normal distribution, then for all samples of size n (no matter how large or small) the numbers

$$t = \frac{\overline{x} - \mu}{s/\sqrt{n}}$$

have the t distribution of Table A.7 of the Appendix, where

$\overline{x}$ = sample mean
s = sample standard deviation
n = number of data points in sample
μ = true mean of population

The use of Table A.7 is valid only when the overall population has the normal distribution. This is due to the fact that Table A.7, the table of percentage points of the t distribution, is derived by theoretical calculations from Table A.4, the table of the normal distribution.

The t distribution has percentage points, just like the standard normal distribution does. There is a different t distribution for each value of n, however. Therefore the percentage points depend on n as well as on α. We denote the percentage points of the t distribution by the symbol $t_{\alpha/2}[df]$ where df is a number involving n. (In this section, $df = n-1$.) The symbol df is an abbreviation for the technical terminology degrees of freedom, which arose from the theoretical notion that the number t depends on $n-1$ algebraically independent quantities.

Even though the numbers

$$t = \frac{\overline{x} - \mu}{s/\sqrt{n}}$$

have the t distribution for all values of n, both large and small, Table A.7 is commonly used only for small samples of $n \leqslant 30$. The reason for this is that, for large values of n, the t distribution is very close to the normal distribution due to the increasing influence of the central limit theorem. Common statistical practice therefore calls for the use of Table A.7 when $n \leqslant 30$ and for Table A.4 when $n > 30$.

Beyond the switch from Table A.4 to Table A.7, there is no recognizable difference in the calculation of confidence intervals. In formula (5.4), we simply replace $z_{\alpha/2}$ by $t_{\alpha/2}[n-1]$, the appropriate number selected from Table A.7. Here the number $n-1$ is called the degrees of freedom and appears in the column headed df in the table. Is it through the degrees of freedom that the number of data points makes its influence felt on the t distribution. We find the proper t value at the intersection of the row corresponding to the number $n-1$ (one less than the number of data points) and the column headed $t_{\alpha/2}$ (for the relevant value of α). The formula for confidence intervals based on small samples, if we do not know the true value of α, is therefore

$$\blacksquare \qquad \overline{x} - t_{\alpha/2}[n-1]\frac{s}{\sqrt{n}} < \mu < \overline{x} + t_{\alpha/2}[n-1]\frac{s}{\sqrt{n}} \qquad (5.6)$$

If we do know the true value of σ, then we use formula (5.4) with s replaced by σ. The following two examples illustrate the use of this formula, particularly how to choose the correct value of $t_{\alpha/2}[n-1]$.

EXAMPLE 5.3. Company Cars. A corporation that maintains a large fleet of company cars for the use of its sales staff needs to determine the average number of miles driven monthly per salesperson. A random sample of 16 monthly car-use records was examined, yielding the information that the 16 salespersons under study had driven a mean of 2,250 miles with a standard deviation of 420 miles. Find a 95% confidence interval for the true mean monthly number of miles per salesperson for the entire sales staff.

SOLUTION. We can summarize the information contained in the sample by recording that

$$n = 16 \qquad \overline{x} = 2,250 \qquad s = 420 .$$

Because $n = 16 < 30$, use of the percentage points of the t distribution, rather than those of the normal distribution, is required. Since we want a 95% confidence interval, we know that $\alpha = .05$ so that $\alpha/2 = .025$. As $n = 16$, we know that $df = n - 1 = 16 - 1 = 15$, so the number we need from Table A.7 is $t_{\alpha/2}[n-1] = t_{.025}[15]$.

We reproduce the relevant portion of Table A.7 in Table 5.6, which shows that $t_{.025}[15] = 2.131$. Now we have in our possession all the components of formula (5.6) for confidence intervals based on small samples, where μ stands for the true mean monthly number of miles per salesperson for the entire sales staff. Therefore, inserting the numerical values for n, $\overline{x}$, s, and $t_{\alpha/2}[n-1]$, we obtain our 95% confidence interval as follows:

TABLE 5.6 A Portion of Table A.7

df	$t_{.025}$
15	2.131

$$2{,}250 - (2.131)\frac{420}{\sqrt{16}} < \mu < 2{,}250 + (2.131)\frac{420}{\sqrt{16}}$$

$$2{,}250 - \frac{(2.131)(420)}{4} < \mu < 2{,}250 + \frac{(2.131)(420)}{4}$$

$$2{,}250 - 223.8 < \mu < 2{,}250 + 223.8$$

$$2{,}026.2 < \mu < 2{,}473.8 .$$

Therefore, the interval 2,026.2 miles to 2,473.8 miles is a 95% confidence interval for the true mean monthly number of miles driven per salesperson for the entire sales staff.

Often in working with small samples of data, the 95% confidence interval will be too wide for practical use. As we have noted in Section 5.B, we can obtain narrower intervals (and, consequently, more precise estimates) if we are willing to give up some confidence. If we drop the confidence level to 90%, or even 80%, the resulting intervals will be narrower.

EXAMPLE 5.4. Manufacture of Antibiotics. Because pharmaceutical companies manufacture antibiotics in large vats for later bottling in smaller units, it is necessary to test potency levels at several locations in the vats before bottling. While it is impossible to achieve complete uniformity in potency levels of the bottled product, such testing is required for the purpose of obtaining the mean potency level of the entire batch. In one such test, the readings at 12 randomly selected locations in the vat were:

8.9   9.0   9.1   8.9   9.1   9.0   9.0   9.0   8.9   8.8   9.1   9.2.

Find a 98% confidence interval for the true mean potency reading for the entire batch.

SOLUTION. In this problem, we know immediately that $n = 12$, but we will have to compute $\bar{x}$ and s from the data. We proceed to make the calculations by the methods introduced in Chapter 2. The basic table of calculations appears in Table 5.7, and shows that $\bar{x} = 9$ and $s = .113$. Because we are looking for a 98% confidence interval, we are using $\alpha = .02$ so that $\alpha/2 = .01$. For $n = 12$, we know that $df = n - 1 = 12 - 1 = 11$, so we obtain from Table A.7 the relevant value of $t_{\alpha/2}[n-1]$, namely $t_{.01}[11] = 2.718$, as shown in Table 5.8. We insert these numbers into formula (5.6), where μ is the true mean potency level of the entire batch.

TABLE 5.7   Calculation of $\bar{x}$ and s for the
Antibiotic Potency Data

x	$x - \bar{x}$	$(x - \bar{x})^2$
8.9	−0.1	.01
9.0	0.0	.00
9.1	0.1	.01
8.9	−0.1	.01
9.1	0.1	.01
9.0	0.0	.00
9.0	0.0	.00
9.0	0.0	.00
8.9	−0.1	.01
8.8	−0.2	.04
9.1	0.1	.01
9.2	0.2	.04
108.0	0.0	.14

$\bar{x} = 108.0/12 = 9.0$

$s = \sqrt{.14/11} = \sqrt{.0127} = .113$

TABLE 5.8   A Portion of Table A.7

df	$t_{.01}$
11	2.718

On the basis of the data collected, we obtain

$$9 - (2.178)\frac{.113}{\sqrt{12}} < \mu < 9 + (2.178)\frac{.113}{\sqrt{12}},$$

$$9 - \frac{(2.178)(.113)}{3.464} < \mu < 9 + \frac{(2.718)(.113)}{3.464},$$

$$9 - .09 < \mu < 9 + .09,$$

$$8.91 < \mu < 9.09.$$

as a 98% confidence interval for the true mean potency level of the entire batch of antibiotic in the vat.

EXERCISES 5.C

5.20. For 18 randomly selected months in the past, the Oryx Veterinary Supply Company (OVS) has sold an average of $16,224.72 worth of goods each month, with a standard deviation of $205.00. Find a 90% confidence interval for the true mean monthly sales level of OVS.

5.21. The Desert Breeze Casualty Insurance Company of Trona wants to estimate the mean number of auto accidents its clients cause per month. They have collected the following data over 12 randomly selected months in the recent past:

Month	No. of Accidents
May 1974	24
July 1974	16
October 1974	20
November 1975	18
January 1976	22
March 1976	8
June 1977	12
July 1978	28
September 1978	20
December 1979	14
April 1980	10
October 1981	12

Find an 80% confidence interval for the true mean number of auto accidents caused per month by the company's clients.

5.22. An auto parts dealer plans on advertising the average lifetime of the auto batteries he sells. To find out what that average lifetime is, he gives away 10 randomly selected batteries to 10 randomly selected customers, with the proviso that the customer use the battery continuously with no maintenance, except for adding water as needed and cleaning the terminals, until it fails. The lifetimes, in years, of the 10 batteries turn out to be as follows:

2.2   1.9   2.0   2.1   1.8   2.1   2.1   1.7   2.0   2.1

   a. Determine a 90% confidence interval for the true mean lifetime of the dealer's batteries.

   b. Determine a 95% confidence interval for the true mean lifetime.

   c. Determine a 99% confidence interval for the true mean lifetime.

5.23. A random sample of 25 cigarettes of the "Rodeo Rider" brand had mean nicotine content of 22 mg., with a standard deviation of 4 mg. Find an 80% confidence interval for the true mean nicotine content of "Rodeo Rider" cigarettes.

5.24. To estimate the length of a typical gas line in the 1970's, a reporter surveyed 9 randomly selected gas stations one morning and found their gas lines to contain the following number of cars:

40 42 35 45 30 60 56 38 41

Find a 90% confidence interval for the true mean number of cars per gas line that morning.

5.25. A stock brokerage firm headquartered in Dallas has a large research staff and maintains well-stocked research libraries in all its major downtown offices so that its brokers may keep informed of current trends in the securities markets. The head office would like to determine the extent to which these libraries are being used. A survey of 6 of the downtown office libraries during the hours of 9 to 10 A.M. found the following numbers of brokers utilizing the facilities:

Office	A	B	C	D	E	F
Number of Brokers	16	5	10	12	8	15

Find an 80% confidence interval for the true mean number of brokers using the research libraries between the indicated hours.

5.26. In 1976, 10 large international banking companies managed the offering of the following amounts of Eurobonds (in billions of dollars):*

2.9   5.9   5.4   2.6   2.7   3.0   3.2   5.5   4.8   3.0

Calculate an 85% confidence interval for the true mean size per bank (in billions of dollars) of the yearly offering.

5.27. Construct a 90% confidence interval for the length of the national mean payback period using the solar heating systems data of Exercise 2.1.

5.28. Find an 80% confidence interval for the true mean executive pay in 1976, based on the data in Exercise 2.5.

5.29. A random sample of nine of the U.S.'s largest corporations yielded the following data on the number of persons employed by each of them:†

Corporation	Thousands of Employees
Timken	23.1
American Airlines	37.3
Scott Paper	21.3
Carnation	22.2
Ohio Edison	6.6
Diamond Shamrock	11.3
R. H. Macy	41.0
Black and Decker	17.2
Walt Disney	18.0

*Business Week, March 14, 1977, p. 63.

†Forbes, May 15, 1978, pp. 242–284.

Find an 80% confidence interval for the true mean number of employees of the U.S.'s largest corporations.

5.30. To obtain an estimate of the true mean freeway speed of a car during the morning rush hour, a delivery service driver recorded her speedometer reading at 12 randomly selected times during a recent trip. These readings in miles per hour were as follows:

41	13	0	50	21	17
18	35	26	8	16	31

Find an 80% confidence interval for the true mean speed during the trip.

5.31. *The Economist* reported in its November 1–7 1980 issue (page 93) the following "latest Tuesday" prime lending rates in a sample of industrial nations:

Nation	Britain	Italy	Switzerland	Japan	U.S.A.
Rate (%)	17	21.5	6.5	8.5	14.5

Find an 80% confidence interval for the true mean national prime lending rate among industrial nations.

SECTION 5.D ESTIMATING STANDARD DEVIATIONS

The problem of estimating the "true" standard deviation σ of a population and the basic principles underlying the idea of confidence intervals for standard deviations are essentially the same as those for the corresponding questions involving means. It is therefore not necessary to redevelop the ideas here, and we will concentrate instead on the technical aspects.

If we have a random sample of n data points, $x_1, x_2, \ldots, x_n$, we have been and will be using formula (5.2), namely

$$s = \sqrt{\frac{\sum(x - \bar{x})^2}{n-1}}$$

to provide an estimate of σ. The number s, as we have already pointed out, is called the sample standard deviation, and we can reasonably expect our value of s to be fairly close to the true standard deviation σ.

Exactly how close s will be to σ can never be known, of course, because we can never know σ exactly (for reasons discussed in Section 5.A in regard to μ). However, in view of some results from theoretical statistics, it turns out that the quantity

$$\chi^2 = \frac{(n-1)s^2}{\sigma^2},$$

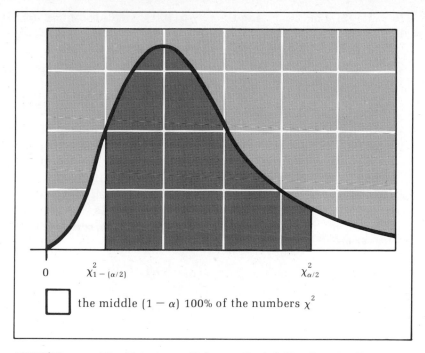

the middle $(1 - \alpha)$ 100% of the numbers χ^2

FIGURE 5.12   The Frequency Polygon (Probability Density Function) of the Chi-Square Distribution

where χ is the Greek letter lowercase *chi* (pronounced "kye"), has the chi-square distribution of Table A.8, with $n-1$ degrees of freedom, if the data come from a population that is normally distributed. The frequency polygon (actually the probability density function—see Section 4.G) of the chi-square distribution, together with a pair of complementary percentage points $\chi^2_{1-\alpha/2}$ and $\chi^2_{\alpha/2}$, is presented in Fig. 5.12. As the picture shows, the numbers $\chi^2_{1-\alpha/2}$ and $\chi^2_{\alpha/2}$ are chosen in such a way that $(1 - \alpha)$100% of the numbers

$$\chi^2 = \frac{(n - 1)s^2}{\sigma^2}$$

fall between them. We can use this information, together with the table of percentage points of the chi-square distribution (Table A.8), in order to calculate confidence intervals for the true population standard deviation σ.

Fig. 5.12 shows that we can be $(1-\alpha)$100% sure that

$$\chi^2_{1-\alpha/2}[n - 1] < \frac{(n - 1)s^2}{\sigma^2} < \chi^2_{\alpha/2}[n - 1]$$

where the $n-1$ in brackets stands for the number of degrees of freedom, *df*. If we apply some basic algebraic techniques to the above inequality, we can derive formula (5.7) below, giving the $(1-\alpha)100\%$ small-sample confidence interval for σ:

$$\sqrt{\frac{(n-1)s^2}{\chi^2_{\alpha/2}[n-1]}} < \sigma < \sqrt{\frac{(n-1)s^2}{\chi^2_{1-\alpha/2}[n-1]}}. \tag{5.7}$$

EXAMPLE 5.5. Manufacture of Antibiotics. In the manufacture of antibiotics in large vats, it is important to know the variability in potency of the product from location to location in the vat. If the variability is too great, the antibiotic could be virtually useless because medical personnel would not be able to judge its effectiveness in any given situation. The standard deviation provides one way of measuring this variability. If potency readings at 12 randomly selected locations in the vat were

8.9   9.0   9.1   8.9   9.1   9.0   9.0   9.0   8.9   8.8   9.1   9.2 ,

find a 90% confidence interval for the true standard deviation in potency of the entire batch.

SOLUTION.   The sample standard deviation of this set of data has been calculated in Table 5.7 to be $s = .113$. For the 90% confidence interval, we have $\alpha = .10$ and so $\alpha/2 = .05$. Here $n = 12$, so that $n-1 = 11$. The chi-square values appearing in the above formula are $\chi^2_{.05}[11] = 19.675$ and $\chi^2_{.95}[11] = 4.575$, as illustrated in Table 5.9, which is a selected portion of Table A.8. Inserting all these numbers into formula (5.7), we obtain that

$$\sqrt{\frac{(12-1)(.113)^2}{19.675}} < \sigma < \sqrt{\frac{(12-1)(.113)^2}{4.575}}$$

$$\sqrt{\frac{(11)(.013)}{19.675}} < \sigma < \sqrt{\frac{(11)(.013)}{4.575}}$$

$$\sqrt{.007} < \sigma < \sqrt{.031}$$

$$.084 < \sigma < .176$$

Therefore we can assert that the interval .084 to .176 is a 90% confidence interval for the true standard deviation of the potency level of the entire batch.

As a general rule, the actual chi-square values are commonly used only for *df*'s of 30 or below. Formula (5.7) can therefore be viewed as the formula

TABLE 5.9   A Small Portion of Table A.8

df	$\chi^2_{.95}$	$\chi^2_{.05}$
11	4.575	19.675

for confidence intervals based on small samples. For large samples, an analogue of the central limit theorem, called the Wilson-Hilferty approximation,* takes over and asserts that we can use the approximation

■ $$\chi^2_{1-\alpha/2}[n-1] = (n-1)\left(1 - z_{\alpha/2}\sqrt{\frac{2}{9(n-1)}} - \frac{2}{9(n-1)}\right)^3 \qquad \textbf{(5.8a)}$$

and

■ $$\chi^2_{\alpha/2}[n-1] = (n-1)\left(1 + z_{\alpha/2}\sqrt{\frac{2}{9(n-1)}} - \frac{2}{9(n-1)}\right)^3 \qquad \textbf{(5.8b)}$$

If formulas (5.8a) and (5.8b) are used as the chi-square values called for in formula (5.7), we will be able to compute the $(1-\alpha)100\%$ large-sample confidence interval for σ. Of course, if the chi-square value is directly available in Table A.8 of the Appendix, the Wilson-Hilferty approximation need not be applied.

EXAMPLE 5.6.   California Grapefruits.   A distributor of citrus fruits judges his stock not only by magnitude of weight, but also by uniformity of weight. The smaller the standard deviation, the less variation there is from the mean, and so the more uniform the weights are. If the distributor selects a random sample of 64 California grapefruits, and these turn out to have (sample) mean 11.36 ounces and (sample) standard deviation 1.76 ounces, find a 95% confidence interval for the true standard deviation of all California grapefruits.

SOLUTION.   We use formulas (5.8a) and (5.8b) because we need a large-sample confidence interval based on $n = 64$ data points. For the 95% confidence interval, we use $\alpha = .05$ so that $z_{\alpha/2} = 1.96$. Further, we have $s = 1.76$. (The sample mean $\overline{x}$ plays no role in the estimation of the standard deviation.) According to formulas (5.8a) and (5.8b), we can consider

$$\chi^2_{.975}[63] = 63\left(1 - 1.96\sqrt{\frac{2}{9(63)}} - \frac{2}{9(63)}\right)^3$$

$$= 63\,(1 - .1164 - .0035)^3 = 63(.8801)^3 = 42.947$$

*See the article by D. C. Hoaglin listed in the bibliography for this chapter.

and

$$\chi^2_{.025}[63] = 63(1 + .1164 - .0035)^3 = 63(1.1129)^3 = 86.838.$$

Using the confidence-interval formula (5.7), we obtain the development:

$$\sqrt{\frac{63(1.76)^2}{86.838}} < \sigma < \sqrt{\frac{63(1.76)^2}{42.947}}$$

$$1.50 < \sigma < 2.13 \ .$$

We can therefore say that the interval 1.50 ounces to 2.13 ounces is a 95% confidence interval for the true standard deviation of the weight of California grapefruits.

EXERCISES 5.D

5.32. Using the data of Exercise 5.21:
   a. Find a 90% confidence interval for the true standard deviation of the number of auto accidents caused per month by the company's clients;
   b. Find a 95% confidence interval for the true standard deviation;
   c. Find a 99% confidence interval for the true standard deviation.

5.33. From the information presented in Exercise 5.23, determine a 90% confidence interval for the true standard deviation of the nicotine content of all "Rodeo Rider" cigarettes.

5.34. Over a period of 50 months, the Desert Sands Casualty Insurance Company of Stovepipe Wells conducted an actuarial study that showed that its clients caused an average of 17 auto accidents per month (as a group, of course, not individually), with a standard deviation of 6 accidents.
   a. Find an 80% confidence interval for the true standard deviation of the number of auto accidents caused per month by the company's clients.
   b. Find a 90% confidence interval for the true standard deviation.
   c. Find a 95% confidence interval for the true standard deviation.

5.35. Using the data of Exercise 5.20, find a 90% confidence interval for the true standard deviation of monthly sales of OVS.

5.36. Find a 90% confidence interval for the standard deviation of the yearly number of worker-days lost due to labor disputes, based on the data of Exercise 2.22.

5.37. Calculate a 95% confidence interval for the national standard deviation of the payback periods, based on the solar heating systems data of Exercise 2.1.

5.38. Construct a 98% confidence interval for the true standard deviation of the number of attempts required by new employees to learn the quality control procedure discussed in Exercise 1.18.

5.39. Find a 97% confidence interval for the true standard deviation of the price forecasts from which the data of Exercise 1.11 were taken.

5.40. Using the Hertz Corporation data of Exercise 5.10, determine a 99% confidence interval for the true standard deviation of the cost per mile of operating the type of vehicle discussed.

SECTION 5.E   ESTIMATING PROPORTIONS

The objective of a marketing survey is to estimate the proportion of shoppers preferring one or another brand or product. The objective of an auditing procedure is to estimate the proportion of a company's records that are correct. And the objective of scholastic and professional competency examinations is to estimate the proportion of subject matter learned by each student or known by each applicant.

When we are estimating a proportion, we collect a set of binomially distributed data having parameters n (the number of data points) and π (the true proportion we want to estimate). We can therefore use the facts about the binomial distribution, recorded in the last paragraph of Section 4.B that $\mu = n\pi$ and $\sigma = \sqrt{n\pi(1-\pi)}$. If we use p to denote the sample proportion based on the data, then the sample mean will be np (to be contrasted with $\mu = n\pi$). If the sample size n is large enough, the normal approximation to the binomial distribution, discussed in Section 4.D, then implies that the z score

$$z = \frac{np - \mu}{\sigma} = \frac{np - n\pi}{\sqrt{n\pi(1-\pi)}} = \frac{p - \pi}{\sqrt{\frac{\pi(1-\pi)}{n}}}$$

has the standard normal distribution. Here the sample proportion p could be, for example, the proportion of sampled shoppers who prefer the brand in question, the proportion of a company's accounts satisfactorily reported on its records, or the proportion of examination questions correctly answered by the student. Using essentially the same algebraic operations that led to formula (5.4), we derive a $(1-\alpha)100\%$ confidence interval formula for the true proportion π:

$$p - z_{\alpha/2} \sqrt{\frac{p(1-p)}{n}} < \pi < p + z_{\alpha/2} \sqrt{\frac{p(1-p)}{n}} \tag{5.9}$$

(Note that π has been replaced by p to get the sample value of σ.) The above formula is usually used for $n > 20$, which generally puts us within the domain of validity of the normal approximation to the binomial distribution. Naturally, the larger the value of n, the more valid is the normal approximation.

EXAMPLE 5.7.   Automobile Loan Approvals.   (This example continues the framework of Examples 4.2 and 4.10.) Of a sample of 700 loan applications, 390 are eventually approved. Find a 95% confidence interval for the true proportion of all loan applications submitted to the bank that are eventually approved.

SOLUTION.   Here we have $\alpha = .05$ so that $z_{\alpha/2} = 1.96$. We have $n = 700$, and we see that the sample proportion eventually approved is $p = 390/700 = .56$.

Therefore, denoting by π the true proportion eventually approved, we can say that, successively,

$$.56 - (1.96) \sqrt{\frac{.56(1-.56)}{700}} < \pi < .56 + (1.96) \sqrt{\frac{(.56)(1-.56)}{700}}$$

$$.56 - (1.96) \sqrt{\frac{(.56)(.44)}{700}} < \pi < .56 + (1.96) \sqrt{\frac{(.56)(.44)}{700}}$$

$$.56 - (1.96) \sqrt{.000352} < \pi < .56 + 1.96 \sqrt{.000352}$$

$$.56 - (1.96)(.0188) < \pi < .56 + (1.96)(.0188)$$

$$.56 - .037 < \pi < .56 + .037$$

We conclude on the basis of the 700 loan applications sampled that the interval .523 to .597, namely 52.3% to 59.7%, is a 95% confidence interval for the true proportion of loans eventually approved by the bank. (The fact that the lower confidence limit is .523 provides a strong hint that a majority of applications is approved.)

EXERCISES 5.E

5.41. The auditors of a medium-sized retail firm would like to know the proportion of accurate accounts receivable based on an audit verification letter sent to customers. On the basis of an experiment in which 130 out of 170 responses were verified as accurate, find a 95% confidence interval for the true proportion of the accounts receivable that will be accurate.

5.42. An insurance company is planning to offer a 10-year term group life insurance policy to all 45-year-old men in a certain occupational grouping. In order to determine a reasonable premium, the company conducts an actuarial study of all men of that age in that occupational grouping. If a survey of 2,000 such men in the past showed that 150 died before reaching age 55, construct a 99% confidence interval for the true proportion of such men who die before age 55.

5.43. A financial analyst would like to predict the percentage of bank presidents who favor a certain controversial change in the banking laws. If a poll of 120 presidents yields 72 who support the change, find a 90% confidence interval for the true percentage of bank presidents favoring the change.

5.44. The personnel officer of a company would like to estimate the proportion of employees within five years of mandatory retirement among the company's employees. Research turns up 45 such employees in a random sample of 195 employees.

    a. Find an 80% confidence interval for the true proportion of employees nearing retirement in the entire company.

    b. Find a 98% confidence interval for the true proportion.

5.45. A marketing analyst studying the market penetration of McDonald's in one major city asks 1,800 randomly selected residents whether or not they have been to McDonald's in the past week. Of those surveyed, 700 responded in the affirmative.

    a. Find a 90% confidence interval for the true percentage of the city residents who have been to McDonald's in the week under study.

    b. Find a 95% confidence interval for that percentage.

    c. Find a 99% confidence interval for that percentage.

5.46. A survey of California families in 1978 revealed an improvement in perceived financial status over the previous year. The California Poll surveyed 1,034 persons. Thirty-four percent of the families interviewed felt better off financially than they had been the previous year. Find a 97% confidence interval for the true percentage of California families feeling better off financially in 1978 than in 1977.

5.47. A study by the California Department of Transportation (Caltrans) revealed that, of 83 randomly selected commuters in the Los Angeles metropolitan area, only 10 traveled by bus or car pool.* Construct a 90% confidence interval for the true proportion of Southern Californians that commute by bus or car pool.

5.48. Of 200 randomly selected tax accountants, 80 had their own individual tax returns for last year audited by the IRS. Find a 95% confidence interval for the true proportion of tax accountants who had their returns audited last year.

SECTION 5.F   CHOOSING THE SAMPLE SIZE

In previous sections, we introduced methods of estimating means and proportions in such a way that we know how accurate our estimates are. Let's look again at the discussion associated with formula (5.5), where we considered the probable bound

$$E = z_{\alpha/2} \frac{\sigma}{\sqrt{n}}$$

on the error of estimation. Roughly speaking, we can be $(1-\alpha)100\%$ sure that the maximum possible error in using $\bar{x}$ as an estimate of μ cannot exceed E.

*Reported in *Orange County Business*, Sept.-Oct., 1977, p. 39.

TABLE 5.10   Relationship Between Sample Size and Probable Bounds on
the Error of Estimation   *(Based on a 95% confidence interval for the true
mean of a population having standard deviation 1.76)*

n	$z_{\alpha/2}$	σ	$E = z_{\alpha/2}\dfrac{\sigma}{\sqrt{n}}$
36	1.96	1.76	.57
64	1.96	1.76	.44
100	1.96	1.76	.34
144	1.96	1.76	.29
196	1.96	1.76	.25
225	1.96	1.76	.23
400	1.96	1.76	.17
900	1.96	1.76	.11
2500	1.96	1.76	.07

A glance at the error formula above reveals the following interesting fact: as n increases, E decreases. In other words, the more data points we use in our sample, the smaller the probable bound on our error of estimation will be. Assuming $\sigma = 1.76$, we have already seen that, for n = 64, we get E = .44. If n = 144, we will have $E = z_{\alpha/2}\,\sigma/\sqrt{n} = (1.96)(1.76)/\sqrt{144} = .29$, and if we use a sample of n = 225 data points, our bound will be $E = (1.96)(1.76)/\sqrt{225} = .23$. Table 5.10 contains a list of sample sizes together with their corresponding probable bounds on error of estimation, so that we can trace a typical relationship between E and n.

Because of the relationship between sample size and probable bound on error of estimation, as expressed by formula (5.5), we can even do better than knowing how accurate our estimates are. We can decide *in advance* how accurate we want to be and then the error formula will tell us how large a sample we need in order to achieve the desired level of accuracy.

EXAMPLE 5.8.   Soft-Drink Machines.   Owners of a chain of coin-operated beverage machines advertise that their machines dispense, on the average, 7 ounces of soft-drink per cup. A consumer testing organization, which is interested in the question of whether a particular machine is meeting its specifications, wants to run a study so that they can be 95% sure of having the true mean amount dispensed by the machine correctly estimated to within an error of .01 ounce (one-hundredth of an ounce). If a preliminary analysis indicates that the amounts dispensed have standard deviation .12 ounce (twelve-hundredths of an ounce), how many sample cupfuls does the organization need to measure in order to achieve the desired level of accuracy?

SOLUTION.   The unknown here is n, the number of data points required for the analysis, in this case the number of sample cupfuls needed. Ordinarily, n is known, along with the standard deviation and α, and it is desired to compute the error. Here, however, the desired error level, $E = .01$, is given, and we want to know instead the number n of data points required. The degree of confidence is 95% so that α $= .05$ and $α/2 = .025$. Therefore $z_{α/2} = z_{.025} = 1.96$. Finally, in place of the sample standard deviation (which is unavailable, for we have not yet taken the actual sample), we use the preliminary estimate that σ $= .12$. Inserting these numbers into their proper places in the error formula (5.5), we have

$$.01 = (1.96)\frac{0.12}{\sqrt{n}} .$$

Successive steps of algebraic operations yield

$$(.01)\sqrt{n} = (1.96)(0.12)$$

$$\sqrt{n} = \frac{(1.96)(0.12)}{.01}$$

$$\sqrt{n} = 23.52 .$$

It then follows that $n = (23.52)' = 553.10$. To achieve the desired level of accuracy, then, would require at least 553.19 data points. Therefore, we would need 554 sample cupfuls for the survey. (Notice that even though 553.19 rounds down to 553, the latter number would leave us just short of the desired accuracy. That 554th data points puts us over the top.)

Using the error formula, we can derive an explicit formula for the required sample size n. Multiplying formula (5.5) through by $\sqrt{n}$ gives us

$$E\sqrt{n} = z_{α/2}\, σ$$

and dividing by E yields

$$\sqrt{n} = \frac{z_{α/2}\, σ}{E} .$$

If we square both sides, we get the *sample size formula* for estimating means,

$$n = \left(\frac{z_{α/2}\, σ}{E}\right)^2 . \tag{5.10}$$

In Example 5.8 we have $z_{\alpha/2} = z_{.025} = 1.96$, $\sigma = .12$, and $E = .01$, so we would have

$$n = \left(\frac{(1.96)(.12)}{.01}\right)^2 = (23.52)^2 = 553.19 \, ,$$

using formula (5.10). This is, of course, the same result we obtained earlier.

We can also use the confidence interval formula (5.9) for proportions in order to determine how many data points are needed to achieve a desired level of accuracy. By analogy with the technique developed for means, we start with the error formula

$$E = z_{\alpha/2} \sqrt{\frac{\pi(1-\pi)}{n}}$$

and solve the error formula for n, to get

$$n = \left(\frac{z_{\alpha/2} \sqrt{\pi(1-\pi)}}{E}\right)^2$$

As we have not yet taken the data (because we are trying to figure out how many data points to use), we do not know what the value of π will be, and so we really cannot use the above formula for n. However, it is a fact of arithmetic that $\pi(1-\pi)$ will *always* be less than 0.25, no matter what number (between 0 and 1, of course) π turns out to be. (Try a few numerical values of π to confirm this fact.) Therefore $\sqrt{\pi(1-\pi)}$ will always be equal to or less than $\sqrt{.25} = 0.5$, so that, no matter what the eventual value of π turns out to be, it will always be sufficient to use

$$n = \left(\frac{(z_{\alpha/2})(0.5)}{E}\right)^2 \qquad\qquad \textbf{(5.11)}$$

data points. Because 0.5 is the *maximum possible* value of $\sqrt{\pi(1-\pi)}$, the number of data points determined using formula (5.11) will generally be quite a bit larger than the minimally required sample size. For example, if we are sure that π cannot exceed 0.10, then $\sqrt{\pi(1-\pi)}$ cannot exceed $\sqrt{(.10)(.90)} = \sqrt{.09} = 0.3$. It will therefore be sufficient to use a sample of size n calculated by replacing the 0.5 in formula (5.11) by 0.3.

EXAMPLE 5.9.   An Auditing Procedure.   (This example is related to Example 4.3.) A public accounting firm, conducting a formal audit, wants to estimate the accuracy of a company's records. If they want to be 98% sure of having the true proportion of individual accounts accurately recorded on the company's books correctly estimated to within .03, how many accounts do they have to examine in detail?

SOLUTION. Here $\alpha = .02$ so that $z_{\alpha/2} = z_{.01} = 2.33$. Furthermore, the desired level of accuracy is given by $E = .03$. Using formula (5.11), we see that a sufficient number of data points for our study would be

$$n = \left(\frac{(z_{\alpha/2})(0.5)}{E}\right)^2 = \left(\frac{(2.33)(0.5)}{.03}\right)^2 = (38.83)^2$$

$$= 1,508.$$

Therefore, a sample of 1,508 accounts examined in detail will be sufficient to guarantee the level of accuracy desired in the audit. On the basis of a sample of that size we can be 98% sure of having the true proportion of accurate accounts correctly estimated to within .03.

As an aside to the conclusion of the above example, it is interesting to note that the Gallup Poll, an organization that undertakes to predict the outcome of elections, uses a sample of 1,500 voters in its surveys.

If we are sure that the true proportion of accounts accurately recorded is at least $\pi = .90$, then the 0.5 in formula (5.11) can be replaced by $\sqrt{\pi(1-\pi)} = \sqrt{(.90)(.10)} = \sqrt{.09} = 0.3$. We could then get away with

$$n = \left(\frac{(2.33)(0.3)}{.03}\right)^2 = 543$$

data points, a considerable saving in cost and time.

EXERCISES 5.F

5.49. A personnel officer of the Beach City Bank & Trust Company of Newport Beach, which has 55 branch offices up and down the state, would like to be 99% sure of having the true mean number of days absent per year per employee correctly estimated to within 0.5 days. A preliminary sampling indicates that the standard deviation of the number of absences is 2.3 days. How many employees' attendance records are needed as a random sample in the personnel officer's study?

5.50. Because weather conditions exert a strong influence on the volume of beer sold locally, the distributor of Bucko Beer would like to be 90% sure of having the true mean daily precipitation in her region measured accurately to within .05 inch. If meteorological history indicates that the standard deviation can reasonably be assumed to be .30 inch, how many observations of daily precipitation are required?

5.51. To determine the optimal length of time to conduct a training program for its employees in the details of a new accounting system recently adopted by the home office, the Pacific Northwest Division of Amalgamated Veneer & Plywood wants to estimate the mean number of half-hour sessions required for its employees to obtain a full working knowledge of the system. Experience with similar training programs suggests that, overall, employees will almost certainly have a standard deviation of 9.2 sessions in this regard. If the company would

like to be 95% sure of having the true mean number of sessions correctly estimated to within two sessions, how many employees are needed for the pilot
project?

5.52. At a General Motors plant in Michigan there are 5,000 assembly line workers.
A financial analyst wants to be 98% sure of having the true mean weekly salary
of the typical assembly line worker to within $10. A pilot study indicated a
standard deviation of $35 for the variable under study. On how many workers
are data needed in order to achieve the desired level of accuracy?

5.53. The Eastingfield Electric Company, which produces light bulbs, is required by
law to estimate statistically the true mean lifetime of its bulbs. In particular, it
has to be 95% sure of having the true mean estimated accurately to within 25
hours. If the lifetimes have standard deviation 200 hours, how many bulbs must
be tested in order to achieve the desired accuracy in estimating the true mean?

5.54. Recall the marketing analyst studying McDonald's in Exercise 5.45. How many
residents must the analyst poll in order to be 95% sure of having the true
percentage estimated correctly to within 1.5%?

5.55. In the insurance situation of Exercise 5.42, records on how many men are
needed in order to be 99% sure of having the true percentage correctly estimated
to within 1%?

5.56. If the auditors of Exercise 5.41 wants to be 95% sure of having the true proportion of accurate accounts correctly estimated to within .02, how many verification responses are required?

5.57. Amalgamated Educators, Inc. (AEI) has been hired by the Board of Education's
contract learning office to develop a pilot program of raising reading scores of
sixth-grade pupils. Preliminary analysis has indicated to AEI researchers that
the true standard deviation of reading scores of such pupils is 22 points. If AEI
has to be 94% sure of having the true mean score of all pupils eventually taking
its program correctly estimated to within five points, how many pupils does it
need for its pilot project?

5.58. If $1.74 is a reasonable estimate for the standard deviation of NYSE corporate
earnings per share, how many stocks do we need to survey in order to be 99%
sure of having the true mean earnings per share correctly estimated to within
20¢ ($0.20)?

5.59. In order to accurately forecast demand for its services, Chick's Quit-Smoking
Health Spa needs to estimate the proportion of smokers who sincerely want to
break the habit. In particular, Chick wants to be 92.5% sure of having that
proportion correctly estimated to within .05. How many smokers does Chick
need for his survey?

5.60. Suppose 11.1 is a good estimate for the standard deviation of the number of
employees (in thousands) of the U.S.'s largest corporations. On how many corporations do we need employment data in order to be 90% sure of having the
true mean number of employees correctly estimated to within two thousand?

5.61. One particular automobile manufacturer wants to estimate the repair and maintenance cost over the first five years to owners of its cars. It is known that the

standard deviation of the five-year cost figure is $400, due to differences in interval vehicle components and personal driving habits. How many automobile records have to be analyzed in order to be 95% sure of having the true five-year cost figure correctly estimated to within $50?

SECTION 5.G   Control Charts

A machine that puts out plastic pipe is, according to contractual specification, supposed to be producing pipe 4 inches in diameter. Pipe that is either too large or too small will not meet the construction requirements. In order to insure that the machine is operating properly (so that the pipe it is producing falls within the contractual specifications), its output must be carefully monitored. This procedure of monitoring the output of an industrial process or machine is called *quality control*. When the output fails to meet specifications, the process or machine is said to be *out of control*.

Because the machine's output can be viewed statistically as a set of data that is generated by successive repetition of the same physical process (and therefore fluctuating about the specified diameter of 4 inches), numerical characteristics of the output (in this case the diameters of the pieces of pipe) are normally distributed. If the set of diameters has true mean μ and true standard deviation σ, the control limit theorem of Section 5.B asserts that the set of *all* possible sample means $\bar{x}$ of size n is *approximately* normally distributed with mean μ and standard deviation $\sigma/\sqrt{n}$ if n is large. However, if the original set of data (namely, the diameters of the pieces of pipe) is itself normally distributed, then it turns out that the set of all possible sample means $\bar{x}$ of size n is *exactly* normally distributed with mean μ and standard deviation $\sigma/\sqrt{n}$, *for any value of n, large or small*. This means that, if we take a sample of n pieces of pipe, no matter what the numerical value of n is, we can be $(1 - \alpha)100\%$ sure that

$$\mu - z_{\alpha/2}\frac{\sigma}{\sqrt{n}} < \bar{x} < \mu + z_{\alpha/2}\frac{\sigma}{\sqrt{n}}.^*$$

In particular, when $z_{\alpha/2} = 3$, a glance at Table A.4 of the Appendix shows that $1 - \alpha/2 = .9987$, so that $\alpha/2 = 1 - .9987 = .0013$ and $\alpha = 2(\alpha/2) = 2(.0013) = .0026$. Therefore, we can be $(1-\alpha)100\% = (1-.0026)100\% = (.9974)100\% = 99.74\%$ sure that

$$\mu - 3\frac{\sigma}{\sqrt{n}} < \bar{x} < \mu + 3\frac{\sigma}{\sqrt{n}} \tag{5.12}$$

---

*By analogy with the confidence intervals of Section 5.B: this statement is an algebraic transformation of formula (5.4).

where

μ = true mean diameter of pipe being produced
σ = true standard deviation of pipe

Formula (5.12) provides the theoretical basis of the quality control monitoring technique known as the *control chart*.

Suppose that the pipe-producing machine *is operating according to specifications* (in particular, μ = 4) and that we know from experience or prior testing that σ = .12. Then, if we look at a sample of five pieces of pipe as they come out of the machine, we can be 99.74% sure that

$$4 - 3\left(\frac{.12}{\sqrt{5}}\right) < \overline{x} < 4 + 3\left(\frac{.12}{\sqrt{5}}\right),$$

where $\overline{x}$ = sample mean diameter of the five pieces. Making the calculations, we see that the chances are 99.74% that

$$4 - \frac{.36}{2.24} < \overline{x} < 4 + \frac{.36}{2.24}$$

$$4 - .16 < \overline{x} < 4 + .16$$

$$3.84 < \overline{x} < 4.16 .$$

This means that it would be extremely rare (if the machine were really operating according to specifications) for us to get a sample of five pieces of pipe having a mean diameter outside the interval 3.84 to 4.16. This is so rare (the chances are only 26/100 of 1%), in fact, that we ordinarily consider it a signal that the machine is *not* operating according to specifications. For example, if a sequence of 5 pieces of pipe coming off the machine were to have a sample mean $\overline{x}$ = 3.77, we would strongly suspect that the production process is out of control; we should stop the process and bring in a technician for a careful inspection and adjustment of the machine.

To monitor the process, control charts of the sort illustrated in Fig. 5.13 are often used, where dots representing the successive sample means are plotted on a graph.* As long as the dots fall in the shaded (under control) region, no problems are indicated. However, when $\overline{x}$ = 3.77 (the sample mean of the fifth sample taken) falls outside the shaded region, the process is considered to be out of control. The control chart of Fig. 5.13 was constructed from the data of Table 5.11.

---

*In the most modern factories, pipe coming off the machine would be measured by a computer programmed to print out the control chart and automatically shut off the machine at the predetermined out-of-control level.

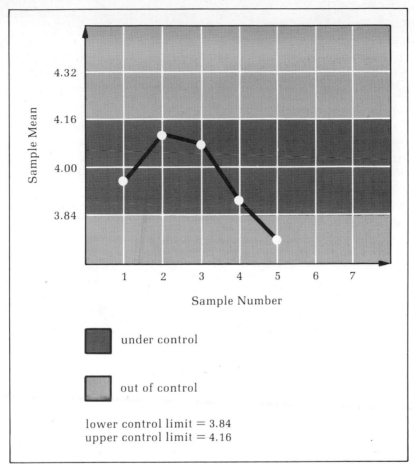

FIGURE 5.13   Control Chart for Pipe Production
(Specification: diameter = 4 inches)

TABLE 5.11   Diameters of Successive Pieces of Pipe   *(In inches, measured as they come off the machine)*

Sample	No. 1	No. 2	No. 3	No. 4	No. 5
	4.01	3.95	4.13	3.80	4.03
	3.90	4.21	3.97	3.91	3.94
	4.02	4.15	4.05	3.80	3.86
	3.92	4.05	4.11	4.07	3.72
	3.95	4.19	3.99	3.87	3.30
Sums	19.80	20.55	20.25	19.45	18.85
Sample Means	3.96	4.11	4.05	3.89	3.77

TABLE 5.12   Monthly Volume of Bad Loans   *(In millions of dollars)*

Month	Volume	Month	Volume	Month	Volume
Jan 1977	7.0	Oct 1977	6.5	Jul 1978	7.2
Feb 1977	7.3	Nov 1977	7.3	Aug 1978	7.1
Mar 1977	7.6	Dec 1977	7.5	Sep 1978	7.3
Apr 1977	7.8	Jan 1978	8.1	Oct 1978	7.1
May 1977	7.4	Feb 1978	8.4	Nov 1978	7.0
Jun 1977	8.1	Mar 1978	8.6	Dec 1978	6.8
Jul 1977	8.2	Apr 1978	8.4	Jan 1979	7.5
Aug 1977	8.3	May 1978	7.7	Feb 1979	7.6
Sep 1977	7.8	Jun 1978	7.4	Mar 1979	7.5

EXAMPLE 5.10.   Bank Defaults.   The Federal Deposit Insurance Corporation (FDIC) supposedly maintains a "watch list" of banks that may be coming dangerously close to insolvency. One of the items the FDIC keeps track of is the average quarterly volume of probably bad loans a bank is maintaining on its list of assets. A reasonable quarterly average for the Bank of Huntington Harbour is 7.2 million dollars with a standard deviation of 0.5 million. Based on the monthly data of Table 5.12, construct a control chart for measuring the bank's potential for insolvency.

SOLUTION.   We use the monthly data in Table 5.12 to construct a table of each quarter's sample mean bad loan volume. We do this in Table 5.13. We determine the upper and lower control limits from formula (5.12) with $\mu = 7.2$ (the reasonable quarterly average), $\sigma = 0.5$, and $n = 3$ (since the sample means are determined by averaging three monthly data points). Therefore the upper control limit is

$$\mu + 3\frac{\sigma}{\sqrt{n}} = 7.2 + 3\left(\frac{.5}{\sqrt{3}}\right) = 7.2 + 0.87 = 8.07$$

and the lower control limit is

$$\mu - 3\frac{\sigma}{\sqrt{n}} = 7.2 - 3\left(\frac{.5}{\sqrt{3}}\right) = 7.2 - 0.87 = 6.33 \ .$$

The control chart appears in Fig. 5.14, and shows that the bank experienced potential for insolvency in quarters No. 3 and No. 5.

TABLE 5.13   Quarterly Sample Means of Bad Loan Volume

Quarter No.	Month	Volume	Quarterly Total	Quarterly Sample Mean
	Jan 1977	7.0		
1	Feb 1977	7.3	21.9	7.30
	Mar 1977	7.6		
	Apr 1977	7.8		
2	May 1977	7.4	23.3	7.77
	Jun 1977	8.1		
	Jul 1977	8.2		
3	Aug 1977	8.3	24.3	8.10
	Sep 1977	7.8		
	Oct 1977	6.5		
4	Nov 1977	7.3	21.3	7.10
	Dec 1977	7.5		
	Jan 1978	8.1		
5	Feb 1978	8.4	25.1	8.37
	Mar 1978	8.6		
	Apr 1978	8.4		
6	May 1978	7.7	23.5	7.83
	Jun 1978	7.4		
	Jul 1978	7.2		
7	Aug 1978	7.1	21.6	7.20
	Sep 1978	7.3		
	Oct 1978	7.1		
8	Nov 1978	7.0	20.9	6.97
	Dec 1978	6.8		
	Jan 1979	7.5		
9	Feb 1979	7.6	22.6	7.53
	Mar 1979	7.5		

EXERCISES   5.G

5.62. Daily closing prices of shares of common stock of the United Garbage Haulers (UGH) Corporation have been very stable over the past several years, being normally distributed with a mean of 16.5 and a standard deviation of 2.5. The apparent stability of the stock has led one investor to sell UGH whenever its weekly (five-day) average closing price exceeds the upper control limit, because he feels that it is overvalued at such a time. On the other hand, he buys the stock heavily whenever its five-day average falls below the lower control limit.

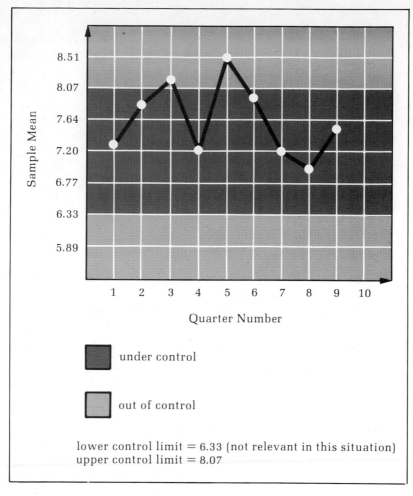

FIGURE 5.14   Control Chart for Bad Loan Volume
(Reasonable quarterly average: 7.2 million dollars)

a. Calculate the upper and lower control limits for the five-day average of closing prices of UGH stock.

b. The following data show the closing prices of UGH stock over a 30-day period:

17.375	14.250	16.500
18.125	13.875	15.250
19.875	12.625	16.375
19.625	13.000	15.875
18.750	12.750	16.000
17.875	13.375	17.125
16.625	14.125	19.875
16.125	15.000	20.125
15.500	15.875	20.500
14.750	16.125	21.875

Construct the control chart for the five-day average of UGH stock, and determine when the investor should buy and when he should sell the stock.

5.63. Chic Engine Bolts, Inc. produces 16 cartons of 1,728 engine bolts each month to fit the Peccarie automobile. As the Peccarie assembly plant uses an effective acceptance sampling scheme, Chic must make certain, before the bolts leave their factory, that all or most of them meet the Peccarie specifications, namely a mean length of 3.15 cm. with a standard deviation of .03 cm. The quality control procedure Chic uses is to calculate the sample mean of every group of 25 bolts as they come down the line to the testers, and then to plot these sample means on a control chart.

a. Determine the upper and lower control limits for the sample means of 25 bolts.

b. The following sample means were recorded for 20 groups of 25 bolts each:

Group No.	Sample Mean (cm.)
1	3.156
2	3.161
3	3.164
4	3.159
5	3.151
6	3.147
7	3.144
8	3.138
9	3.135
10	3.131
11	3.130
12	3.135
13	3.141
14	3.148
15	3.153
16	3.156
17	3.158
18	3.155
19	3.152
20	3.149

Draw the control chart, and note the times that the production process was out of control.

5.64. An analyst in the county property-tax assessor's office is studying the assessed values of single-family homes to determine which ones, if any, are underassessed by the office and which are overassessed. One particular long block consists of 80 similar homes whose assessed values should be normally distributed with a mean of $45,000 and a standard deviation of $2,000. The analyst calculated the following averages of values of groups of eight consecutive address numbers:

Group	Average Value
A	43,747
B	42,556
C	44,954
D	45,632
E	46,184
F	46,996
G	47,429
H	47,105
I	46,650
J	44,625

Construct the control chart and find out whether or not over- or underevaluation is occurring.

5.65. So far this section has dealt only with control charts for sample means. It is also often useful to be able to construct control charts for sample proportions. The upper and lower control limits in this kind of situation are based on the confidence interval formula (5.9) of Section 5.E. If the ideal proportion to be controlled is denoted as p, then the upper control limit for successive samples of size n is

$$\text{UCL} = p + 3\sqrt{\frac{p(1-p)}{n}}$$

while the lower control limit is

$$\text{LCL} = p - 3\sqrt{\frac{p(1-p)}{n}}.$$

Let's consider the following alternative quality control procedure for the Chic Engine Bolts of Exercise 5.63. The production process is to be considered out of control if the proportion of defective bolts produced exceeds .06, namely 6%. Samples of 400 bolts each are to be inspected and the proportion of defective bolts in each such sample is to be noted.

a. Determine the upper and lower control limits for the proportion of defective bolts in the samples of size 400.

b. Twelve consecutive samples of 400 bolts each have the following proportions of defectives. Draw the control chart, and determine when, if at all, the production process was out of control.

Sample No.	Proportion
1	.0735
2	.0882
3	.0890
4	.0950
5	.0975
6	.0903
7	.0865
8	.0824
9	.0835
10	.0750
11	.0770
12	.0738

5.66. An auditor verifying the outstanding credit card balances due by customers of Nickle's Department Stores, Inc., considers the overall credit system to be under control if the proportion of accounts containing minor errors is $p = .10$.

a. Determine the upper and lower control limits if the auditor uses samples of size $n = 120$ from the alphabetized list of Nickle's customers.

b. Twenty consecutive samples of 120 accounts each had the following proportions containing minor errors:

Sample No.	Proportion	Sample No.	Proportion
1	.0871	11	.1063
2	.0850	12	.1144
3	.0895	13	.1186
4	.0964	14	.1234
5	.0997	15	.1289
6	.1078	16	.1305
7	.1056	17	.1255
8	.1031	18	.1207
9	.1054	19	.1186
10	.1028	20	.1122

Construct the control chart and find out when the credit system is out of control.

SECTION 5.H   RANDOM SAMPLING

At the foundation of all the techniques of statistical inference, in particular the estimation methods we discussed in this chapter, lies the concept of a *random sample*. As we mentioned in Section 5.A, a random sample is a sample chosen according to the following principle: each possible combination of n elements of the population, whether it be a population of grapefruits, banking corporations, bolt diameters, etc., has an equal chance to be selected as the sample. Whole books, even several-volume series, have been written on the subject of how to guarantee that the sample we select is truly representative of the population. (Some of these books are listed in the bibliography at the end of this chapter.) Our objective in this section is to present one method, the method of random numbers, which can often be used to attain a degree of randomness sufficient for management decision making.

Random Numbers

Consider the population consisting of the top 160 stock market performers, ranked according to their 5-year price changes over 1973–77, as reported in the *Forbes'* article, "Who's Where in the Stock Market," in the issue of January 9, 1978. This information is reproduced in the three pages of Table 5.14. Suppose we would like to estimate μ, the mean number of shares outstanding for 1977 of the top 160 companies, and we want to do it with a random sample of size 16. Let's now find out how to choose a random sample of 16 of the 160 banking corporations.

We first turn to Table A.9 of the Appendix, the Table of Random Digits (also called a table of random numbers). This table consists of 10 columns and 50 rows of 5-digit numbers, generated in such a way as to imitate the ideal concept of randomness as closely as is humanly possible. In order to use this table to impose a measure of randomness on our real-life situation, we let the table direct us to the banking corporations that should comprise our random sample.

Next we must pick a starting point within the table of random digits. It does not really matter how we do this, for no matter where we start the internal structure of the table itself will guarantee that successive digits appear in a random manner. So suppose we start with the number 37262 at the intersection of the seventh column and the twenty-second row. We reproduce in Table 5.15 the portion of Table A.9 below and to the right of the number 37262.

Now, because we are choosing a random sample from a population consisting of 160 ranked members, we need only three of the five digits of 37262. (This is true because the highest-ranking member of the population can be described by a three-digit number.) It really doesn't matter which three we take, so let's look at 372. We now record in Table 5.16 the three-digit random numbers of Table 5.15 below 372 and in the corresponding locations of the

TABLE 5.14 Who's Where in the Stock Market
(Reprinted by permission of FORBES *Magazine* from the January 9, 1978 issue.)

Who's Where in the Stock Market

Five-Year Change in Consumer Price Index: 44.9%

Rank '77	'76	Company	5-Year Price Change	1973-77 Price Range	Recent Price	Shares Outstanding (millions)	Total Market Value (millions)	Latest 12-Month Earnings Per Share	Price/ Earnings Ratio	Indicated Annual Dividend	Current Yield
1	*	Utd Energy Resources	425.9%‡‡	35½– 3	35½	1.8	$ 419	$6.12	6	$1.72	4.8%
2	1	Moore McCormack Res	422.9	38¼– 3½	31⅜	.7	179	5.24	6	0.90	2.9
3	5	Westmoreland Coal	380.6	65¼– 7⅞	41¾	5.8	285	3.43	12	1.60	3.8
4	6	E-Systems	320.5	33 – 4⅜	26⅞	3.7	99	4.75	6	1.20	4.5
5	8	Earth Resources	300.0	21¾– 4½	18½	5.1	95	2.70	7	1.00	5.4
6	91	Pneumo Corp	284.9	19¾– 2½	15⅞	2.5	39	3.45	5	1.00	6.3
7	14	Teledyne	266.1	78⅞– 6⅞	64¼	1.8	757	13.66	5	none	0.0
8	*	Wyman-Gordon	262.9	34 – 7½	31¾	4.3	136	4.66	7	1.20	3.8
9	*	Service Merchandise	258.7	18⅝– ¾	15½	8.2	127	1.66	9	0.10	0.6
10	64	Raymond Intl	255.4	28⅜– 4⅛	24⅞	4.3	107	4.55	5	1.00	4.0
11	323	Great Western United	242.3	46⅝– 1½	22¼	2.1	47	2.62	8	none	0.0
12	74	Dean Foods	237.5	27 – 5⅜	22½	2.0	45	3.54	6	0.80	3.6
13	18	Athlone Industries	231.1	16⅜– 3⅜	14⅝	3.6	52	2.70	5	0.80	5.5
14	480	Amtel	227.5	17 – 4½	16⅜	4.7	77	1.24	13	0.44	2.7
15	7	Archer Daniels Midland	217.4	29⅝– 6⅛	20½	31.3	641	1.92	11	0.20	1.0
16	37	A E Staley Mfg	212.1	32½– 5¼	22⅝	11.1	251	2.20	10	1.00	4.4
17	156	Foster Wheeler	204.9	32⅜– 6¼	30⅜	8.2	246	3.02	10	1.00	3.3
18	427	Inmont	203.9	29¾– 4⅞	29⅝	7.9	234	3.10	10	1.00	3.4
19	60	MGM	202.4	23⅞– 2⅜	23¼	14.2	331	2.23	10	1.00	4.3
20	40	Crane	190.2	39 – 6⅝	28¾	10.2	288	6.64	4	1.40	5.0
21	13	Diamond Shamrock	181.0	40 – 8½	29½	36.7	1,082	4.16	7	1.40	4.7
22	26	Koppers	179.4	30¼– 7¾	24⅝	24.9	612	2.34	11	1.10	4.5
23	94	American Standard	175.5	36½– 7¾	36½	12.4	454	5.15	7	2.00	5.5
24	54	Harnischfeger	175.0	24⅝– 5⅝	17⅞	8.8	157	2.49	7	1.00	5.6
25	29	Amsted	174.3	59⅝– 16	57¼	5.4	308	6.62	9	2.60	4.5
26	*	Northwest Energy	173.2‡‡	43⅞– 11⅞	36⅞	4.3	158	5.49	7	2.20	6.0
27	16	Northrop	172.4	25 – 4¾	21	13.8	291	4.68	4	1.20	5.7
28	3	NVF	163.9	8¼– 1⅜	5⅜	11.0	59	1.20	4	none	0.0
29	12	McKee	159.6	20⅝– 4¾	16⅞	3.2	54	2.82	6	1.00	5.9
30	*	Consolidated Papers	158.6	34¼– 11	32	5.2	166	4.47	7	2.00	6.3

* Not ranked last year. ‡‡ Three-year price change.

(continued)

TABLE 5.14 (continued)

Rank '77	Rank '76	Company	5-Year Price Change	1973-77 Price Range	Recent Price	Shares Outstanding (millions)	Total Market Value (millions)	Latest 12-Month Earnings Per Share	Price/ Earnings Ratio	Indicated Annual Dividend	Current Yield
31	50	**Missouri Pacific Corp**	**156.8%**	50⅞– 14	47½	13.4	**$638**	$8.30	6	$2.10	4.4%
32	125	**Boise Cascade**	**151.7**	33¾– 8¼	28	27.0	**755**	3.90	7	1.10	3.9
33	62	**Cone Mills**	**151.6**	30¾– 7⅜	30½	5.8	**176**	6.35	5	1.60	5.2
34	*	**Parker Pen**	**150.2**	20⅜– 4¾	20⅜	7.9	**160**	2.01	10	0.48	2.4
35	11	**Cooper Industries**	**147.3**	49¼– 10¼	45¾	11.2	**511**	4.59	10	1.08	2.4
36	151	**Super Valu Stores**	**145.8**	29⅝– 5⅞	29½	8.9	**264**	2.94	10	0.86	2.9
37	10	**St Joe Minerals**	**141.6**	50 – 12¼	33⅜	22.3	**745**	3.24	10	1.30	3.9
38	*	**Chic & North Western**	**141.2‡‡**	13½– 3½	10¼	4.4	**45**	3.99	3	none	0.0
39	100	**Super Food Services**	**140.0**	16⅝– 3¾	13⅞	1.1	**15**	2.15	6	0.40	2.9
40	282	**Globe-Union**	**139.9**	25½– 3⅞	24¾	6.2	**153**	3.64	7	1.00	4.1
41	9	**Cameron Iron Works**	**138.5**	49 – 10½	31	9.5	**295**	3.04	10	0.40	1.3
42	51	**UV Industries**	**138.3**	24¼– 7	21¼	8.6	**184**	3.99	5	1.00	4.7
43	172	**Belco Petroleum**	**138.1**	34⅜– 8⅝	32⅞	7.6	**248**	5.46	6	1.00	3.0
44	22	**Potlatch**	**136.9**	39⅝– 9¾	30½	15.1	**461**	4.07	7	1.00	3.3
45	*	**Olin Corp**	**134.2‡‡**	22⅝– 6½	17⅛	24.1	**413**	3.23	5	0.88	5.1
46	365	**Ingredient Technology**	**134.2**	12⅝– 3⅜	12	1.8	**22**	2.94	4	0.50	4.2
47	44	**Boeing**	**134.0**	30 – 5⅞	29¼	42.6	**1,245**	3.82	8	1.00	3.4
48	197	**Chemetron**	**129.6**	52 – 13½	48½	4.0	**195**	2.16	22	1.50	3.1
49	19	**Colt Industries**	**127.1**	59¼– 13	48¼	7.8	**376**	7.47	6	2.75	5.7
50	34	**Bucyrus-Erie**	**125.5**	30¼– 9	22⅝	20.3	**460**	2.55	9	0.80	3.5
51	71	**Niagara Frontier Svc**	**121.9**	19 – 5	17¾	2.2	**39**	3.46	5	0.70	3.9
52	216	**Kaiser Steel**	**120.7**	47½– 8	24	7.0	**168**	4.41	5	1.50	6.3
53	33	**MCA**	**120.7**	44¼– 8⅜	35¾	18.6	**663**	4.53	8	1.00	2.8
54	130	**Midland-Ross**	**116.4**	33½– 8⅝	33	5.4	**177**	4.21	8	1.80	5.5
55	79	**Allis-Chalmers**	**115.3**	33¾– 6⅛	26⅜	10.7	**282**	5.34	5	1.30	4.9
56	112	**Occidental Petroleum**	**114.9**	31⅛– 7⅜	25¼	66.7	**1,684**	3.05	8	1.25	5.0
57	363	**Alco Standard**	**114.1**	23⅞– 6¼	22¾	10.1	**229**	4.14	5	1.01	4.4
58	17	**Intl Minerals & Chem**	**111.8**	48½– 15¼	41½	17.9	**745**	6.41	6	2.60	6.3
59	137	**Dravo Corp**	**111.7**	33 – 12½	29⅜	5.0	**146**	3.10	9	1.05	3.6
60	105	**Pittston**	**110.9**	47 – 9⅛	24	37.7	**906**	2.88	8	1.45	6.0
61	472	**Twentieth Century-Fox**	**109.8**	25⅞– 4½	24⅛	7.7	**185**	5.90	4	0.70	2.9
62	452	**Babcock & Wilcox**	**109.3**	60⅝– 11½	56½	12.2	**688**	4.42	13	1.50	2.7
63	529	**National Industries**	**108.6**	11⅝– 2½	9⅛	5.8	**53**	1.59	6	0.40	4.4
64	138	**Raytheon**	**108.2**	35 – 10	35	30.8	**1,079**	3.37	10	1.00	2.9
65	70	**H K Porter**	**108.0**	60 – 24	52	1.2	**62**	9.20	6	2.00	3.8
66	*	**Big Three Industries**	**107.8**	39¾– 10½	34	20.0	**681**	2.33	15	0.40	1.2
67	*	**Mervyn's**	**105.1**	40½– 5½	40½	4.7	**191**	3.37	12	none	0.0
68	447	**SCOA Industries**	**104.2**	18⅞– 4¼	18⅜	3.4	**62**	3.02	6	1.00	5.4
69	2	**MAPCO**	**104.1**	50⅛– 11¼	37¼	18.2	**677**	3.32	11	1.10	3.0
70	90	**Santa Fe Intl**	**102.8**	28 – 6½	26¾	20.8	**556**	4.11	7	0.60	2.2
71	53	**Univar**	**102.8**	15¾– 3⅜	9⅛	6.7	**61**	–0.19	def	0.56	6.1
72	202	**Mead**	**102.5**	24 – 7⅞	20¼	22.4	**453**	4.03	5	1.00	4.9
73	217	**Airco**	**100.7**	35⅞– 10	35⅛	11.8	**414**	4.80	7	1.35	3.8
74	742	**Columbia Pictures**	**100.0**	20½– 1⅝	19	8.4	**159**	2.39	8	none	0.0
74	35	**General Dynamics**	**100.0**	65 – 13½	50¾	10.9	**551**	9.40	5	none	0.0

(continued)

76	96	Eastern Gas & Fuel	97.2	29⅛	- 5⅝	19¼	21.8	420	1.98	10	0.80	4.2
77	45	Entex	96.7	26	- 5⅝	25	8.?	205	3.68	7	1.20	4.8
78	*	North American Coal	95.9	29	- 7½	23⅞	3.?	80	2.12	11	0.50	2.1
79	106	Federal Co	95.2	32½	- 9	30½	2.0?	88	2.36	13	1.80	5.9
80	707	Memorex	93.9	34⅞	- 1⅜	31¾	5.?	167	5.08	6	none	C.0
81	95	Nash Finch	93.8	31	- 12	31	1.?	53	5.19	6	1.85	6.0
82	43	Dresser Industries	93.6	47⅛	- 16½	45¼	39.?	1,765	4.56	10	0.88	1.9
83	15	Houston Natural Gas	93.5	36⅝	- 7⅝	29⅝	39.?	1,160	3.33	9	0.80	2.7
84	84	Handy & Harman	92.4	23⅜	- 9¾	22⅝	3.?	75	3.25	7	1.00	4.5
85	66	Reynolds Metals	92.2	44⅞	- 12	30¾	18.3	577	4.46	7	1.50	4.9
86	214	Joy Manufacturing	91.5	50⅞	- 12⅜	33⅞	12.2	413	3.92	9	1.50	4.4
87	93	Northern Natural Gas	91.0	53	- 17	40	22.2	889	5.87	7	2.40	6.0
88	103	Universal Leaf	90.7	33⅜	- 10⅞	33⅜	4.7	157	4.18	8	1.88	5.6
89	*	Itel	90.7	21⅞	- 2½	20½	8.7	178	2.71	8	0.45	2.2
90	368	American Stores	90.1	36⅜	- 11	30¾	.3	160	4.49	7	2.10	6.9
91	321	Grumman	89.8	22½	- 7¼	17¼	.9	136	3.78	5	1.00	5.8
92	357	Harris Corp	89.0	44½	- 6⅝	42⅞	12.3	526	3.56	12	1.00	2.3
93	136	Tyler	87.7	24¾	- 5¾	22⅞	5.8	132	3.61	6	0.60	2.6
94	169	Skaggs Companies	87.1	28⅛	- 5⅝	26½	3.0	211	2.46	11	0.70	2.6
95	205	Northwest Industries	84.2	60⅞	- 14¾	56⅝	2.2	693	7.93	7	2.85	5.0
96	20	Schlumberger	84.0%	74	- 32½	74	5.9	$6,359	$4.34	17	$1.10	1.5%
97	880	Foodarama Supermkts	81.8	8	- 2	7½	1.4	10	1.87	4	0.40	5.3
98	204	Ogden	81.1	29¼	- 11½	25⅝	9.3	234	4.95	5	1.40	5.6
99	4	Natl Semiconductor	80.9	55⅝	- 6¾	20½	13.0	266	0.94	22	none	0.0
100	491	Lockheed	80.8	19⅛	- 2⅜	16½	11.4	187	3.71	4	none	0.0
101	*	Bruno's	80.8	12¾	- 2⅝	11¾	2.1	24	1.79	7	0.40	3.4
102	*	Olinkraft	80.2‡‡	39¾	- 10⅝	26⅝	8.8	231	3.80	7	1.20	4.6
103	113	Carpenter Technology	80.0	21¾	- 7⅞	21⅜	8.5	182	3.47	6	1.20	5.6
104	209	Lear Siegler	79.2	17⅜	- 3⅜	16⅛	12.3	199	2.91	6	0.60	3.7
105	126	Beech Aircraft	78.8	31¾	- 6⅜	28½	7.2	204	3.27	9	1.10	3.9
106	27	Air Prods & Chems	77.8	40⅝	- 15⅛	27⅞	28.2	780	2.40	12	0.40	1.4
107	25	Fluor	77.3	49	- 15	39⅝	15.5	605	4.37	9	1.00	2.6
108	314	Morrison-Knudsen	77.0	37⅞	- 10¾	33½	2.9	97	5.17	6	1.10	3.3
109	335	Carborundum	76.4	63¾	- 12⅜	61¾	8.0	491	4.56	14	1.24	2.0
110	110	Washington Post	76.3	33½	- 7⅞	33½	8.2	276	3.68	9	0.36	1.1
111	293	Carson Pirie Scott	75.4	25¼	- 6¾	25	2.3	58	4.59	5	1.30	5.2
112	92	Albertson's	75.3	27¼	- 10	26½	7.3	194	3.06	9	0.80	3.0
113	89	Butler Manufacturing	73.8	26¼	- 11⅛	23¾	5.8	137	3.30	7	0.80	3.4
114	*	DPF	72.7	9¾	- 2⅜	9½	4.2	40	1.23	8	none	0.0
115	691	Ruddick	71.4	7½	- 1⅜	7½	2.3	17	1.50	5	0.35	4.7
116	41	United Technologies	71.4	41¼	- 10⅜	38⅝	26.6	1,012	5.48	7	1.80	4.7
117	432	Interpace	71.3	22⅞	- 7	22⅞	3.7	85	3.17	7	1.22	5.3
118	753	Alterman Foods	70.4	18½	- 6	17¼	1.4	24	2.00	9	0.5?	2.9
119	83	Standard Oil Ohio	70.2	91¼	- 37⅝	78⅝	38.6	3,032	3.88	20	1.36	1.7
120	31	Inland Container	70.1	34	- 10½	25⅜	8.0	204	2.89	9	1.10	4.3

* Not ranked last year. ‡‡ Three-year price change. def Deficit.

TABLE 5.14 (continued)

Rank '77	Rank '76	Company	5-Year Price Change	1973-77 Price Range	Recent Price	Shares Outstanding (millions)	Total Market Value (millions)	Latest 12-Month Earnings Per Share	Price/ Earnings Ratio	Indicated Annual Dividend	Current Yield
121	199	**SmithKline Corp**	**69.6**	49½– 15⅛	49½	29.9	**1,481**	2.69	18	1.10	2.2
122	56	**Getty Oil**	**68.6**	212 – 92½	158½	20.5	**3,255**	14.78	11	2.80	1.8
123	46	**Reserve Oil & Gas**	**68.1**	21⅜– 3¾	14½	13.1	**190**	1.27	11	0.20	1.4
124	102	**Hughes Tool**	**67.7**	52½– 18⅜	35¾	10.8	**385**	3.34	11	0.70	2.0
125	393	**Dayton-Hudson**	**67.0**	44¼– 6½	44¼	15.9	**703**	4.60	10	1.40	3.2
126	127	**Kaiser Alum & Chem**	**66.9**	40⅛– 12	29⅝	19.7	**585**	3.89	8	1.40	4.7
126	21	**Pioneer Corp**	**66.9**	39¼– 11¼	30¼	9.3	**282**	4.13	7	1.50	5.0
128	183	**Signal Companies**	**66.5**	34⅝– 12⅜	32⅜	19.9	**643**	4.66	7	1.36	4.2
129	143	**Eagle-Picher Inds**	**66.0**	22 – 6	20⅜	9.9	**202**	2.47	8	0.68	3.3
130	63	**Digital Equipment**	**65.5**	60⅞– 15⅛	50⅝	39.3	**1,987**	3.01	17	none	0.0
131	194	**Harsco**	**65.4**	28⅜– 8¾	27½	9.7	**266**	4.36	6	1.60	5.8
132	98	**Central Steel & Wire**	**64.6**	137 – 76	130	0.3	**44**	34.78	4	7.00	5.4
133	637	**Revere Copper**	**63.9**	21¾– 5⅞	14¾	5.6	**83**	1.78	8	none	0.0
134	59	**Vulcan Materials**	**63.8**	29¼– 10¼	22⅝	11.4	**259**	3.33	7	1.10	4.9
135	*	**Pioneer Hi-Bred Intl**	**62.8‡**	24⅛– 10⅜	20¾	15.9	**330**	2.18	10	0.68	3.3
136	49	**Stauffer Chemical**	**62.7**	54 – 17⅝	37⅝	21.8	**821**	4.94	8	1.80	4.8
137	238	**Castle & Cooke**	**62.6**	19⅛– 7½	18¾	19.0	**356**	2.10	9	0.80	4.3
138	159	**Ark Louisiana Gas**	**62.4**	36⅛– 12⅝	35¼	12.6	**443**	3.74	9	2.00	5.7
139	554	**Fairchild Industries**	**62.3**	16⅞– 3¾	15⅝	4.6	**72**	1.64	10	0.50	3.2
140	232	**Foxboro**	**62.1**	54 – 18½	47	5.3	**251**	5.84	8	1.50	3.2
141	68	**Denny's**	**62.0**	29 – 5⅞	28¾	8.6	**247**	2.59	11	0.60	2.1
142	*	**Pennzoil**	**61.5‡‡**	35¾– 12¾	29⅞	32.3	**964**	4.18	7	1.80	6.0
143	57	**Anderson-Clayton**	**61.3**	26¼– 7¼	21⅜	13.7	**293**	2.85	8	1.00	4.7
144	361	**American Bakeries**	**61.3**	17¾– 3⅛	16⅛	1.7	**28**	2.85	6	1.20	7.4
145	144	**Abbott Laboratories**	**61.1**	57¼– 15¼	56⅜	29.5	**1,665**	3.74	15	1.20	2.1
146	86	**Dillon Companies**	**60.3**	36⅜– 13¾	31⅝	9.7	**306**	2.77	11	1.20	3.8
147	622	**Johnson Controls**	**60.0**	29⅜– 3⅜	28½	8.6	**245**	2.31	12	0.80	2.8
147	548	**Del E Webb**	**60.0**	14¼– 2	12	8.6	**103**	1.08	11	0.20	1.7
149	260	**Ralph M Parsons**	**59.2**	28¾– 6⅛	28	3.5	**97**	4.57	6	1.00	3.6
150	891	**Bausch & Lomb**	**58.3**	60¼– 17⅛	43⅛	5.8	**249**	4.06	11	1.40	3.2
151	461	**Automation Inds**	**58.1**	12⅝– 1¾	12¼	5.0	**62**	1.79	7	0.40	3.3
152	817	**Ampex**	**58.0**	11¼– 2	9⅞	10.9	**108**	1.05	9	none	0.0
153	253	**Campbell Taggart**	**56.6**	24¼– 7¼	21¾	10.8	**235**	2.30	9	0.88	4.0
154	81	**Tektronix**	**56.0**	40 – 9⅛	39	17.7	**689**	2.68	15	0.48	1.2
155	251	**Edison Bros Stores**	**56.0**	74¼– 16¾	71¾	4.0	**284**	8.28	9	2.70	3.8
156	134	**Pillsbury**	**55.7**	45 – 15¾	40⅞	16.2	**662**	3.67	11	1.28	3.1
157	259	**Emhart**	**55.6**	32⅞– 9½	31⅞	10.9	**349**	5.16	6	1.80	5.6
158	295	**Allied Chemical**	**54.7**	54¼– 23	44⅞	28.1	**1,260**	4.85	9	2.00	4.5
159	24	**Smith International**	**53.5**	43⅛– 10⅛	35⅞	9.5	**342**	3.62	10	0.72	2.0
160	164	**Hanes**	**53.5**	29 – 6	27⅝	4.3	**117**	4.55	6	1.52	5.5

* Not ranked last year. ‡ Four-year price change. ‡‡ Three-year price change.

TABLE 5.15   Random Numbers Below and to the Right of 37262

51603	39632	09308	18023
31684	07421	74411	94873
58786	99347	38137	71018
33858	41254	05114	18748
85496	98215	14644	54949
30520	61690	34076	13101
26638	73767	84644	13066
06484	11383	42522	86489
87211	89188	87472	62569
12988	91659	20840	41839
10249	46573	41177	77616
18511	48170	54274	35223
37156	21654	73948	39454
26487	94169	94770	79757
86393	14702	96954	81180
15887	35938	57377	50222
75668	91909	48839	71560
55650	85352	05221	74645
43724	69633	48123	83852
03992	00524	33338	63320
28300	07758	88565	99968
93215	20689	95890	85617
62406	07600	31083	04375
85595	52459	21400	40246
11207	28485	48862	11813
51181	98961	30460	40066
79229	03874	00138	55781
06125	59984	42227	94356

other columns to the right. We next cross out all numbers of Table 5.16 that exceed 160. The numbers that remain (listed in Table 5.17) are a random selection of numbers from 1 (rather 001) to 160 and can therefore be viewed as a random selection of ranks. The banking corporations having the first 16 of these randomly selected ranks can therefore be considered as constituting a random sample of the top 160 companies.

Let's see how this actually works. The first 16 randomly selected ranks are listed in Table 5.18. For each of these ranks, we look up in Table 5.14 the company having that rank. This random selection of 16 of the top 160 companies appears in Table 5.19, together with the number of shares outstanding of the companies selected.

TABLE 5.16   A List of Random 3-Digit Numbers

516	396	093	180
316	074	744	948
587	993	381	710
338	412	051	187
854	982	146	549
305	616	340	131
266	737	846	130
064	113	425	864
872	891	874	625
129	916	208	418
102	465	411	776
185	481	542	352
371	216	739	394
264	941	947	797
863	147	969	811
158	359	573	502
756	919	488	715
556	853	052	746
437	696	481	838
039	005	333	633
283	077	885	999
932	206	958	856
624	076	310	043
855	524	214	402
112	284	488	118
511	989	304	400
792	038	001	557
061	599	422	943

TABLE 5.17   A Random Selection of Numbers from 001 to 160   (Numbers remaining in Table 5.16)

064	074	093	131
129	113	051	130
102	147	146	043
158	005	052	118
039	077	001	
112	076		
061	038		

TABLE 5.18   Sixteen Randomly Selected
Numbers from 001 to 160

064	039	147	093
129	112	005	051
102	061	077	
158	074	076	
	113	038	

TABLE 5.19   Sixteen Randomly Selected Companies from *Forbes'* List

Rank	Company	Number of Shares Outstanding (Millions)
64	Raytheon	30.8
129	Eagle-Picher Inds	9.9
102	Olinkraft	8.8
158	Allied Chemical	28.1
39	Super Food Services	1.1
112	Albertson's	7.3
61	Twentieth Century Fox	7.7
74	Columbia Pictures	8.4
113	Butler Manufacturing	5.8
147	Del E Webb	8.6
5	Earth Resources	5.1
77	Entex	8.2
76	Eastern Gas & Fuel	21.8
38	Chic & Northwestern	4.4
93	Tyler	5.8
51	Niagara Frontier Svc	2.2

Once we have in our possession the 16 randomly selected values of the 1977 number of shares outstanding, as presented in Table 5.19, we can calculate their sample mean (it is $\bar{x} = 10.25$) and their sample standard deviation (it is $s = 8.76$) as estimates of the true parameters μ and σ. If it is reasonable to assume that the overall population of values of 1977 number of shares outstanding of the various companies has the normal distribution, we can use our small-sample methods to find confidence intervals for these parameters. The key is the ability to select a random sample of data from a population. If the population cannot be considered to have the normal distribution, then our small-sample methods will not apply. If we have a random sample, however, we will still be able to find nonparametric methods (see Chapters 12 and 13) that will allow us to draw inferences about the population as a whole.

Other Sampling Methods

Not all statistical sampling is done in the random manner described above. Sometimes it is useful to apply techniques such as *stratified* or *cluster* sampling. In stratified sampling, the objective is to make sure that all possible subdivisions or strata in the population are appropriately represented in the sample. Suppose we want to choose a sample of 40 stocks from the massive portfolio held by a major mutual fund. If we were to choose a random sample, we would be taking the risk that all insurance, automotive, utility, or other stocks would be omitted from the sample. By selecting a stratified sample, we guarantee that the sample will contain a cross section of all types of stocks held by the fund.

Cluster sampling is extensively used in marketing surveys because of the difficulty and expense required to obtain a random sample of shoppers nationwide. Rather than interview two shoppers in Cincinnati, three in Los Angeles, five in New York, two in Seattle, one in Omaha, etc. for 80 more cities, it is more efficient to contact 50 shoppers in each of 10 major cities. These groups of 50 shoppers are called clusters.

Technical details of these and other sampling methods are available in many specialized books, including the ones listed in the bibliography for this chapter.

EXERCISES 5.H

5.67. Use the data from Table 5.14 to carry out the following operations.

    a. Beginning with the middle three digits of the numbers below 03797 (11th row, 4th column), construct a random sample of 10 of the 160 top companies.

    b. Calculate the sample mean indicated annual dividend, as an estimate of the true mean indicated annual dividend of the 160 companies.

    c. Find the sample standard deviation of the indicated annual dividends as an estimate of the true standard deviation.

    d. Based on the random sample, estimate the median indicated annual dividend of the 160 top companies.

5.68. Start with the last three digits of the numbers following 22169 (2nd row, 8th column), and construct a random sample of size 15 of the 160 top companies. Based on this random sample, obtain an estimate of the mean price/earnings ratio for these companies.

5.69. Estimate, using a properly selected random sample of size 18, the mean recent prices of the stocks of the 160 top companies.

5.70. Based on a properly selected random sample of size 30, estimate the median latest 12-month earnings per share.

SUMMARY AND DISCUSSION

This chapter has been the first chapter of the portion of this book dealing with statistical inference, the science of drawing conclusions from information contained in a sample of data.

The major problem of statistical inference is the attempt to determine to what extent a sample can be considered to be representative of the underlying population. Using two spectacular theoretical results, the law of large numbers and the central limit theorem, we have developed the method of confidence intervals to solve that problem. We have presented confidence interval formulas that show the extent to which the sample mean represents the true population mean, the sample standard deviation represents the true population standard deviation, and the sample proportion represents the true population proportion.

The concept of a control chart, used to monitor quality control in industrial processes, was also developed as a consequence of the theory underlying confidence intervals.

CHAPTER 5   BIBLIOGRAPHY

STATISTICAL SAMPLING

Cochran, W. A., Sampling Techniques, (2nd ed.), 1963, New York: John Wiley.

Deming, W. E., Some Theory of Sampling. 1950, New York: John Wiley.

Raj, D., The Design of Sample Surveys. 1972, New York: McGraw-Hill.

Williams, W. H., "How bad can 'good' data really be?" The American Statistician, May 1978, pp. 61–65.

VARIATIONS ON CONFIDENCE INTERVALS

Burstein, H., "Finite population correction for binomial confidence limits," Journal of the American Statistical Association, March 1975, pp. 67–69.

Crisman, R., "Shortest confidence interval for the standard deviation," Journal of Undergraduate Mathematics, September 1975, pp. 57–62.

Guenther, W. C., "Shortest confidence intervals," The American Statistician, February 1969, pp. 22–25.

Guenther, W. C., "Unbiased confidence intervals," The American Statistician, February 1971, pp. 51–53.

Hoaglin, D. C., "Direct approximations for chi-squared percentage points," Journal of the American Statistical Association, September 1977, pp. 508–515.

Machel, R. E. and J. Rosenblatt, "Confidence intervals based on a single observation," Proceedings of the IEEE, August 1966, pp. 1087–1088.

Morrison, D. G., "On forming confidence intervals for certain Poisson ratios," Decision Sciences, January–April 1970, pp. 234–236.

Tummins, M. and R. H. Strawser, "A confidence limits table for attribute analysis," The Accounting Review, October 1976, pp. 907–912.

WILSON, P. D. and J. TONASCIA, "Tables for the shortest confidence intervals on the standard deviation and variance ratio from normal distributions," *Journal of the American Statistical Association*, December 1971, pp. 909–912.

THE PROPER SAMPLE SIZE

BECKWITH, R. E., "Bounds on sample size in modified Bernoulli sampling: with applications in opinion surveys," *Decision Sciences*, January 1973, pp. 31–39.
BURSTEIN, H., *Sample-Size Tables for Quality Control and Auditing*. 1973, Westport, Conn.: Redgrave Information Resources.
PENTICO, D. W., "On the determination and use of optimal sample sizes for estimating the difference in means," *The American Statistician*, February 1981, pp. 40–42.
PUSHKIN, A. B., "Presenting beta risk to students," *The Accounting Review*, January 1980, pp. 117–122.

CONTROL CHARTS AND APPLICATIONS

HAYYA, J. C., R. M. COPELAND, and K. H. CHAN, "On extensions of probabilistic profit budgets," *Decision Sciences*, January 1975, pp. 106–119.
HUBBARD, C. L., "Statistical control charts for administrative decision," *Decision Sciences*, January–April 1970, pp. 163–173.
MEDDAUGH, E. J., "The bias of cost control charts toward Type II errors," *Decision Sciences*, April 1975, pp. 376–382.

CHAPTER 5   SUPPLEMENTARY EXERCISES

5.71. Proposed SEC regulations state that an auditor using random sampling must be 99% sure of having one class of companies' mean amount of accounts receivable estimated correctly to within $5.00. If experience has shown that accounts receivable of this type of company generally have standard deviation of $60.00, how many accounts does the auditor have to include in his random sample in order to achieve the specified accuracy?

5.72. In connection with a study on consumer buying behavior, a department store analyst would like to choose a random sample of credit card holders in order to estimate the mean yearly income level for that class of customers. It is known from earlier studies of this kind that such incomes have standard deviation $3,000. If the analyst wants to be 90% sure of having the true mean income estimated correctly to within $100, how many customers does she need for the random sample?

5.73. The research branch of a major automobile manufacturer has been working on an engine that is supposed to be able to get more miles per gallon of gasoline. In 9 trial runs during a preliminary testing period, the engine was used on a 100-mile track, and the amount of gasoline (in gallons) needed to cover the route was recorded each time, as follows:

$$2.5 \quad 3.0 \quad 2.0 \quad 2.4 \quad 2.9 \quad 2.6 \quad 2.7 \quad 2.1 \quad 2.3.$$

a. Find a 90% confidence interval for the engine's true mean consumption of gasoline per 100 miles.

b. Find a 90% confidence interval for the engine's true standard deviation of consumption of gasoline per 100 miles.

c. If it is reasonable to use 0.35 as an estimate of the standard deviation, how many trial runs would be needed to be able to estimate the true mean consumption to within 0.1 gallons per 100 miles with 98% confidence?

d. Using 0.35 as an estimate of the standard deviation, how sure can we be of having the true mean consumption correctly estimated to within 0.1 gallons, if we base our estimate on a sample of 40 trial runs?

5.74. A 1977 Harris poll of consumers claimed that 61% of those questioned felt that the quality of most products has deteriorated over the past 10 years. To check on these disturbing results, a consumer analyst for a large appliance manufacturer conducted his own poll of his company's recent customers. Of 100 randomly selected customers, 70 agreed that product quality has indeed declined. Find a 95% confidence interval for the true proportion of the company's recent customers who feel that way.

5.75. A toy store's survey involving 400 randomly selected families in one section of a large city showed that the average family surveyed had 2.37 children, with a standard deviation of 1.80 children. Find a 95% confidence interval for the true mean number of children per family in that section of the city.

5.76. As part of a study of highway safety, a random sample of 25 automobiles from the freeways of Los Angeles were tested for their stopping ability after application of the brakes. The 25 cars required a mean distance of 148 feet to stop, with a standard deviation of 21 feet. Find a 90% confidence interval for the true mean stopping distance of all cars on Los Angeles freeways.

5.77. In connection with an efficiency study of the productivity of candy factory workers, 60 employees were able to produce per day an average of 14.8 pounds of chocolate-covered cherries, with a standard deviation of 6.5 pounds.

a. Find a 90% confidence interval for the true standard deviation of the number of pounds of chocolate-covered cherries produced.

b. Find a 98% confidence interval for the standard deviation.

5.78. As part of an analysis of the relationship between smoking and lung disease, a corporation's physician conducts a survey of company employees with lung disorders, with the goal of finding out what percentage of such employees are heavy smokers.

a. Of 200 such patients, the physician determines that 120 are heavy smokers. Find a 95% confidence interval for the true percentage of lung patients that are heavy smokers.

b. If an expanded survey reveals that 1,200 out of 2,000 patients are heavy smokers, what would be a 95% confidence interval for the true percentage?

c. If the physician wants to be 95% sure of having the true percentage estimated correctly to within 1%, how many patients must he include in his survey?

5.79. Of 300 licensed drivers selected at random from Department of Motor Vehicles records, it is learned that 45 are problem drinkers. Construct a 90% confidence interval for the true proportion of licensed drivers that are problem drinkers.

5.80. Prior to a union election in which Able is running against Baker, a survey of

1,500 union members shows that 780 intend to vote for Able.

   a. Find an 80% confidence interval for the true proportion of the vote going to Able.

   b. Find a 90% confidence interval for the true proportion of the vote going to Able.

   c. With what probability can we assert that Able will win a majority of the votes?

5.81. To estimate the proportion of a company's accounts that have defaulted over the years, an accountant chooses a random sample of 2,000 accounts and discovers that 60 have defaulted.

   a. Find a 95% confidence interval for the true proportion of defaulting accounts.

   b. If it were necessary to be 99.9% sure of having the true proportion estimated to within .01, how large a random sample would be needed to guarantee that level of accuracy?

5.82. As part of an analysis of financial transactions conducted in the nation's smaller stock and commodity exchanges, it was found that the dollar volume of shares traded on the Boston Stock Exchange averaged $78,000 per day, with a standard deviation of $5,000, over a period of 40 trading days.

   a. Find a 90% confidence interval for the true mean daily dollar volume on the Boston Stock Exchange.

   b. Suppose it is reasonable, based on this analysis, to assume that the standard deviation is $5,000. For how many days must a study be conducted in order to be 90% sure of having the true mean dollar volume correctly estimated to within $500?

5.83. A sociological study conducted in 1980 of the relationship between unemployment and crime showed that, of 150 prisoners surveyed, 90 were employed full-time at the time of their arrest.

   a. Find an 80% confidence interval for the true proportion of prisoners who were employed full-time at the time they were arrested.

   b. How many prisoners' employment records must be investigated in order to be 95% sure of having the true proportion correctly estimated to within .02 or 2%?

5.84. As part of a study of projected claims arising from a group medical insurance policy to be offered to one state's public school teachers, an actuary discovers that the number of days per year a teacher is absent from work due to illness is a random variable having standard deviation 2.2 days. If she would like to be 90% sure of having the mean number of days absent correctly estimated to within 0.1 day (one-tenth of a day), how many teachers' records does she need to check as part of her survey?

5.85. Recent experience leads the Desert Breeze Casualty Insurance Company of Trona to consider 17 as a reasonable number for the mean amount of auto accidents its clients cause per month, and 5.9 for the standard deviation. To determine whether or not its underwriting procedures are out of control, the company observes the average monthly number of accidents over several four-month periods as follows:

Period	No. of Accidents	Period	No. of Accidents
Jan–Apr 1978	19	Jan–Apr 1980	25
May–Aug 1978	21	May–Aug 1980	23
Sep–Dec 1978	22	Sep–Dec 1980	23
Jan–Apr 1979	27	Jan–Apr 1981	26
May–Aug 1979	18	May–Aug 1981	21
Sep–Dec 1979	15	Sep–Dec 1981	17

a. Calculate the upper and lower control limits.

b. Construct the control chart and determine which periods, if any, indicate that the underwriting procedures may be out of control.

5.86. The Quick Cow, a chain of fast-service mini-markets specializing in dairy products, figures that only 15% of its milk cartons ought to remain on the shelves longer than two days. It would like to conduct a test of whether or not its inventory stocking schedule is under control. Careful records on 20 days indicate the following percentages of milk cartons remaining from two days earlier:

Percentage	Percentage	Percentage	Percentage
11	21	8	20
10	9	12	22
12	14	16	11
18	17	15	7
23	22	18	15

a. Determine the upper and lower control limits.

b. Draw the control chart and note when it indicates that the inventory stocking schedule is out of control.

5.87. An October, 1977, survey of bank economists showed that of 51 economists contacted, 39 thought that borrowing costs for business loans would increase moderately by the middle of 1978.* Find a 95% confidence interval for the true proportion of bank economists having that opinion.

5.88. Consider the data of Table 5.14, entitled "Who's Where in the Stock Market" and taken from *Forbes*, January 9, 1978, pages 201–203.

a. Select a random sample of size 12 from among the 160 companies listed.

b. Use the random sample to find an 80% confidence interval for the true mean current yield of the top 160 companies in five-year stock price change, as measured by *Forbes*.

c. Find an 80% confidence interval for the true mean standard deviation of the current yields of the 160 companies.

d. Select a random sample of size 25, and use it to find an 80% confidence interval for the true proportion of companies whose current yield exceeds 5.0%.

*U.S. News & World Report, October 31, 1977, p. 82.

6

Testing Statistical Hypotheses

One of the fundamental principles of the scientific method of increasing our knowledge of the world around us is the formulation of a hypothesis. Today this and other principles of the scientific method have just as much validity for business and the management sciences as they do for the natural sciences. When we formulate a hypothesis, we define very clearly the problem under discussion and we provide the framework for gathering the relevant facts. An analysis of the accumulated facts (or data) then leads us to accept or reject the hypothesis.

For example, a recent Texas business school graduate concluded, on the basis of extensive market research and statistical analysis, that the frozen yogurt business was here to stay. After she decided to open her own store dealing in that particular product, a number of questions arose regarding factors that determine an optimal location for such a store. In particular, she wanted to test the *hypothesis*

> H: Yogurt sales *will not* be greater in neighborhoods where median household income exceeds $20,000 than in neighborhoods having median income below that level.

against the *alternative*

> A: Yogurt sales *will* be greater in neighborhoods where median household income exceeds $20,000.

To choose between the hypothesis H and the alternative A, the first step would be to conduct a market research study comparing projected sales of frozen yogurt in neighborhoods of various income levels. Analysis of the results of the marketing study would then lead the decision maker to reject or accept the hypothesis H.

It should be noticed that the hypothesis H is formulated in a negative sort of way. Rather than asserting that frozen yogurt will sell better in certain neighborhoods than in others, it states that it actually will not sell better. Such a formulation of the hypothesis H has the effect of placing the "burden of proof" of its popularity in certain neighborhoods on the yogurt itself. In order for us to decide that high-income neighborhoods really prefer more frozen yogurt, the data must move us to make the decision to reject H. Unless the projected sales figures in high-income areas are substantially above those of low-income areas, we will merely accept H. Because the logic of the situation often requires H to be structured in this negative way, we usually refer to H as the *null hypothesis*.

Our objective in this chapter is to provide a framework for managerial decision making by showing how to formulate a hypothesis and how to use statistical methods of data analysis for the purpose of deciding whether to reject or accept that hypothesis.

SECTION 6.A.   THE STATISTICAL DECISION-MAKING PROCESS

Suppose, while we are traveling (first-class, of course) on the over-the-pole flight from San Francisco to London, an indicator light on the pilot's instrument panel warns that the left wing is about to fall off. Mentally, the pilot immediately formulates the hypothesis

H: Left wing is about to fall off.

and the alternative

A: Indicator light isn't working properly.

We can run some visual, mechanical, and electrical checks in this and any other such situation, and these checks provide a set of data upon which we can base our analysis of the problem. After all the data are assembled, our situation fits into one of the following categories:

1. Left wing is almost certain to fall off.
2. Left wing will probably fall off.
3. Evidence is inconclusive.

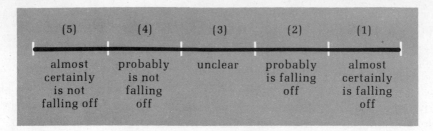

FIGURE 6.1   Status of Airplane in Regard to State of Left Wing

4. Indicator light probably isn't working properly.
5. Indicator light almost certainly isn't working properly.

There is no such thing as absolute certainty in a situation like this until it is almost too late. Fig. 6.1 illustrates the situation.

Two Types of Errors

After the pilot arrives at his decision regarding the true state of the left wing, he may have committed one of two types of errors. These types of errors are formally designated *Type I* and *Type II*.

TYPE I ERROR
We reject H when it is really true (in this case, we decide that the indicator light is malfunctioning when, in fact, the left wing is about to fall off.)

TYPE II ERROR
We accept H when it is really false (in this case, we decide that the left wing is about to fall off when, in fact, the indicator light is malfunctioning).

Of course, we might possibly arrive at the correct decision: we could reject H when it's really false or accept it when it's really true.

There are two questions we have to settle in connection with the Type I and Type II errors: (1) which of the two errors is the worse, and (2) how often are we willing to make the worse of the two errors?

Regarding the indicator light problem, we would probably agree that the Type I error is the worse. If we make a Type I error, we are risking the lives of the passengers and crew, while if we make a Type II error, we are merely wasting the passengers' time, the airline's money, and the Arctic rescue crew's time, money, and effort.

How often would we be willing to make a Type I error? Well, I don't know about you, but if I were the pilot, I'd prefer never to make a Type I error. What decision strategy can we follow that would guarantee our never making a Type I error? If you think about it, you'll see that the only way we can be 100% sure of never making a Type I error would be to always decide that the left wing is about to fall off and then to attempt an emergency landing on the polar ice cap. This way, we can be sure of never rejecting H when it's really true, because we will never reject H—never when it's true and never when it's false.

Imagine the situation for a moment. The indicator light comes on, and before another minute passes, we radio the Arctic rescue patrol that our plane can be found sticking tail up out of the polar ice. Although we are guaranteed never to make a Type I error, you'll have to agree that this emergency operating plan is somewhat impractical. Why is this procedure impractical? Primarily because we will be ditching the plane many, many times when the only problem is a crossed wire in the indicator light. In our zeal to avoid Type I errors, we will be committing Type II errors whenever it is possible to do so. If we consider the fact that there is a real problem relatively rarely, even when the indicator light comes on, it's truly ridiculous to ditch the plane at every sign of a malfunction.

Well, now we're back where we started. We agree that we cannot ditch the plane every time, and therefore we're going to have to fly on to London with the indicator light flashing on some occasions. Yet, every time we decide to fly on to London instead of landing on the polar ice, we're risking a Type I error. Even if we fly on to London in only that case when we are most absolutely sure that there is an electrical malfunction in the indicator light, there is even then some chance that the left wing is actually falling off. What do we do? How often do we fly on to London? How often do we land on the polar ice?

The problem revolves around the difficulty of deciding upon an acceptable level of Type I errors. How often are we willing to make a Type I error; i.e., how many times when the left wing is really (unknown to us at the time) falling off are we willing to fly on to London?

Of course, there is no real answer to this question. This is a subjective decision that must take many factors into account. If we denote by α (pronounced "alpha") our probability of making a Type I error and by β (pronounced "bayta" but spelled "beta") our probability of making a Type II error, there is an inverse relationship between α and β. If α is very low, we would not be making a Type I error too often. Therefore, we would not be rejecting H (and consequently asserting A to be true) unless it were fairly certain that H was false. It would, therefore, be necessary to accept H in some questionable cases, and so our probability of a Type II error would be high. So a low value of α would almost automatically lead to a high value of β. The reverse is also true: a high value of α means we are rejecting H in questionable cases, so that we accept H only when it is probably true and β is therefore low. It is therefore necessary for us to weigh the relative seriousness of the two errors, not merely to decide separately on an acceptable level of Type I errors. As we have seen in the indicator light example above, the insistence on a low incidence of Type I errors almost certainly leads to an unacceptably high level of Type II errors. On the other hand, we can't really accept too high a level of Type I errors either.

To make the best of a bad situation, we use the following procedure for handling statistical hypotheses:

1. We decide which of the two errors, Type I or Type II, is the worse.
2. If we think the Type I error is the worse, we choose a testing procedure that makes α acceptably low, and we agree to be satisfied with whatever the value of β turns out to be.
3. If we think the Type II error is the worse, we choose a testing procedure that makes α high, and then we can expect β to be correspondingly low.

Despite the general rule of the inverse behavior of α and β, it unfortunately turns out that there is usually not a simple algebraic relationship between α and β. This is too bad, but not very much can be done about it, for the computation of the exact value of β in many applied situations is a difficult theoretical problem and depends on the relationships among many unknown quantities. The only thing we can really be sure of is that, as α increases from low to high, β will decrease from high to low. For this reason, it is advisable whenever possible to construct H and A in such a way that the Type I error will be the worse, and then our ability to choose α will give us some control over that error.

The Level of Significance

Our choice of α can therefore be viewed as a subjective measure, on our part, of the relative seriousness of the two types of errors. For this reason, α is said to be the *significance level* of the test. When a resaercher or manager chooses a particular significance level α as part of the analysis of a problem, he is morally (although not legally) obligated to state at the outset what level of α is being used. Any decision that is made on the basis of a statistical analysis must include a statement of the numerical value of α, for only in this way can a reader of the report know how the writer viewed the relative seriousness of the two errors. Sometimes the reader may not have the same subjective viewpoint as the writer and may therefore disagree strenuously with the writer's choice of α. In borderline cases, a change in the numerical value of α may very well result in a change in the decision. (Some practical examples of this will be illustrated in the remaining sections of this chapter.) Therefore, the choice of α could be the determining factor in making the decision, and so it must be reported as an integral part of the research study.

You have probably heard or read somewhere the allegation, "anything can be proved using statistics." The greater part of whatever truth there may be to this allegation arises when opposing parties disagree as to what level of significance should be used in analyzing the data. As a prime example of this sort of situation, we can look at the recent history of disputes between environmentalists and economic growth advocates over offshore oil drilling in locations such as California's Santa Barbara Channel. The hypothesis to be tested is

> H: Offshore oil drilling *does not* cause significant environmental damage.

and the alternative is

> A: Offshore oil drilling *does* cause significant environmental damage.

In this situation, the significance level would be

$$\begin{aligned} \alpha &= P(\text{Type I error}) \\ &= P(\text{rejecting H when it's really true}) \\ &= \text{probability of asserting that offshore oil drilling } does \text{ cause significant environmental damage } when, \text{ in fact, it } does\ not \end{aligned}$$

while

$$\begin{aligned} \beta &= P(\text{Type II error}) \\ &= P(\text{accepting H when it's really false}) \\ &= \text{probability of asserting that offshore oil drilling } does\ not \text{ cause significant environmental damage } when, \text{ in fact, it } does \end{aligned}$$

From the environmentalists' point of view, the Type II error would be the worse, for the environmentalist believes that the possibility of significant damage to the environment outweighs any economic advantage to be gained from offshore oil drilling. In the opinion of advocates of economic growth, the Type I error is the worse of the two because such an error would tend to stagnate economic growth, leading to loss of jobs and a lowered standard of living, even if there were no real evidence that significant environmental damage would occur. It is not that environmentalists are necessarily out to

TABLE 6.1   Choice of Level of Significance α—Environmentalists vs. Economic Growth Advocates   *(H: Drilling does not cause environmental damage; A: Drilling does cause environmental damage)*

Primary Concern	Worse Error	Significance Level α	β
Environment	Type II	High	Low
Economy	Type I	Low	High

wreck the economy, nor are economic growth advocates deliberately out to destroy the environment. What is really the case is that, if an error (Type I or Type II) is to be made, the environmentalist would rather err on the side of the environment (while willing to risk some economic dislocation) and the economic growth advocate would rather err on the side of economic well-being (while willing to risk some environmental problems). Environmentalists would then base a statistical analysis on a high level of significance α, because they would insist that $\beta = P(\text{Type II error})$ be held as low as possible. Growth advocates, on the other hand, would favor a low level of significance, for they would feel that $\alpha = P(\text{Type I error})$ itself should be low. We summarize this discussion in Table 6.1.

One more comment before leaving this section: What do we mean by *low* α, and what do we mean by *high* α? Because the choice of α is a subjective decision to be made by the researcher or manager on the scene, this question cannot be answered using a statistical formula. An examination of the business and management publications listed in the bibliography indicates that a high level of α usually means a value between .02 and .10, while a low level of α is taken to be between .005 and .02. If the Type I error is quite a bit worse, the typical value of α is .01, while if the Type II error is also very bad, α is typically taken to be .05.

EXERCISES   6.A

6.1. Many statistical studies have been conducted in an attempt to find out whether or not smoking really causes cancer. To put the burden of proof on those who feel that it does, we can formulate the hypothesis

H: Smoking does not cause cancer.

and the alternative

A: Smoking causes cancer.

a. The cigarette industry would be unfairly hurt by a definitive decision that smoking causes cancer if the opposite were, in fact, true. To protect the cigarette industry, should α be chosen high or low?

b. The American Cancer Society wants to eliminate all possible causes of cancer even if their connection with cancer is not conclusively proved beyond all doubt. Should the Society argue for a high or low value for α?

6.2. A jury is supposed to convict an individual on trial for a string of murders and assaults if the individual is guilty beyond a reasonable doubt. The word "reasonable" is related to the numerical value of α in testing the hypothesis

   H: The individual is innocent.

against the alternative

   A: The individual is guilty.

a. To avoid convicting an innocent person, should the jury choose a high value for α or a low value?

b. To avoid releasing a guilty person to possibly continue the wave of attacks, should the jury select a high value for α or a low value?

6.3. A business executive has been assigned the task of deciding whether or not to invest several thousand dollars in a new business venture. He has to choose between the hypothesis

   H: The venture will succeed.

and the alternative

   A: The venture will fail.

a. If there is a shortage of investment capital, so that only the safest investments should be undertaken, should the executive use a high value or a low value of α?

b. If investment capital is plentiful, allowing the funding of some possibly risky ventures, should α be chosen as high or low?

6.4. An investor doesn't know whether a stock she plans to buy will go up or down. She has to choose between the hypothesis

   H: The stock will go down in the near term.

and the alternative

   A: The stock will go up in the near term.

a. If she feels she may need the money in the near future and so doesn't want to take too great a risk, should she choose α high or low?

b. If she is fairly certain that she won't need the cash any time soon and is therefore willing to gamble on the chance of a long-term gain, should she choose α high or low?

SECTION 6.B.   TESTING FOR THE MEAN OF A POPULATION

Because of the expense involved in manufacturing Nitro-Plus, a new chemical fertilizer, the fertilizer must produce an average yield of more than 20,000 pounds of tomatoes per acre within a specified growing period in order to be economically feasible. An independent agricultural testing organization chose 40 acres of farmland at random in various geographical regions, fertilized each with Nitro-Plus, and planted tomatoes intensively. The results show that the 40 acres had mean yield 20,400 pounds, and the yield per acre varied widely, having a standard deviation of 1,200 pounds. On the basis of the test results involving 40 randomly selected acres, can we conclude that the *true* mean yield really exceeds 20,000 pounds per acre?

In order to put the burden of proof on the fertilizer, we denote the true mean yield by the symbol μ, and we define the hypothesis to be

H: μ = 20,000 (the true mean yield fails to exceed 20,000)

and the alternative

A: μ > 20,000 (the true mean yield exceeds 20,000)

What is a reasonable significance level for this test? Well, in this case, the Type I error (rejecting H when it's really true) would be to conclude that the fertilizer produced the desired level of yield when, in fact, it did not. The Type II error, on the other hand, would be to say that Nitro-Plus does not meet the specifications when, in fact, it really does. We might be able to agree that the Type I is the worse error from the farmer's viewpoint, for it would eventually lead to financial failure of the operation, while the Type II would only prevent us from taking advantage of an effective new fertilizer. Suppose, therefore, we set the significance level low, say α = .01.

Large Samples

Now, how do we use the data to decide whether to accept or reject the hypothesis H? If we recall the statement of the central limit theorem that appears in Section 5.B, we know that the collection of all possible sample means $\bar{x}$ of size n = 40 comprise a set of normally distributed data having mean μ and standard deviation $\sigma/\sqrt{n}$. More precisely, we know that for all possible samples composed of n = 40 data points, the numbers

$$z = \frac{\bar{x} - \mu}{s/\sqrt{n}},$$

where we have used the sample standard deviation s to estimate σ, have approximately the standard normal distribution. Assuming that $\mu = 20{,}000$ (namely, that H is really true), under what circumstances would we be misled into rejecting H? We would tend to reject H if the sample mean $\bar{x}$ came out to be substantially larger than $\mu = 20{,}000$.

What we do we mean by "substantially larger"? Well, this is the point at which the significance level of the test enters the picture. As we have noted that a reasonable level of significance here would be $\alpha = .01$, this means that we want to reject H only 1% of the time when H is really true. Because we would tend to reject H if $\bar{x}$ turned out to be much larger than $\mu = 20{,}000$, the rejection of H would be indicated by a large value of

$$z = \frac{\bar{x} - 20{,}000}{s/\sqrt{n}},$$

because z would be large when $\bar{x}$ is much larger than 20,000. (z is actually a standardized measure of the difference between $\bar{x}$ and 20,000.) Now, when H is really true, there is a 1% chance that z would exceed $z_{.01} = 2.33$. Therefore, if we agree to reject H in case z turns out to be larger than 2.33, we will have only a 1% chance of rejecting the hypothesis H that $\mu = 20{,}000$ when it is really true. This means that our test of whether or not the true mean yield per acre of tomatoes, using Nitro-Plus fertilizer, really exceeds 20,000 pounds will have significance level $\alpha = .01$.

The set of values of z which leads us to reject H is called the *rejection region* of the test. Therefore, the rejection region of this test is the region where $z > 2.33$. A diagram with the rejection region shaded appears in Fig. 6.2. We can formalize the hypothesis testing mechanism for this problem as follows:

H: $\mu = 20{,}000$
A: $\mu > 20{,}000$

Compute

$$z = \frac{\bar{x} - 20{,}000}{s/\sqrt{n}}$$

and reject H at level $\alpha = .01$ if $z > z_{.01} = 2.33$. To complete this problem and to make our decision regarding the economic feasibility of the Nitro-Plus fertilizer, it remains only to calculate z and to decide whether or not it exceeds 2.33.

From the data accumulated by the independent testing organization, a study of $n = 40$ randomly selected acres gave a sample mean yield of $\bar{x} = $

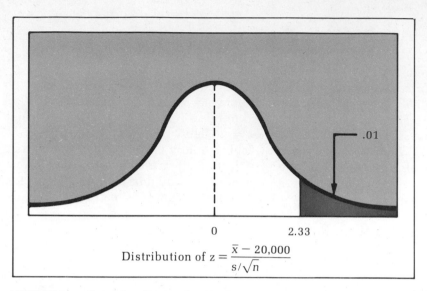

$$\text{Distribution of } z = \frac{\bar{x} - 20{,}000}{s/\sqrt{n}}$$

FIGURE 6.2   Rejection Region for H when α = .01
(Tomatoes-fertilizer example)

20,400 with a sample standard deviation of s = 1,200. Inserting these numbers into the formula for z, we find that

$$z = \frac{\bar{x} - 20{,}000}{s/\sqrt{n}} = \frac{20{,}400 - 20{,}000}{1{,}200/\sqrt{40}} = \frac{400}{1{,}200/6.325} = \frac{400}{189.72} = 2.11 \ .$$

As we agreed to reject H if z > 2.33, but z actually turned out to be 2.11 (which is less than 2.33), we *cannot* reject H. Therefore, at significance level α = .01, we cannot conclude that the true mean yield exceeds 20,000 pounds per acre. Our experimental yields have not been large enough to provide convincing proof of the economic feasibility of Nitro-Plus.

Suppose we have looked at the problem from a different point of view. Suppose the economic goal of running a profitable operation had been outweighed by the emergency necessity of producing a larger food supply as soon as possible. In such a situation, we would probably be less concerned with the precise level of mean yield and more concerned merely with the ability of Nitro-Plus to produce a large yield. In testing H against A we would therefore be likely to consider the Type II error (accepting H when it's really false) to be the worse, for it would lead us into rejecting the use of Nitro-Plus in some cases when the fertilizer would be truly effective.* In our desire to avoid a Type II error, we would then be willing to ease up somewhat on the probability of a Type I error, perhaps by raising the significance level from α

---

*The Type II error will be studied in greater detail in Section 6.C.

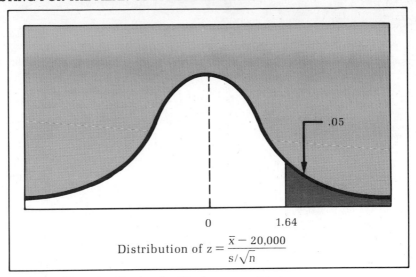

FIGURE 6.3   Rejection Region for H when α = .05
(Tomatoes-fertilizer example)

= .01 to α = .05. At significance level α = .05, our rejection rule would be
to reject H in favor of A if z > $z_{.05}$ = 1.64. The rejection region is illustrated
in the diagram of Fig. 6.3.

In viewing the problem from this different point of view, the only thing
that has changed is the level of significance α. The data, of course, remain
the same. (Our opinion of the problem will have no effect on the data, for the
data record only how many pounds of tomatoes are actually produced using
the Nitro-Plus fertilizer.) Therefore we still have n = 40, $\bar{x}$ = 20,400, and s
= 1,200, so that z = 2.11 as before. Now, however, we have agreed to reject
H if z > 1.64. Because z = 2.11 has turned out to be larger than 1.64, we
would therefore make the decision to reject the hypothesis H in favor of the
alternative A.

In the circumstances of the decline in importance of the precise economic "break-even" point and the rise in importance of the food output itself,
our conclusion would then be that Nitro-Plus increases tomato output to an
extent sufficient to justify its use. At significance level α = .05, our experimental results indicate that Nitro-Plus will be able to produce yields in excess
of 20,000 pounds per acre. A tabular comparison betwen the cases α = .01
and α = .05 is presented in Table 6.2.

Our statistical analysis of the tomatoes-fertilizer example has given us
the following decision rule: Reject H (and consider the fertilizer to be effective) if α = .05, but accept H (thereby considering the fertilizer to be ineffective) if α = .01.

The decision rule turned out this way because z = 2.11 is larger than
$z_{.05}$ = 1.64, but smaller than $z_{.01}$ = 2.33. It would be interesting and useful
to know exactly which significance level α would be the dividing line

TABLE 6.2 Decision Making in the Tomatoes-Fertilizer Example
($H:\mu = 20{,}000$ vs. $A:\mu > 20{,}000$)

Significance Level	Rejection Rule	Calculated Value of z	Decision
$\alpha = .01$	$z > 2.33$	$z = 2.11$	Accept H
$\alpha = .05$	$z > 1.64$	$z = 2.11$	Reject H

between rejecting and accepting H. This cut-off level is called the *P-value* of the number z. More precisely, the P-value of $z = 2.11$ is the significance level α for which $z_\alpha = 2.11$.

If we look up $z = 2.11$ in the table of the normal distribution (Table A.4 of the Appendix), we find its corresponding proportion to be .9826, as shown in Table 6.3. As illustrated in Fig. 6.4, it follows that $\alpha = 1.0000 - .9826 = .0174$, when $z_\alpha = 2.11$. Therefore the P-value of $z = 2.11$ is $\alpha = .0174$. This means that we should reject H if we feel that α ought to be larger than

TABLE 6.3 A Small Portion of Table A.4

z	proportion
2.11	.9826

FIGURE 6.4 The P-Value of $z = 2.11$

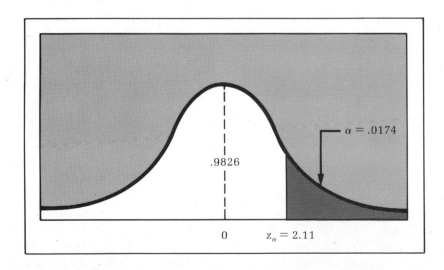

.0174, but we should accept H if we feel that α ought to be smaller than .0174. In the particular cases we looked at, $\alpha = .01$ was smaller than .0174, so it led us to accept H, while $\alpha = .05$ was larger than .0174, so it led us to reject H.

The statistical test we have just carried out is called the *one-sample z test*. It is a "one-sample" test because only one set of data is used in the calculations, and it is a "z test" because the availability of more than 30 data points (here n = 40) allows us to apply the central limit theorem, using percentage points of the standard normal distribution. One-sample tests for the means of a population are characterized by the use of three distinct means: the true mean μ of the population, the sample mean $\bar{x}$ based on the data, and finally the *hypothesized mean*, which we shall denote by the symbol μ_0 ("mew-zero"). In the tomatoes-fertilizer example, the hypothesized mean is $\mu_0 = 20,000$, for the hypothesis reads

H: $\mu = 20,000.$

The statement does not really say that the true mean μ is 20,000, but merely that we are hypothesizing that it is. The sample mean is $\bar{x} = 20,400$, and the numerical value of the true mean μ is, as always, unknown to us.

The format of the one-sample z test is often as it was in the tomatoes-fertilizer example. In particular, when we test the hypothesis

H: $\mu = \mu_0$

against the alternative

A: $\mu > \mu_0$

we compute

■
$$z = \frac{\bar{x} - \mu_0}{s/\sqrt{n}} \tag{6.1}$$

and we reject H at level α if $z > z_\alpha$. Sometimes, however, the alternative is formulated differently, due to the nature of the question being asked. The following example provides an illustration.

EXAMPLE 6.1. Calculator Battery Lifetimes. One manufacturer of pocket calculators advertises that its battery pack allows the calculator to operate continuously for 22 hours, on the average, without recharging. A prospective corporate customer tested a sample of 50 calculators, and these turned out to have mean 21.8 hours with a standard deviation of .9 hour. If a significance level of $\alpha = .10$ seems to be appropriate, is there sufficient evidence to indicate that the true mean duration of continuous operation (without recharging) is actually less than the 22 hours claimed by the manufacturer? Also, calculate the P-value of the test.

SOLUTION. Denoting by μ the true mean duration of continuous operation (without recharging) of the battery pack, we have to test the hypothesis

H: $\mu = 22$

vs.

A: $\mu < 22$

at the level of significance $\alpha = .10$. Because we have a single sample of data and $n = 50 > 30$ data points, the appropriate method of analysis is again the one-sample z-test. We compute, as before,

$$z = \frac{\overline{x} - 22}{s/\sqrt{n}}$$

and we have to determine which values of z will lead us to reject H in favor of A. We will be inclined to reject H if $\overline{x}$ turns out to be considerably smaller than 22 because, if our sample mean comes out quite a bit lower than 22, we'd have a hard time convincing ourselves that the true mean was really 22. Yet, as $\alpha = .10$ here, we would be making a Type I error that 10% of the time z actually falls below $-z_{.10} = -1.28$ even if μ is really 22. Therefore, if we decide to reject H in favor of A in case z turns out to be less than (even more negative than) $-z_{.10} = -1.28$, our test would have significance level $\alpha = .10$, indicating a 10% chance of a Type I error. We therefore agree to reject H at level $\alpha = .10$ in this situation if $z < -1.28$. A diagram of the rejection region is presented in Fig. 6.5.

It now remains only to calculate the numerical value of z and to determine whether or not it is more negative than -1.28. From the data, we see that $n = 50$, $\overline{x} = 21.8$, and $s = .9$. We therefore calculate z as follows:

$$z = \frac{\overline{x} - 22}{s/\sqrt{n}} = \frac{21.8 - 22}{.9/\sqrt{50}} = \frac{-.2}{.9/7.07} = \frac{(-.2)(7.07)}{.9} = -1.57$$

As $z = -1.57 < -1.28$, we have to reject H in favor of A. At level $\alpha = .10$, our conclusion is that the mean duration of continuous operation (without recharging) of the battery pack is actually *less than* the 22 hours claimed by the manufacturer.

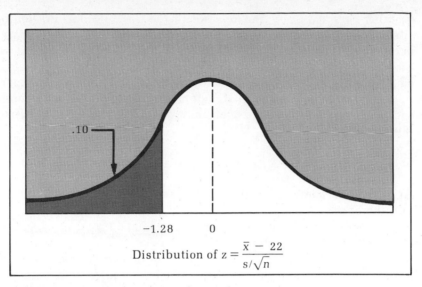

FIGURE 6.5   Rejection Region for H when α = .10
(Calculator battery example)

Finally, a glance at Table A.4 of the Appendix shows that the P-value of z = −1.57 is the listed proportion .0582. Therefore, we would reject H (concluding the manufactuer's claim is not valid) for all significance levels greater than α = .0582, while we would accept H for levels below that number.

Small Samples

The two situations above called for the use of the one-sample z test. The "z" in the one-sample z test refers to the fact that more than 30 data points are available, so that we can base our analysis of the problem on the z scores of the normal distribution. What happens in a one-sample problem where fewer than 30 data points are available? A brief glance at the discussion near the beginning of Section 5.C should provide the answer. In situations where the underlying population (from which the sample of data points is randomly selected) is normally distributed, we can replace the rejection regions based on the z scores of Table A.4 of the Appendix by rejection regions based on the t distribution of Table A.7. (If it's not normally distributed, see Chapter 12.)

EXAMPLE 6.2.   Apartment Occupancy Rates.   A real estate invest-ment company with primary holdings in Chicago is considering

whether or not to buy and operate properties in Houston. Before it
commits itself to investments in the Houston area, it wants to know how
occupancy rates for apartment complexes in Houston compare with
those in Chicago. Presently, buildings that the company operates in
Chicago have an average occupancy rate of 96.1 (per cent). A market
research study conducted in Houston recently showed that a random
sample of eight comparable apartment buildings there had occupancy
rates as follows:

95.4   95.0   97.5   95.6   93.3   94.2   94.9   97.3

At significance level α = .05, do the results of the study support the
opinion of one of the company's officers that occupancy rates in Houston are really lower than their counterparts in Chicago?

SOLUTION.   Because n = 8 < 30, the solution of this problem requires
the use of the one-sample t test. If we denote by μ the true mean occupancy rate of comparable apartment buildings in Houston, we want to
test the hypothesis

H: μ = 96.1 (average occupancy rate in Houston is the
            same as it is in Chicago),

against the alternative

A: μ < 96.1 (average occupancy rate in Houston is below
            that of Chicago)

at level of significance α = .05. If we were using the z test, we would
reject H if z < $-z_{.05}$ = -1.64. However, as we are using the t test, we
replace $z_{.05}$ by $t_{.05}$ $[n-1]$, where $n-1$ is the degrees of freedom required
in using Table A.7. Therefore, as n = 8, we have df = $n-1$ = $8-1$ =
7. We compute

$$t = \frac{\bar{x} - \mu_0}{s/\sqrt{n}} \tag{6.2}$$

where μ_0 = 96.1 here, and we reject H in favor of A if $t < -t_{.05}$ [7] =
-1.895. We illustrate the relevant portion of Table A.7 in Table 6.4 and
the rejection region in Fig. 6.6.
It remains, then, only to compute the value of

$$t = \frac{\bar{x} - 96.1}{s/\sqrt{n}}$$

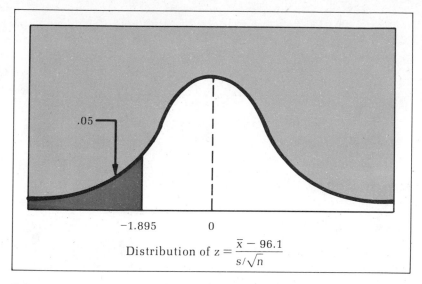

FIGURE 6.6   Rejection Region for H when $\alpha = .05$
(*Occupancy rate example*)

TABLE 6.4   A Small Portion of Table A.7

df	$t_{.05}$
7	1.895

and to check whether or not $t < -1.895$. We can insert the value $n = 8$ immediately, but the numerical values of $\bar{x}$ and s will have to be computed directly from the data. As in Chapters 2 and 5, we will set up a table of calculations to do the job. The calculations, which can be found in Table 6.5, show that $\bar{x} = 95.4$ and $s = 1.43$. Substitution of these numbers into the formula for t yields

$$t = \frac{\bar{x} - 96.1}{s/\sqrt{n}} = \frac{95.4 - 96.1}{1.43/\sqrt{8}} = \frac{-.7}{1.43/2.83} = \frac{(-.7)(2.83)}{1.43} = -1.39$$

Now recall that we agreed to reject H in favor of A if t turned out to be less than -1.895. Well, as it turns out, $t = -1.39$ is *not* less than -1.895, because it is less negative and therefore larger. So, since $t = -1.39 > -1.895$, we *cannot* reject H in favor of A. We conclude, at level $\alpha = .05$, that the data *do not support* the opinion that corresponding occupancy rates are lower, on the average, in Houston than in Chicago.

TABLE 6.5   Calculations Leading to the Mean
and Standard Deviation of the Occupancy Rate
Data

x	$x - \bar{x}$	$(x - \bar{x})^2$
95.4	0.0	0.00
95.0	−0.4	0.16
97.5	2.1	4.41
95.6	0.2	0.04
93.3	−2.1	4.41
94.2	−1.2	1.44
94.9	−0.5	0.25
97.3	1.9	3.61
763.2	0.0	14.32

$$n = 8 \qquad \bar{x} = \frac{\Sigma x}{n} = 763.2/8 = 95.4$$

$$s = \sqrt{\frac{\Sigma(x - \bar{x})^2}{n - 1}} = \sqrt{14.32/7} = \sqrt{2.05} = 1.43$$

Our conclusion to this question does not, of course, imply that occupancy rates in Houston are higher than, or even equal to, those in Chicago. It merely says that the data do not provide sufficient evidence for
us to assert with confidence that Houston rates are really lower than
Chicago rates.

EXAMPLE 6.3.   Beer Distribution.   A major beer producer would like
to know if it would be worthwhile to establish an internal weather forecasting group within the company. The idea would be that specific
weather information could then be fed into a computer model that
would forecast daily beer sales for various regional distribution centers.
Inaccurate or vague temperature and precipitation forecasts would
result in a disparity between local beer demand and supply—too much
beer shipped to Cleveland, for example, might force an undersupply in
Washington, D.C. To find out whether or not an internal group could be
expected to develop accurate forecasts, 10 randomly selected meteorologists had been asked to predict the total rainfall in Cleveland for the
next weekend. Their predictions, in tenths of an inch, were as follows:

2.1   1.8   2.3   2.3   2.6   2.5   2.3   2.5   2.1   2.5

As it turned out, 2.5 tenths of an inch fell in Cleveland that weekend. At level α = .05, do the results of the study indicate that, on the average, meteorologists' predictions accurately estimate the actual rainfall?

SOLUTION. A one-sample t test is required for the solution of this problem, in view of the fact that we have fewer than 30 data points available, namely 10. We use μ to represent the true mean prediction of the rainfall by the entire population in question, which is probably all meteorologists available for employment by the producer. We then test the hypothesis

H: μ = 2.5 (on the average, meteorologists can accurately estimate the rainfall),

vs.

A: $\mu \neq$ 2.5 (on the average, meteorologists have difficulty accurately estimating the rainfall).

Under what circumstances would we feel compelled to reject H in favor of A? Well, as $\bar{x}$ is an approximation of μ, we would be inclined to reject H if the number

$$t = \frac{\bar{x} - 2.5}{s/\sqrt{n}}$$

turned out to be either too large or too small, because both of these extremes would tend to indicate that μ is relatively far from 2.5.

For a significance level of α = .05, we are willing to make a Type I error 5% of the time; this means that a rejection region of the sort pictured in Figs. 6.2, 6.3, 6.5, and 6.6 must have a shaded region representing a probability of .05. In the present example, however, the rejection region would have to appear in two parts, one part at each extreme of the graph of the t distribution. It would make sense to organize the rejection region in such a way that the far right part and the far left part each account for 2.5% of the possible values of t. (The alternative A: $\mu \neq$ 2.5 does not allow us to favor one extreme over the other.) We would therefore be using the 2.5% percentage point of the t distribution, the number $t_{.025} [n - 1]$ = $t_{.025} [9]$ = 2.262, to mark our points of division between the rejection and acceptance regions. A small portion of Table A.7 is reproduced in Table 6.6 to point up the selection of $t_{.025} [9]$.

We would therefore reject H at level α = .05 if either t were more negative than $-t_{.025} [9]$ = -2.262 or if t were greater than $t_{.025} [9]$ = 2.262. In other words, we would compute t and we would reject H in

TABLE 6.6   A Small Portion of Table A.7

df	$t_{.05}$	$t_{.025}$
9	1.833	2.262

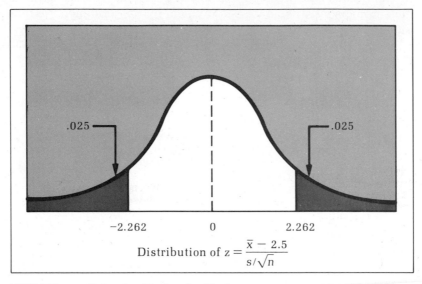

FIGURE 6.7   Rejection Region for H when α = .05
(Beer distribution example)

favor of A if $t < -2.262$ or if $t > 2.262$. We can express the two-part rejection rule more compactly in terms of the absolute value of t: we reject H if $|t| > 2.262$. The rejection region is illustrated in the diagram of Fig. 6.7.

The preliminary calculations appearing in Table 6.7 are aimed at obtaining the numerical values of $\bar{x}$ and s directly from the meteorologists' 10 data points. From those calculations, we discover that $\bar{x}$ = 2.3 and s = .245. You should observe that we have used for variety the "short-cut formula" for computing the standard deviation, namely formula (5.3), which was introduced in Section 5.B. Inserting these numbers into the formula for t, we get

$$ t = \frac{\bar{x} - 2.5}{s/\sqrt{n}} = \frac{2.3 - 2.5}{.245/\sqrt{10}} = \frac{-.2}{.245/3.162} = \frac{-.2}{.0775} = -2.58. $$

TABLE 6.7   Calculations Leading to the Mean
and Standard Deviations of the Beer
Distribution Data   *(Short-cut formula)*

x	x^2
2.1	4.41
1.8	3.24
2.3	5.29
2.3	5.29
2.6	6.76
2.5	6.25
2.3	5.29
2.5	6.25
2.1	4.41
2.5	6.25
23.0	53.44

$n = 10$

$$\bar{x} = \frac{\Sigma x}{n} = 23/10 = 2.3$$

$$s = \sqrt{\frac{n\Sigma x^2 - (\Sigma x)^2}{n(n-1)}} = \sqrt{\frac{10(53.44) - (23)^2}{(10)(9)}} - \sqrt{\frac{534.3 - 529}{90}} = \sqrt{5.4/90} = \sqrt{.06} = .245.$$

We agreed to reject H if $|t| > 2.262$. Because $t = -2.58$, it follows that $|t| = 2.58$, which indeed exceeds 2.262, and we should therefore reject H. We conclude, at level $\alpha = .05$, that meteorologists generally have difficulty, on the average, in accurately estimating rainfall. Our analysis of the forecasts reveals that, for a weekend on which 2.5 tenths of an inch of rain actually fell, the average meteorologist will tend to estimate the rainfall as being significantly different from 2.5.

The beer distribution example has provided an illustration of what is called a *two-tailed t test*. A two-tailed test is a statistical test having a two-part rejection region of the sort pictured in Fig. 6.7. Such a rejection region arises in tests of population means when the alternative is of the form A: $\mu \neq \mu_0$, as opposed to A: $\mu > \mu_0$ or A: $\mu < \mu_0$. For a two-tailed test having significance level α, we reject H if $|t| > t_{\alpha/2}[n - 1]$, using $\alpha/2$ rather than α to locate the rejection region.

For quick reference in dealing with situations like those discussed in the present section and also those of the next two sections, we can collect

TABLE 6.8   Rejection Rules in the Presence
of Various Alternatives for Tests Involving
Population Means $(H: \mu = \mu_0)$

Alternative	Reject H in Favor of A at Significance Level α if:
A: $\mu > \mu_0$	$z > z_\alpha$
A: $\mu < \mu_0$	$z < -z_\alpha$
A: $\mu \neq \mu_0$	$|z| > z_{\alpha/2}$

Note: While the above information is given in the language
of the one-sample z test, it is necessary merely to replace
the z_α and $z_{\alpha/2}$ by $t_\alpha [n - 1]$ and $t_{\alpha/2} [n - 1]$ in order to obtain
the rejection rules for the one-sample t test.

the various possible alternatives together with the rejection rules corresponding to them. This listing of possible alternatives appears in Table 6.8, and use
of that table will make it unnecessary to draw diagrams like those of Figs.
6.2, 6.3, 6.5, 6.6, and 6.7 for each and every problem that comes up. Table 6.8
incorporates the ideas that went into the determination of each of the rejection regions pictured above. We shall have occasion to refer to Table 6.8
during the next two sections of the present chapter and also when we study
some sections of Chapters 12 and 13.

EXERCISES   6.B

6.5. In the 1950s, the average number of children per family in the United States
was 2.35. The president of Amalgamated Baby Foods, Inc., wants to find out
whether the number of children per family has declined from that level during
the past decade. He selects a random sample of 81 families from around the
nation and discovers that they had 2.17 children, on the average, with a standard
deviation of 0.9 children. Do the data support, at level $\alpha = .05$, the contention
that families of the 1970s had, on the average, fewer children than families of
the 1950s?

6.6. An automobile manufacturer claims that his cars use an average of 4.50 gallons
of gasoline for each 100 miles. A consumer organization tests 36 of the cars and
finds that a mean of 4.65 gallons, with a standard deviation of 0.36 gallons, is
used for each 100 miles by the cars tested. At significance level $\alpha = .01$, do the
data provide evidence that the car really uses more than 4.50 gallons, on the
average, per 100 miles?

6.7. A machine that turns out plastic piping is, according to contractual specifications, supposed to be producing pipe of 4 inches in diameter. Piping that is
either too large or too small will not meet the construction requirements. As
part of a routine quality control test, a random sample of 50 pieces was carefully

measured, and their diameters had mean 4.05 inches and standard deviation .12 inch. At level $\alpha = .05$, can we conclude that the true mean diameter differs significantly from 4 inches?

6.8. Over a period of 50 months, the Desert Sands Casualty Insurance Company of Stovepipe Wells conducted an actuarial study, which showed that, over the 50-month period, its clients as a group caused an average of 17 auto accidents per month, with a standard deviation of 6. Using a significance level of $\alpha = .05$, should the company reconcile itself to the fact that the true mean number of accidents its clients cause per month is above 15?

6.9. A tobacco processor certified that his cigarettes contained no more than 21 mg of nicotine, on the average. A random sample of 25 cigarettes was analyzed by an independent medical researcher and found to have a mean of 22.6 mg of nicotine, with a standard deviation of 3 mg. At level $\alpha = .05$, do the data provide enough evidence to assert that the processor's certification was fallacious?

6.10. A small factory produces timing devices under subcontract to a major aerospace company. Ideally, the devices have a mean timing advance of 0.00 seconds per week. (A *negative* time advance is called a time lag.) A sample of six devices, tested as they came off the assembly line, had the following advances:

Serial no. of device	00693A	01014F	01983B	04441W	15672P	18449G
Time advance as tested	0.1	0.2	0.1	0.0	0.1	0.0

At significance level $\alpha = .10$, do the data provide evidence that the true mean advance is different from 0.00, so that the production process requires some adjustment?

6.11. If a new, slightly cheaper process for mining copper is adopted, it will not be economical if it yields fewer than 50 tons of ore per day in an experimental mine. The results of a five-day trial period are summarized by the following daily production figures (in tons):

$$50 \quad 47 \quad 53 \quad 51 \quad 52$$

At significance level $\alpha = .01$, do the results indicate that the mean production does not fall below 50 tons per day, and so that the new process is economical?

6.12. A manufacturer of chemicals for use in the pharmaceutical industry conducts a study of the time it takes, after receipt of an order, for the chemicals to reach the customer's plant. The following data give the transit times from the manufacturer to the customer of six randomly selected shipments of the past month:

Shipment no.	017	009	021	058	031	002
Transit time (days)	6.0	5.0	3.0	4.5	4.0	5.5

Can the manufacturer be truthful, at significance level $\alpha = .05$, in asserting that the true mean shipping time does not differ from 4.5 days?

6.13. One brand of automobile, the sleek and sophisticated Peccarie, is advertised as being able to get 52 miles per gallon at freeway speeds. In test runs over a standard freeway mock-up, 100 randomly selected specimens got an average of 46 miles per gallon, with a standard deviation of 10. At level $\alpha = .01$, can we consider the advertised statement to be correct?

6.14. Based on the *U.S. News & World Report* unemployment rate data appearing in Exercise 5.17, can we assert that the nationwide average urban unemployment rate was below 6.5% at the time the survey was taken? Use a level of significance of $\alpha = .05$.

6.15. The following data list the number of employees of 12 randomly selected member corporations of the Fortune 500 largest U.S. industrial corporations:*

Corporation	Rank by Sales Volume	Number of Employees
American Petrofina	212	3,209
Cluett, Peabody	330	24,265
Hobart	415	12,056
Mead	141	26,200
Springs Mills	311	20,100
Weyerhauser	75	47,211
Phelps Dodge	240	14,700
Liggett Group	283	6,800
Gold Kist	251	6,726
Frederick & Herrud	469	1,300
Bendix	70	79,700
National Gypsum	314	12,738

At level $\alpha = .10$, do the sample data indicate that the average number of employees of Fortune 500 industrial corporations is below 20,000?

6.16. A newspaper columnist asserted that the mean 1976 chief executive pay of major United States corporations significantly exceeded $300,000 for the year. Is this assertion substantiated at level $\alpha = .10$ by the data of Exercise 2.5?

SECTION 6.C   ANALYSIS OF THE TYPE II ERROR

Let's look again at the Nitro-Plus situation of Section B, where we were testing the hypothesis

     H: $\mu = 20,000$,

*Fortune, May 1977, pp. 366–385.

against the alternative

A: $\mu > 20,000$.

Here μ represented the true mean yield (in pounds of tomatoes) per acre of farmland treated with Nitro-Plus fertilizer. Our data on $n = 40$ randomly selected acres showed a sample mean yield of $\bar{x} = 20,400$ pounds per acre with a sample standard deviation of $s = 1,200$ pounds. Our one-sample z statistic turned out to have the numerical value

$$z = \frac{\bar{x} - \mu_0}{s/\sqrt{n}} = \frac{20,400 - 20,000}{1,200/\sqrt{40}} = 2.11,$$

permitting us to reject H at significance level $\alpha = .05$, but not at significance level $\alpha = .01$.

The significance level of the test is, as discussed in Section 6.A, the probability of making a Type I error, namely the probability of rejecting H when H is really true. When we use $\alpha = .05$ as the significance level, we are agreeing to reject H 5% of the time when it is really true. If we lower α to .01, we want to reject H only 1% of the time when it's really true. When $\alpha = .05$, we should reject H whenever $z > 1.64$, while, when $\alpha = .01$, we should reject H whenever $z > 2.33$. By raising α to .05 from .01, we are increasing our probability of making a Type I error by making it easier to reject H. Therefore, in borderline cases where it is not too clear that H is either definitely true or definitely false, we could reject H when $\alpha = .05$ but accept it when $\alpha = .01$.

Now, the probability β of making a Type II error is the probability of accepting H when H is really false. Unfortunately, the statement that "H is really false" is considerably more vague than the statement that "H is really true," because there are many ways for H to be false, but only one way for it to be true. For example, H is false if $\mu = 20,500$ or $\mu = 21,000$ or $\mu = 21,435$, or any one of an infinite number of possible values, but H is true *only* if $\mu = 20,000$. Therefore it is impossible to calculate the probability β of accepting H when H is really false. The best we can do is to calculate the probability of accepting H for each possible value of μ ($> 20,000$) separately.

Let's first work with the case of the significance level $\alpha = .05$, so that we ought to reject H in favor of A if $z > 1.64$. We should therefore accept H if $z \leq 1.64$. By analogy with our procedure of denoting the hypothesized mean by μ_0, we will use the symbol μ_A to denote the various alternative values of $\mu > \mu_0$. If we borrow the language of conditional probability, we can represent for each possible value of μ_A the probability of making a Type II error as

$$\beta(\mu_A) = P(\text{Type II error if } \mu = \mu_A)$$

$$= P(\text{accept H}| \mu = \mu_A) = P(z \leq 1.64| \mu = \mu_A).$$

In the above expression, z will have the same formula as before, namely formula (6.1). However, under the assumption that $\mu = \mu_A$, this value of z will no longer have the standard normal distribution. Instead, the quantity

$$z_A = \frac{\overline{x} - \mu_A}{\sigma/\sqrt{n}}$$

will be a standard normal random variable. (We must replace μ in the statement of the central limit theorem by μ_A, because we are assuming that $\mu = \mu_A$, while the true standard deviation σ remains unchanged.)

To make a typical Type II error calculation, we can set $\mu_A = 20,500$, which is one of the possible values of μ when H is false. Our decision is still based on the same data as before, so that $n = 40$, $\overline{x} = 20,400$, and $s = 1,200$. We use the following formula (where we have replaced σ by the best estimate we have for it, namely $s = 1,200$):

$$\beta(\mu_A) = P(z \leq z_\alpha | \mu = \mu_A) = P\left(\frac{\overline{x} - \mu_0}{s/\sqrt{n}} \leq z_\alpha \,\middle|\, \mu = \mu_A\right)$$

$$= P\left(\frac{\overline{x} - \mu_0}{s/\sqrt{n}} + \frac{\mu_0 - \mu_A}{s/\sqrt{n}} \leq z_\alpha + \frac{\mu_0 - \mu_A}{s/\sqrt{n}} \,\middle|\, \mu = \mu_A\right)$$

$$= P\left(\frac{\overline{x} - \mu_A}{s/\sqrt{n}} \leq \frac{\mu_0 - \mu_A}{s/\sqrt{n}} + z_\alpha \,\middle|\, \mu = \mu_A\right)$$

$$= P\left(\text{SNRV} \leq \frac{\mu_0 - \mu_A}{s/\sqrt{n}} + z_\alpha\right) \qquad (6.3)$$

where SNRV represents a standard normal random variable. We therefore simply look up the proportion in Table A.4 corresponding to a z value of $(\mu_0 - \mu_A)/(s/\sqrt{n}) + z_\alpha$, where α is the significance level of the test.

Setting $\mu_A = 20,500$, along with the known values of $\mu_0 = 20,000$, $s = 1,200$, $n = 40$, and $z = 1.64$ (since $\alpha = .05$), we get

$$\beta(\mu_A) = P(\text{SNRV} \leq \frac{20,000 - 20,500}{1,200/\sqrt{40}} + 1.64) = P(\text{SNRV} \leq \frac{-500}{1,200/6.32} + 1.64)$$

$$= P(\text{SNRV} \leq -2.63 + 1.64) = P(\text{SNRV} \leq -0.99) = 0.1611$$

referring to Table A.4. This means that, if the true mean yield were actually 20,500 pounds, and we wanted to test H:$\mu = 20,000$ against A:$\mu > 20,000$ at significance level (probability of Type I error) $\alpha = .05$, we would make a Type II error about .1611 = 16.11% of the time. That is, our chances of being fooled into thinking that $\mu = 20,000$ when actually $\mu = 20,500$ are .1611. (The fact that the significance level is $\alpha = .05$ means that our chances of being fooled into thinking that $\mu > 20,000$ when actually $\mu = 20,000$ are .05 = 5%.)

What about the Type II error when $\alpha = .01$? When $\mu_A = 20,500$ and $z_\alpha = z_{.01} = 2.33$, we have that

$$\beta(\mu_A) = P(\text{SNRV} \leq \frac{20,000 - 20,500}{1,200/\sqrt{40}} + 2.33)$$

$$= P(\text{SNRV} \leq -2.63 + 2.33)$$

$$= P(\text{SNRV} \leq -0.30) = .3821$$

according to Table A.4. This calculation shows that, if we desire to lower our probability of making a Type I error from .05 down to .01, our probability of making a Type II error rises from .1611 up to .3821. This inverse relationship between α and β must be carefully taken into acount when we are making our choice of α.

Using formula (6.3), we can determine the Type II error probability $\beta(\mu_A)$ for any desired alternative value μ_A of the true mean. The simplest way to report such information would be by way of a table of the sort illustrated in Table 6.9. In addition to showing that the Type II error probabilities are considerably greater when α, the Type I error probability, is smaller, Table 6.9 also shows that the Type II error probabilities are rather substantial when μ_A is very close to $\mu_0 = 20,000$. We should really have expected this, because if the true mean is very close to the hypothesized mean, it would be unlikely that our statistical test could distinguish between the two. We would be strongly inclined to accept H, thereby committing a Type II error.

Because $\beta(\mu_A)$ is the probability of accepting H when the actual value of μ is μ_A, there is a strong analogy between Table 6.9 and the acceptance sampling probabilities of Tables 3.4 and 3.5 in Section 3.D. We can construct an operating characteristic (OC) curve analogous to that of Fig. 3.13 and based

TABLE 6.9   Type II Error Probabilities in the Nitro-Plus Example
(H: $\mu = 20,000$ vs. A: $\mu > 20,000$)

μ_A	$z_A = \dfrac{\mu_0 - \mu_A}{s/\sqrt{n}}$	$\alpha = .05$ $z_\alpha = 1.64$	$\alpha = .01$ $z_\alpha = 2.33$	$\beta(\mu_A)$ when $\alpha = .05$	$\beta(\mu_A)$ when $\alpha = .01$
20,100	-0.53	1.11	1.80	.8665	.9641
20,200	-1.05	0.59	1.28	.7224	.8997
20,300	-1.58	0.06	0.75	.5239	.7734
20,400	-2.11	-0.47	0.22	.3192	.5871
20,500	-2.63	-0.99	-0.30	.1611	.3821
20,600	-3.16	-1.52	-0.83	.0643	.2033
20,700	-3.69	-2.05	-1.36	.0202	.0869
20,800	-4.21	-2.57	-1.88	.0051	.0301
20,900	-4.74	-3.10	-2.41	.0010	.0080
21,000	-5.27	-3.63	-2.94	.0002	.0016

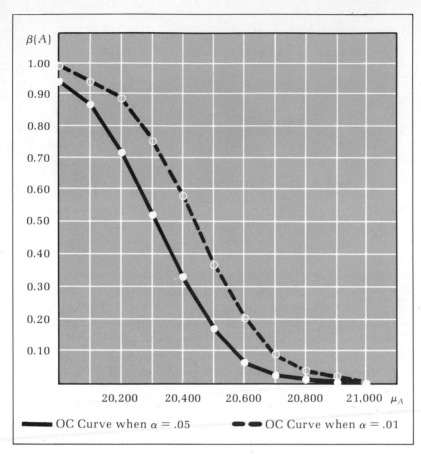

FIGURE 6.8 Operating Characteristic (OC) Curve for Type II Errors
(Nitro-plus example)

on the acceptance probabilities of Table 6.9. We do this in Fig. 6.8, where the
solid line traces $\beta(\mu_A)$ when $\alpha = .05$ and the broken line illustrates $\beta(\mu_A)$
when $\alpha = .01$.

The quantity $\pi(\mu_A) = 1 - \beta(\mu_A)$ is the probability of *avoiding* a Type
II error, because $\beta(\mu_A)$ is the probabiity of *making* a Type II error. The probability $\pi(\mu_A)$ is called the *power* of the test against the alternative $\mu = \mu_A$,
since it signifies the ability of the test to fight off a Type II error. Because the
relationship between $\pi(\mu_A)$ and $\beta(\mu_A)$ is so simple, it is easy to tabulate the
various power values in a form such as appears in Table 6.10. On the basis
of Table 6.10, we can construct *power curves* of the sort illustrated in Fig.
6.9. Because the power curves rise as μ_A gets farther and farther away from
μ_0, we see that the test grows more resistant to the Type II error.

Before closing this section, we present formulas analogous to formula
(6.3), to be used for computing $\beta(\mu_A)$ in case of alternatives A: $\mu < \mu_0$ and A:

TABLE 6.10 Power of the Test in the Nitro-Plus Example

μ_A	$\alpha = .05$		$\alpha = .01$	
	$\beta(\mu_A)$	$\pi(\mu_A)$	$\beta(\mu_A)$	$\pi(\mu_A)$
20,100	.8665	.1335	.9641	.0359
20,200	.7224	.2776	.8997	.1003
20,300	.5239	.4761	.7734	.2266
20,400	.3192	.6808	.5871	.4129
20,500	.1611	.8389	.3821	.6179
20,600	.0643	.9357	.2033	.7967
20,700	.0202	.9798	.0869	.9131
20,800	.0051	.9949	.0301	.9699
20,900	.0010	.9990	.0080	.9920
21,000	.0002	.9998	.0016	.9984

FIGURE 6.9 Power Curves in the Nitro-Plus Example

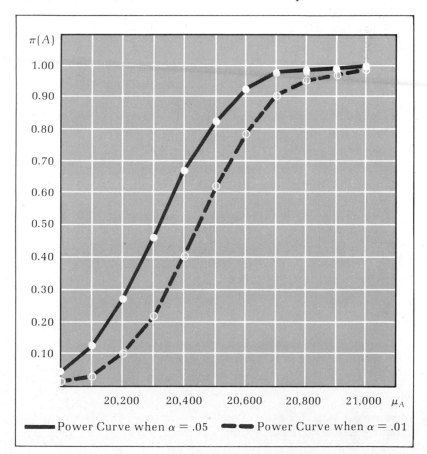

Power Curve when $\alpha = .05$ ● ● Power Curve when $\alpha = .01$

$\mu \neq \mu_0$. When we are testing H: $\mu = \mu_0$ against A: $\mu < \mu_0$, the Type II error probability against $\mu = \mu_A$ is given by

$$\blacksquare \qquad \beta(\mu_A) = P\left(SNRV \geq \frac{\mu_0 - \mu_A}{s/\sqrt{n}} - z_\alpha\right) \qquad (6.4)$$

and when we are testing H: $\mu = \mu_0$ against A: $\mu \neq \mu_0$, we use the formula

$$\blacksquare \qquad \beta(\mu_A) = P\left(\frac{\mu_0 - \mu_A}{s/\sqrt{n}} - z_{\alpha/2} \leq SNRV \leq \frac{\mu_0 - \mu_A}{s/\sqrt{n}} + z_{\alpha/2}\right). \qquad (6.5)$$

The exercises following this section will provide examples where formulas (6.4) and (6.5) should be used.

Before we close this section, we should mention the effect that sample size has on the probability of making a Type II error. Take a look, for example, at formula (6.5). If the sample size n is very large, then the term

$$\frac{\mu_0 - \mu_A}{s/\sqrt{n}} = (\mu_0 - \mu_A) \div \frac{s}{\sqrt{n}} = (\mu_0 - \mu_A)\left(\frac{\sqrt{n}}{s}\right) = \left(\frac{\mu_0 - \mu_A}{s}\right)\sqrt{n}$$

is correspondingly very large in magnitude (in either the positive or negative direction, depending upon the sign of $\mu_0 - \mu_A$). Therefore the interval

$$\frac{\mu_0 - \mu_A}{s/\sqrt{n}} - z_{\alpha/2} \leq SNRV \leq \frac{\mu_0 - \mu_A}{s/\sqrt{n}} + z_{\alpha/2}$$

would be found in one of the extreme tails of the bell-shaped curve, and the probability associated with it would consequently be very small. This small probability is $\beta(\mu_A)$. To summarize, we can decrease our probability of making a Type II error by working with as large a sample as is practical to select.

EXERCISES 6.C

6.17. In Exercise 6.5, the significance level $\alpha = .05$ was used to test whether or not families of the 1970s had, on the average, fewer children than families of the 1950s.

   a. Calculate the Type II error probabilities and power of the test against various alternatives.

   b. Construct the operating characteristic and power curves of the test.

6.18. In Exercise 6.6, the significance level $\alpha = .01$ was used to test whether or not the car uses more than 4.50 gallons, on the average, per 100 miles.

   a. Compute the Type II error probabilities of the test against several alternatives.

   b. Draw the OC curve of the test.

6.19. Exercise 6.7 asked for a test at level $\alpha = .05$ of whether or not the true mean diameter differs significantly from 4 inches.

    a. Determine the power of the test against a range of alternatives.

    b. Graph the power curve of the test.

6.20. In the actuarial study of Exercise 6.8, a significance level of α = .05 was used in a test of whether or not the average number of accidents caused per month exceeds 15.

    a. Calculate the Type II error probabilities of the test against various alternatives.

    b. Construct the OC curve of the test.

6.21. Example 6.1 of Section B dealt with calculator battery lifetimes. A level of α = .10 was used to test whether or not the true mean duration was less than 22 hours.

    a. Compute the power of the test against several alternatives.

    b. Draw the power curve of the test.

SECTION 6.D   TESTING FOR THE DIFFERENCE BETWEEN MEANS

A local manufacturer in Milwaukee, with a view toward eventually cutting costs, conducts a comparison test of the length of time light bulbs last before requiring replacement. A sample of 40 randomly selected Universal Electric bulbs, the brand currently used, has a mean lifetime of 1,110 hours, with a standard deviation of 20 hours. A cheaper brand (Light, Ltd.), which is under consideration as a replacement, is also tested, and a random sample of 60 bulbs turn out to have a mean lifetime of 1,081 hours, with a standard deviation of 18 hours. The company's buyer would like to use the cheaper bulbs unless the results of the comparison test indicate strongly that the cheaper ones are significantly worse.

Large Samples

In answering this question, it is apparent that we are dealing with two distinct samples of data—the 40 Universal Electric bulbs and the 60 Light, Ltd., bulbs. We are comparing these two samples against each other, rather than against a hypothesized mean as was done in Section 6.B. Because a large number of data points is available, the appropriate statistical test to use is the *two-sample z test*. If we set

μ_U = true mean lifetime of *Universal* Electric bulbs

μ_L = true mean lifetime of *Light*, Ltd., bulbs

then the company's buyer wants to test the hypothesis

H: $\mu_L = \mu_U$ (there is no real difference between the
cheaper and the more expensive bulbs),

vs.

A: $\mu_L < \mu_U$ (the cheaper bulbs are really worse in length
of lifetime).

The formula for the two-sample z test is

■

$$z = \frac{\overline{x}_L - \overline{x}_U}{\sqrt{\dfrac{s_L^2}{n_L} + \dfrac{s_U^2}{n_U}}}$$

(6.6)

where

n_L = number of Light, Ltd., bulbs in sample
n_U = number of Universal Electric bulbs in sample
$\overline{x}_L$ = sample mean of Light, Ltd., bulbs
$\overline{x}_U$ = sample mean of Universal Electric bulbs
s_L = sample standard deviation of Light, Ltd., bulbs
s_U = sample standard deviation of Universal Electric bulbs

(Here the denominator of z, a combination of the standard deviations of the
two samples of data, plays the role of $s/\sqrt{n} = \sqrt{s^2/n}$.) By analogy with the
second in the list of alternatives presented in Table 6.8, we should reject H:
$\mu_L = \mu_U$ in favor of A: $\mu_L < \mu_U$ at significance level α if $z < -z_\alpha$.
    What would be an appropriate level of significance for this problem?
Well, because the Light, Ltd., bulbs are cheaper, the company would prefer
to use that brand unless it was convincingly demonstrated that they are really
worse. This means that they would not like to reject H unless it was really
false. Therefore, a test procedure with a very low probability of making a
Type I error would be quite appealing to the buyer. Since the significance
level α is the probability of a Type I error, α should be chosen to be very
small, say $\alpha = .005$. For this value of α, we would have $z_\alpha = z_{.005} = 2.57$. It
follows that we should reject H in favor of A if $z < -2.57$.
    From the data, we know that

$$n_L = 60 \qquad \overline{x}_L = 1081 \qquad s_L = 18$$

$$n_U = 40 \qquad \overline{x}_U = 1110 \qquad s_U = 20$$

Inserting these numbers into their proper places in the formula for z, we get

$$z = \frac{\overline{x}_L - \overline{x}_U}{\sqrt{\dfrac{s_L^2}{n_L} + \dfrac{s_U^2}{n_U}}} = \frac{1081 - 1110}{\sqrt{\dfrac{(18)^2}{60} + \dfrac{(20)^2}{40}}} = \frac{-29}{\sqrt{\dfrac{324}{60} + \dfrac{400}{40}}} = \frac{-29}{\sqrt{5.4 + 10.0}}$$

$$= -29/\sqrt{15.4} = -29/3.92 = -7.4.$$

We agreed to reject H if $z < -2.57$. As it turned out, $z = -7.4 < -2.57$, so we must reject H. Our conclusion is that, at level $\alpha = .005$, the cheaper Light, Ltd., bulbs are really worse than the Universal Electric bulbs, in the sense that they last significantly fewer hours. On the basis of the information gained from this study, we would advise the company to continue using Universal Electric light bulbs.

Small Samples

EXAMPLE 6.4.   Collection Methods.   The Mainstreet, a regional chain of department stores located primarily in small rural towns, wanted to find out which of two suggested methods of collecting accounts receivable is most effective. They first divided a sample of their customers by random assignment into two groups. To one group of six customers, they mailed a series of past due notices at the end of 30 days, 60 days, and 90 days beyond the invoice data. To a second group of nine customers they made telephone calls at the same time intervals. From the six customers in the mail group, payment was received, respectively, 62, 68, 106, 47, 94, and 43 days after the invoice date. The nine customers in the telephone group paid their bills, respectively, 81, 31, 93, 44, 49, 38, 62, 34, and 36 days past the invoice date. At level of significance $\alpha = .10$, do the sample results indicate that the telephone method is more effective than the mail method in speeding up collection of accounts receivable?

SOLUTION.   We must use the two-sample t test because of the relatively small number of data points available, six in one sample and nine in the other. We define the following symbols for the true means relevant to the problem:

μ_M = true mean number of days until payment when *mail* method is used.

μ_T = true mean number of days until payment when *telephone* method is used.

We then want to test the hypothesis

H: $\mu_M = \mu_T$ (there is no difference in effectiveness between the two collection methods),

vs.

A: $\mu_M > \mu_T$ (the telephone method is more effective).

The number of degrees of freedom (df) appropriate to the two-sample t test is

$df = n_M + n_T - 2$, where
n_M = number of data points in mail sample
n_T = number of data points in telephone sample.

In this problem $n_M = 6$ and $n_T = 9$, so that

$df = n_M + n_T - 2 = 6 + 9 - 2 = 13$.

Because our alternative is A: $\mu_M > \mu_T$, a glance at Table 6.8 shows that we should reject H in favor of A if

$t > t_\alpha [n_M + n_T - 2] = t_{.10} [13] = 1.350$,

in view of the fact that our significance level here is $\alpha = .10$.

The only task remaining in the solution of this problem is the computation of the numerical value of t, which we have to compare with $t_{.10} [13] = 1.350$. The value of t in the two-sample t test is calculated using the formula

$$t = \frac{\overline{x}_M - \overline{x}_T}{\sqrt{\dfrac{\Sigma(x_M - \overline{x}_M)^2 + \Sigma(x_T - \overline{x}_T)^2}{n_M + n_T - 2} \left(\dfrac{1}{n_M} + \dfrac{1}{n_T} \right)}} \qquad (6.7)$$

where

$\overline{x}_M$ = sample mean number of days until payment when mail method is used, and
$\overline{x}_T$ = sample mean number of days until payment when telephone method is used.

The quantities $\Sigma(x_M - \overline{x}_M)^2$ and $\Sigma(x_T - \overline{x}_T)^2$ are the sums of squared deviations from the mean, analogous to the number computed by summing the third column of Table 6.5, for example. The square root

$$\sqrt{\frac{\Sigma(x_M - \overline{x}_M)^2 + \Sigma(x_T - \overline{x}_T)^2}{n_M + n_T - 2}}$$

is called the *pooled standard deviation* of the two sets of data. We calculate the components of the formula for t by constructing two tables like that of Table 6.5, one such table for each of the two samples of data. The calculations are presented in Table 6.11. From the information pre-

TABLE 6.11 Calculations Leading to the Two-Sample t Statistic (Collection methods data)

	Mail Method			Telephone Method	
x_M	$x_M - \bar{x}_M$	$(x_M - \bar{x}_M)^2$	x_T	$x_T - \bar{x}_T$	$(x_T - \bar{x}_T)^2$
62	−8	64	81	29	841
68	−2	4	31	−21	441
106	36	1296	93	41	1681
47	−23	529	44	−8	64
94	24	576	49	−3	9
43	−27	729	38	−14	196
			62	10	100
			34	−18	324
			36	−16	256
420	0	3198	468	0	3912

$n_M = 6$ $\Sigma(x_M - \bar{x}_M)^2 = 3{,}198$ $\bar{x}_T = 468/9 = 52$

$\bar{x}_M = 420/6 = 70$ $n_T = 9$ $\Sigma(x_T - \bar{x}_T)^2 = 3{,}912$

sented in Table 6.11, we see that the components of the formula for t are:

$$n_M = 6 \qquad \bar{x}_M = 70 \qquad \Sigma(x_M - \bar{x}_M)^2 = 3{,}198$$

$$n_T = 9 \qquad \bar{x}_T = 52 \qquad \Sigma(x_T - \bar{x}_T)^2 = 3{,}912$$

Inserting these numbers into their proper places in formula (6.7), we obtain that

$$t = \frac{\bar{x}_M - \bar{x}_T}{\sqrt{\dfrac{\Sigma(x_M - \bar{x}_M)^2 + \Sigma(x_T - \bar{x}_T)^2}{n_M + n_T - 2}\left(\dfrac{1}{n_M} + \dfrac{1}{n_T}\right)}}$$

$$= \frac{70 - 52}{\sqrt{\dfrac{3{,}198 + 3{,}912}{6 + 9 - 2}\left(\dfrac{1}{6} + \dfrac{1}{9}\right)}} = \frac{18}{\sqrt{\left(\dfrac{7{,}110}{13}\right)\left(\dfrac{15}{54}\right)}}$$

$$= \frac{18}{\sqrt{106{,}650/702}} = \frac{18}{\sqrt{151.92}} = \frac{18}{12.33} = 1.46.$$

We agreed to reject H in favor of A if t turned out to be larger than $t_{.10}[13] = 1.350$. Because $t = 1.46 > 1.350$, we must therefore reject H.

Our conclusion, at level $\alpha = .10$, is that the telephone method of collection is significantly more effective than the mail method.

EXAMPLE 6.5.   Restaurant Locations.   A restaurateur, holding the rights to a local franchise of the Ms. Neptune's Scrumptious Sardine Sandwich chain of fast-food restaurants, has to determine which of two prospective locations is potentially more desirable. She studied a report in a local business magazine containing published accounts of seven randomly selected fast-food restaurants on the east side of town, and found that they had mean sales of $6 thousand per week with a standard deviation of $1.5 thousand. She then compared this with the mean sales of $5.5 thousand and the standard deviation of $1.4 thousand for a random sample of eight restaurants on the west side of town. At level $\alpha = .05$, can she conclude from the data that there is a significant difference between the mean weekly sales of fast-food restaurants on the east and west sides of town?

SOLUTION.   The problem calls for a two-sample t test of the hypothesis

H: $\mu_E = \mu_W$

against the alternative

A: $\mu_E \neq \mu_W$

where

μ_E = true mean weekly sales on east side of town, and
μ_W = true mean weekly sales on west side of town.

We have used the alternative A: $\mu_E \neq \mu_W$ because the restaurateur wants to know only whether or not mean weekly sales differed in the two locations. She is not interested at present in finding out which of two locations has the larger mean, but only whether the two means were different.

The number of degrees of freedom for this problem is $df = n_E + n_W - 2 = 7 + 8 - 2 = 13$, and Table 6.8 requires us to reject H at level $\alpha = .05$ if

$|t| > t_{\alpha/2} [df] = t_{.025} [13] = 2.160,$

because our alternative is A; $\mu_E \neq \mu_W$. It now remains only to compute the value of

$$t = \frac{\bar{x}_E - \bar{x}_W}{\sqrt{\dfrac{\Sigma(x_E - \bar{x}_E)^2 + \Sigma(x_W - \bar{x}_W)^2}{n_E + n_W - 2}\left(\dfrac{1}{n_E} + \dfrac{1}{n_W}\right)}}$$

and to check to see whether or not $|t| > 2.160$.

From the information given in the statement of the problem, we know the following quantities:

$$n_E = 7 \qquad \bar{x}_E = 6 \qquad s_E = 1.5$$

$$n_W = 8 \qquad \bar{x}_W = 5.5 \qquad s_W = 1.4$$

Inserting as many of these numbers into the formula for t as will fit, we obtain that

$$t = \frac{6 - 5.5}{\sqrt{\dfrac{\Sigma(x_E - \bar{x}_E)^2 + \Sigma(x_W - \bar{x}_W)^2}{7 + 8 - 2}\left(\dfrac{1}{7} + \dfrac{1}{8}\right)}}.$$

Unfortunately, we cannot complete the calculation of t because we were not given the numerical values of $\Sigma(x_E - \bar{x}_E)^2$ and $\Sigma(x_W - \bar{x}_W)^2$. How are we to find these numbers?

Well, how did we manage to do it in the last two examples? In those examples, we had access to all the original data points, so we merely computed the sums of squared deviations from the mean directly. That computation was carried out explicitly in Table 6.11. This time, we unfortunately do not know the original data points, but instead we know only how many there were and what their sample mean and standard deviation were. What are we going to do?

The quantities we need to know, namely $\Sigma(x_E - \bar{x}_E)^2$ and $\Sigma(x_W - \bar{x}_W)^2$, are mathematically related to the sample standard deviations by the formulas

$$s_E = \sqrt{\frac{\Sigma(x_E - \bar{x}_E)^2}{n_E - 1}}$$

and

$$s_W = \sqrt{\frac{\Sigma(x_W - \bar{x}_W)^2}{n_W - 1}}$$

for those formulas would be used to compute s_E and s_W if we did know or were able to compute $\Sigma(x_E - \bar{x}_E)^2$ and $\Sigma(x_W - \bar{x}_W)^2$. Now, however, the shoe is on the other foot: we know s_E and s_W, but we want to know $\Sigma(x_E - \bar{x}_E)^2$ and $\Sigma(x_W - \bar{x}_W)^2$. Therefore we develop a formula for calculating $\Sigma(x_E - \bar{x}_E)^2$ and $\Sigma(x_W - \bar{x}_W)^2$ based upon our knowledge of the values of s_E and s_W.

In the above formulas for s_E and s_W, we can square both sides, obtaining thereby

$$s_E^2 = \left(\sqrt{\frac{\Sigma(x_E - \bar{x}_E)^2}{n_E - 1}} \right)^2 = \frac{\Sigma(x_E - \bar{x}_E)^2}{n_E - 1} \text{, and}$$

$$s_W^2 = \left(\sqrt{\frac{\Sigma(x_W - \bar{x}_W)^2}{n_W - 1}} \right)^2 = \frac{\Sigma(x_W - \bar{x}_W)^2}{n_W - 1}.$$

It follows algebraically from the above results that

$$\Sigma(x_E - \bar{x}_E)^2 = (n_E - 1)s_E^2$$

and

$$\Sigma(x_W - \bar{x}_W)^2 = (n_W - 1)s_W^2. \qquad (6.8)$$

Using these expressions, we can indirectly calculate the numerical values of $\Sigma(x_E - \bar{x}_E)^2$ and $\Sigma(x_W - \bar{x}_W)^2$ that we need in order to complete the computation of t. In particular, we get that

$$\Sigma(x_E - \bar{x}_E)^2 = (n_E - 1)s_E^2 = (7 - 1)(1.5)^2 = (6)(2.25) = 13.5, \text{and}$$

$$\Sigma(x_W - \bar{x}_W)^2 = (n_W - 1)s_W^2 = (8 - 1)(1.4)^2 = (7)(1.96) = 13.72.$$

Inserting these numbers into their proper places in the expression for t, we see that

$$t = \frac{6 - 5.5}{\sqrt{\frac{13.5 + 13.72}{7 + 8 - 2}\left(\frac{1}{7} + \frac{1}{8}\right)}} = \frac{0.5}{\sqrt{\left(\frac{27.22}{13}\right)\left(\frac{15}{56}\right)}}$$

$$= \frac{.5}{\sqrt{408.3/728}} \doteq .5/\sqrt{.561} = .5/.749 = .67.$$

The numerical value of t therefore turned out to be $t = .67$, while, as you will recall, we agreed to reject H if $|t| > 2.160$. In view of the fact that $|t| = .67 < 2.160$, we cannot reject H at level of significance $\alpha = .05$. We therefore conclude that the data does not reveal a significant difference between the true mean sales of fast-food restaurants on the two sides of town.

Conditions on the Use of Two-Sample Tests

With this example, we have completed our discussion of two-sample tests for the difference of means. Before we go on to another topic, we should point out the conditions under which we may validly use these tests for the solution of applied problems. Those conditions are as follows.

When large samples of data ($n > 30$) are available, the central limit theorem discussed in Section 5.B allows us to use the percentage points of the normal distribution in making a statistical decision. When only small samples are available, we have to restrict ourselves to normally distributed data as we did in Section 5.C. This restriction is necessary only because the table of the t distribution is derived from the table of the normal distribution for the expressed purpose of taking care of small-sample situations. In connection with the two-sample tests, two further restrictions are required: (1) it must be reasonable for us to assume that both samples of data have approximately the same "true" standard deviation, and (2) both samples of data must be selected independently of each other; i.e., individual members of one sample must bear no relation to individual members of the other sample.

CRITERIA FOR USE OF THE TWO-SAMPLE t TEST

We use a two-sample t test for the difference of means when the following four criteria are met:

1. Fewer than 32 data points (total of both samples) are available;

2. Both sets of data come from normally distributed populations;

3. Both sets of data have the same "true" population standard deviation; and

4. The sets of data are independent of each other.

Whenever a problem seems to call for the two-sample t test, then, the above criteria must be carefully observed so that we can be sure that our solution of the problem is based on sound statistical principles rather than on the shifting sands of convenience. In future sections and chapters, we therefore must also study ways of checking whether or not these criteria are satisified in any particular problem of interest. Criterion 1, of course, can be checked merely by counting the number of available data points. For criterion 2, we will study in Section 13.C the chi-square test of normality, the objective of which is to decide whether or not a set of randomly selected data has come from a population that is normally distributed. In Section 6.F, we will show how to check criterion 3 by testing for the difference of two standard deviations. Criterion 4 can be analyzed only by thinking about the nature of the problem under discussion and the way in which the data were gathered.

What do we do in two-sample problems when one or more of these criteria are not met? If criterion 1 is not met, but all the others are, we can use the two-sample z test. If criteria 2 or 3 or both are not met, we must use a nonparametric test such as the sign test or the Mann-Whitney test to study the differences between the sets of data. These nonparametric tests will be studied in detail in Chapter 12. Finally, when criterion 4 is not satisfied, we must use the paired-sample t test, which we will present in the next section.

EXERCISES 6.D

6.22. A new chemical fertilizer, Nitro-Plus, yielded 20,400 pounds of tomatoes, on the average, with a standard deviation of 1,200 pounds on 40 randomly selected acres of farmland. On another 100 randomly selected acres, the standard organic fertilizer produced a mean yield of 19,000 pounds, with a standard deviation of 1,000 pounds. At significance level $\alpha = .01$, do the results of the comparison indicate that the chemical fertilizer really produces larger yields than the organic fertilizer?

6.23. Two methods of teaching a complicated assembly-line component operation to new employees, the Chicago method and the Boston method, were applied to several randomly selected new employees divided into two groups. On an efficiency test given to both groups at the conclusion of the training program, the 60 employees exposed to the Chicago method scored a mean of 20 with a standard deviation of 7, while the 50 employees exposed to the Boston method had an average score of 22 with a standard deviation of 8. At level $\alpha = .05$, can we say that the two methods differ significantly in effectiveness?

6.24. Last year a sample of 100 new college graduates accepted jobs paying an average of $1,050 per month, with a standard deviation of $40. This year, 50 recently hired new graduates will start at an average pay level of $1,070 per month, with a standard deviation of $30. At level $\alpha = .01$, do the data indicate that starting pay levels are rising for new college graduates?

6.25. An economist wants to know whether Commonwealth trade arrangements result in lower prices in Canada for British imports. As an example, he obtains price data on English cheese in six major metropolitan areas in the U.S. and six in Canada. The data, in cents per pound, follow.

U.S. Prices	250	260	280	270	260	240
Canada Prices	255	260	250	240	265	255

At significance level $\alpha = .10$, can the economist conclude that English cheese sells at lower prices in Canada?

6.26. A housing study conducted in 1975 showed that Honolulu had an average monthly rent of $258 per dwelling unit (the highest among U.S. cities), with a standard deviation of $30, based on an analysis of 12 randomly selected units. In 1978, a random sample of 16 units had a mean rental price of $280, with a standard deviation of $35. Does the study contain enough information to assert at level $\alpha = .10$ that Honolulu rents increased significantly in the three-year period?

6.27. School buildings can be built more economically if state regulations on the required number of windows per classroom are relaxed. In an era of increasing costs, combined with an eroding financial structure, the only thing standing between this fact and its immediate implementation is a psychological theory that the lack of windows in the classroom tends to increase anxiety among pupils (presumably because there is no escape). To check out the theory, an educational psychologist prepared a test of anxiety and administered it to a class of 15 pupils who had spent a month attending classes in an experimental windowless classroom, and also to a class of 20 pupils (acting as controls) who had spent the same month studying the same material in an ordinary classroom. Those pupils in the windowless classroom had a mean anxiety level of 105 with a standard deviation of 25, while those in the ordinary room had a mean anxiety level of 95 with a standard deviation of 30. At level $\alpha = .01$, do the test results indicate that lack of windows in the classroom leads to a significant increase in anxiety among schoolchildren?

6.28. A study conducted by I. P. Sharp Associates comparing the operating expenses of the Boeing 747 and the McDonnell-Douglas DC-10 yielded the following information for six airlines using the 747 and five using the DC-10, where the numbers represent the total operating expenses in dollars per block hour per passenger:*

747 Data		DC-10 Data	
Airline	Expenses	Airline	Expenses
Trans World	9.71	United	9.60
Northwest	9.86	Continental	10.21
American	7.34	Western	9.24
Braniff	7.37	National	9.06
United	7.34	American	7.74
Delta	10.06		

At level of significance $\alpha = .05$, do the data indicate that mean operating expenses differ significantly between the two types of airplanes?

6.29. A study determined the family income required in various metropolitan areas around the nation to maintain 319 goods and services considered important to an "average" way of life.† The results on five areas in the East and 6 in the West follow (in thousands of dollars):

---

*Based on data reported in *Aviation Week & Space Technology*, May 16, 1977, pp. 34–35.

†Reported in *U.S. News & World Report*, Feb. 20, 1978, p. 69.

| | East | | West | |
Area	Income Required	Area	Income Required
Pitts	24.4	Seat	22.9
Wash	26.7	Hous	23.5
Chi	25.9	SF	26.9
NY	29.0	LA	26.5
St.L	23.0	Denv	23.2
		Phoe	21.6

At level $\alpha = .05$, do the results of the study indicate that a higher family income is required in the East?

6.30. Let's try to find out which is the more profitable industry: alcoholic beverages or energy. The following data compare the price per earnings ratios of the stocks of various alcoholic beverage and energy corporations for 1977.*

Alcoholic Beverage Corporations	P-E Ratio	Energy Corporations	P-E Ratio
Olympia Brewing	8	Westmoreland Coal	12
Heublein	14	Belco Petroleum	6
Seagram	9	Continental Oil	8
Pabst	11	Phillips Petroleum	10
Jos. Schlitz Brew.	14	Consol Natural Gas	6
Adolph Coors	7	Combustion Engineer.	10
National Distillers	7	Houston Natural Gas	9
		Marathon Oil	7

At level $\alpha = .05$, can we detect from the data a significant difference between the mean price per earnings rations of the two industrial groupings?

6.31. A recent survey of 40 savings and loan associations on the East Coast showed that they were charging an average interest rate of 9.25 percent on home mortgages, with a standard deviation of 0.65 per cent. At the same time, a survey of 50 S & L's on the West Coast indicated an average rate of 9.75 percent, with a standard deviation of 0.45 percent. At level $\alpha = .05$, does the survey indicate that mortgage rates are significantly higher on the West Coast?

6.32. The Knitwhitt Corporation, a major textiles firm, surveyed its personnel for their preference ratings of two competing office redecorating plans. On a scale of 0 to 100, 13 employees rated the Innerspace Corp.'s plan and 8 other employees rated the Ambience Corp.'s plan. The ratings are listed below:

Innerspace	45	92	68	47	31	19	63	29	53	64	71	81	65
Ambience	96	49	18	68	73	74	85	57					

*Randomly selected from Forbes, January 9, 1978, pp. 201–218.

The Ambience plan is considerably more expensive. At level $\alpha = .05$, is there any reason to believe that the Knitwhitt employees would give that plan a significantly higher mean rating?

6.33. Top management of a national chain of Do-It-Yourself Massage Parlors wants to buy the best quality neon lights it can find, as long as they don't cost too much. The most expensive brand, Ace Flaming Signatures, and a cheaper brand, Zappo, offer to compete with each other for the lucrative contract in the booming new fad. The following data reveal the length of time that four samples of the Ace brand and five samples of the Zappo brand lasted until failing:

Brand	Consecutive Hours				
Ace	220	160	200	220	
Zappo	180	150	210	160	200

At significance level $\alpha = .05$, do the test results indicate that the more expensive brand actually lasts longer, on the average?

6.34. Blockade-Run Imported Junque (BRIJ), Inc., is considering sending its entire 2,000-strong sales force to one of two nationally-recognized training programs. To find out which one would be more appropriate in its particular field, BRIJ sent 60 randomly selected employees to the Deuce Sales School and enrolled another 50 in the Western Imaginative Marketing Program. Results showed that those who attended the Deuce School increased their volume over the next six months by $32,000, on the average, with a standard deviation of $8,600. Comparable figures for the Western attendees were a mean of $30,100 and a standard deviation of $3,800. At significance level $\alpha = .03$, is the evidence sufficient to conclude that the Deuce Sales School offers superior training?

SECTION 6.E.   A TEST FOR PAIRED SAMPLES

In this section, our goal will be to introduce the *paired-sample t test*, which tests the difference between two sets of data in cases where the individual members of one sample are *directly related* ("paired") to corresponding individual members of the other sample. As you can recall from the last few paragraphs of the previous section, the two-sample t test can be used only if the individual members of one sample are independent of the individual members of the other sample. Therefore, the paired-sample t test can be viewed as a replacement for the two-sample t test when the two samples of data are not independent of each other.

We will first show why we really need the paired-sample t test, that is, why the two-sample t test cannot be applied to paired data. The best way to illustrate this is to present an actual example of a real two-sample problem that cannot be solved using the two-sample t test. By the words, "cannot be solved," we *do not mean* that we are unable to compute the numerical value of t by formula (6.7), but *rather* that the results we get by carrying out this

TABLE 6.12   Before and After Weights of Ten Dieting Colleagues

	Drink-More Diet			Eat-Less Diet	
Colleague	Weight Before	Weight After	Colleague	Weight Before	Weight After
A	150	150	F	150	141
B	160	160	G	160	150
C	170	170	H	170	160
D	180	179	I	180	170
E	190	141	J	190	179

process do not make sense. In Example 6.6 following, we have a perfectly logical problem involving the effectiveness of two different reducing diets. When we apply the two-sample t test, however, the answers we get are obviously nonsensical. Then, after we reassess the problem, we will be able to see exactly why the two-sample t test failed to give an adequate solution and exactly what adjustments are required in order to correctly solve the problem.

EXAMPLE 6.6.   Weight-Reducing Diets.   A business executive who feels the need to lose some weight is considering going on one of the two famous reducing diets, the Drink-More diet and the Eat-Less diet, but only if statistical analysis shows that one or both of these diets are effective in inducing weight loss. As it turns out, five colleagues have tried the Drink-More diet and another five have tried the Eat-Less diet. The weights of the 10 colleagues, both before beginning and after completing their diet programs, are recorded in Table 6.12. The question we have to answer is this: "Are either of the diets really effective, at significance level $\alpha = .05$, in producing weight loss?"

DISCUSSION OF THE PROBLEM.   As we can discover by looking at the data of Table 6.12, the Drink-More diet is virtually ineffective in the great majority (80%) of cases studied. The Eat-Less diet, on the other hand, seems to cause everybody to lose approximately 10 pounds. On the surface, it therefore seems that the Eat-Less diet is effective in producing weight loss, while the Drink-More diet is really not effective. This distinction between the diets ought to be reflected in our statistical analysis of the problem.

ATTEMPTED SOLUTION USING THE TWO-SAMPLE t TEST.   We first test the effectiveness of the Drink-More diet, using the two-sample t test. If we use the symbols

μ_B = true mean weight of group *before* dieting,
μ_A = true mean weight of group *after* dieting.

TABLE 6.13   Calculations Leading to the Two-Sample
t Statistic   (Drink-more diet data)

	Weight Before			Weight After	
x_B	$x_B - \bar{x}_B$	$(x_B - \bar{x}_B)^2$	x_A	$x_A - \bar{x}_A$	$(x_A - \bar{x}_A)^2$
150	-20	400	150	-10	100
160	-10	100	160	0	0
170	0	0	170	10	100
180	10	100	179	19	361
190	20	400	141	-19	361
850	0	1000	800	0	922

$n_B = 5$ $\qquad\qquad\qquad \Sigma(x_B - \bar{x}_B)^2 = 1000 \qquad\qquad \bar{x}_A = 800/5 = 160$

$\bar{x}_B = 850/5 = 170 \qquad\qquad\qquad n_A = 5 \qquad\qquad \Sigma(x_A - \bar{x}_A)^2 = 922$

we want to test the hypothesis

H: $\mu_B = \mu_A$ (the diet is not effective),

vs.

A: $\mu_B > \mu_A$ (the diet is effective).

The alternative is structured in such a way as to read that the diet is
effective if a significant amount of weight is lost, namely if mean weight
before dieting exceeds mean weight after dieting.
    Using the two-sample t test, we agree to reject H in favor of A at signif-
icance level $\alpha = .05$ if

$$t > t_\alpha [n_B + n_A - 2] = t_{.05} [8] = 1.860,$$

because $n_B = 5$ and $n_A = 5$. Table 6.13 shows the usual calculations for
the two-sample t test. Upon inserting the components of t into their
proper places in the formula, we obtain

$$t = \frac{\bar{x}_B - \bar{x}_A}{\sqrt{\dfrac{\Sigma(x_B - \bar{x}_B)^2 + \Sigma(x_A - \bar{x}_A)^2}{n_B + n_A - 2} \left(\dfrac{1}{n_B} + \dfrac{1}{n_A}\right)}}$$

$$= \frac{170 - 160}{\sqrt{\dfrac{1{,}000 + 922}{5 + 5 - 2}\left(\dfrac{1}{5} + \dfrac{1}{5}\right)}} = \frac{10}{\sqrt{\left(\dfrac{1{,}922}{8}\right)\left(\dfrac{2}{5}\right)}}$$

$$= 10/\sqrt{96.1} = 10/9.803 = 1.02 \ .$$

TABLE 6.14 Calculations Leading to the Two-Sample t Statistic   *(Eat-less diet data)*

x_B	$x_B - \bar{x}_B$	$(x_B - \bar{x}_B)^2$	x_A	$x_A - \bar{x}_A$	$(x_A - \bar{x}_A)^2$
150	-20	400	141	-19	361
160	-10	100	150	-10	100
170	0	0	160	0	0
180	10	100	170	10	100
190	20	400	179	19	361
850	0	1000	800	0	922

$n_B = 5$ $\Sigma(x_B - \bar{x}_B)^2 = 1{,}000$ $\bar{x}_A = 800/5 = 160$

$\bar{x}_B = 850/5 = 170$ $n_A = 5$ $\Sigma(x_A - \bar{x}_A)^2 = 922$

Because it turned out that $t = 1.02 < 1.860$, we cannot reject H. The two-sample t test therefore asserts at level $\alpha = .05$ that the Drink-More diet is not effective.

So far so good. The result we have just obtained confirmed our original suspicion that the Drink-More diet could not be counted upon to give an individual a reasonable amount of weight loss. Now let's apply the same procedure to analyze the Eat-Less diet (which from direct observation of the data seems to almost guarantee each participant a weight loss of 10 pounds). We are still testing the hypothesis

H: $\mu_B = \mu_A$

vs.

A: $\mu_B > \mu_A$

and we still reject H at level $\alpha = .05$ when $t > 1.860$, because $df = n_B + n_A - 2 = 5 + 5 - 2 = 8$ as before. It therefore remains only to compute the value of t and to note whether or not it exceeds 1.860. The relevant calculations appear in Table 6.14 and they show that

$$t = \frac{\bar{x}_B - \bar{x}_A}{\sqrt{\dfrac{\Sigma(x_B - \bar{x}_B)^2 + \Sigma(x_A - \bar{x}_A)^2}{n_B + n_A - 2}\left(\dfrac{1}{n_B} + \dfrac{1}{n_A}\right)}}$$

$$= \frac{170 - 160}{\sqrt{\dfrac{1{,}000 + 922}{5 + 5 - 2}\left(\dfrac{1}{5} + \dfrac{1}{5}\right)}} = \frac{10}{\sqrt{\left(\dfrac{1{,}922}{8}\right)\left(\dfrac{2}{5}\right)}}$$

$$= 10/\sqrt{96.1} = 10/9.803 = 1.02 \ .$$

Again $t = 1.02 < 1.860$, and so we again cannot reject H in favor of A in the case of the Eat-Less diet. Therefore the two-sample t test gives us the same result regarding the Eat-Less diet as it did regarding the Drink-More diet, namely that, at level $\alpha = .05$, the Eat-Less diet cannot be considered an effective way to lose weight.

ANALYSIS OF THE ATTEMPTED SOLUTION.   Despite the fact that there exist important differences between the two diets, the two-sample t test has decided, at level $\alpha = .05$, that neither diet can be considered significantly effective in producing weight loss. More surprising than the fact that the two-sample t test judged both diets to be similarly ineffective is the fact that the numerical values of t turned out to be exactly the same in both cases. In fact, all calculations leading to the numerical value of t were also identical in both cases. A comparison of Tables 6.13 and 6.14 reveals that they would be carbon (or photo-) copies of one another except for the positions of the numbers 141, -19, and 361 in the "Weight After" portion of the tables. In Table 6.13, these numbers appear at the bottom of the "Weight After" columns, while in Table 6.14 they appear at the top. We can therefore assert that the two-sample t test is incapable of detecting the distinctions between the obviously ineffective Drink-More diet and the obviously effective Eat-Less diet.

Why did the two-sample t test fail to give a logical solution to this problem? Well, let's review the basic principles underlying the two-sample t test. We tested, in both cases, the hypothesis

H: $\mu_B = \mu_A$

against the alternative

A: $\mu_B > \mu_A$.

In other words, we tested whether or not the overall average (true mean) of the group's weight before undertaking the diet was significantly greater than the overall average of the group's weight after completing the diet. In the case of the Drink-More diet, the group weighed a total of 850 pounds (for an average of 170) before dieting, and a total of 800 (for an average of 160) after dieting, as we can see from Table 6.13. Therefore, participants in the Drink-More diet lost 10 pounds, on the average. Table 6.15 shows that exactly the same average results have been obtained in the case of the Eat-Less diet. Participants in that dieting program weighed an average of 170 pounds before and 160 pounds after for an average loss of 10 pounds. It is this *average* loss for which the two-sample t test looks, and that is what it has found. Because the *average* weight loss is the same for both diets, the two diets are equally effective (in this case, equally ineffective) in the eyes of the two-sample *t* test.

But is it the overall average weight loss that we are primarily interested in? We can answer this question by going back to the reasoning that led us to believe that the Drink-More diet was not really effective, while the Eat-Less diet was. As demonstrated in the Drink-More data of Table 6.12, acquaintances A, B, C, and D lost virtually no weight, while acquaintance E lost enough weight for the entire group to have an average loss of 10 pounds. The Eat-Less diet of the same table, on the other hand, shows that each participant individually lost approximately 10 pounds, so the diet could reasonably be said to guarantee everyone a loss of around 10 pounds. Therefore it was the *individual* changes in weight from "before" to "after," not the overall group *averages*, that influenced us to think that the Eat-Less diet was effective in producing weight loss, while the Drink-More diet was not. It is these individual differences that are not reflected at all in the calculations for the two-sample *t* test, for those calculations take only overall averages into account. Therefore, because the question of the effectiveness of the diets hinges on the individual rather than the overall average weight loss, the two-sample *t* test is *not* an appropriate vehicle for answering the question.

TABLE 6.15   Calculations Leading to the Paired-Sample
t Statistic   (*Drink-more diet data*)

Colleague	Weight Before x_B	Weight After x_A	$d = x_B - x_A$	d^2
A	150	150	0	0
B	160	160	0	0
C	170	170	0	0
D	180	179	1	1
E	190	141	49	2,401
Sums			50	2,402

$n = 5$

$\overline{d} = \Sigma d/n = 50/5 = 10$

$$s_d = \sqrt{\frac{n\Sigma d^2 - (\Sigma d)^2}{n(n-1)}} = \sqrt{\frac{5(2,402) - (50)^2}{(5)(4)}} = \sqrt{\frac{12,010 - 2,500}{20}}$$

$$= \sqrt{\frac{9,510}{20}} = \sqrt{475.5} = 21.81$$

What we require in order to make a proper analysis of the diets is a method in which the calculations are based on *individual* weight losses. Such a method is the "paired-sample *t* test."

CORRECT SOLUTION USING THE PAIRED-SAMPLE *t* TEST. We first test the effectiveness of the Drink-More diet using the paired-sample *t* test. We define the quantity

μ_d = true mean *individual* *difference* between "before" weight and "after" weight.

and we want to test the hypothesis

H: $\mu_d = 0$ (diet is not effective)

vs.

A: $\mu_d > 0$ (diet is effective).

The alternative is structured so that the diet is judged to be effective if the average individual loses a positive amount of weight; namely, if the average individual's weight before exceeds his or her weight after the diet. To carry out the paired-sample *t* test, we would reject H in favor of A at level α if $t > t_\alpha [n - 1]$, where

$$t = \frac{\overline{d}\sqrt{n}}{s_d} \tag{6.9}$$

in which

n = number of "before-after" pairs
$\overline{d}$ = sample mean of individual differences
s_d = sample standard deviation of individual differences.

It is usually more convenient, when working with the paired-sample *t* test, to compute s_d using the short-cut formula for the standard deviation given by

$$s_d = \sqrt{\frac{n\Sigma d^2 - (\Sigma d)^2}{n(n - 1)}},$$

recalling formula (5.3) of Section 5.B. Here

Σd = sum of individual differences
Σd^2 = sum of squares of individual differences

Taking $\alpha = .05$ as our level of significance and noting that there are $n = 5$ before-after pairs comprising the Drink-More data of Table 6.12, we agree to reject

H: $\mu_d = 0$

in favor of

A: $\mu_d > 0$

if $t > t_{.05}$ [4] $= 2.132$. The computations needed for the numerical determination of t are presented in Table 6.15, and they yield that

$$t = \frac{\overline{d}\sqrt{n}}{s_d} = \frac{(10)\sqrt{5}}{21.81} = \frac{(10)(2.236)}{21.81} = 1.025.$$

Therefore t has turned out to be 1.025, which is not larger than 2.132. As we have agreed to reject H in favor of A if $t > 2.132$, we cannot reject H under the circumstances. We, therefore, conclude at level $\alpha = .05$, using the paired-sample t test, that the Drink-More diet is not effective in producing individual weight loss.

Now that the paired-sample t test has confirmed our earlier judgment that the Drink-More diet does not produce significant individual weight losses, we await its decision on the Eat-Less diet. It was the latter diet that provided us with the first intimation that the analysis based on the two-sample t test was breaking down. It remains to be seen if the paired-sample t test will be able to detect the substantial individual weight losses that we know are present in the data. We are still testing the hypothesis

H: $\mu_d = 0$

vs.

A: $\mu_d > 0$,

and we still will reject H at level $\alpha = .05$, if $t > 2.132$ because $df = n - 1 = 5 - 1 = 4$ as before. Table 6.16 shows the calculations used in finding the numerical value of t. Inserting the numerical values $n = 5$, $\overline{d} = 10$, and $s_d = .71$ (obtained from the calculations of Table 6.16) into the formula for t, we get that

$$t = \frac{\overline{d}\sqrt{n}}{s_d} = \frac{10\sqrt{5}}{.71} = \frac{(10)(2.236)}{.71} = 31.5 .$$

TABLE 6.16   Calculations Leading to the Paired-Sample
t Statistic   (Eat-less diet data)

Colleague	Weight Before x_B	Weight After x_A	$d = x_B - x_A$	d^2
F	150	141	9	81
G	160	150	10	100
H	170	160	10	100
I	180	170	10	100
J	190	179	11	121
Sums			50	502

$n = 5$ $\bar{d} = d/n = 50/5 = 10$

$$s_d = \sqrt{\frac{n\Sigma d^2 - (\Sigma d)^2}{n(n-1)}} = \sqrt{\frac{5(502) - (50)^2}{(5)(4)}} = \sqrt{\frac{2{,}510 - 2{,}500}{20}} = \sqrt{10/20} = \sqrt{.5} = .71$$

We have agreed to reject H if $t > 2.312$. As it turned out, $t = 31.5$, a number substantially in excess of 2.312, so that we can proceed with the greatest confidence to the rejection of H. In the eyes of the paired-sample t test, then, there is little doubt that the Eat-Less diet is effective at level $\alpha = .05$ (or any other level, for that matter). The numerical value $t = 31.5$ is so overwhelming that we must conclude, at any reasonable level α, that the Eat-Less diet is truly effective in producing individual weight loss.

The conclusions based on the use of the paired-sample t test have therefore provided us with statistical justification of the effectiveness of the Eat-Less diet and the ineffectiveness of the Drink-More diet. Because of its emphasis on *individual* differences between related pairs of data points rather than on *group averages*, the paired-sample t test has successfully distinguished between the fundamental characteristics of the two diets under study. The lesson to be learned from all this is that the two-sample t test should be used only when we want to investigate the difference between the means of two independent sets of data, while the paired-sample t test should be used whenever consideration of individual differences between related pairs of data points is paramount.

EXAMPLE 6.7.   Sales of Farm Equipment.   A regional wholesaler of farm equipment, parts, and supplies wants to investigate whether or not he really needs a staff of salespeople to personally present products at

TABLE 6.17   Farm Equipment Sales Data   (Thousands of dollars)

Week Beginning	Norfolk (mail order)	Fairbury (sales staff)
Aug 15	1.5	1.4
Apr 18	1.4	1.4
Jan 10	2.2	2.6
Nov 14	2.6	2.5
Jun 20	3.9	4.8
Feb 28	4.5	4.3
Jul 4	3.7	4.0
Dec 12	3.4	3.5
Jan 24	4.0	3.9
May 16	2.5	2.6
Mar 21	1.9	1.7
Oct 3	1.5	1.4

the customer's own location. Although he has always had a sales staff, he is wondering whether a change to a mail and phone ordering method would significantly reduce his sales volume. Last year he therefore conducted an experiment to compare sales in Fairbury, where he used his sales staff, and Norfolk, where he operated with mail and phone orders only. These two Nebraska cities were chosen because they were comparable in size and sales volume generated by the wholesaler. A random sample of weekly sales figures for the year, in thousands of dollars, is recorded in Table 6.17. At level $\alpha = .10$, do the results of the experiment indicate that the switch to a mail and phone operation in Norfolk led to a significant reduction in weekly sales volume?

SOLUTION.   We use the paired-sample t test in preference to the two-sample t test because the columns of data are paired by week. We would not want to conclude that the switch in sales methods was a bad mistake merely on the basis of a large decline in sales for one or two weeks. We would want instead to establish a pattern of decreased sales regularly throughout the year. The only way we can address ourselves to this question is to apply the paired-sample t test.

We want to test the hypothesis

H: $\mu_d = 0$ (switching to mail-order *does not* decrease sales)

vs.

A: $\mu_d < 0$ (switching to mail-order *does* decrease sales)

where

μ_d = true mean difference between sales volume in mail-order region and sales volume in sales staff region.

The alternative A: $\mu_d < 0$ has been selected (rather than A: $\mu_d > 0$) because, as the calculations in Table 6.18 show, the individual difference d is given by the subtraction

d = Norfolk sales minus Fairbury sales
   = mail-order sales minus sales staff sales.

The alternative A: $\mu_d < 0$ therefore expresses the statement that switching to mail order decreases sales, for it means that the bulk of the d's are negative and, consequently, Norfolk sales are usually below Fairbury sales. Using Table 6.8 to help us determine the rejection rule, we then see that we should reject H at level α if $t < -t_\alpha$ [$n - 1$]. Here $\alpha = .10$

TABLE 6.18   Calculations Leading to the Paired-Sample
t Statistic   (Farm equipment sales data)

| Week beginning | thousands of dollars | | $d = x_M - x_S$ | d^2 |
	Mail Order x_M	Sales Staff x_S		
Aug. 15	1.5	1.4	0.1	0.01
April 18	1.4	1.4	0.0	0.00
Jan. 10	2.2	2.6	−0.4	0.16
Nov. 14	2.6	2.5	0.1	0.01
June 20	3.9	4.8	−0.9	0.81
Feb. 28	4.5	4.3	0.2	0.04
July 4	3.7	4.0	−0.3	0.09
Dec. 12	3.4	3.5	−0.1	0.01
Jan. 24	4.0	3.9	0.1	0.01
May 16	2.5	2.6	−0.1	0.01
March 21	1.9	1.7	0.2	0.04
Oct. 3	1.5	1.4	0.1	0.01
Sums			−1.0	1.20

$n = 12$ $d = \Sigma d/n = -1/12 = -.083$

$$s_d = \sqrt{\frac{n\Sigma d^2 - (\Sigma d)^2}{n(n-1)}} = \sqrt{\frac{12(1.2) - (-1)^2}{(12)(11)}} = \sqrt{\frac{14.4 - 1}{132}}$$

$$= \sqrt{\frac{13.4}{132}} = \sqrt{.102} = .32$$

and $n = 12$ (for there is one pair for each month), so that we agree to reject H in favor of A if $t < -t_{.10}[11] = -1.363$. All that remains now is to determine the numerical value of t, the preliminary calculations for which appear in Table 6.18, and show that $n = 12, \bar{d} = -.083$, and $s_d = .32$. Inserting these numbers into the expression for t, we have that

$$t = \frac{\bar{d}\sqrt{n}}{s_d} = \frac{(-.083)\sqrt{12}}{.32} = \frac{(-.083)(3.464)}{.32} = -.90 .$$

Therefore, $t = -.90 > -1.363$ (because $-.90$ is *less negative* than -1.363). As we agreed to reject H if $t < -1.363$, the results do not permit us to reject H. At level $\alpha = .10$, we then conclude that the experiment seems to show that the switch to mail order does *not* significantly decrease weekly sales volume.

Before we close this section, it would be instructive to comment on the origin of the formula

$$t = \frac{\bar{d}\sqrt{n}}{s_d}$$

used in the paired-sample t test. Toward this end, let's take a look at Tables 6.7 and 6.18 with a view toward comparing them. If we compare Table 6.7 with the last two columns of Table 6.18, we see that the tables have exactly the same structure. The only differences are the numerical values of the data points and the fact that the columns of Table 6.7 are headed by x's, while the columns of Table 6.18 are headed by d's. Other than that, we are really carrying out the same set of calculations. Because Table 6.7 contains calculations used in connection with the one-sample t test, it seems that Table 6.18 could also be viewed in terms of the one-sample t test. That is, in fact, exactly what the paired-sample t test is—a one-sample t test where the one sample is the set of difference (d's). To test the hypothesis H: $\mu_d = 0$ using the one-sample t test on the d's would call for the formula

$$t = \frac{\bar{x} - \mu_0}{s/\sqrt{n}} = \frac{\bar{d} - 0}{s_d/\sqrt{n}} .$$

But a little algebra shows that

$$t = \frac{\bar{d} - 0}{s_d/\sqrt{n}} = \frac{\bar{d}}{s_d/\sqrt{n}} = \left(\frac{\bar{d}}{1}\right)\left(\frac{\sqrt{n}}{s_d}\right) = \frac{\bar{d}\sqrt{n}}{s_d} ,$$

which is the formula for the paired-sample t statistic. Therefore, the paired-sample t test can be considered as a special case of a one-sample t test. The one sample of data is obtained by computing the pair-by-pair differences of two original sets of data, rather than being observed directly from an experiment or survey. Therefore the requirements for valid use of the paired-sample *t* test are the same as those for the one-sample *t* test: it must be reasonable to assume that each pair is independent of any other pair and that the pairwise differences come from a normally distributed population.

Finally, if there are more than 30 pairs of data points, we would use the paired-sample *z* test having the same formula. However, we would base our rejection rules on percentage points of the normal distribution rather than on the *t* distribution.

EXERCISES 6.E

6.35. Some psychologists feel that there is a statistical correlation between smoking and absenteeism. The management of a leather goods factory has under consideration a stop-smoking incentive plan, and they would therefore be interested in knowing whether employees who have stopped smoking have better absenteeism records than they had while they were smoking. Nine such employees are randomly selected, and their absenteeism records before and after they have stopped smoking are compared. The data follow:

Employee	A	B	C	D	E	F	G	H	I
Days absent per year while smoking	20	30	14	6	42	19	18	12	24
Days absent per year after stopping smoking	10	20	16	5	40	15	22	10	20

At significance level $\alpha = .05$, does the comparison tend to support the theory that employees have less absenteeism after they stop smoking?

6.36. One financial analyst thinks that the closing price of a stock is generally lower than its midrange price for the day. To test his theory, he obtained the following information on the closing and midrange prices of eight randomly selected stocks one recent day.

Stock	Midrange Price	Closing Price
HMW	3.5	3.5
DAL	36.1	35.875
LHD	8.0	8.125
EDS	15.7	15.5
USS	46.6	46.5
SVC	15.4	15.5
WPS	58.5	59.0
RXD	30.5	30.75

At level $\alpha = .05$, do the results indicate that closing prices are generally lower than midrange prices?

6.37. Nine randomly selected automobile manufacturing companies (foreign and domestic) had the following production records in 1977 and 1978:

Company	Millions of Vehicles Produced	
	1977	1978
A	2.4	2.1
B	1.5	1.5
C	0.5	0.5
D	2.2	2.2
E	1.3	1.0
F	1.0	0.9
G	1.1	0.8
H	0.3	0.3
I	0.8	0.7

At significance level $\alpha = .05$, do the data indicate that the 1978 production levels of automobile companies were significantly below the 1977 levels?

6.38. Environmental Nucleonics of Barstow instituted a time-consuming and rather expensive industrial safety program, the goal of which was to reduce the number of worker-hours lost due to on-the-job accidents. The following data were recorded for six of their branches before and after institution of the safety program:

Branch	Hours Lost per Employee per Quarter	
	Before Program	After Program
Barstow, Ca.	46	43
Hoboken, N.J.	72	66
Ensenada, B.C.	50	51
Elko, Ne.	38	37
Rawlins, Wy.	78	73
Vancouver, B.C.	66	60

At level $\alpha = .01$, decide whether or not the data indicate that the safety program is effective in reducing time lost due to industrial accidents.

6.39. An agricultural experiment station tested the comparative yields of two varieties of corn, a new experimental variety and a standard variety, to find out whether or not the new variety would permit greater food production on the same land. Each of seven farmers was asked to grow both varieties on similar plots of land and to give each exactly the same amount of treatment, water, fertilizer, etc. The yields, in bushels, are recorded below:

Farmer	A	B	C	D	E	F	G
New variety	28.2	24.6	29.7	20.5	34.6	27.1	31.4
Standard variety	27.2	24.3	29.0	22.5	34.2	26.8	30.4

At level $\alpha = .05$, do the results of the experiment confirm the claim that the new variety is significantly more productive than the standard variety?

6.40. A recent study on "how taxes and inflation erode an 8% raise" compared the after-tax income of 13 individuals in 1976 with their after-tax income, after an 8% raise and adjusted for inflation, in 1977.* The data follow in dollars:

Person	1976	1977
A	$13,339	$13,541
B	17,454	17,614
C	21,335	21,439
D	24,990	25,112
E	28,636	28,668
F	32,237	32,162
G	35,850	35,439
H	38,880	38,518
I	41,981	42,078
J	45,112	44,106
K	17,303	47,641
L	50,579	50,863
M	53,731	53,736

Using a level of significance $\alpha = .10$, decide whether or not the 8% raise seems to induce a significant difference in after-tax income.

6.41. A corporation psychologist has been assigned the task of finding out whether executives hired from outside the company feel any more uneasy in their jobs than do their colleagues who have been promoted from within. She pairs 14 randomly selected executives, based on their length of service as an executive, and gives each a psychological test of uneasiness. A high score on the test indicates "very uneasy." The scores follow:

Years as Executive	Hired from Outside	Promoted from Within
2	80	50
5	60	40
8	40	40
10	40	30
11	20	10
14	10	5

*Reported in *Business Week*, September 12, 1977, p. 83.

At level $\alpha = .05$, does there seem to be evidence that executives hired from the outside are more uneasy than those promoted from within?

6.42. In an attempt to determine the most economical lodging facilities for its employees to use on overnight business trips, the Amalgamated National Termite Supply (ANTS) Company compared two motel chains' nightly rates at various locations across the country:

Location	"Sleepy Six" Rate (dollars)	"Lazy Eight" Rate (dollars)
Cleveland	19	20
Seattle	16	18
Boston	19	20
Minneapolis	17	18
Dallas	19	22
St. Louis	18	18
Denver	19	18

At level $\alpha = .05$, do the chains seem to differ significantly in their rates?

6.43. One of the modern economic indicators is turning out to be the revenue-passenger-miles (RPM) accumulated each month by regional and national airlines. The following data* compare RPM figures for September 1979 and 1980 operations of six randomly selected regional airlines.

Airline	Air Calif	Pied- mont	Texas Int'l	Aloha	Alaska	Fron- tier
1979 RPM	78	151	173	32	69	234
1980 RPM	93	181	168	22	79	229

At significance level $\alpha = .01$, can we assert that RPM for a typical regional airline has increased over the period of time studied?

SECTION 6.F.   TESTS FOR THE STANDARD DEVIATION

Up to this point, we have concentrated on statistical tests of hypotheses involving the mean of a population or the difference between means of two populations. In this section, our objective will be to study analogous problems involving standard deviations.

EXAMPLE 6.8. Apartment Occupancy Rates. Recall the market research study of Example 6.2, which showed that eight randomly

---

*Based on a report in *Aviation Week & Space Technology*, October 20, 1980, page 35.

selected apartment buildings in Houston had occupancy percentage rates as follows:

95.4   95.0   97.5   95.6   93.3   94.2   94.9   97.3

In attempting to forecast rental income, it would be useful for the company to know how variable apartment occupancy rates are. Do the results of the market research study indicate, at level α = .05, that occupancy rates in Houston have standard deviation below 2.0 percent?

SOLUTION.   The procedure we follow in solving this problem is called the "one-sample chi-square test for the population standard deviation." As in the case of the small-sample confidence intervals for the population standard deviation (studied in Section 5.D), the use of the chi-square distribution is valid only in situations where the underlying population is normally distributed. Therefore, only when the population has the normal distribution can we apply the small-sample methods (both one-sample and two-sample) of this section.

Returning now to the problem at hand, we look at the general structure of the one-sample chi-square test: we test the hypothesis

H: $\sigma = \sigma_0$

vs.

A: $\sigma < \sigma_0$,

where

σ = true population standard deviation

and

σ_0 = hypothesized standard deviation.

We then reject H in favor of A at significance level α if

$$\chi^2 < \chi^2_{1-\alpha}[n-1]$$

where

n = number of data points in the sample,

and

■

$$\chi^2 = \frac{(n-1)s^2}{\sigma_0^2} = \frac{\Sigma(x-\bar{x})^2}{\sigma_0^2} \qquad\qquad \textbf{(6.10)}$$

TABLE 6.19   A Small Portion of Table A.8

df	$\chi^2_{.95}$
7	2.167

We have encountered all of these quantities before, in connection with confidence intervals for standard deviations. Here s is the sample standard deviation that we have used many times before, and it is an algebraic fact that $(n - 1)s^2 = \Sigma(x - \bar{x})^2$ as shown in formula (6.8). Because $n = 8$ and $\alpha = .05$, we agree to reject H in favor of A if

$$\chi^2 < \chi^2_{.95}[7] = 2.167$$

as in Table A.8 of the Appendix. The relevant portion of Table A.8 is reproduced in Table 6.19. It therefore remains only to compute the numerical value of

$$\chi^2 = \frac{\Sigma(x - \bar{x})^2}{\sigma_0^2}.$$

Well, we know from the formulation of the hypothesis $\sigma_0 = 2.0$, and we can refer back to the calculations of Table 6.5 where it was shown that $\Sigma(x - \bar{x})^2 = 14.32$. It follows that

$$\chi^2 = \frac{\Sigma(x - \bar{x})^2}{\sigma_0^2} = \frac{14.32}{(2.0)^2} = 14.32/4 = 3.58 .$$

Naturally, if Table 6.5 had not already been in existence, we would have had to construct one like it in order to find out what $\Sigma(x - \bar{x})^2$ is. As we have agreed to reject H in favor of A if $\chi^2 < 2.167$, we note that $\chi^2 = 3.58 > 2.167$ so we cannot reject H. At level $\alpha = .05$, there is not enough evidence to infer conclusively that the true standard deviation of occupancy rates is below 2.0 percent.

As we have discussed earlier, there are three possible alternatives in one-sample tests involving the hypothesis H: $\sigma = \sigma_0$. These alternatives are A: $\sigma > \sigma_0$, A: $\sigma < \sigma_0$, and A: $\sigma \neq \sigma_0$. By analogy with Table 6.8 (involving means), we record in Table 6.20 the rules for rejecting H: $\sigma = \sigma_0$ in favor of each alternative in cases involving standard deviations.

EXAMPLE 6.9. Calculator Battery Lifetimes. One manufacturer of pocket calculators advertises that its battery pack allows the calculator to operate continuously for 22 hours, on the average, without recharging. (See Example 6.1) A prospective corporate customer, however, is often interested in the uniformity of the product as well, for it would be quite

TABLE 6.20 Rejection Rules in the Presence of Various Alternatives for One-Sample Tests of Population Standard Deviations ($H: \sigma = \sigma_0$)

Alternative	Reject H in Favor of A at Significance Level α If:
A: $\sigma > \sigma_0$	$\chi^2 > \chi^2_\alpha[n-1]$
A: $\sigma < \sigma_0$	$\chi^2 < \chi^2_{1-\alpha}[n-1]$
A: $\sigma \neq \sigma_0$	$\chi^2 < \chi^2_{1-\alpha/2}[n-1]$ or $\chi^2 > \chi^2_{\alpha/2}[n-1]$

disconcerting to order a supply of 100 calculators for company use and then to find out that half operate continuously for 38 hours, while the other half operate continuously only for 6 hours. Such a shipment would certainly satisfy the criterion of an average operating time of 22 hours, but the tremendous variation in quality would considerably reduce the ultimate value to the user of the total package of 100 calculators. Suppose, then, that the corporate customer insists on a uniformity of operation requiring that the standard deviation of battery pack lifetimes not exceed one-half hour (.5 hour). Based on a sample of 50 calculators whose battery lifetimes turn out to have mean 21.8 hours and standard deviation .9 hour, can the customer conclude at level $\alpha = .01$ that the true standard deviation of battery lifetimes is in excess of the acceptable .5 hour?

SOLUTION. We note in this problem that $\sigma_0 = .5$, $s = .9$, $n = 50$, and the sample mean 21.8 is not relevant to the question. We are therefore testing the hypothesis

H: $\sigma = .5$ (battery pack uniformity is acceptable)

against the alternative

A: $\sigma > .5$ (battery pack lifetimes vary too much).

Because $n = 50 > 30$, we could use a large-sample, rather than a small-sample, test. We compute χ^2 from formula (6.10) as before, and then we calculate the chi-square percentage point using the Wilson-Hilferty approximation

$$\blacksquare \qquad \chi^2_\alpha[n-1] = (n-1)\left(1 + z_\alpha \sqrt{\frac{2}{9(n-1)}} - \frac{2}{9(n-1)}\right)^3 \qquad (6.11)$$

that we first mentioned in Section 5.D, formulas (5.8a) and (5.8b).

The rejection rule is based on the appropriate percentage points of the chi-square distribution as listed in Table 6.20. We therefore reject H in favor of A at level α if $\chi^2 > \chi^2_\alpha[n-1]$. Since $n = 50$, $\alpha = .01$, and

$z_\alpha = z_{.01} = 2.33$, we agree to reject H if $\chi^2 > \chi^2_{.01}[49]$. Using formula (6.11), we get that

$$\chi^2_{.01}[49] = 49\left(1 + 2.33\sqrt{\frac{2}{9(49)}} - \frac{2}{9(49)}\right)^3$$

$$= 49\,(1 + (2.33)(.06734) - .004535)^3$$

$$= 49\,(1 + .1569 - .004535)^3 = 49\,(1.1524)^3 = 74.99.$$

From the data, we know that $n = 50$ and $s = .9$, and from the formulation of the hypothesis that $\sigma_0 = .5$. Therefore

$$\chi^2 = \frac{(n-1)s^2}{\sigma_0^2} = \frac{(49)(.9)^2}{(.5)^2} = \frac{(49)(.81)}{.25} = 158.76\;.$$

As we have agreed, we must reject H in favor of A because $\chi^2 = 158.76 > 74.99$. We then conclude, at level $\alpha = .01$, that the true standard deviation of the battery pack lifetimes is significantly larger than .5 hour, and they are consequently too variable to meet the customer's specifications.

The two-sample test for the difference of standard deviations introduces a new distribution into our discussion. This new distribution is the F distribution of Table A.10 of the Appendix. (As well as seeing action in the remainder of this section, the F distribution will appear throughout the entirety of Chapter 7.)

EXAMPLE 6.10. Supermarket Prices. It is known that some supermarkets give trading stamps with purchases, while others do not. By and large, most people realize that those supermarkets that give trading stamps generally post slightly higher prices, in order to cover the additional cost of the stamps. On the other hand, one consumer organization feels that stores that do not give stamps are prone to offer "loss leaders" or other price adjustment gimmicks rather than to give uniformly lower prices. In other words, the consumer organization feels that prices are more variable among supermarkets not offering trading stamps than among those who do. To test out its guess, the organization checks out watermelon prices (per pound) at four supermarkets that give trading stamps and six that do not. The results of the survey follow:

Prices per pound of watermelons (cents)					
At supermarkets that give stamps	12	18	14	16		
At supermarkets that give no stamps	10	16	16	12	8	10

At level $\alpha = .05$, do the data tend to support the allegation that prices are more variable among supermarkets not giving trading stamps?

SOLUTION. The appropriate way to answer this question is by means of the *two-sample F test for the difference of standard deviations*. We define the quantities

σ_T = true standard deviation of watermelon prices at supermarkets that give *trading* stamps,

σ_N = true standard deviation of watermelon prices at supermarkets that do *not* give trading stamps.

We then want to test the hypothesis

H: $\sigma_N = \sigma_T$ (watermelon prices are equally variable),

against the alternative

A: $\sigma_N > \sigma_T$ (watermelon prices are more variable at supermarkets not offering trading stamps).

To test H against A, we compute

s_T = sample standard deviation of watermelon prices at supermarkets that give trading stamps

and

s_N = sample standard deviation of watermelon prices at supermarkets that do not give trading stamps,

and we note the numerical values of

n_T = number of data points in sample of supermarkets that give trading stamps,

n_N = number of data points in sample of supermarkets that do not give trading stamps.

We then calculate the value of

$$F = \frac{s_N^2}{s_T^2} \tag{6.12}$$

and we reject H at significance level α if

$$F > F_\alpha [n_N - 1; n_T - 1].$$

Degrees of Freedom of the F Distribution

A word about the F distribution is now in order. The percentage points of the F distribution are of the form

$$F_\alpha [dfn\,; dfd],$$

where

  dfn = degrees of freedom in the numerator

and

  dfd = degrees of freedom in the denominator.

The fact that the percentage points of the F distribution are classified by two indices of degrees of freedom contrasts markedly with those of the t and chi-square distributions that are classified by only one index of degrees of freedom. In the F test for the difference of standard deviations, each index of degrees of freedom is the number of data points involved, minus one. For example, dfn equals one less than the number of data points used in computing the numerator of F. In the watermelon price example, therefore,

$$dfn = n_N - 1 = 6 - 1 = 5$$

because 6 data points were used in the computation of the numerator s_N^2. By a similar reasoning process, we see that

$$dfd = n_T - 1 = 4 - 1 = 3.$$

Therefore, at level α = .05, we agree to reject

  H: $\sigma_N = \sigma_T$

in favor of

  A: $\sigma_N > \sigma_T$

if

$$F > F_{.05}\,[5;3] = 9.01\,.$$

In Table 6.21, we exhibit a small portion of Table A.10, showing where to find $F_{.05}\,[5;3]$.

  It now remains only to calculate the value of F and to observe whether or not the value so calculated exceeds 9.01. To calculate F, we must first calculate s_N^2 and s_T^2. It may look to you as though we have to calculate s_N and s_T first, and then square the resulting numbers. However, the nature of the standard deviation allows us to short-cut this procedure somewhat. Because

$$s_N = \sqrt{\frac{\Sigma(x_N - \bar{x}_N)^2}{n_N - 1}}$$

TABLE 6.21   A Small Portion of Table
A.10   (Values of $F_{.05}$)

	Degrees of Freedom for Numerator	
Degrees of Freedom for Denominator		5
	3	9.01

we can square both sides of this equation to get

$$s_N^2 = \frac{\Sigma(x_N - \bar{x}_N)^2}{n_N - 1}.$$

Similarly

$$s_T^2 = \frac{\Sigma(x_T - \bar{x}_T)^2}{n_T - 1}.$$

We can use these formulas to calculate the numerator and denominator of the F statistic when the actual values of the data points are known to us. The calculational details involving the watermelon price data are presented in Table 6.22.

TABLE 6.22   Calculations Leading to the Two-Sample
F Statistic   (Watermelon price data)

Supermarkets Not Offering Stamps			Supermarkets Offering Stamps		
x_N	$x_N - \bar{x}_N$	$(x_N - \bar{x}_N)^2$	x_T	$x_T - \bar{x}_T$	$(x_T - \bar{x}_T)^2$
10	-2	4	12	-3	9
16	4	16	18	3	9
16	4	16	14	-1	1
12	0	0	16	1	1
8	-4	16			
10	-2	4			
72	0	56	60	0	20

$\bar{x}_N = 72/6 = 12$

$s_N^2 = \dfrac{\Sigma(x_N - \bar{x}_N)^2}{n_N - 1} = \dfrac{56}{6 - 1} = 56/5 = 11.2$

$F = \dfrac{s_N^2}{s_T^2} = 11.2/6.67 = 1.68$

$\bar{x}_T = 60/4 = 15$

$s_T^2 = \dfrac{\Sigma(x_T - \bar{x}_T)^2}{n_T - 1} = \dfrac{20}{4 - 1} = 20/3 = 6.67$

The calculations made in Table 6.22 show that $F = 1.68$. As we have agreed to reject H in favor of A at level $\alpha = .05$ if $F > 9.01$, our calculation that $F = 1.68 < 9.01$ shows that we cannot reject H. On the basis of the given data, therefore, we do not have enough evidence to assert, at level $\alpha = .05$, that watermelon prices are more variable at supermarkets that do not give trading stamps.

Rejection Rules for the *F* Test

Before we present a second example illustrating the application of the two-sample F test for the difference of standard deviations, let's set up a table of rejection rules analogous to those of Tables 6.8 and 6.20. It turns out that the F test has a curious characteristic in this regard: there are only two possible alternatives, rather than the three appearing in Tables 6.8 and 6.20. Taking as our hypothesis

$$\text{H:} \quad \sigma_1 = \sigma_2 \, ,$$

the only alternatives that the F test, as we have organized it, knows how to handle are

$$\text{A:} \quad \sigma_1 > \sigma_2$$

and

$$\text{A:} \quad \sigma_1 \neq \sigma_2.$$

Because of the complicated nature of the F distribution, involving as it does two indices of degrees of freedom, it turns out that we cannot use the percentage points presented in Table A.10 to test H against the alternative

$$\text{A:} \quad \sigma_1 < \sigma_2 \, .$$

A simple reformulation of that alternative, however, will turn it into

$$\text{A:} \quad \sigma_2 > \sigma_1$$

and this pattern is acceptable to the F test. In short, the F test can be used only if the alternative contains the symbol $>$ or $\neq$. If $<$ is present in the statement of the alternative the populations must be interchanged so that the $<$ is reversed. For example, in the question concerning the variability of watermelon prices, we want to test the hypothesis that σ_T and σ_N were the same, as opposed to the alternative that σ_T was smaller than σ_N. If we had set up our hypothesis testing problem as testing

$$\text{H:} \quad \sigma_T = \sigma_N$$

TABLE 6.23 Rejection Rules for the Two Possible Alternatives for the Two-Sample F Test of Population Standard Deviations $(H: \sigma_1 = \sigma_2)$

Alternative	Reject H in Favor of A at Significance Level α if:	F
A: $\sigma_1 > \sigma_2$	$F > F_\alpha[n_1 - 1;\, n_2 - 1]$	$F = \dfrac{s_1^2}{s_2^2}$
A: $\sigma_1 \neq \sigma_2$	$F > F_{\alpha/2}[n_L - 1;\, n_S - 1]$	$F = \dfrac{\text{larger } s^2}{\text{smaller } s^2}$

Notes: For the alternative A: $\sigma_1 \neq \sigma_2$, the test statistic is

$$F = \frac{\text{larger } s^2}{\text{smaller } s^2},$$

where

larger s^2 = larger of the two squared sample standard deviations s_1^2 and s_2^2, and smaller s^2 = smaller of the two squared sample standard deviations s_1^2 and s_2^2.

Therefore, the degrees of freedom for the numerator and denominator are, respectively, $n_L - 1$ and $n_S - 1$, where

n_L = number of data points in sample having the larger squared standard deviation, and
n_S = number of data points in sample having the smaller squared standard deviation.

vs.

A: $\sigma_T < \sigma_N$,

we could not make use of Table A.10 in applying the F test. But all that is necessary to solve the problem is to reformulate the hypothesis and alternative by interchanging the two populations. After the interchange, we get

H: $\sigma_N = \sigma_T$

vs.

A: $\sigma_N > \sigma_T$,

and in this formulation the problem can be solved by the methods illustrated in Example 6.10. With these facts in mind, we present the rejection rules for the two-sample F test in Table 6.23.

EXAMPLE 6.11. Management Aptitude Scores. A corporation psychologist wants to check on the theory that college graduates with different major fields vary to different extents in their aptitude for corporate management. Some experts hold that the degree of variation among liberal arts graduates does not differ significantly from that among business school graduates, while others believe that there is a substantial difference in variation within the graduates of each group. The psychologist therefore conducts the following experiment: he randomly selects a group of 20 liberal arts graduates and 25 business school graduates, and gives them each a standardized management aptitude test of the sort generally used by corporation personnel departments. The 20 liberal arts graduates obtained a standard deviation of 7 on their scores, while the 25 business school graduates had a standard deviation of 11. At level $\alpha = .02$, do the results of the study indicate the existence of a significant difference in variability of management aptitude scores within the two groups?

SOLUTION.   We apply a two-sample F test of the hypothesis

H: $\sigma_L = \sigma_B$ (variation does not differ between groups)

vs.

A: $\sigma_L \neq \sigma_B$ (variation does differ between groups)

where

σ_L = true standard deviation of liberal arts graduates

and

σ_B = true standard deviation of business school graduates.

Using symbols with analogous meanings, we summarize the data as

$$n_L = 20 \qquad n_B = 25$$

$$s_L = 7 \qquad s_B = 11$$

According to the rejection rules presented in Table 6.23, we should reject H at level $\alpha = .02$ if

$$F > F_{\alpha/2}[n_L - 1; n_S - 1] = F_{.01}[n_B - 1; n_L - 1]$$

$$= F_{.01}[24; 19] = 2.92,$$

because s_B^2 is the larger s^2 and s_L^2 is the smaller. The numerical value of $F_{.01}[24; 19]$ has been read off from Table A.10, more precisely from the portion of it appearing in Table 6.24. The numerical value of F is

$$F = \frac{\text{larger } s^2}{\text{smaller } s^2} = \frac{s_B^2}{s_L^2} = \frac{(11)^2}{(7)^2} = 121/49 = 2.47 .$$

TABLE 6.24   A Small Portion
of Table A.10   (Values of $F_{.01}$)

		Degrees of Freedom for Numerator	
Degrees of Freedom for Denominator		24	
	19	2.92	

As we have agreed to reject H in favor of A if $F > 2.92$, the result that $F = 2.47 < 2.92$ does not lead us to reject H. We therefore conclude at level $\alpha = .02$ that the data presented does not provide convincing evidence of differing degrees of variability in management aptitude between the two groups.

One final note on the F test: as in the cases of the t and chi-square tests before it, the F test is valid only when it is reasonable to assume that both sets of data come from normally distributed populations. This restriction is necessary due to the fact that the same technical mathematics involved in the construction of Table A.4 (the table of the standard normal distribution) is used in the construction of Table A.10 (the table of the F distribution).

EXERCISES   6.F

6.44. In Exercise 6.7, contractual specifications require a machine to produce plastic piping of 4 inches diameter. A random sample of 50 pieces coming off the assembly line had standard deviation .12 inch. Can we conclude from the sample, at level $\alpha = .05$, that the diameter of the piping produced has standard deviation in excess of .10 inch and is therefore too variable to be useful?

6.45. A manufacturer of chemicals for use in the pharmaceutical industry conducts a study of the time it takes, after receipt of an order, for the chemicals to reach the customer's plant. The following data give the transit times from the manufacturer to the customer of six randomly selected shipments of the past month:

Shipment number	017	009	021	058	031	002
Transit time (days)	6.0	5.0	3.0	4.5	4.0	5.5

At level $\alpha = .01$, is it reasonable to assert that the transit times have standard deviation that does not exceed 1.0 day?

6.46. From the data of Exercise 6.10, dealing with the time advance of aerospace timing devices, can we say, at level $\alpha = .05$, that the time advance has standard deviation larger than 0.05 seconds?

6.47. The following table gives the producer price index and the index of industrial production in the United States on July 1 of each year for a five-year period:

Year	Producer Prices	Industrial Production
1	80	110
2	80	90
3	90	80
4	80	100
5	70	120

Do the data indicate, at level $\alpha = .05$, that the index of industrial production is significantly more variable than the producer price index?

6.48. In Exercise 6.25, an economist obtained data on the price of English cheese in various metropolitan areas of the U.S. and Canada. Can he conclude from the data, at level $\alpha = .10$, that the prices vary to approximately the same extent in the two countries?

6.49. Based on the training program data of Exercise 6.23, is it reasonable to conclude at level $\alpha = .01$ that the Chicago and Boston methods are equally consistent (namely that they lead to the same standard deviation of scores)?

6.50. Do the Honolulu rental data of Exercise 6.26 indicate, at level $\alpha = .10$, that rentals in 1978 varied significantly more in price than those in 1975?

6.51. On the basis of the airline operating expense data of Exercise 6.28, does it appear that the 747 and DC-10 expenses are equally variable? Use $\alpha = .01$.

SECTION 6.G   TESTS FOR PROPORTIONS

As has been pointed out in Section 4.B and Section 5.E, many questions involving proportions can be better understood if they are structured according to the format of the binomial distribution. As we will soon see from the examples, only the small-sample tests make explicit use of the binomial distribution of Table A.2. The large-sample tests of proportions are based on percentage points of the normal distribution, because for large numbers of data points the normal approximation to the binomial distribution comes into play.

Small Samples

EXAMPLE 6.12.   Income Levels.   In the opinion of the research staff attached to a congressional subcommittee on the economics of urban areas, 30% of full-time employed persons in the Dallas metropolitan area have incomes below $8,000 per year. Conflicting testimony given the committee by a consulting sociologist claims that the true percentage in question is substantially larger than 30%. To resolve the disagreement, the committee appoints an independent researcher to study the situation. The researcher surveys a random sample of 17 persons employed full-time in the Dallas metropolitan area and finds that seven have incomes below $8,000 per year. Does her survey support, at level $\alpha = .05$, the testimony that the true percentage in question is in excess of 30%?

SOLUTION.   We have to choose between the hypothesis

H: $\pi = .30$ (committee staff opinion is correct)

and the alternative

A: $\pi > .30$ (consulting sociologist is correct),

where

$\pi =$ true proportion of full-time employed persons in
      Dallas metropolitan area having incomes below $8,000
      per year.

Because we are using a level of significance of $\alpha = .05$, we want the probability of a Type I error, namely the probability of rejecting H when it's really true, to be no more than 5%. Now, in testing H vs. A, we would be inclined to reject H if substantially more than 30% of the 17 persons sampled have incomes below $8,000 per year. How are we to determine the meaning of the phrase "substantially more?"

   In the case of the *small-sample test of proportion*, we reason as follows: if the hypothesis H: $\pi = .30$ were really true, then, as we are dealing with a random sample of 17 persons, the random variable

B = number of persons in sample having incomes below $8,000 per
      year

has the binomial distribution with parameters $\pi = .30$ and $n = 17$. (Perhaps, at this point, you should glance back to Section 4.B to remind yourself of the details of the binomial distribution.) For easy reference, we reproduce in the first three columns of Table 6.25 the portion of Table A.2 concerned with the binomial distribution having parameters $\pi = .30$ and $n = 17$. As we have noted in the above paragraph, we would be inclined to reject H if the random variable B turned out to be very large. Again, what does "very large" mean? Well, as we have been

TABLE 6.25   A Small Portion of Table A.2   (With adjustment)

		Values of $P(B \leq k)$	Values of $P(B > k) = 1 - P(B \leq k)$
n	k	$\pi = .30$	
17	0	.0023	.9977
	1	.0193	.9837
	2	.0774	.9226
	3	.2019	.7981
	4	.3887	.6113
	5	.5968	.4032
	6	.7752	.2248
	7	.8954	.1046
	8	.9597	.0403
	9	.9873	.0127
	10	.9968	.0032
	11	.9993	.0007
	12	.9999	.0001
	13	1.0000	.0000
	14	1.0000	.0000
	15	1.0000	.0000
	16	1.0000	.0000
	17	1.0000	.0000

doing throughout this whole chapter, we define "very large" by referring to the significance level α. The number represents the probability of our rejecting H when H (namely, $\pi = .30$) is really true. As $\alpha = .05$ for this example, we want to find the numerical value of k such that the probability is .05 that B is larger than k.

If we find this numerical value of k, then we will agree to reject H in favor of A at level $\alpha = .05$ if $B > k$, because our probability of rejecting H when it's really true would be .05. This value of k is called the "critical value of k." Therefore, if $B > k$, we would reject H in favor of A and we would assert, at level $\alpha = .05$, that the true percentage of persons employed full-time, having incomes below $8,000, in the Dallas metropolitan area is indeed in excess of 30%.

We look for the critical value of k by using the fourth column of Table 6.25. While the third column gives the number $P(B \leq k)$, we have adjusted these numbers (by subtracting them from 1) to record the items $P(B > k)$ in the fourth column. We seek the critical value of k, namely the value of k for which $P(B > k) = .05$, as follows: halfway down the fourth column of Table 6.25, we discover the facts that

$P(B > 7) = .1046$, and
$P(B > 8) = .0403$.

From those facts, we see that the critical value of k for $\alpha = .05$ should be somewhere between 7 and 8. However, there are unfortunately no *possible values* of k between 7 and 8, since the binomial distribution is a discrete distribution, as discussed in Section 4.B. We therefore come up with the unpleasant conclusion that there is no rejection rule having level of significance $\alpha = .05$. The two rejection rules having level of significance just above and just below $\alpha = .05$ are:

1. Reject H at level $\alpha = .1046$ if $B > 7$.
2. Reject H at level $\alpha = .0403$ if $B > 8$.

We have now to choose the more appropriate of these two significance levels, and we would probably agree that it should be $\alpha = .0403$, for that is the one closest to .05. We therefore agree to reject H at level $\alpha = .0403$ if $B > 8$.

It now remains only to find the numerical value of B, and this value is evident from the data. It is clear from the data that 7 of the 17 persons studied have incomes below \$8,000. Therefore $B = 7$. As we have agreed to reject H in favor of A at level $\alpha = .0403$ if $B > 8$, and it turned out that $B = 7 < 8$, we cannot reject H. Our conclusion at level $\alpha = .0403$ is then that the independent researcher's survey does *not* support the assertion that the percentage of full-time employed persons having incomes below \$8,000 significantly exceeds 30%.

As it turned out, in the above example we would have failed to reject H at level $\alpha = .1046$ also. This is due to the fact that the rejection rule told us to reject H if $B > 7$, while B turned out to equal 7 exactly. Therefore $B > 7$ did not occur, and so we do not reject H. Because we did not reject H at level $\alpha = .0403$ and we also did not reject it at level $\alpha = .1046$, it is reasonable to conclude that we also would not reject H at level $\alpha = .05$. This kind of argument can be advanced only because .05 is sandwiched between two significance levels, at both of which H was not rejected.

If we recall the discussion of P-values in Section 6.B, we can state that the P-value of $B = 7$ is $P(B > 6) = .2248$, according to Table 6.25. We would therefore reject H at only those levels of α that are above .2248. For any $\alpha < .2248$, we would not reject H.

EXAMPLE 6.13.   Accuracy of Financial Statements.   In order to make an informed decision on whether or not to purchase an oil drilling tool manufacturing company, a group of prospective buyers has to check on the accuracy of the numbers appearing on the company's financial statements. The present owners say that the accounts receivable are at least 95% accurate. A check of 20 randomly selected accounts receivable turns up 3 that are incorrect. At level $\alpha = .10$, do the prospective buyers have reason to believe that the true percentage of accuracy is below 95%?

SOLUTION. If we set

π = true proportion of statements that are accurate,

then we want to test the hypothesis

H: $\pi = .95$

against the alternative

A: $\pi < .95$.

We must therefore carry out the small-sample test of proportion, using the binomial distribution with parameters $\pi = .95$ and $n = 20$, as the random sample is composed of 20 accounts receivable. If the hypothesis H is really true, the random variable

A = number of statements that are accurate

has the binomial distribution with parameters $\pi = .95$ and $n = 20$. We would be inclined to reject H in favor of A if the numerical value of A turned out to be small. Since our probability of rejecting H is $\alpha = .10$, we would reject H if A fell below the critical value of k for which

$P(A < k) = .10$.

From Table A.2, for $\pi = .95$ and $n = 20$, we find that

$P(A \leqslant 17) = .0755$
$P(A \leqslant 18) = .2642$

Because "$A \leqslant 17$" means the same thing as "$A < 18$," it follows that

$P(A < 18) = .0755$.

Therefore, we have discovered that, at level $\alpha = .0755$, the critical value of k is 18.

It would then be logical to agree to reject H in favor of A if $A < 18$, for the significance level (the probability of rejecting H when it's really true) would then be $\alpha = .0755$, as close to $\alpha = .10$ as we could expect to get in this discrete situation.

Looking at the data, we see that the sample has detected errors in 3 out of the 20 accounts sampled. Therefore $A = 17 < 18$, so we must reject H at level $\alpha = .0755$. The same decision holds true at level $\alpha = .2642$ as well, because $A = 17 < 19$, too. At level $\alpha = .10$, we are

confident in asserting that the percentage of accounts receivable statements that are correct is actually below 95%.

If, in this example, the appropriate level of significance was $\alpha = .05$,
we would be unable to make a decision. This is due to the fact that Table A.2
shows $P(A \leq 16) = .0159$, so that we should reject H at level $\alpha = .0159$ if A
< 17. Since A = 17 exactly, we see that A < 17 does not occur so we cannot
reject H. As we must reject H at level $\alpha = .0755$, but we cannot do it at level
$\alpha = .0159$, the discrete nature of the situation does not allow us to make a
decision at level $\alpha = .05$. In fact, Table A.2 shows that the P-value of A = 17
is $P(A < 18) = .0755$, so we can reject H for certain only when $\alpha > .0755$. It
is customary, in situations like this, to accept H whenever we do not have
sufficient evidence to reject it. At level $\alpha = .05$, the best we could do would
be to say that the data do not support the belief that fewer than 95% of the
accounts receivable are incorrect.

Large Samples

EXAMPLE 6.14.   A Corporate Takeover. General Nucleonics, a
major energy corporation, is attempting to take over a smaller concern,
Sergeant Nucleonics. General's securities consultant reports that 60%
of Sergeant's shareholders support the takeover bid. To be sure of this,
General's president requests a telephone survey of a random sample of
Sergeant's shareholders. The staff polls 1,500 shareholders and finds
that 784 support the takeover bid. At level $\alpha = .01$, does the telephone
poll tend to refute the consultant's report?

SOLUTION.   We use the *one-sample z test for the proportion*. Denote
by π the true proportion of shareholders who support the takeover bid.
We want to test the hypothesis

H: $\pi = .60$ (true proportion is 60%)

against the alternative

A: $\pi < .60$ (less than 60% support takeover bid).

In accordance with the principles of the normal approximation to the
binomial distribution (since n = 1,500 > 30), we compute

$$z = \frac{p - \pi_0}{\sqrt{\dfrac{\pi_0(1 - \pi_0)}{n}}} \qquad (6.13)$$

where

π_0 = hypothesized proportion supporting takeover bid,
p = sample proportion supporting takeover bid,
n = number of shareholders involved in poll.

(This formula is reminiscent of the one used in the one-sample z test for the mean of population.)

We then reject H at level α if $z < -z_\alpha$. This rejection rule has been chosen on the basis of the three options presented in Table 6.8, upon replacing the μ's by π's. We therefore agree to reject H in favor of A at level $\alpha = .01$ if $z < -z_{.01} = -2.33$. It now remains only to compute the numerical value of z. From the data accumulated in the poll, we know that

$n = 1,500$
$p = 784/1,500 = .5227$,

and, on the basis of the question asked, we see that the hypothesized proportion is $\pi_0 = .60$. It follows automatically that

$$ z = \frac{p - \pi_0}{\sqrt{\dfrac{\pi_0(1 - \pi_0)}{n}}} = \frac{.5227 - .60}{\sqrt{\dfrac{(.60)(.40)}{1,500}}} $$

$$ = -.0773/\sqrt{.00016} = -.0773/.01265 = -6.114 . $$

As we agreed to reject H if z turned out to be less than (more negative than!) -2.33, the result that $z = -6.114 < -2.33$ leads us to reject H. At level $\alpha = .01$, the president's poll therefore tends to refute the consultant's report that 60% of the shareholders support the takeover bid. The poll indicates that the true percentage supporting the takeover bid is significantly below 60%.

EXAMPLE 6.15. Advertising Effectiveness. The advertising department of a major Southern California automobile dealership, which has branches in Orange and Los Angeles Counties, wants to know whether its advertising is really effective in influencing prospective customers. In particular, the department wants to find out if more station wagons can be sold by saturating an area with advertising urging customers to buy a station wagon. To test out the effectiveness of such an advertising campaign, the dealership for one week floods Orange County with advertising, while maintaining a low profile in Los Angeles County. Over the next two months 120 of 400 cars it sells at the Orange County branch are station wagons, while 150 of the 600 cars it sells at the L.A.

branch are station wagons. Do the results of the study tend to indicate that a proportionately higher number of station wagons was sold in Orange County, where the advertising was concentrated? Use a level of significance of $\alpha = .05$.

SOLUTION.   We use the *two-sample z test for the difference of proportions*. We define some symbols as follows:

π_H = true proportion of station wagons sold in regions of *heavy* advertising.

and

π_N = true proportion of station wagons sold in region of *normal* advertising.

We want to test the hypothesis

H: $\pi_H = \pi_N$ (advertising does not affect sales of station
             wagons)

against the alternative

A: $\pi_H > \pi_N$ (heavy advertising increases sales of station
             wagons).

From the data, we can determine the items

n_H = number of data points in the region of heavy advertising,
p_H = sample proportion of station wagons sold in the region of heavy advertising,
n_N = number of data points in the region of normal advertising,
p_N = sample proportion of station wagons sold in the region of normal advertising,
p = sample proportion of station wagons sold in both regions combined.

Sales records indicate that

$$n_H = 400 \qquad\qquad\qquad n_N = 600$$

$$p_H = 120/400 = .30 \qquad\qquad p_N = 150/600 = .25$$

$$p = \frac{120 + 150}{400 + 600} = 270/1{,}000 = .27 \ .$$

In accordance with the rejection rules of Table 6.8, we calculate

■ $$z = \frac{p_H - p_N}{\sqrt{p(1-p)\left(\frac{1}{n_H} + \frac{1}{n_N}\right)}} \qquad (6.14)$$

(this formula might remind you of the one for the two-sample t test for the difference of means), and then we reject H in favor of A at level α if $z > z_\alpha$. Here $\alpha = .05$, so we agree to reject H if $z > z_{.05} = 1.64$. Upon inserting the numerical values of the components of the formula for z, we get that

$$z = \frac{p_H - p_N}{\sqrt{p(1-p)\left(\frac{1}{n_H} + \frac{1}{n_N}\right)}} = \frac{.30 - .25}{\sqrt{(.27)(.73)\left(\frac{1}{400} + \frac{1}{600}\right)}}$$

$$= \frac{.05}{\sqrt{(.197)\left(\frac{1,000}{240,000}\right)}} = \frac{.05}{\sqrt{(.197)(.004167)}} = \frac{.05}{\sqrt{.00082}}$$

$$= .05/.0286 = 1.75.$$

Therefore $z = 1.75 > 1.64$, so we have to reject H in favor of A at level $\alpha = .05$. We conclude that, at level $\alpha = .05$, the advertising campaign can be considered to have been effective in increasing the proportion of station wagons sold.

One final remark on the above example: if the proper level of significance had been $\alpha = .01$, in the opinion of the advertising department, the advertising campaign would have been judged noneffective. This is due to the fact that the rejection rule would have indicated the rejection of H when $z > z_{.01} = 2.33$. Since z turned out to be 1.75, H could not validly have been rejected and we would have to say that the advertising did not seem to affect sales. We have here one more example of the importance of making an appropriate subjective choice of the significance level α, the probability of rejecting H when it's really true. According to Table A.4, the P-value of $z = 1.75$ is $\alpha = .0401$, so we would reject H for choices of $\alpha > .0401$, but accept H if $\alpha < .0401$.

EXERCISES 6.G

6.52. To find out whether or not a particular coin is really fair (equally likely to fall heads or tails), the coin is tossed several times and the number of heads resulting is carefully noted.

a. If four out of ten tosses result in heads, can we assert at level $\alpha = .05$ that the coin is fair?

b. If 40 out of 100 tosses result in heads, can we conclude at level $\alpha = .05$ that the coin is fair?

6.53. An executive in charge of new product development at a local dairy feels that at least 10% of the population would be willing to try a carton of chocolate-flavored buttermilk, while the other 90% would not be at all interested. Is her opinion substantiated at level $\alpha = .10$ by a survey of 50 randomly selected persons, which turned up four who would be interested?

6.54. The manufacturer of a new brand of soap guarantees that, with the aid of her unique marketing strategies, at least 80% of the soap marketed will sell quickly. Does an experiment in which 10 of 17 bars marketed sell quickly tend to refute her claims at level $\alpha = .05$?

6.55. A straw poll of 14 randomly selected members of the various state legislatures reveals seven who favor a certain controversial change in the banking laws.

a. At level $\alpha = .10$, do the results of the poll conflict with a news reporter's assertion that at least 60% of the membership supports the change?

b. At level $\alpha = .10$, do the results of the poll conflict with a banking lobbyist's assertion that no more than 40% of the membership supports the change?

6.56. A straw poll of 140 randomly selected state legislators turns up 70 who favor a certain controversial change in the banking laws.

a. At level $\alpha = .10$, do the results of the poll conflict with a news reporter's assertion that at least 60% of the membership supports the change?

b. At level $\alpha = .10$, do the results of the poll conflict with a banking lobbyist's assertion that no more than 40% of the membership supports the change?

6.57. A 1977 Harris poll of consumers claimed that 61% of those questioned felt that the quality of most products has deteriorated over the past 10 years. To check on these disturbing results, a consumer analyst for a large appliance manufacturer conducted his own poll of his company's recent customers. Of 100 randomly selected customers, 70 agreed that product quality has indeed declined. At significance level $\alpha = .05$, can the analyst be sure that the results of his own poll show that his company's recent customers are even more disappointed than the general public surveyed in the Harris poll?

6.58. A survey of 50 licensed drivers living in an urban area turns up 8 who have a drinking problem, while a survey of 30 drivers in a rural area reveals that 4 are problem drinkers. Can we conclude at level $\alpha = .05$ that the proportion of problem drinkers differs significantly between urban and rural areas?

6.59. A tight race is developing for the presidency of a major national labor union between the incumbent and the challenger.

a. A preelection poll of 100 union members nationwide concludes that 53% of those surveyed support the incumbent. At level $\alpha = .05$, does this indicate that the incumbent will obtain more than 50% of the vote?

b. A preelection poll of 1000 union members nationwide concludes that 53% of those surveyed support the incumbent. At level $\alpha = .05$, does this indicate that the incumbent will obtain more than 50% of the vote?

6.60. A survey of 70 men and 60 women taking the College Entrance Examinations showed that 10 of the men and 7 of the women intended to major in business and economics. At level $\alpha = .05$, is there a significant difference between the proportions of men and women interested in that subject area?

6.61. A recent survey of attorneys in Orange County, California, showed that 16 out of 23 opposed the 1977 Supreme Court decision allowing lawyers to advertise their services.* Do the survey results indicate that significantly more than 65% of Orange County lawyers opposed the Supreme Court decision? Use $\alpha = .05$.

6.62. Rule 85-P, recently promulgated by the SEC at the urging of the AICPA, drew the support of 90 of 600 management accountants attending an accounting convention in Phoenix, and the support of 80 of 400 auditors attending. Does this indicate, at level $\alpha = .02$, that auditors favor the proposal in a significantly higher proportion than do management accountants?

6.63. *Forbes* Magazine (in its May 15, 1978, issue, pp. 129–130) reported that over 50% of the players in the National Basketball Association earn over $100,000 per year. Can this assertion be substantiated on the basis of a survey of 35 such players of whom 20 earn over $100,000? Use a level of significance of $\alpha = .10$.

SUMMARY AND DISCUSSION

In this chapter, we have introduced the procedure of formulating a question in statistical terms and then answering it using available data. As we have seen, the first step is setting up a hypothesis and an alternative, and the second is using the data to help us choose between them.

We have illustrated the technique of *testing statistical hypotheses* in examples involving means, standard deviations, and proportions.

The general problem of statistical inference that we first encountered in our study of confidence intervals in Chapter 5, namely, the fact that we cannot be 100% sure of our decision, is still with us. It is reflected in the fact that each statistical decision is made in terms of a level of significance, which is our probability of rejecting the hypothesis we formulated when that hypothesis is actually true. Having chosen a level of significance gives us an idea of our chances of making an incorrect choice between hypothesis and alternative.

Our study of testing statistical hypotheses does not end here, but continues into Chapters 7, 8, 9, 12, and 13, as we continue to ask important questions and to answer them on the basis of a collection of statistical data.

---

*Orange County Business, November/December 1977, p. 39.

CHAPTER 6   BIBLIOGRAPHY

SIGNIFICANCE LEVELS

COHEN, J. *Statistical Power Analysis for the Behavioral Sciences*. 1969, New York: Academic Press.

FEINBERG, W. E. "Teaching the Type I and Type II Errors: The Judicial Process," *The American Statistician*, June 1971, pp. 30–32.

FRIEDMAN, H., "Trial by Jury: Criteria for Convictions, Jury Size and Type I and Type II Errors," *The American Statistician*, April 1972, pp. 21–23.

MORRISON, D. E., and R. E. HENKEL (eds.), *The Significance Test Controversy—A Reader*. 1970, Chicago: Aldine.

TSANG, H. H., "The Effects of Changing Sample Size on the Alpha and Beta Errors: A Pedagogic Note," *Decision Sciences*, October 1977, pp. 757–759.

WEBSTER, G. W., and T. A. CLEARY, "A Proposal for a New Editorial Policy in the Social Sciences," *The American Statistician*, April 1970, pp. 16–19.

SPECIAL TESTING SITUATIONS

DAUGHERTY, T. F., "Two apparent contradictions in testing hypotheses," *Journal of Quality Technology*, July 1980, pp. 154–157.

EBERHARDT, K. R., and M. A. FLIGNER, "A comparison of two tests for equality of two proportions," *The American Statistician*, November 1977, pp. 151–155.

FOSTER, T. W., III, and D. VICKREY, "The Information Content of Stock Dividend Announcements," *The Accounting Review*, April 1978, pp. 360–370.

GIBBONS, J. D., and J. W. PRATT, "P-values: Interpretation and Methodology, *The American Statistician*, February 1975, pp. 20–25.

HERREY, E. M. J., "Percentage Points of H-Distribution for Computing Confidence Limits or Performing *t*-tests by the Mean Absolute Deviation," *Journal of the American Statistical Association*, March 1971, pp. 187–188.

SOLOMON, S. L., A Decision Model for Selecting Alternative Hypotheses, *Decision Sciences*, July 1975, pp. 581–589.

SELECTED PROFESSIONAL PUBLICATIONS REPORTING STATISTICAL ANALYSES

The Accounting Review
The American Economic Review
The American Statistician
Bell Journal of Economics
Business and Society
Canadian Journal of Economics
Columbia Journal of World Business
Decision Sciences
Econometrika
Economic Geography
Financial Management
Industrial & Labor Relations Review
Journal of Accounting Research
Journal of the American Statistical Association

Journal of Economics and Business
Journal of Finance
Journal of Financial and Quantitative Analysis
Journal of Marketing Research
Management Science
National Tax Journal
Personnel Psychology
Review of Economics and Statistics
Transportation Journal
Urban Affairs Quarterly
University of Michigan Business Review

CHAPTER 6   SUPPLEMENTARY EXERCISES

6.64. The manufacturer of a hospital room humidifier advertises that the control dial on the device will maintain a mean room humidity of 80%, with a standard deviation of 2.0%. Its performance was carefully observed during a one-day test at 60 different times throughout the day. The mean humidity level turned out to be 78.3 with a standard deviation of 2.9.

a. At level $\alpha = .01$, do the observed data tend to contradict the manufacturer's claim that the true mean humidity level will be 80?

b. Construct the power curve of the test in part a.

c. At level $\alpha = .05$, does the one-day test indicate that the true standard deviation is somewhat larger than the advertised value of 2.0?

6.65. In a routine check of drug packaging, the FDA counts the contents of 16 bottles of aspirin labeled as containing 500 tablets each. The 16 bottles have mean contents 494 aspirins with a standard deviation of 30.

a. Can the FDA assert, at level $\alpha = .05$, that the bottles are underfilled?

b. Can they assert, at level $\alpha = .05$, that the true standard deviation of such bottles is significantly above 15?

6.66. A company that markets canned carrots advertises that their 15-ounce cans actually have mean weight 15.1 ounces, with a standard deviation of .1 ounce, so that a substantial percentage of cans exceed the stated contents (15.0 ounces). To test the validity of the company's assertion, a state consumer agency selects a random sample of 11 cans and carefully weighs the contents of each. The 11 weights are as follows:

15.1   15.0   15.2   14.9   15.0   15.1   15.1   14.9   15.1   15.0   14.6

a. At significance level $\alpha = .05$, can the agency assert that the true mean contents are below the 15.1 ounces claimed by the company?

b. At level $\alpha = .05$, do the data indicate that the true mean contents are below the 15.0 ounces printed on the label?

c. At level $\alpha = .05$, can the agency conclude from the data that the true standard deviation exceeds .1 ounce?

6.67. A truck farmer decides to test a new nonpolluting insecticide that also, according to the inventor, definitely reduces the loss attributable to a common pest. The farmer has a feeling that the new spray is really worse than the standard (polluting) spray so he treats 80 acres with the new spray and 100 with the standard spray throughout the growing season. For the new insecticide, the mean yield per acre turned out to be 980 pounds, with a standard deviation of 60 pounds, while for the standard spray there was a mean yield of 1,040 pounds, with a standard deviation of 50 pounds.

a. At level $\alpha = .01$, do the results of the test provide evidence that the new spray is significantly worse than the standard one?

b. At level $\alpha = .05$, can we conclude that the new spray is more variable in its performance than is the standard spray, in the sense of having a larger yield standard deviation?

6.68. In order to compare the merits of two short-range ground-to-ground antitank weapons, a Defense Department analyst conducts a test of 8 of the conventional type and 10 of the more expensive wire-guided rockets. The conventional type have mean target error of 5.2 feet with a standard deviation of 1.8 feet, while the wire-guided ones have mean target error of 3.6 feet with a standard deviation of 1.5 feet. At level $\alpha = .10$, can we say that the more expensive rockets are really more accurate?

6.69. A candidate for the presidency of a statewide labor union feels that she can lose approximately 60% of the vote in the northern part of the state and still accumulate enough votes in the south to put together a winning margin. A recent poll of 180 union members in the north revealed 117 who planned to vote against her. At level $\alpha = .01$, does the poll indicate that she is losing significantly more than 60% of the vote?

6.70. An insurance industry pamphlet claims that the average yearly cost of a $50,000 five-year renewable term life insurance policy to a 25-year-old male is below $200 per year. A random sample of 10 major insurance companies shows that the costs of their programs (in dollars) are as follows:

220   230   190   200   190   190   190   160   150   230

At level $\alpha = .01$, do the sample data tend to confirm the industry claim that the true mean cost falls significantly below 200?

6.71. It is the job of a tea tester for an English tea-importing company to decide which tea leaves are most suitable for "English Breakfast" tea. The tea tester has to decide between two lots of leaves, an inexpensive lot and an expensive lot. The tea tester and his laboratory staff come up with the following levels of impurities in eight randomly selected leaves of each lot:

Inexpensive lot	3	4	25	9	11	4	12	20
Expensive lot	6	5	7	10	6	8	5	4

The tea tester knows, of course, that he cannot validly use the t test to decide whether or not both lots of leaves have the same average impurity levels unless both populations (inexpensive and expensive impurity levels) have the same standard deviation. At significance level $\alpha = .02$, do the data indicate that the two standard deviations are the same?

6.72. The research department of a company that produces industrial string proposes a new technique for strengthening the string and making it able to withstand stronger forces. To find out whether the new technique really does strengthen the string, five lengths of string are produced by the standard technique and five by the new technique, and the samples are compared for their respective breaking points, in pounds of force. The breaking points are as follows:

Standard technique	144	131	155	126	134
New technique	139	154	132	143	147

a. At level $\alpha = .10$, does the comparison indicate that the breaking points of both groups have the same standard deviation?

b. At level $\alpha = .10$, does the comparison support the contention of the research department that the new technique produces stronger string?

6.73. Several insurance adjusters were concerned about the unusually high repair estimates they seemed to be getting from Fosbert's U-Bet Repair Station. To test their suspicions, they brought each of eight damaged cars to Fosbert's and also to the auto repair shop of Nickle's Department Store, a concern generally regarded as reliable. They obtained the following estimates, in hundreds of dollars:

Car number	#1	#2	#3	#4	#5	#6	#7	#8
Fosbert estimate	2.1	4.5	6.3	3.0	1.2	5.4	7.3	9.3
Nickle estimate	2.0	3.8	5.9	2.8	1.3	5.0	6.5	8.6

At level $\alpha = .01$, can the adjusters conclude from their survey that Fosbert's estimates are significantly higher than Nickle's?

6.74. A committee of personnel officers was assigned the task of comparing young employees with older employees in their ability in the quantitative skills area. Twenty employees, 10 selected from among young employees and the other 10 from among the older employees, were paired on the basis of job classification, educational level, and general health, and then the committee administered the same quantitative skills test to all 20. The scores follow:

Pair	A	B	C	D	E	F	G	H	I	J
Young employees	100	50	10	60	80	60	80	40	20	20
Older employees	90	60	20	60	70	70	90	50	10	40

Do the test scores indicate, at level $\alpha = .05$, that the committee should report that young employees are less effective than older employees in the quantitative skills area?

6.75. A compendium of actuarial tables published in 1958 shows that of all 45-year-old men in a certain occupational grouping 7% die before reaching age 55. A more recent survey of 2,000 men in the age and occupational group reveals that percentage to be 5%. At level $\alpha = .01$, does the recent survey indicate that the probability of death in that group has been significantly reduced since 1958?

6.76. One bank loan examiner claims that 55% of the adult population of a particular major city have applied for a bank loan at one time or another.

a. Does a random sample of 18 residents, of whom eight have applied for a loan, tend to refute the claim at level of significance $\alpha = .05$?

b. Does a random sample of 160 residents, of whom 80 have applied, tend to refute the claim at level $\alpha = .05$?

6.77. As part of an analysis of overdue accounts that eventually require legal action to force payment, a corporation accountant discovers that 45 out of 130 corporate accounts required legal action in the past, while 214 of 1,070 individual accounts required such action. Can the accountant conclude that a corporate account is significantly more likely to require legal action than is an individual account? Use level $\alpha = .01$.

6.78. A political poll concerning a union election in which Able is running against Baker shows that 520 of 1,000 urban members favor Able, while 240 of 500 rural members favor Able. Can we conclude from the results of the poll that Able is significantly more popular, at level $\alpha = .05$, in urban areas than in rural areas?

6.79. An investment counselor, studying the per-share dividends of major U.S. corporations, selects a random sample of companies from the *Forbes* list of about 800 of the biggest corporations and comes up with the following 1976 information on per-share yearly dividends in dollars:*

Company	Dividend	Company	Dividend
Hanna Mining	1.65	Cone Mills	0.95
American Express	0.85	Olin Corp.	1.37
Dufouu	1.10	Waldbaum	0.00
Niagara Mohawk	1.24	First Hawaiian	1.50
Florida Power	2.15	Eli Lilly	1.25
Potlatch	0.78	J. P. Stevens	1.10
Timken	2.20	Allis-Chalmers	0.50
Bache Group	0.40	Halliburton	0.80
Nash Finch	1.60	Scott Paper	0.72
Owens-Illinois	0.94	Jefferson-Pilot	0.78
Esmark	1.58	Chubb	1.42
Vornado	0.00	Mobil	3.50
Kaiser Steel	1.50	Fidelcor	2.20
Bendix	1.79	Coca-Cola	2.65
Dillingham	0.48	General Mills	0.72
Fruehauf	1.80	Northrop	1.35
Richmond Corp.	0.80	Travelers	1.08
Jim Walter	1.10	Paine Webber	0.30

a. At level $\alpha = .01$, do the investment counselor's data indicate that the true mean per-share dividends paid by major U.S. corporations exceed $1?

b. What about at level $\alpha = .05$?

c. Construct the power curve of the test when $\alpha = .01$.

d. Construct the power curve of the test when $\alpha = .05$.

*Forbes, May 15, 1977, pp. 202–243.

6.80. In order to find out if there is a significant difference between banking companies and utilities in regard to average 12-month earnings per share, a financial analyst selected a random sample of 12 banking companies and 12 utilities; and discovered that their respective earnings per share were as follows for the 12 months between October 1, 1976, and September 30, 1977:*

Banking Companies	Earnings per Share	Utilities	Earnings per Share
Security Pacific	4.45	Long Island Lighting	2.58
Mellon National	7.08	Ohio Edison	2.17
First Bank System	4.63	Rochester Telephone	1.98
BancOhio	3.44	Texas Eastern	5.15
Wachovia	1.95	Neptune International	1.66
Citicorp	3.18	Communic. Satellite	3.62
First Wisconsin	2.93	Arizona Public Service	3.14
Texas Commerce	3.61	Western Union	2.48
Manufacturers Hanover	5.23	Oklahoma Natural Gas	5.59
Charter New York	5.09	Duke Power	2.39
Rainier Bancorp.	3.23	Allegheny Power	2.44
Crocker National	4.05	So. Calif. Edison	3.98

At level α = .02, do the data indicate that there is a significant difference between the average earnings per share of the two industry groupings?

6.81. During the weeks ending, respectively, December 3, 1976, and December 2, 1977, U.S. auto manufacturers produced the following numbers of cars of eight randomly selected models:*

Model	Week Ending Dec. 3, 1976	Week Ending Dec. 2, 1977
Monte Carlo	8,708	10,500
Pacer	1,385	1,260
Firebird	3,191	4,552
Monaco	1,579	1,750
Skylark	2,598	2,001
Mark V	2,311	1,833
LTD II	5,722	4,048
Cougar	4,347	3,766

At level α = .10, do the data seem to show that auto production in the U.S. differed significantly between the weeks studied?

*Business Week, November 14, 1977, pp. 107, 134, and 138.

*The Wall Street Journal, December 2, 1977, p. 7.

6.82. Amalgamated Motor Fuzzwalls (AMF), Inc., manufactures a sound-absorption screen that is advertised to reduce audible engine noise by an average of 10 decibels. The Federal Institute of Quality of Life (FIQL) tested 82 such screens and found that they had a mean noise reduction level of 9.4 decibels with a standard deviation of 1.2 decibels. Using a significance level of $\alpha = .08$, can FIQL assert that the true mean noise reduction is below the 10-decibel figure that AMF advertises?

6.83. The four Marx Brothers made a number of very successful movies back in the 1930s and 1940s. So successful, in fact, that when they were shown on commercial television during the past decade, a TV magazine found that, in 17 randomly selected hours, an average of 22 minutes of time was sold to advertisers, with a standard deviation of 3.1 minutes. At level $\alpha = .01$, can we assert that more than 20 minutes of each hour, on the average, was occupied by commercial messages?

7

Analysis
of Variance

In our discussion of testing statistical hypotheses in Chapter 6, we studied one-sample tests and two-sample tests about population means. We introduced some techniques valid for large samples and some for small samples, and we allowed for two independent samples and for paired samples.

As you may have noticed, we did not analyze a situation involving three or four or more samples; yet, it is often necessary to run an experiment involving several samples. For example, a manufacturer often wants to know whether or not sales of a new product will vary in several different regions of the country. The way to find out would be to select sales offices in several metropolitan areas in each of four regions and then compare the four mean sales levels.

As a second example, an office executive might be considering five competing computerized storage systems for his company's office records, and he would want to know whether they are equally efficient or not. The problem could be analyzed by solving some sample information retrieval problems in each of the five systems and then comparing the five mean efficiency levels. As it turns out, none of the techniques studied in Chapter 6 is effective in studying either of these and similarly structured problems. We need a completely new method of analysis.

Before we proceed with the details of this new method, called the *analysis of variance*, let's take a brief look at the reasons why our previous methods are inadequate for the study of these problems. Suppose we are working with the new product sales problem just mentioned. We denote the true mean sales of the new product in the various regions as follows:

μ_E = true mean sales per sales office in the East
μ_S = true mean sales per sales office in the South
μ_M = true mean sales per sales office in the Midwest
μ_W = true mean sales per sales office in the West

We would then want to test, at significance level α, the hypothesis

H: $\mu_E = \mu_S = \mu_M = \mu_W$

against the alternative

A: these true means are not all the same.

If we were to apply our previous techniques to this problem, the simplest thing we could do would be to run each of the following three tests, using a two-sample test on each one:

1. H: $\mu_E = \mu_S$

vs.

   A: $\mu_E \neq \mu_S$

2. H: $\mu_S = \mu_M$

vs.

   A: $\mu_S \neq \mu_M$

3. H: $\mu_M = \mu_W$

vs.

   A: $\mu_M \neq \mu_W$

If we were to reject H in any one of these three tests, we would be concluding that the true means are not all the same, and we would therefore have to reject the overall hypothesis

H: $\mu_E = \mu_S = \mu_M = \mu_W$

in favor of the alternative

A: these true means are not all the same.

So far so good, but let's now talk about the significance levels of the tests.

If we ran each of the three two-sample tests at significance level α = .05, we would have a probability .05 of making a Type I error and a probability

.95 of avoiding a Type I error on each one. This means that we would have a 5% chance of rejecting each of the three H's if they were really true. In other words, we would have a 95% chance in each case of asserting that the two means in question were the same, when they really were. Now suppose the overall hypothesis H were really true. What would be the probability of avoiding a Type I error here, namely what would be the probability of our deciding that H was really true?

Well, to accept the overall H, on the basis of the three two-sample tests, we would have to accept those three H's. But, we have a 95% chance of accepting each H, because all three would really be true if the overall H were. To be fair, all three two-sample tests would have to be conducted independently of each other, and so we would see that, if the overall H were really true.

$$
\begin{aligned}
P(\text{accepting overall H}) &= P(\text{accepting all three two-sample H's}) \\
&= P(\text{accepting first H}) \times P(\text{accepting second H}) \\
&\quad \times P(\text{accepting third H}) \\
&= (.95)(.95)(.95) = .857 = 85.7\%
\end{aligned}
$$

(using properties of independent events discussed in Section 3.C.). Therefore we have an 85.7% chance of accepting the overall H when it's really true, if we use three two-sample tests. It automatically follows that we have a 14.3% chance of rejecting the overall H when it's really true. This means that the probability of a Type I error, the significance level of our test, will be .143.

We began with three two-sample tests, each having significance level $\alpha = .05$, and we would up with a test of our overall hypothesis H having significance level $\alpha = .143$. If we intend to set up a test of level $\alpha = .05$, the resulting level $\alpha = .143$ is an unacceptably high level of significance, of course, for it indicates that we will be making a Type I error nearly one-seventh (rather than one-twentieth) of the time when H is really true. This is what invalidates the method of compounding two-sample tests—it unfortunately compounds the probability of error. (To guarantee a significance level of .05 for the test of the overall hypothesis, it would be necessary to use a level $\alpha = .017$ on each of the three two-sample tests. Such a low level of α would probably push β, the probability of a Type II error, to an unacceptable level.) Therefore we need a method of testing the equality of several means by a single several-sample test. That method, the analysis of variance, is our subject for this chapter.

SECTION 7.A.   COMPONENTS OF THE VARIANCE

If we market a new product in five randomly selected sales offices in metropolitan areas of each of four regions, we will have a total of 20 pieces of sales information to analyze. Even if there is no real difference between the sales potential in differing regions, it would be unreasonable to expect all 20 metropolitan areas to give exactly the same figures. As we have discussed earlier in Chapter 5, for example, this would be as unreasonable as expecting a sample mean to be exactly the same as a true mean. Therefore, we would expect to have some variation among the sales figures, and it is the extent of this variation that leads us to decide whether or not there is a significant difference in sales potential among different regions.

Suppose our marketing experiment results in the data presented in Table 7.1. How are we to decide whether or not the true mean sales are the same for the four regions? We first have to discuss two pertinent questions:

1. To what extent do the four sample means differ among themselves?

2. To what extent do the four sample means represent the true means of their respective groups?

Our decision to accept or reject the hypothesis that the four true means are equal will be made on the basis of the answers to the above two questions. We will be inclined to reject H if the sample means differ greatly among themselves and we think that they are accurate representations of the true means. Otherwise, we will probably accept H. In Table 7.2, we illustrate the relationship between the answers to questions 1 and 2 above and our decision regarding the hypothesis H.

Let's deal first with question 1, which involves the extent to which the four sample means differ among themselves. As can be seen from the data of Table 7.1, we have the sample means 8, 8, 6, and 5, each based on one of the four columns of data. The extent to which these sample means differ among themselves can be measured by their variation from *their own* mean 6.75,

TABLE 7.1   Daily Sales of New Product by Sales Offices in Four Regions   (*Thousands of dollars*)

	East	South	Midwest	West	Entire Set of Data
	13	8	3	4	
	6	2	5	8	
	5	10	10	5	
	6	9	6	6	
	10	11	6	2	
Sums	40	40	30	25	135
Sample Means	8	8	6	5	6.75

TABLE 7.2   When Should the Hypothesis H: $\mu_E = \mu_S = \mu_M = \mu_W$ Be Rejected in Favor of A: These True Means Are Not All the Same?

Extent to Which Sample Means Represent True Means	Extent to Which Sample Means Differ Among Themselves	
	Little	Much
Little	Cannot Reject H	Cannot Reject H
Much	Cannot Reject H	Should Reject H

which is also the "grand" mean of all 20 data points. We can denote this measure of variation by

$$VB = (8-6.75)^2 + (8-6.75)^2 + (6-6.75)^2 + (5-6.75)^2$$

$$= (1.25)^2 + (1.25)^2 + (-0.75)^2 + (-1.75)^2$$

$$= 1.5625 + 1.5625 + 0.5625 + 3.0625 = 6.75$$

where VB stands for *variation between* the samples. (The fact that VB also turned out to be 6.75 is pure coincidence, not related in any way to the fact that the sample means had mean 6.75.)

Before we investigate the significance of VB in detail, let's work with question 2, which concerns the extent to which the four sample means represent the true means of their groups. To study this question, we can compute the *variation within* each sample from *its own* mean as follows (using the squared deviations from the mean, as we have been doing since Section 2.C).

$$VW_E = (13-8)^2 + (6-8)^2 + (5-8)^2 + (6-8)^2 + (10-8)^2$$

$$= 25 + 4 + 9 + 4 + 4 = 46$$

$$VW_S = (8-8)^2 + (2-8)^2 + (10-8)^2 + (9-8)^2 + (11-8)^2$$

$$= 0 + 36 + 4 + 1 + 9 = 50$$

$$VW_M = (3-6)^2 + (5-6)^2 + (10-6)^2 + (6-6)^2 + (6-6)^2$$

$$= 9 + 1 + 16 + 0 + 0 = 26$$

$$VW_W = (4-5)^2 + (8-5)^2 + (5-5)^2 + (6-5)^2 + (2-5)^2$$

$$= 1 + 9 + 0 + 1 + 9 = 20$$

The *variation within* the samples, denoted by VW, can then be calculated as

$$VW = VW_E + VW_S + VW_M + VW_W$$

$$= 46 + 50 + 26 + 20 = 142.$$

Before proceeding, let's take a look at what we have done so far. We have four samples of data, and we have defined two measures of variation:

VB = variation between the samples
VW = variation within the samples

Now if H (the hypothesis that the four true means are equal) is true, there should be little variation *between* the samples, while if H is false, there should be a lot of variation *between* the samples. What does this have to do with variation *within* the samples? Well, as it turns out, there are only two types of variation—variation between samples and variation within samples. These two types of variation must add up to the total variation of the data points, namely the sum of squared deviations of all the data points from the grand mean. Therefore, when VB is small, VW is large, and when VB is large, VW is small.

BASIC PRINCIPLE OF ANALYSIS OF VARIANCE
When the hypothesis H is true, VB is small and VW is large, and when the alternative A is true, VB is large and VW is small.

To see how this principle works in the example data of Table 7.1, it is necessary to compute the *total variation*, denoted by TV, of the data and then to compare it with VB and VW. In Table 7.3, we compute the total variation of the data, and we come up with the result that TV = 175.75.

If we set n = number of data points in each sample, then the relationship between TV, VB, and VW is given by the equation

■ $$TV = nVB + VW, \tag{7.1}$$

which is called the *components of the variance formula*. For the data of Table 7.1, we have shown that

$$TV = 175.75 \qquad n = 5 \qquad VB = 6.75 \qquad VW = 142$$

The components of the variance formula, $TV = nVB + VW$, then becomes the assertion that

$$175.75 = (5)(6.75) + 142,$$

which is a true statement since $(5)(6.75) = 33.75$.

In the remainder of this section, we shall complete the discussion of the new product sales problem by formally testing the hypothesis

$$H: \quad \mu_E = \mu_S = \mu_M = \mu_W$$

TABLE 7.3   Calculation of the Total
Variation   *(New product sales data)*

x	$x - \bar{x}$	$(x - \bar{x})^2$
13	6.25	39.0625
6	−0.75	0.5625
5	−1.75	3.0625
6	−0.75	0.5625
10	3.25	10.5625
8	1.25	1.5625
2	−4.75	22.5625
10	3.25	10.5625
9	2.25	5.0625
11	4.25	18.0625
3	−3.75	14.0625
5	−1.75	3.0625
10	3.25	10.5625
6	−0.75	0.5625
6	−0.75	0.5625
4	−2.75	7.5625
8	1.25	1.5625
5	−1.75	3.0625
6	−0.75	0.5625
2	−4.75	22.5625
135	0	175.7500

$\bar{x} = 135/20 = 6.75$ TV = 175.75

against the alternative

A: these true means are not all the same.

To test H against A, we use the F test. This is the same F test that we used in
Section 6.F to study the differences between standard deviations. It is reasonable to use the F test here because the quantities of TV, VB, and VW are
basically the squares of standard deviations. (In fact, the word "variance" of
a set of data is simply the technical name for the square of the standard
deviation.) All that is needed to make them into exact squares of standard
deviations is to divide them by the proper numbers.

The rejection rule is as follows: we reject H in favor of A at level α if

$$F > F_\alpha [m - 1; m(n - 1)]$$

where

■
$$F = \frac{nVB/(m-1)}{VW/m(n-1)}.$$

(7.2)

In the expression for F, we have used the symbols

m = number of samples involved
n = number of data points in each sample.

If we use a level of significance $\alpha = .05$, then we agree to reject H if

$$F > F_{.05}[3; 16] = 3.24,$$

because $m = 4$ and $n = 5$ here.* Now, because

$$nVB = (5)(6.75) = 33.75 \qquad m-1 = 4-1 = 3$$

$$VW = 142 \qquad\qquad m(n-1) = 4(5-1) = 16$$

we have

$$F = \frac{nVB/(m-1)}{VW/m(n-1)} = \frac{(33.75)/3}{(142)/16} = \frac{(33.75)(16)}{(142)(3)} = 1.27.$$

We agreed to reject H if $F > 3.24$, but actually it turned out that $F = 1.27 < 3.24$; so we cannot reject H at level $\alpha = .05$. We therefore conclude that the data supports the hypothesis that the mean sales volume of the new product does not differ significantly among the 4 regions.

How were the components of the variance involved in the solution to the above problem? Well, the final decision was that F was not large enough to permit us to reject H. Looking at the formula for F, we can see that this means that VB was not large enough relative to VW to warrant the rejection of H in favor of A. In other words, the variation between the sample means was not much greater than the variation within the samples, so it was not large enough to indicate the existence of a significant difference among the true means. We were then led to believe that the sample means did not exhibit inordinately large differences among themselves. Therefore, we could not conclude that variations among the data points could be attributed largely to variations between the true means.

EXAMPLE 7.1. **Word Processing Systems.** In Table 7.4, we have listed the weekly costs of operating each of five competing word processing (WP) systems in offices of a major steel corporation. The data were taken on randomly selected weeks during a six-month trial period. Costs

---

*You might find it useful at this point to refer to the material on the degrees of freedom of the F distribution in Section 6.F.

TABLE 7.4 Weekly Costs of Different WP Systems (Data given in hundreds of dollars)

	United Electronics	Verbiage Control	Western Words	X-Pensive Palaver	Yorktown Machine	Entire Set of Data
	12	12	3	14	16	
	13	10	6	18	21	
	7	6	10	18	15	
	8	4	13	14	20	
Sums	40	32	32	64	72	240
Sample Means	10	8	8	16	18	12

are one factor influencing the corporation's choice of a WP system, so the corporation would like to know whether or not costs run about the same for each of the five systems. Use a level of significance of $\alpha = .05$.

SOLUTION. We use the following symbols to denote the true means involved in this example:

μ_U = true mean cost of United Electronics system;
μ_V = true mean cost of Verbiage Control system;
μ_W = true mean cost of Western Words system;
μ_X = true mean cost of X-Pensive Palaver system;
μ_Y = true mean cost of Yorktown Machine system.

We want to test the hypothesis

H: $\mu_U = \mu_V = \mu_W = \mu_X = \mu_Y$

against the alternative

A: these true means are not all the same.

Recalling that

m = number of samples involved
n = number of data points in each sample

we see that $m = 5$ and $n = 4$. Since the general rule is to reject H in favor of A at level α if

$F > F_\alpha [m-1; m(n-1)]$,

we should reject H at level $\alpha = .05$ if

$$F > F_{.05} [4; 15] = 3.06 .$$

Therefore it remains only to compute

$$F = \frac{nVB/(m-1)}{VW/m(n-1)} = \frac{4VB/4}{VW/15} = 15VB/VW.$$

In the above formula

$$VB = (\bar{x}_U - \bar{x})^2 + (\bar{x}_V - \bar{x})^2 + (\bar{x}_W - \bar{x})^2 + (\bar{x}_X - \bar{x})^2 + (\bar{x}_Y - \bar{x})^2$$

$$= (10-12)^2 + (8-12)^2 + (8-12)^2 + (16-12)^2 + (18-12)^2$$

$$= 2^2 + 4^2 + 4^2 + 4^2 + 6^2 = 4 + 16 + 16 + 16 + 36$$

$$= 88$$

and

$$VW = VW_U + VW_V + VW_W + VW_X + VW_Y$$

where

$$VW_U = (12-10)^2 + (13-10)^2 + (7-10)^2 + (8-10)^2$$

$$= 2^2 + 3^2 + 3^2 + 2^2 = 4 + 9 + 9 + 4 = 26$$

$$VW_V = (12-8)^2 + (10-8)^2 + (6-8)^2 + (4-8)^2$$

$$= 4^2 + 2^2 + 2^2 + 4^2 = 16 + 4 + 4 + 16 = 40$$

$$VW_W = (3-8)^2 + (6-8)^2 + (10-8)^2 + (13-8)^2$$

$$= 5^2 + 2^2 + 2^2 + 5^2 = 25 + 4 + 4 + 25 = 58$$

$$VW_X = (14-16)^2 + (18-16)^2 + (18-16)^2 + (14-16)^2$$

$$= 2^2 + 2^2 + 2^2 + 2^2 = 4 + 4 + 4 + 4 = 16$$

$$VW_Y = (16-18)^2 + (21-18)^2 + (15-18)^2 + (20-18)^2$$

$$= 2^2 + 3^2 + 3^2 + 2^2 = 4 + 9 + 9 + 4 = 26 \ .$$

It therefore follows that

$$VW = VW_U + VW_V + VW_W + VW_X + VW_Y$$

$$= 26 + 40 + 58 + 16 + 26$$

$$= 166$$

and so

$$F = \frac{15VB}{VW} = \frac{(15)(88)}{166} = 7.95 \ .$$

Because we have agreed to reject H in favor of A at level of significance $\alpha = .05$ if F turned out to be greater than 3.06, the result that $F = 7.95$

> 3.06 leads us to reject H. Therefore, at level α = .05, we conclude that the costs of the five WP systems are not all the same.

The material dealing with the components of the variance that we have presented in this chapter provides the philosophical and mathematical basis of a group of statistical procedures known collectively as *the analysis of variance*. In the next two sections, we shall study in detail the way in which two of these procedures are widely used in statistical analysis of applied systems. Before we proceed, however, let's take a look at the conditions under which the analysis of variance procedures are valid.

By analogy with the t distribution that we discussed in Chapters 5 and 6, it is proper to use the F distribution when working with a set of normally distributed data. In particular, the valid application of the F test to the solution of analysis of variance problems requires that the populations (all m of them!) from which the random samples were selected be normally distributed with the same standard deviation. The data points must be independently selected, and the central limit theorem operates if the n's are larger than 5.* It is *not* required, however, that the number of data points (what we have been representing by n) be the same for each sample, although unequal values of n require the use of a different set of computing formulas to be described in the next section.

These conditions are reminiscent of those we met in our discussion of the two-sample t test in Section 6.D, where both sets of data were required to come from populations having normal distributions with the same standard deviation. It was not necessary that both sets have the same number of data points.

EXERCISES 7.A

7.1. An individual considering purchase of a $5,000 life insurance policy wants to find out whether there are substantial differences between insurance premiums of the various types of policies available. In particular, he investigates the prices for the four types of policies listed below, and he obtains price quotations from six different competing underwriters. These price quotations are as follows:

Price Quotations for Monthly Premium (dollars)			
Individual Renewable Term	Individual Convertible Term	Group Renewable Term	Government Group Insurance
2.50	2.00	2.00	1.00
3.00	2.20	1.50	1.20
2.20	1.80	1.80	1.00
2.80	2.50	2.00	1.10
2.50	3.00	2.00	1.50
3.00	2.00	1.80	1.40

*In working the exercises, we will assume that all the required conditions are satisfied.

a. Calculate the components of the variance, and write down the numbers appearing in the components of the variance formula TV = nVB + VW.

b. Decide at level of significance α = .05 whether or not there are significant differences among the average monthly premiums of the 4 types of policies under study.

7.2. A stock brokerage firm, members of the New York Stock Exchange and other major exchanges, thinks it has a psychological test that can pick out those individuals who will turn out to be successful stockbrokers. As the firm usually recruits its employees from outside the securities industry, it would like to know in which occupations it can expect to find those with a high level of aptitude, as determined by the psychological test. Before the current recruiting drive begins, the firm therefore decides to give its psychological test to six randomly selected individuals in each of six occupational classifications. The aptitude scores follow:

Aptitude Scores for Persons in Six Occupational Classifications

Commission Salesperson	Office Clerk	Teacher	Civil Servant	Career Military	Pro Athlete
80	70	80	40	50	70
70	80	60	90	10	60
80	60	90	50	50	30
90	30	70	40	60	50
60	50	80	30	30	80
40	70	10	20	40	70

a. Calculate the components of the variance.

b. At level α = .01, can the stock brokerage firm conclude that there is a significant difference in aptitude among persons in the six occupational classifications?

7.3. An ichthyologist employed by the New England Fishers Co-op wants to find out whether or not temperature changes in the ocean's water produced by coastal nuclear generating facilities will have a significant effect on the growth of fish indigenous to the region. He sets up an experiment involving four groups of seven recently hatched specimens each of the same species of fish. Each group is placed in a simulated ocean environment in which all factors are controlled and identical, with the exception of the temperature of the water. Six months later, the 28 specimens are weighed, and their weights are recorded in the following table:

Weights of Specimens from Each of Four Water-Temperature Groups (ounces)

40°F	42°F	44°F	46°F
20	18	24	16
18	25	17	17
23	16	14	26
16	20	22	19
15	24	25	14
25	16	27	20
16	21	18	21

a. Calculate the components of the variance.

b. At level $\alpha = .05$, do the results of the experiment indicate that the temperature differences tested have a significant effect on the average weight of fish in the region?

7.4. As part of a study of the manner in which pork-barrel projects are funded, a political scientist gathered the following data from eight randomly selected federal budgets over the past 40 years. The data points are the total appropriation in the budgets for pork-barrel projects in each of the three socio-politico-economic subdivisions of the nation: urban, suburban, and rural.

Appropriation for Each Socio-Politico-Economic Subdivision (millions of dollars)

Urban	Suburban	Rural
8	4	11
10	8	10
12	9	8
15	10	9
18	12	10
24	17	12
30	25	16
27	35	20

a. Calculate the components of the variance.

b. At level $\alpha = .01$, can we conclude from the data that the average amounts of federal money dispensed to each of the three subdivisions are approximately the same?

7.5. An analysis of business conditions across the country, region by region, yielded the following data on 21 cities, seven in each region of the country, on the prevailing unemployment rate in October, 1977:*

Eastern Cities	Central Cities	Western Cities
6.9	4.7	3.0
7.7	5.5	6.6
6.8	3.7	7.3
5.2	6.5	7.3
5.6	5.5	7.0
4.7	4.9	4.8
7.2	4.3	8.7

a. Determine the components of the variance.

b. At level $\alpha = .01$, do the data indicate that unemployment rates were virtually the same, on the average, in cities of each of the three regions?

* U.S. News & World Report, October 24, 1977, pp. 73–75.

7.6. A stockbroker observes the following data as part of a study of the yearly earnings per share made by companies in each of six industrial groupings:[†]

12-Month Earnings per Share (dollars)					
Chemicals	Electrical	Metals	Fuel	Paper	Textiles
4.80	1.72	5.28	5.45	4.47	3.18
2.81	3.86	2.97	1.98	3.51	1.57
2.96	4.62	1.78	6.62	3.55	4.54
1.72	0.94	4.02	3.08	3.59	2.10
3.21	2.59	3.20	4.13	2.37	2.75

a. Calculate the components of the variance.
b. At level $\alpha = .05$, can the stockbroker draw the conclusion that there are significant differences among the industrial groupings in average earnings per share?

SECTION 7.B.   ONE-WAY ANALYSIS OF VARIANCE

We now proceed to the development of the techniques of *one-way analysis of variance*, a method of organizing the data and computations needed in carrying out the F test for the equality of several means.

Let's begin by taking another look at Example 7.1 and the data of Table 7.4. In order to test the hypothesis

$$\text{H:} \quad \mu_U = \mu_V = \mu_W = \mu_X = \mu_Y$$

against the alternative

$$\text{A:} \quad \text{these true means are not all the same,}$$

it was necessary to compute the number

$$F = \frac{nVB/(m-1)}{VW/m(n-1)} .$$

From the data of Table 7.4, it was easy to see that $n = 4$ and $m = 5$, and it was not really difficult to calculate $VB = 88$. The calculation of VW, however, was another story. That required us to do six separate calculations, one each of VW_U, VW_V, VW_W, VW_X, and VW_Y, and one summing all of the latter quantities. You must admit that the work required to calculate VW was somewhat out of line with that needed for the remainder of the problem, and that a

[†]*Business Week*, November 14, 1977, pp. 106–138.

short-cut method of computing VW would be much appreciated by all those working in the field (even by those having access to expensive calculators!). The level of general appreciation would increase as the number of samples involved grew, for the work required to compute VW by the method of Section 7.A would grow correspondingly.

As it turns out, there is a very simple way of getting the value of VW. It is based on an algebraic inversion of the components of the variance formula (7.1), asserting that

$$VW = TV - nVB.$$

As we have already pointed out, n and VB are relatively easy to compute, so if we can come up with a quick way of getting TV, we will have done the job.

Fortunately, there is a quick way of calculating TV that is well-suited to both machine and hand computation. The explicit formula for TV is similar to the short-cut formula (5.3) for the standard deviation that we first discussed way back in Chapter 2 and later made extensive use of in Chapters 5 and 6. In particular, TV is given by the formula

$$TV = \Sigma x^2 - \frac{(\Sigma x)^2}{N},$$

where

$$N = \text{total number of data points in all samples together}$$
$$\Sigma x^2 = \text{sum of squares of all } N \text{ data points}$$
$$\Sigma x = \text{sum of all } N \text{ data points.}$$

In Table 7.5, we set up the data of Example 7.1 (Table 7.4) for the computation of TV. (A somewhat related construction in Table 7.3 gave a slightly longer calculation of TV for a different set of data.) The computations in Table 7.5 show that

$$N = 20 \qquad \Sigma x = 240 \qquad \Sigma x^2 = 3,398$$

for the WP systems data, from which it follows that

$$TV = \Sigma x^2 - \frac{(\Sigma x)^2}{N} = 3,398 - \frac{(240)^2}{20}$$

$$= 3,398 - \frac{57,600}{20} = 3,398 - 2,880 = 518$$

As long as we are on the subject of short-cut formulas, let's complete the story by presenting and showing how to use a short-cut formula for computing the term nVB. Our calculational task will be immediately shortened by one step

TABLE 7.5   Calculation of the
Total Variation (TV)   (WP systems data)

x	x²
12	144
13	169
7	49
8	64
12	144
10	100
6	36
4	16
3	9
6	36
10	100
13	169
14	196
18	324
18	324
14	196
16	256
21	441
15	225
20	400
Sums 240	3,398

$N = 20$ $\Sigma x^2 = 3,398$

$\Sigma x = 240$ $TV = \Sigma x^2 - \dfrac{(\Sigma x)^2}{N} = 3,398 - \dfrac{(240)^2}{20} = 518$

simply because we will be getting the term nVB in its entirety, rather than merely VB alone. The short-cut formula for nVB is

$$nVB = \Sigma\left(\frac{S_i^2}{n_i}\right) - \frac{(\Sigma x)^2}{N}$$

where S_i is the sum of the n_i data points of sample i. Note that the sizes n_i of the samples need not all be the same. Referring to Table 7.4, we can see that

$S_U = 40$ $S_X = 64$
$S_V = 32$ $S_Y = 72$
$S_W = 32$

TABLE 7.6   One-Way Analysis of Variance Calculations
(WP systems data)

x^2	x	S	S^2	n	$\dfrac{S^2}{n}$
144	12				
169	13	40	1,600	4	400
49	7				
64	8				
144	12				
100	10	32	1,024	4	256
36	6				
16	4				
9	3				
36	6	32	1,024	4	256
100	10				
169	13				
196	14				
324	18	64	4,096	4	1,024
324	18				
196	14				
256	16				
441	21	72	5,184	4	1,296
225	15				
400	20				
Sums 3,398	**240**			20	3,232

$$TV = \Sigma x^2 - \frac{(\Sigma x)^2}{N} = 3,398 - \frac{(240)^2}{20} = 518$$

$$nVB = \Sigma \left(\frac{S^2}{n}\right) - \frac{(\Sigma x)^2}{N} = 3,232 - \frac{(240)^2}{20} = 352$$

$$VW = TV - nVB = 518 - 352 = 166$$

Because $\Sigma x = 240$, $N = 20$, and $n = 4$ as before, the short-cut formula for nVB yields that

$$nVB = \left(\frac{40^2}{4} + \frac{32^2}{4} + \frac{32^2}{4} + \frac{64^2}{4} + \frac{72^2}{4}\right) - \frac{(240)^2}{20}$$

$$= \left(\frac{1,600}{4} + \frac{1,024}{4} + \frac{1,024}{4} + \frac{4,096}{4} + \frac{5,184}{4} \right) - \frac{57,600}{20}$$

$$= (400 + 256 + 256 + 1,024 + 1,296) - 2,880$$

$$= 3,232 - 2,880 = 352$$

Because TV = 518 and nVB = 352, it follows that

$$\text{VW} = \text{TV} - \text{nVB} = 518 - 352 = 166,$$

which is, of course, the same result we obtained in the last section for the WP systems data.

In Table 7.6, we illustrate how to combine the calculations made in Tables 7.4 and 7.5 into a single table and thereby obtain a more direct computation of TV, nVB, and ultimately VW.

A computer program, written in the BASIC language, for handling the one-way analysis of variance situation is listed in Table 7.7. After the data are entered in response to the computer's prompting, the solution is printed out as in Table 7.8.

EXAMPLE 7.2.   Gasoline Efficiency.   A motorist would like to find out whether or not he gets the same gasoline efficiency (in miles per gallon), regardless of the brand of fuel he uses. He chooses five brands of gasoline available in his locality, and he records the number of miles per gallon he obtained on several tankfuls of each brand. In Table 7.9, his recorded mileage (mpg) for each tankful is listed. Use a level of significance of $\alpha = .01$ to test whether or not gasoline efficiency is the same for all five brands.

SOLUTION.   The items of primary interest are the following five true means:

μ_T = true mean number of miles per gallon of Thrifty gasoline,
μ_B = true mean number of miles per gallon of Broadway gasoline,
μ_F = true mean number of miles per gallon of Federated gasoline,
μ_G = true mean number of miles per gallon of Gibraltar gasoline,
μ_H = true mean number of miles per gallon of Holiday gasoline.

We want to test the hypothesis

H: $\mu_T = \mu_B = \mu_F = \mu_G = \mu_H$

TABLE 7.7   BASIC Computer Program for One-Way Analysis of Variance

```
10        REMARK: ONE-WAY ANALYSIS OF VARIANCE
20        DIM X(10,10),S1(10),N(10),S2(10)
30        REMARK: M = NUMBER OF SAMPLES
40        REMARK: N(I) = NUMBER OF DATA POINTS IN ITH SAMPLE
50        REMARK: X(I,J) = JTH DATA POINT OF ITH SAMPLE
60        PRINT "HOW MANY SAMPLES ARE THERE";
70        INPUT M
80        PRINT
90        REMARK: N = TOTAL NUMBER OF DATA POINTS
100       N = 0
110       FOR I = 1 TO M
120          REMARK: S1(I) = RUNNING SUM OF ITH SAMPLE
130          REMARK: S2(I) = RUNNING SUM OF SQUARES
140          S1(I) = 0
150          S2(I) = 0
160          PRINT "HOW MANY DATA POINTS ARE IN SAMPLE";I;
170          INPUT N(I)
180          N = N + N(I)
200          FOR J = 1 TO N(I)
210             READ X(I,J)
220             S1(I) = S1(I) + X(I,J)
230             S2(I) = S2(I) + (X(I,J)**2)
240             NEXT J
250          NEXT I
260          PRINT
270       REMARK: B = VARIATION BETWEEN SAMPLES
280       REMARK: W = VARIATION WITHIN SAMPLES
290       T = 0
300       S = 0
310       Q = 0
320       FOR I = 1 TO M
330          T = T + (S1(I)**2)/N(I)
340          S = S + S1(I)
350          Q = Q + S2(I)
360          NEXT I
370       PRINT "SIGMA X =";S
380       PRINT "SIGMA X2 =";Q
390       PRINT "N =";N
400       PRINT "SIGMA(S2/N) =";T
410       PRINT
420       B = T - (S**2)/N
430       W = Q - T
440       PRINT "VARIATION BETWEEN SAMPLES IS";B
450       PRINT "VARIATION WITHIN SAMPLES IS";W
460       PRINT
470       F = ((N-M)*B)/((M-1)*W)
480       PRINT "THE F-STATISTIC HAS VALUE";F
490       PRINT
500       PRINT "DEGREES OF FREEDOM FOR NUMERATOR IS";M-1
510       PRINT "DEGREES OF FREEDOM FOR DENOMINATOR IS";N-M
530       PRINT "WHAT IS LEVEL OF SIGNIFICANCE ALPHA";
540       INPUT A
550       PRINT "LOOK UP THE VALUE OF F";A;"(";M-1;";";N-M;").";
560       PRINT "WHAT IS IT";
570       INPUT F1
580       PRINT
590       IF F > F1 THEN 640
600       PRINT "WE CANNOT REJECT H; OUR INFORMATION DOES NOT"
610       PRINT "CONTRADICT THE HYPOTHESIS THAT THE TRUE MEANS"
620       PRINT "ARE ALL THE SAME."
630       STOP
640       PRINT "WE REJECT H, CONCLUDING THEREBY THAT THE"
650       PRINT "TRUE MEANS ARE NOT ALL THE SAME."
700       REMARK: LINES 700-720 CONTAIN DATA POINTS BY SAMPLE
800       END
```

TABLE 7.8   Computer Printout of Solution: WP Systems Example

```
701      DATA 12,13,7,8
702      DATA 12,10,6,4
703      DATA 3,6,10,13
704      DATA 14,18,18,14
705      DATA 16,21,15,20


run


HOW MANY SAMPLES ARE THERE ? 5

HOW MANY DATA POINTS ARE IN SAMPLE 1   ?4
HOW MANY DATA POINTS ARE IN SAMPLE 2   ?4
HOW MANY DATA POINTS ARE IN SAMPLE 3   ?4
HOW MANY DATA POINTS ARE IN SAMPLE 4   ?4
HOW MANY DATA POINTS ARE IN SAMPLE 5   ?4

SIGMA X = 240
SIGMA X2 = 3398
N = 20
SIGMA(S2/N) = 3232

VARIATION BETWEEN SAMPLES IS 352
VARIATION WITHIN SAMPLES IS 166

THE F-STATISTIC HAS VALUE 7.95181

DEGREES OF FREEDOM FOR NUMERATOR IS 4
DEGREES OF FREEDOM FOR DENOMINATOR IS 15
WHAT IS LEVEL OF SIGNIFICANCE ALPHA ?.05
LOOK UP THE VALUE OF F .05 ( 4 ; 15 ).
WHAT IS IT ? 3.06

WE REJECT H, CONCLUDING THEREBY THAT THE
TRUE MEANS ARE NOT ALL THE SAME.
```

TABLE 7.9   Efficiency of Different Brands of Gasoline   (Data in miles per gallon)

Brand:	Thrifty	Broadway	Federated	Gibraltar	Holiday
	26	29	29	31	28
	25	27	29	28	23
	28	26	26	26	24
	24	30	24	28	28
	29	25	25	26	25
	24	25	23	27	22
	26		26	28	25
				30	
n	7	6	7	8	7

against the alternative

A: these true means are not all the same.

In one-way analysis of variance, we compute

$$F = \frac{nVB/(m-1)}{VW/(N-m)} \tag{7.3}$$

and we reject H in favor of A at level α if

$$F > F_{\alpha}[m-1; N-m].* \tag{7.4}$$

In this gasoline efficiency example

m = number of samples = 5
N = total number of data points
  = $n_T + n_B + n_F + n_G + n_H$ = 7 + 6 + 7 + 8 + 7 = 35

Therefore the degrees of freedom for the appropriate percentage point
of the F distribution are

$m - 1$ = $5 - 1$ = 4 (for the numerator)
$N - m$ = $35 - 5$ = 30 (for the denominator).

Because the level of significance is α = .01 for this problem, we agree
to reject H in favor of A if

$$F > F_{.01}[4; 30] = 4.02.$$

It therefore remains only to calculate the numerical value of F and to
observe whether or not it exceeds 4.02. Inserting the numbers m = 5
and N = 35, we have that

$$F = \frac{nVB/(m-1)}{VW/(N-m)} = \frac{nVB/4}{VW/30} = \frac{30nVB}{4VW}$$

where

$$nVB = \Sigma\left(\frac{S_i^2}{n_i}\right) - \frac{(\Sigma x)^2}{N}$$

and

---

*If all n's are equal, then $N = mn$, so that $N - m = mn - m = m(n-1)$, the number that appeared
in formula (7.2).

$$TV = \Sigma x^2 - \frac{(\Sigma x)^2}{N}$$

so that

$$
\begin{aligned}
VW &= TV - nVB \\
&= \left[\Sigma x^2 - \frac{(\Sigma x)^2}{N} \right] - \left[\Sigma \left(\frac{S_i^2}{n_i} \right) - \frac{(\Sigma x)^2}{N} \right] \\
&= \Sigma x^2 - \Sigma \left(\frac{S_i^2}{n_i} \right)
\end{aligned}
$$

In Table 7.10, we present the calculations required to obtain the value of F, especially the calculation of

$$nVB = \Sigma \left(\frac{S_i^2}{n_i} \right) - \frac{(\Sigma x)^2}{N} \tag{7.5}$$

and

$$VW = \Sigma x^2 - \Sigma \left(\frac{S_i^2}{n_i} \right) \tag{7.6}$$

We see from the results of Table 7.10 that $nVB = 38.57$ and $VW = 130$, from which it follows that

$$F = \frac{30nVB}{4VW} = \frac{(30)(38.57)}{(4)(130)} = 2.23.$$

The above information may be summarized in the ANOVA (ANalysis Of VAriance) table of Table 7.11. The ANOVA table illustrates how the total variance is distributed among its components.

We have agreed to reject the hypothesis

H: $\mu_T = \mu_B = \mu_F = \mu_G = \mu_H$

in favor of the alternative

A: these true means are not all the same

at level $\alpha = .01$ if F turned out to be larger than 4.02. Therefore, the result that

$$F = 2.23 < 4.02$$

indicates that we should not reject H. At significance level $\alpha = .01$, then, we conclude that the data do not reveal significant differences in

TABLE 7.10 One-Way Analysis of Variance Calculations (Gasoline efficiency data)

	x^2	x	S	S^2	n	S^2/n
	676	26				
	625	25				
	784	28				
	576	24	182	33,124	7	4,732
	841	29				
	576	24				
	676	26				
4,754						
	841	29				
	729	27				
	676	26	162	26,244	6	4,374
	900	30				
	625	25				
	625	25				
4,396						
	841	29				
	841	29				
	676	26				
	576	24	182	33,124	7	4,732
	625	25				
	529	23				
	676	26				
4,764						
	961	31				
	784	28				
	676	26				
	784	28	224	50,176	8	6,272
	676	26				
	729	27				
	784	28				
	900	30				
6,294						
	784	28				
	529	23				
	576	24				
	784	28	175	30,625	7	4,375
	625	25				
	484	22				
	625	25				
4,407						
Sums	24,615	925			35	24,485

$$nVB = \Sigma\left(\frac{S^2}{n}\right) - \frac{(\Sigma x)^2}{N} = 24,485 - \frac{(925)^2}{35} = 38.57$$

$$VW = \Sigma x^2 - \Sigma\left(\frac{S^2}{n}\right) = 24,615 - 24,485 = 130$$

TABLE 7.11 One-Way ANOVA Table: Gasoline Efficiency Example

Source of Variation	Degrees of Freedom	Sum of Squares	Mean Sum of Squares	F
Between samples	$m-1$	nVB	$MSB = \dfrac{nVB}{m-1}$	$\dfrac{MSB}{MSW}$
Within samples	$N-m$	VW	$MSW = \dfrac{VW}{N-m}$	*
Total	$N-1$	TV	*	*

Source of Variation	Degrees of Freedom	Sum of Squares	Mean Sum of Squares	F
Between samples	4	38.57	9.64	2.23
Within samples	30	130	4.33	*
Total	34	168.57	*	*

mileage per gallon between the five brands of gasoline tested. (We have not conclusively proved that the gasolines are equivalent in efficiency, however. All we have said is that we cannot validly infer from our data that there are significant differences among them.)

A computer printout of the solution, using the BASIC program of Table 7.7, appears in Table 7.12.

Before we go on to the next section, it would be useful to explain what we have done in terms of the traditional language and vocabulary of analysis of variance. Referring again to the basic gasoline efficiency data of Table 7.9, we can view the five different brands of gasoline as "treatments" affecting the car's mileage per gallon. Under the analysis of variance restrictions discussed at the conclusion of Section 7.A, we operate under the assumption that each treatment yields a set of data whose underlying population is normally distributed, and all such populations have the same standard deviation σ. The hypothesis

H: $\mu_T = \mu_B = \mu_F = \mu_G = \mu_H$

asserts that all such populations also have the same mean μ (which is, of course, the common numerical value of μ_T, μ_B, μ_F, μ_G, and μ_H), and therefore that *all* the data points come from a *single* normally distributed population having mean μ and standard deviation σ.

TABLE 7.12   Computer Printout of Solution:
Gasoline Efficiency Example

```
701     DATA 26,25,28,24,29,24,26
702     DATA 29,27,26,30,25,25
703     DATA 29,29,26,24,25,23,26
704     DATA 31,28,26,28,26,27,28,30
705     DATA 28,23,24,28,25,22,25


run


HOW MANY SAMPLES ARE THERE ? 5

HOW MANY DATA POINTS ARE IN SAMPLE 1   ?7
HOW MANY DATA POINTS ARE IN SAMPLE 2   ?6
HOW MANY DATA POINTS ARE IN SAMPLE 3   ?7
HOW MANY DATA POINTS ARE IN SAMPLE 4   ?8
HOW MANY DATA POINTS ARE IN SAMPLE 5   ?7

SIGMA X = 925
SIGMA X2 = 24615
N = 35
SIGMA(S2/N) = 24485

VARIATION BETWEEN SAMPLES IS 38.5714
VARIATION WITHIN SAMPLES IS 130

THE F-STATISTIC HAS VALUE 2.22527

DEGREES OF FREEDOM FOR NUMERATOR IS 4
DEGREES OF FREEDOM FOR DENOMINATOR IS 30
WHAT IS LEVEL OF SIGNIFICANCE ALPHA ?.01
LOOK UP THE VALUE OF F .01 ( 4 ; 30 ).
WHAT IS IT ? 4.02

WE CANNOT REJECT H; OUR INFORMATION DOES NOT
CONTRADICT THE HYPOTHESIS THAT THE TRUE MEANS
ARE ALL THE SAME.
```

Looked at from this point of view, the quantity

$$nVB = \Sigma\left(\frac{S_i^2}{n_i}\right) - \frac{(\Sigma x)^2}{N},$$

which we have been referring to as the "variation between the samples"
can be considered as a measure of variation between the treatments.
Traditionally, then, the quantity nVB has been called the *treatment sum
of squares* and has often been denoted by the symbols SST. What about
the quantity VW? This number, which we have been calling the "variation within the samples," represents the squared error made by using

the mean of each sample as an estimate of the true mean of the corresponding population. Historically, then,

$$VW = \Sigma x^2 - \Sigma \left(\frac{S_i^2}{n_i} \right)$$

has been called the *error sum of squares* and has often been denoted by the symbol SSE. In the language of "treatment sum of squares" and "error sum of squares," the F statistic used to test the hypothesis in question would be

$$F = \frac{SST/(m-1)}{SSE/(N-m)} .$$

We would then reject H, concluding that treatment effects differ among themselves, at level α if $F > F_\alpha [m-1; N-m]$, namely if the treatment sum of squares is substantially larger than the error sum of squares.

EXERCISES   7.B

7.7. In the situation of Exercise 7.1, use the procedures of one-way analysis of variance to decide at level $\alpha = .05$ whether or not there are significant differences between the monthly premiums of the four competing plans.

7.8. Carry out a one-way analysis of variance, based on the data of Exercise 7.2, in order to decide at level $\alpha = .01$ whether or not there are significant differences in aptitude among persons in the six occupational classifications.

7.9. Apply the one-way analysis of variance technique to the ichthyological data of Exercise 7.3 to find out, at level $\alpha = .05$, whether temperature changes affect the weights of fish.

7.10. Using the data of Exercise 7.4, run a one-way analysis of variance to answer the question as to whether the average appropriations to each of the three subdivisions are basically the same. Use level $\alpha = .01$.

7.11. Apply the one-way analysis of variance procedure to the data of Exercise 7.5 in order to test at level $\alpha = .01$ whether or not average unemployment rates in cities of the three regions were virtually the same.

7.12. Use the one-way analysis of variance method on the data of Exercise 7.6 to test whether significant differences exist among the industrial groupings in average earnings per share. Take $\alpha = .05$.

7.13. A department store with several outlets in North Dakota, Oklahoma, and Wyoming wants to find out, as part of an inventory study, whether or not its private-label refrigerators sell equally well in each of the three states. The following data list the number of units (in hundreds) sold so far this year in several randomly selected branches in each of the three states:

North Dakota	Oklahoma	Wyoming
2	1	2
3	4	2
5	8	4
6	1	2
1	6	3
	3	

At level $\alpha = .05$, do the three states seem to differ significantly in the number of refrigerators sold per branch?

7.14. The following data record percentages that represent, for various companies, their 1976 net income as a percentage of stockholders' equity.* This information is useful to banking officials, who have to make judgments about the soundness of a company's financial structure.

Energy Companies	Steel Companies	Paper Companies	Drug Companies
14.3	8.0	13.9	6.8
17.0	8.8	11.7	19.9
8.9	0.9	10.7	23.2
8.0	21.3	9.5	22.2
8.8	5.0	18.6	18.6
13.9	9.7	16.9	10.4
14.6	6.2		14.1
15.7			
19.5			

Use a level of significance of $\alpha = .01$ in testing whether or not there are significant differences among the various industry groups.

SECTION 7.C   TWO-WAY ANALYSIS OF VARIANCE

During the course of its work on medication for common burns, the J&L Pharmaceutical Company came up with a new compound, MM-77, which it feels will offer substantial relief. To find out whether or not MM-77 is truly effective, the consumer testing department chooses an appropriate analysis-of-variance procedure based on a randomized block design.

A *randomized block design* is a statistical structure for testing the performance of, say MM-77, against competing medications, in the presence of

---

*Data selected randomly from *Fortune*, May 1977, pp. 366–385.

various external conditions. The external conditions could be, for example, several different causes of the burns. A random sample of 30 individuals with burns is selected in such a way that, for each of the five types of burns, there are six individuals with that type. Among each group of six individuals, six possible medications (one of which is MM-77) are assigned randomly, one medication to an individual. The resulting data, which appear in Table 7.13, record the length of time it takes for the various treatments to yield noticeable relief.

We define the following notation:

μ_M = true mean relief time when treated by MM-77
μ_A = true mean relief time when treated by aerosol spray
μ_P = true mean relief time when treated by "pump" spray
μ_C = true mean relief time when treated by cold water
μ_J = true mean relief time when treated by petroleum jelly
μ_N = true mean relief time when not treated at all

We can test the hypothesis that the six treatments do not differ significantly in mean relief time, namely

H_T: $\mu_M = \mu_A = \mu_P = \mu_C = \mu_J = \mu_N$

against the alternative

A_T: These true mean relief times are not all the same.

We use the symbols H_T and A_T to indicate that these are the *treatment hypothesis* and *treatment alternative*, respectively. (The *failure* to reject H in favor of A would be evidence supporting an assertion that MM-77 is *not* effective, for it would mean that all the remedies, MM-77 included, are just as effective as no treatment at all, the last of the six treatments considered in the study.)

TABLE 7.13   Effectiveness of New Compound MM-77 in the Medication of Common Burns   (*Data in hours until occurrence of noticeable relief*)

Type of Burn		Treatment				
	MM-77	Aerosol Spray	"Pump" Spray	Cold Water	Petroleum Jelly	No Treatment
Sunburn	35	30	25	41	37	42
Kitchen grease	38	28	23	29	30	38
Flame	41	32	24	32	28	47
Electrical	29	35	24	36	25	37
Hot water	37	25	29	42	30	47

In addition to the hypothesis of equal treatment effects, the two-way structure of Table 7.13 provides us with enough information to deal with another question, the question of whether or not the five types of burns differ in duration of time until relief begins. For this question, we test the hypothesis

H_B: $\mu_S = \mu_K = \mu_F = \mu_E = \mu_H$

against the alternative

A_B: these true mean relief times are not all the same,

where

μ_S = true mean time until relief of sunburn
μ_K = true mean time until relief of kitchen grease burn
μ_F = true mean time until relief of flame burn
μ_E = true mean time until relief of electrical burn
μ_H = true mean time until relief of hot water burn

Here we have used the symbols H_B and A_B to indicate that these are the *block hypothesis* and the *block alternative*, respectively. In two-way analysis of variance, we refer to the categories along the top of the table as the "treatments" and the categories along the side as the "blocks." Then we view the data points in the body of Table 7.13 as depending, to various extents, on *treatment effects* and *block effects*.

In this language, we can reformulate the treatment hypothesis as

H_T: all treatments have equal effects on relief time (so that MM-77 is not particularly effective),

and the treatment alternative as

A_T: the treatment effects are not all the same (some treatments are better than others).

Similarly, we can express the *block hypothesis* as

H_B: all blocks have equal effects on relief time (so that all burn types have the same average pain duration),

and the *block alternative* as

A_B: the block effects are not all the same (pain of some burn types lasts longer than that of others).

Now that we have formulated the questions to be answered on the basis of the data of Table 7.13, we proceed to develop the two-way analysis of variance techniques used toward their solution. As in one-way analysis of variance, we use an F test to test each of the hypotheses H_T and H_B. Here, as in the previous section, the F test for H_T involves a treatment sum of squares SST and an error sum of squares SSE. By analogy, the F test for H_B involves a *block sum of squares* SSB and the error sum of squares SSE. In particular, if we set

m = number of treatments
n = number of blocks,

then we reject H_T in favor of A_T at level α if

$$F_T > F_\alpha [m - 1; (n-1)(m-1)] \tag{7.7}$$

where

$$F_T = \frac{\text{SST}/(m-1)}{\text{SSE}/(n-1)(m-1)} = \frac{(n-1)\text{SST}}{\text{SSE}} \tag{7.8}$$

and we reject H_B in favor of A_B at level α if

$$F_B > F_\alpha [n - 1; (n-1)(m-1)] \tag{7.9}$$

where

$$F_B = \frac{\text{SSB}/(n-1)}{\text{SSE}/(n-1)(m-1)} = \frac{(m-1)\text{SSB}}{\text{SSE}} \tag{7.10}$$

If we denote by N the total number of data points ($N = mn$), then we have the following formulas that are used in computing F_T and F_B:

$$\text{SST} = \frac{\Sigma S_T^2}{n} - \frac{(\Sigma x)^2}{N} \tag{7.11}$$

$$\text{SSB} = \frac{\Sigma S_B^2}{m} - \frac{(\Sigma x)^2}{N} \tag{7.12}$$

and

$$\text{SSE} = \Sigma x^2 - \frac{\Sigma S_T^2}{n} - \frac{\Sigma S_B^2}{m} + \frac{(\Sigma x)^2}{N} \tag{7.13}$$

In the above computing formulas, S_T stands for the column (treatment) sums and S_B stands for the row (block) sums. In Table 7.14, we illustrate the form a typical two-way analysis of variance table will take.

TABLE 7.14 The Form of a Two-Way Analysis of Variance Table   (*For use in calculations*)

| Blocks | Treatments | | | | | |
	Treat-ment 1	Treat-ment 2	...	Treat-ment m	Block Sums	Squares of Block Sums
Block 1	X_{11}	X_{12}	...	X_{1m}	S_{B1}	S_{B1}^2
Block 2	X_{21}	X_{22}	...	X_{2m}	S_{B2}	S_{B2}^2
⋮	⋮	⋮	.⋮.	⋮	⋮	⋮
Block n	X_{n1}	X_{n2}	...	X_{nm}	S_{Bn}	S_{Bn}^2
Treatment Sums	S_{T1}	S_{T2}	...	S_{Tm}	Σx	ΣS_B^2
Squares of Treatment Sums	S_{T1}^2	S_{T2}^2	...	S_{Tm}^2	ΣS_T^2	

We now proceed to the solution of the two-way analysis of variance problem based on the data of Table 7.13. Using a level of significance of $\alpha = .05$, we agree to reject H_T in favor of A_T if, using formula (7.7),

$$F_T > F_{.05}[5; 20] = 2.71$$

because $m = 6$ (number of treatments) and $n = 5$ (number of blocks). And, also at level $\alpha = .05$, we agree to reject H_B in favor of A_B if, using formula (7.9),

$$F_B > F_{.05}[4; 20] = 2.87.$$

It remains therefore only to compute the numerical values of F_T and F_B and to determine whether or not they exceed 2.71 and 2.87, respectively. The calculations leading to the computation of these numbers require a table having the form of Table 7.14. They are presented in Table 7.15, using the data of Table 7.13. The table at the bottom of Table 7.15 contains the square of each data point in Table 7.13. Its only objective is to calculate the numerical value of Σx^2 (the sum of squares of all the data points), which appears in the formula for SSE.

Using the results of the calculation appearing at the bottom of Table 7.15, we see that

$$SST = \frac{\Sigma S_T^2}{n} - \frac{(\Sigma x)^2}{N} = \frac{169,946}{5} - \frac{(996)^2}{30}$$

$$= 33,989.2 - \frac{992,016}{30} = 33,989.2 - 33,067.2 = 922$$

$$SSB = \frac{\Sigma S_B^2}{m} - \frac{(\Sigma x)^2}{N} = \frac{199,008}{6} - 33,067.2$$

$$= 33,168 - 33,067.2 = 100.8$$

TABLE 7.15   Calculations for Two-Way Analysis of Variance   *(New compound for treating common burns data)*

Block	Treatment M	A	P	C	J	N	Block Sums	Squares of Block Sums
S	35	30	25	41	37	42	210	44,100
K	38	28	23	29	30	38	186	34,596
F	41	32	24	32	28	47	204	41,616
E	29	35	24	36	25	37	186	34,596
H	37	25	29	42	30	47	210	44,100
Treatment Sums	180	150	125	180	150	211	996	199,008
Squares of Treatment Sums	32,400	22,500	15,625	32,400	22,500	44,521	169,946	

Calculation of Σx^2, the Sum of Squares of All Data Points

Block	Treatment M	A	P	C	J	N	Σ
S	1,225	900	625	1,681	1,369	1,764	7,564
K	1,444	784	529	841	900	1,444	5,942
F	1,681	1,024	576	1,024	784	2,209	7,298
E	841	1,225	576	1,296	625	1,369	5,932
H	1,369	625	841	1,764	900	2,209	7,708
Σ	6,560	4,558	3,147	6,606	4,578	8,995	34,444

$$\Sigma x = 996 \qquad m = 6 \qquad \Sigma S_T^2 = 169,946$$

$$\Sigma x^2 = 34,444 \qquad n = 5 \qquad \Sigma S_B^2 = 199,008$$

$$N = 30$$

$$SSE = \Sigma x^2 - \frac{\Sigma S_T^2}{n} - \frac{\Sigma S_B^2}{m} + \frac{(\Sigma x)^2}{N}$$

$$= 34,444 - 33,989.2 - 33,168 + 33,067.2$$

$$= 354.$$

From these calculations, it then follows that

$$F_T = \frac{(n-1)SST}{SSE} = \frac{(4)(922)}{354} = 10.42$$

TABLE 7.16   Two-Way ANOVA Table: Common Burns Example

Source of Variation	Degrees of Freedom	Sum of Squares	Mean Sum of Squares	F
Treatments	$m-1$	SST	$MST = \dfrac{SST}{m-1}$	$\dfrac{MST}{MSE}$
Blocks	$n-1$	SSB	$MSB = \dfrac{SSB}{n-1}$	$\dfrac{MSB}{MSE}$
Error	$(n-1)(m-1)$	SSE	$MSE = \dfrac{SSE}{(n-1)(m-1)}$	*
Total	$nm-1$	TV	*	*

Source of Variation	Degrees of Freedom	Sum of Squares	Mean Sum of Squares	F
Treatments	5	922	184.4	10.42
Blocks	4	100.8	25.2	1.42
Error	20	354	17.7	*
Total	29	1,376.8	*	*

and

$$F_B = \frac{(m-1)SSB}{SSE} = \frac{(50)(100.8)}{354} = 1.42.$$

The above information may be summarized in the ANOVA table of Table 7.16. The ANOVA table indicates the distribution of the total variance among its components.

We are now ready to draw our conclusions. We agreed to reject H_T (all treatments have equal effects) in favor of A_T (the treatment effects are not all the same) if F_T turned out to be greater than 2.71. And $F_T = 10.42 > 2.71$, so we do reject H_T. This means that, at level $\alpha = .05$, we conclude from the data of Table 7.13 that the treatment effects are not all the same. We therefore seem to have some statistical evidence that the various medications studied do have a detectable effect on the waiting time until noticeable relief occurs, at least when compared with no treatment at all. This conclusion would indicate to the company that a further look at the data is warranted, since MM-77 seems to have different effects from those of some standard medications.

In regard to the blocks, we have agreed to reject H_B (all burn types have the same average pain duration) in favor of A_B (pain of some burn types lasts longer than others) at level $\alpha = .05$ if $F_T > 2.87$. As it turned out, however, $F_T = 1.42 < 2.87$, so we cannot reject H. We therefore assert, at level $\alpha = .05$, that the data of Table 7.13 indicate that all burn types studied have the same average pain duration, in the sense that mean times until relief of the various burn types do not differ significantly among themselves.

A BASIC computer program for solving two-way analysis of variance problems appears in Table 7.17. The computer solution to the burn relief example is printed out in Table 7.18.

TABLE 7.17   BASIC Computer Program for Two-Way
Analysis of Variance

```
 10      REMARK: TWO-WAY ANALYSIS OF VARIANCE
 20      DIM X(10,10),B1(10),B2(10),T1(10),T2(10)
 30      REMARK: N = NUMBER OF BLOCKS
 40      REMARK: M = NUMBER OF TREATMENTS PER BLOCK
 50      REMARK: X(I,J) = JTH DATA POINT OF ITH TREATMENT
 60      PRINT "HOW MANY BLOCKS ARE THERE";
 70      INPUT N
 80      PRINT "HOW MANY TREATMENTS ARE THERE PER BLOCK";
 90      INPUT M
100      MAT X = ZER(N,M)
110      REMARK: THE DATA MATRIX IS LISTED IN LINES 111 TO 119
120      MAT READ X
130      FOR I = 1 TO N
140          B1(I) = 0
150          B2(I) = 0
160          FOR J = 1 TO M
170              B1(I) = B1(I) + X(I,J)
180              B2(I) = B2(I) + (X(I,J)**2)
190              NEXT J
200          NEXT I
210      FOR J = 1 TO M
220          T1(J) = 0
230          T2(J) = 0
240          FOR I = 1 TO N
250              T1(J) = T1(J) + X(I,J)
260              T2(J) = T2(J) + (X(I,J)**2)
270              NEXT I
280          NEXT J
290      PRINT
300      PRINT "DEGREES OF FREEDOM FOR DENOMINATOR IS";(M-1)*(N-1)
310      PRINT "TEST FOR SIGNIFICANT DIFFERENCES AMONG TREATMENTS"
320      PRINT "DEGREES OF FREEDOM FOR NUMERATOR IS ";M-1
330      PRINT "WHAT IS THE LEVEL OF SIGNIFICANCE ALPHA";
340      INPUT A1
350      PRINT "LOOK UP VALUE OF F";A1;"(";M-1;";";(M-1)*(N-1);")"
360      PRINT "WHAT IS IT";
370      INPUT F8
380      T = 0
390      B = 0
400      Q1 = 0
410      Q2 = 0
420      S1 = 0
430      S2 = 0
440      FOR J = 1 TO M
450          S1 = S1 + T1(J)
460          T = T + (T1(J)**2)
470          S2 = S2 + T2(J)
480          NEXT J
490      FOR I = 1 TO N
500          Q1 = Q1 + B1(I)
510          B = B + (B1(I)**2)
520          Q2 = Q2 + B2(I)
530          NEXT I
540      REMARK: SSE = ERROR SUM OF SQUARES
550      E = S2 - (T/N) - (B/M) + ((S1**2)/(M*N))
560      PRINT "SSE =";E
570      REMARK: SST = TREATMENT SUM OF SQUARES
580      U = (T/N) - ((S1**2)/(M*N))
590      PRINT "SST =";U
600      F1 = (U/E)*(N-1)
610      PRINT "TREATMENT F HAS VALUE";F1
620      PRINT
630      IF F1 > F8 THEN 680
```

(Continued)

```
640     PRINT "WE CANNOT REJECT TREATMENT H; OUR INFORMATION"
650     PRINT "DOES NOT CONTRADICT THE HYPOTHESIS THAT"
660     PRINT "TREATMENT MEANS ARE ALL THE SAME."
670     GO TO 700
680     PRINT "WE REJECT TREATMENT H, CONCLUDING THEREBY"
690     PRINT "THAT TREATMENTS SEEM TO DIFFER IN EFFECT."
700     PRINT
710     PRINT "TEST FOR SIGNIFICANT DIFFERENCES AMONG BLOCKS"
720     PRINT "DEGREES OF FREEDOM FOR NUMERATOR IS";N-1
730     PRINT "WHAT IS THE LEVEL OF SIGNIFICANCE ALPHA";
740     INPUT A2
750     PRINT "LOOK UP VALUE OF F";A2;"(";N-1;";";(M-1)*(N-1);")."
760     PRINT "WHAT IS IT";
770     INPUT F9
780     REMARK: SSB = BLOCK SUM OF SQUARES
790     V = (B/M) - ((S1**2)/(M*N))
800     PRINT "SSB =";V
810     F2 = (V/E)*(M-1)
820     PRINT "BLOCK F HAS VALUE";F2
830     PRINT
840     IF F2 > F9 THEN 890
850     PRINT "WE CANNOT REJECT BLOCK H; OUR INFORMATION DOES NOT"
860     PRINT "CONTRADICT THE HYPOTHESIS THAT BLOCK MEANS ARE ALL"
870     PRINT "THE SAME."
880     GO TO 999
890     PRINT "WE REJECT BLOCK H, CONCLUDING THEREBY THAT THE"
900     PRINT "BLOCKS SEEM TO DIFFER IN EFFECT."
999     END
```

TABLE 7.18   Computer Printout of Solution: Common Burns Example

```
111     DATA 35,30,25,41,37,42
112     DATA 38,28,23,29,30,38
113     DATA 41,32,24,32,28,47
114     DATA 29,35,24,36,25,37
115     DATA 37,25,29,42,30,47


run


HOW MANY BLOCKS ARE THERE ?5
HOW MANY TREATMENTS ARE THERE PER BLOCK ?6

DEGREES OF FREEDOM FOR DENOMINATOR IS 20
TEST FOR SIGNIFICANT DIFFERENCES AMONG TREATMENTS
DEGREES OF FREEDOM FOR NUMERATOR IS  5
WHAT IS THE LEVEL OF SIGNIFICANCE ALPHA ?.05
LOOK UP VALUE OF F .05 ( 5 ; 20 )
WHAT IS IT ? 2.71
SSE = 354
SST = 922
TREATMENT F HAS VALUE 10.4181

WE REJECT TREATMENT H, CONCLUDING THEREBY
THAT TREATMENTS SEEM TO DIFFER IN EFFECT.

TEST FOR SIGNIFICANT DIFFERENCES AMONG BLOCKS
DEGREES OF FREEDOM FOR NUMERATOR IS 4
WHAT IS THE LEVEL OF SIGNIFICANCE ALPHA ?.05
LOOK UP VALUE OF F .05 ( 4 ; 20 ).
WHAT IS IT ? 2.87
SSB = 100.8
BLOCK F HAS VALUE 1.42373

WE CANNOT REJECT BLOCK H; OUR INFORMATION DOES NOT
CONTRADICT THE HYPOTHESIS THAT BLOCK MEANS ARE ALL
THE SAME.
```

EXERCISES   7.C

7.15. Accident figures accumulated by the National Safety Council over a three-day holiday weekend for five randomly selected years show the following number of fatalities in each of three cities of comparable size:

Year	Number of Fatalities		
	Baltimore	Detroit	San Diego
1962	20	22	16
1973	18	21	14
1964	25	28	15
1978	28	30	20
1969	29	29	25

a. At level $\alpha = .05$, do the data indicate that the number of fatalities differs significantly among the three cities?

b. At level $\alpha = .01$, do the data indicate that the number of fatalities differs significantly from year to year?

7.16. As part of a study of the role of the coffee break in the proper functioning of the American economy, a corporation psychologist selected 20 employees of a major company, who were classified according to position and number of cups of coffee drunk per day, and rated each on a job performance scale. The ratings are:

Cups of Coffee Drunk per Day	Position			
	Secretary	Production Supervisor	Accountant	Division Manager
No coffee	15	10	9	11
1–3 cups	8	9	11	14
4–6 cups	12	13	8	15
7–9 cups	18	12	17	17
10 or more cups	12	12	15	18

a. At level of significance $\alpha = .05$, do the various positions seem to have differing ratings?

b. At level $\alpha = .01$, do differences in the number of cups of coffee consumed lead to significant differences in job performance ratings?

7.17. Some economists believe that in a period of simultaneous inflation and recession the real growth of the GNP, the overall price increase, and the unemployment rate will be nearly identical (when each is expressed as a percentage). The following data list the economic forecasts for a recent year that were made by 24 leading economists.

| | Forecasts | | |
Economist	Real GNP Growth	Price Increase	Unemployment Rate
R.D.	6.7	6.7	6.9
A.T.S.	6.9	6.1	8.0
K.B.S.	7.1	5.9	7.8
R.O.	6.7	6.0	7.6
D.A.H.	5.9	7.3	7.7
D.S.A.	6.6	6.5	7.8
R.J.	6.0	6.7	7.6
P.J.M.	7.1	5.1	7.8
R.G.D.	6.0	6.0	7.6
F.H.S.	6.0	6.2	7.8
J.R.F.	5.5	6.5	7.8
B.A.G.	5.9	6.0	7.7
P.L.B.	6.4	5.5	7.5
A.G.H.	5.6	5.8	7.8
H.E.N.	6.0	5.5	8.0
A.G.M.	5.4	5.9	7.7
M.C.	5.9	5.4	7.8
D.R.C.	5.4	5.6	7.9
W.C.F.	5.0	5.5	7.6
R.E.	5.4	5.3	8.0
R.H.P.	4.9	7.7	7.8
G.W.M.	5.7	4.8	8.1
J.J.O.	5.1	4.9	8.4
A.G.S.	4.9	4.1	7.5

a. At level $\alpha = .01$, do the forecasts of the various economic quantities seem to differ significantly on the average?

b. At level $\alpha = .05$, do the data indicate that the 24 economists have significantly differing forecasts?

7.18. A recent study of world bauxite production and the possibility that cartel pricing may arise in the bauxite industry brought out the following data on the trend of bauxite production in various nations.* The quantities listed are in thousands of metric tons.

| | Country | | | |
Year	Guyana	France	Jamaica	Surinam
1961	2,374	2,148	6,663	3,351
1964	2,468	2,387	7,811	3,926
1968	3,490	2,756	8,391	5,484
1971	3,757	3,066	12,565	6,162
1973	3,224	3,084	13,385	6,580
1975	3,200	2,500	11,400	4,900

*Appearing in the article, "Cartel pricing and the structure of the world bauxite market," by R. S. Pindyck in The Bell Journal of Economics, vol. 8, no. 2, Autumn 1977, p. 350.

a. At level $\alpha = .05$, do the figures show that bauxite production differs significantly among the four nations surveyed?

b. Do the data indicate at level $\alpha = .05$ that there were significant differences in bauxite production among the years listed?

7.19. The following data present the total percentage return to investors in 1976 on their investments in several randomly selected corporations, classified according to region in which the head office is located and type of industry the corporation engages in.*

Industry	Northeast	Midwest	Southeast	West Coast
Energy	26.98	45.28	48.17	30.96
Food	14.93	24.15	58.66	17.80
Aerospace	37.36	42.86	32.09	21.27
Electronics	28.04	39.88	9.36	11.46
Textiles	12.87	51.44	27.63	37.42
Drugs	8.54	21.18	23.65	24.88
Metals	19.87	30.98	44.18	79.53
Paper	17.72	28.89	35.84	31.36
Chemicals	17.63	14.19	43.92	36.31
Minerals	28.70	18.03	54.68	44.34

a. At level $\alpha = .01$, can we conclude from the data that there are significant differences in percentage return to investors among the geographical regions in which the head office is located?

b. At level $\alpha = .05$, do the various industry groups seem to differ significantly in percentage return to investors?

SUMMARY AND DISCUSSION

In this chapter, we have extended Chapter 6's discussion of one-sample and two-sample tests for the difference of means to the case of more than two samples. Using one-way analysis of variance, we were able to consider several samples simultaneously, testing the hypothesis that all their population means are equal, against the alternative that significant differences exist between the means. The statistical analysis was based on the relationship between the variation between the sample means and the variation of the data points within each sample. The ratio between these two variations is similar to the ratio between two variances and therefore has the F distribu-

*Fortune, May 1977, pp. 366–385.

tion. It followed, then, that an F test of the sort used in Chapter 6 to test the difference of standard deviations can be applied in analysis of variance to test the difference of several means. Finally, we have discussed the process of two-way analysis of variance, which allows us to test the effects of two types of influences (called "treatments" and "blocks") on the underlying populations. The techniques of two-way analysis of variance were not new, but were merely minor revisions of those used in one-way analysis of variance. BASIC computer programs were presented for solving both the one-way and two-way analysis of variance problems.

CHAPTER 7   BIBLIOGRAPHY

DETAILED TREATMENT OF ANALYSIS OF VARIANCE

GUENTHER, W. C., *Analysis of Variance.* 1964, Englewood Cliffs, N.J.: Prentice-Hall.

SPECIAL TOPICS

D'AGOSTINO, R. B., "Relation between the Chi-Squared and ANOVA Tests for Testing the Equality of k independent Dichotomous Populations," *The American Statistician*, June 1972, pp. 30–32.

FRIEDMAN, H. H. and W. S. DIPPLE, JR., "The Effect of Masculine and Feminine Brand Names on the Perceived Taste of a Cigarette," *Decision Sciences*, July 1978, pp. 467–471.

HARTER, H. L., "Multiple Comparison Procedures for Interactions," *The American Statistician*, December 1970, pp. 30–32.

PATTON, J. M., "An Experimental Investigation of Some Effects of Consolidating Municipal Financial Reports," *The Accounting Review*, April 1978, pp. 402–414.

SIROTNIK, K., "On the Meaning of the Mean in ANOVA (of the Case of the Missing Degree of Freedom)," *The American Statistician*, October 1971, pp. 36–37.

ZIKMUND, W. G., R. F. CATALANELLO, and S. M. WEGENER, "The Accounting Student's Job-Rating Criteria: An Experiment," *The Accounting Review*, July 1977, pp. 729–735.

CHAPTER 7   SUPPLEMENTARY EXERCISES

7.20. To attempt to determine the effects of the various types of pollutants in the air, an EPA team studied the growth rate of mice in the four different atmospheres in the table below. Twenty-eight mice born at the same time were fed exactly the same diet in order to prepare them for the experiment. All other living conditions were identical, except that the mice were divided into four groups according to the type of pollution in their atmosphere. A biologist participating in the study collected the following data on weight gains of the mice over a period of time:

| Weight Gains of Mice (grams) | | | |
Atmosphere with Cigarette Smoke	Atmosphere with Auto Exhaust	Atmosphere with Industrial Smoke	Clean Air
6	9	10	11
7	7	9	8
6	9	8	10
10	8	5	9
8	5	10	8
5	10	8	6
7	8	6	11

a. Calculate the components of the variance.

b. Using the components of the variance, decide at level $\alpha = .05$ whether or not there are significant differences in weight gains of mice in the four types of atmosphere.

c. Carry out a one-way analysis of variance to answer the question of part b.

7.21. It is the job of a tea tester for an English tea-importing company to decide which tea leaves are most suitable for English Breakfast tea. The tea tester must decide among three lots of leaves: an inexpensive lot, a moderately priced lot, and an expensive lot. The tea tester and his laboratory staff discover the following levels of impurities in eight randomly selected leaves of each lot.

Inexpensive lot	3	4	25	9	11	4	12	20
Moderately priced lot	4	9	16	9	12	6	4	8
Expensive lot	6	5	7	10	6	8	5	4

a. Calculate the components of the variance.

b. Use the components of the variance to test at level $\alpha = .01$ whether or not all three lots of leaves have the same average impurity levels.

c. Apply the one-way analysis of variance technique to test the same hypothesis as in part b.

7.22. Peanuts are known to be an excellent source of protein, and they are therefore a good substitute for meat and fish when prices are high. To find out whether storage of peanuts for a length of time tends to change their protein content, a peanut distributor selected several bags of peanuts from among those that had been stored in her warehouse for various periods of time, and then she measured the protein content of each bag. The data, in grams of protein per bag, follow.

Fresh Peanuts	Peanuts Stored 6 Months	Peanuts Stored 12 Months	Peanuts Stored 18 Months	Peanuts Stored 24 Months
10	8	5	10	7
8	6	8	6	8
6	5	9	9	7
10	10	8	10	11
8	7	8	6	5
10	9	6	8	9
6	6	10	7	6
8		5	7	8
6		7		10
10		5		
4				

At level $\alpha = .05$, do the data indicate that storage of peanuts tends to alter their protein content?

7.23. A factory, in the process of assessing its energy costs, conducted a study of how much it costs to run each of four types of machines that it uses with each of five different energy sources. The monthly costs, in hundreds of dollars, are listed below:

| Type of Machine | Energy Source | | | | |
	Elec-tricity	Natural Gas	Oil	Coal	Manual
Drill press	6	5	8	4	5
Metal stamping	4	3	6	2	4
Sorter	3	3	4	2	2
Conveyor	2	3	3	3	6

a. At level $\alpha = .01$, do the data provide evidence sufficient to assert that different energy sources result in different operating costs, on the average, for the types of machines under study?

b. At level $\alpha = .01$, can we conclude from the data that the four types of machines differ significantly among themselves in energy costs?

7.24. An operations analyst for the airline industry conducted a study of seven domestic trunk lines, eight international carriers, and six local service lines in order to determine whether or not passenger load factors differ significantly among the three types of air carrier service. The following data list the various passenger load factors as percentages.[*]

---

[*] *Aviation Week & Space Technology*, October 24, 1977, p. 34.

Domestic Trunks	International	Local Service
5.0	4.8	7.2
16.9	1.4	23.2
8.8	0.7	4.0
5.1	2.3	5.0
6.7	6.2	18.1
10.1	6.9	21.8
6.9	2.5	
	11.7	

Can the analyst conclude, at level $\alpha = .01$, that the three types of service differ significantly in passenger load factors?

7.25. The trust company that manages the employee pension fund for the Amalgamated Glue Corporation wants to find out if it would be reasonable to assume that banks, insurance companies, and utilities have, on the average, the same total dollar value of all their outstanding shares of common stock. The data, in billions of dollars worth of common stock, follow.*

Banks	Insurance Companies	Utilities
0.373	1.902	0.685
0.396	0.420	2.372
0.982	0.381	0.442
0.623	1.591	0.771
0.656	0.602	0.866
0.906	0.491	0.816
0.218	0.496	1.302
0.145	0.335	1.106
0.242	0.347	0.290
1.165		0.511
2.177		

At level of significance $\alpha = .05$, can the trust company conclude that the average dollar value of all common stock outstanding is basically the same for corporations in all three industry groups?

7.26. A nationwide chain of real estate offices surveyed a number of major real estate markets in order to study the behavior of housing prices in 1980. The data below show the typical percentage price increase of each of four types of housing in each of six geographical areas:

*Selected from a detailed report in *Forbes*, May 15, 1977, pp. 216–243.

Housing	So. Calif.	No. Calif.	New England	Pacif. N.W.	Florida	Texas
Suburban houses	14	13	6	4	4	9
Downtown condominiums	18	16	4	3	2	8
Downtown apartments	9	10	10	5	5	10
Mobile homes	10	12	4	11	8	12

a. At level $\alpha = .05$, do the data indicate that percentage price increases differ significantly among the six geographic areas surveyed?

b. At level $\alpha = .01$, can the real estate chain assert that the four types of housing differ significantly in their percentage price increases?

7.27. The securities department of a Cleveland savings and loan wanted to study the variation in money market rates so that it could plan a long-term investment strategy. Looking up the rates for three major money market instruments during six critical weeks in the past year, it found the following data in percentages:

Instrument	Week 1	Week 2	Week 3	Week 4	Week 5	Week 6
Federal funds	6.13	6.00	5.20	6.50	6.46	5.00
Treasury bills	5.88	5.58	5.10	6.15	6.28	4.93
NYC Bank CD's	6.13	5.95	5.30	6.65	6.80	5.10

a. Test, at level $\alpha = .05$, whether or not money market rates vary significantly over periods of time.

b. Test, at level $\alpha = .05$, whether or not the three money market instruments vary significantly in rate.

7.28. A recent study of live TV coverage of congressional hearings over the past several years revealed the following information:*

Year	Total Hours of Live Coverage			
	ABC	CBS	NBC	PBS
1966	22.60	27.55	69.15	1.50
1968	1.00	1.57	10.33	2.50
1969	0.00	0.00	5.58	6.50
1971	0.00	0.00	0.00	5.00
1972	1.50	0.50	0.50	3.00
1974	17.22	27.15	25.70	116.50
1976	0.00	0.00	0.00	3.78

a. At level $\alpha = .05$, do the data indicate that the networks differ significantly in average number of hours of live coverage of congressional hearings?

b. At level $\alpha = .01$, does there seem to be a significant difference in hours of live coverage among the years surveyed?

*The Wilson Quarterly, Autumn 1977, p. 41.

IV

PREDICTION AND FORECASTING

8

Linear Regression and Correlation

In a study of the efficiency of automobile engines with respect to consumption of gasoline, one particular 3,990-pound car was operated at various speeds, and its gasoline consumption in miles per gallon at each speed level was carefully measured. The data of Table 8.1 express the results of the experiment. From a brief glance at the data, we can immediately draw one conclusion: at higher speeds, the car gets fewer miles per gallon. From a detailed analysis of the data, however, we might be able to answer the following four questions with some respectable degree of accuracy.

1. Is there a simple relationship (namely, a relationship expressible by a simple formula) between speed in miles per hour and efficiency in miles per gallon?
2. If so, what is this relationship?
3. How much of the variation in efficiency levels can be attributed to variations in speed? (That is to say, among all those factors that affect gasoline efficiency, how important a role is played by the speed of the car?)
4. How can we use the data to predict efficiency at other speeds, for example at 25, 55, or 80 miles per hour, and how accurate will our predictions be?

How to obtain answers to questions such as these is the subject matter of this chapter.

TABLE 8.1   Efficiency of an Automobile Engine

Speed (miles per hour)	Efficiency (miles per gallon)
30	20
40	18
50	17
60	14
70	11

SECTION 8.A.   THE EQUATION OF THE STRAIGHT LINE

Suppose we put the data of Table 8.1 on a scattergram of the sort introduced in Chapter 1. The scattergram of Fig. 8.1 communicates to the viewer information beyond that contained in our earlier observation that, "at higher speeds, the car gets fewer miles per gallon." From the scattergram, we can see that there seems to be a definite linear trend that, while not perfect, does describe the general characteristics of the relationship between speed and efficiency.

FIGURE 8.1   Scattergram of Speed vs. Efficiency

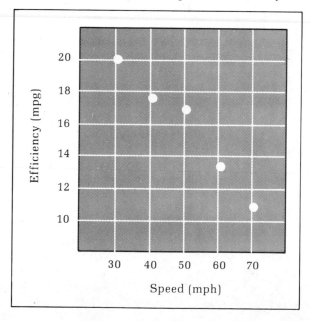

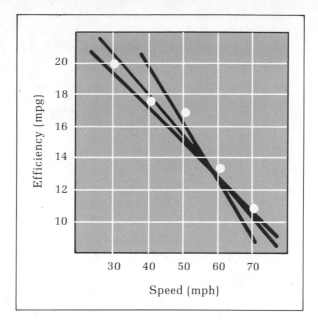

FIGURE 8.2   Possible Best-Fitting Lines

The next logical step is to specify the unique straight line that most closely approximates all the data points simultaneously. Fig. 8.2 shows three possible candidates for the straight line that best fits the data. To select the one best-fitting line, it is necessary to analyze more carefully the numerical relationship between speed and efficiency that is indicated by the data.

As it turns out, every straight line has its own equation, which is a description of the line's slope and location. For example, let's consider the line that joins the two points in Fig. 8.3. One of the points is labeled (1, 2) to indicate that it lies at a perpendicular distance 1 to the *right* of the vertical axis and a perpendicular distance 2 *above* the horizontal axis. The other is labeled (5, 4) because it lies 5 units to the *right* of the vertical axis and 4 units *above* the horizontal axis. A point labeled (−4, −3) would lie 4 units to the *left* of the vertical axis and 3 units *below* the horizontal axis.

The line joining (1, 2) with (5, 4) also contains several other points—infinitely many, in fact. A line can be considered as a collection of points, all "glued" together in a special configuration. A typical point is labeled (x, y) when it lies x units to the right of the vertical axis and y units above the horizontal axis. (If x is negative and if we proceed "x units to the right," we actually wind up to the left of the vertical axis.) Some of these points (x, y) happen to fall exactly on the line, and some do not.

To find out which points (x, y) lie on a particular straight line and which ones do not, we use the equation of the straight line: the equation of a straight

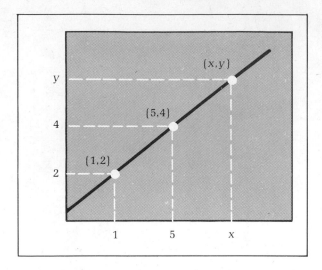

FIGURE 8.3   A Typical Straight Line

line is the relationship between x and y that specifies which (x, y)'s lie on the line and which lie off it.

We can construct the equation of the line joining (1, 2) with (5, 4) as follows: first we observe from Fig. 8.4 that to get to (5, 4) from (1, 2), we have to move 4 units to the *right* and 2 units *up*. If now we are at (5, 4) and again move 4 units to the right and 2 units up, we will reach (9, 6), which is also on the same line. In fact, anytime we move up an amount equal to half the

FIGURE 8.4   Points on the Line

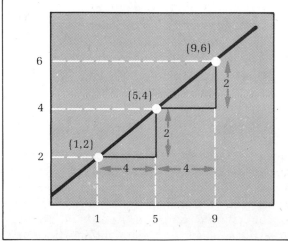

distance we move to the right, we will always arrive at a point that is on the line (as long as we started from a point on the line, of course).

Now look again at Fig. 8.3. To get from (1, 2) to (x, y), we have to move a distance $x - 1$ to the right. Therefore, to arrive at a point on the line, we must move a distance $\frac{1}{2}(x - 1)$ up. But, according to Fig. 8.3, this distance up can also be expressed as $y - 2$. It follows that

$$y - 2 = \frac{1}{2}(x - 1)$$

and algebraically in sequence

$$y - 2 = \frac{1}{2}x - \frac{1}{2} - .5x - .5$$
$$y = 2 + .5x - .5 = 1.5 + .5x$$

The last equation just above, $y = 1.5 + .5x$, is referred to in statistical work as *the* equation of the straight line joining (1, 2) with (5, 4).

Now that we have the equation, we can determine which points are on the line and which are not. For example, (9, 6) is on the line because $6 = 1.5 + (.5)(9)$, and (11, 8) is not on the line because $8 \neq 1.5 + (.5)(11)$.

In the same way, we could start with any two points and work out the equation of the line joining them. Because the algebraic techniques used would be the same, we would come up with an equation of the following form.

THE EQUATION OF THE STRAIGHT LINE

$$y = a + bx$$

where a and b are constants characteristic of the line.

Conversely, from any equation of the form $y = a + bx$, we can draw the corresponding straight line. For example, consider the equation $y = -2 + 3x$. If we can find any two points on the line represented by this equation, we can connect them in order to obtain the graph of the line. We can find points by picking x values at random and then working out the corresponding y values. If we choose $x = 0$, then $y = -2 + (3)(0) = -2 + 0 = -2$, so the point $(0, -2)$ is on the line. If we next select $x = 2$, then $y = -2 + (3)(2) = -2 + 6 = 4$, so that $(2, 4)$ lies on the line. Connecting the points $(0, -2)$ and $(2, 4)$ as in Fig. 8.5, we get a picture of the line. Notice that an individual choosing $x = 1$ and $x = 3$, instead of $x = 0$ and $x = 2$, would obtain the same line by use of the two points $(1, 1)$ and $(3, 7)$.

The number that plays the role of b can be interpreted as follows: b is the amount by which y increases (if $b > 0$), or decreases (if $b < 0$), when x

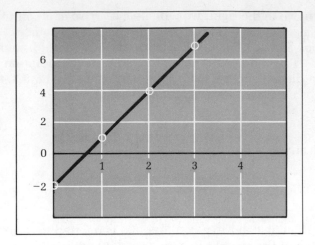

FIGURE 8.5   The Line y = −2 + 3x

increases by one unit. The meaning of *a* is the value of *y* when *x* is zero. We refer to *b* as the *slope* of the straight line and to *a* as the *y-intercept*, for it is the point at which the line crosses the vertical axis.

EXERCISES 8.A

8.1. For each of the following pairs of points, determine the equation of the straight line passing through them.

    a. (1, 2) and (5, 3)

    b. (0, −2) and (2, 1)

    c. (1, 9) and (5, 5)

8.2. For each of the following pairs of points, determine the equation of the straight line passing through them.

    a. (1, 2) and (5, 6)

    b. (0, −2) and (2, 5)

    c. (1, 9) and (5, 8)

8.3. For each of the following equations, find two points on the corresponding straight line and draw the line.

    a. y = 6 + 2x.

    b. y = 8 − 3x.

    c. y = −3 + 4x.

8.4. For each of the following equations, find two points on the corresponding straight line and draw the line.

    a. $y = 8 + 3x$.

    b. $y = 6 - 2x$.

    c. $y = -4 + 3x$.

8.5. Determine the equation of the straight line passing through the points (8.1, 2.3) and (3.2, 1.8), and find a third point lying on that line.

8.6. Find two points lying on the straight line having equation $y = 7.8 - 2.2x$, and draw the line.

SECTION 8.B   THE REGRESSION LINE AND PREDICTION

The line that best fits the points of Fig. 8.1 is called the *regression line* of the points. We know from Section 8.A that in order to find the equation of the regression line, we have only to determine the numerical values of a and b.

    Let's first specify exactly what we mean by the "best-fitting line," one that would "best fit" the purposes for which we want to eventually use it. The data of Table 8.1 relate the gasoline efficiency of the engine to the speed of the car. The regression line would be a relationship between speed and efficiency that we would most likely use to predict efficiency at various speeds. It is traditional, algebraically speaking, to label speed as x (the *independent variable*), and efficiency as y (the *dependent variable*). Then the equation $y = a + bx$ expresses the manner in which efficiency "depends" on speed. A statement such as this does not necessarily imply any causal dependence, like a cause-and-effect relationship, but merely an algebraic connection between the numbers representing efficiency and the numbers representing speed. *In general, we shall always use y to denote the variable to be predicted.*

The Sum of Squared Deviations

A prediction of efficiency, based on the regression line $y = a + bx$, will be as accurate as possible only if we choose a and b in such a way that some measure of the vertical distances between the data points and the line (known as *vertical deviations*) are collectively as small as possible. This situation is illustrated in Fig. 8.6, where the original data points are compared with the locations predicted for them by a possible regression line $y = a + bx$.

    By analogy with the development of the mean and standard deviation in Chapter 2, we consider the regression line as a sort of "mean" or "balancing line" of the data points. Since the positive and negative vertical deviations of the data points from the line should cancel each other, the real deviation of the points from the line will have to be calculated using absolute values or squares in order to eliminate the cancellation effects of points falling above and below the line. As in the case of the mean absolute deviation discussed

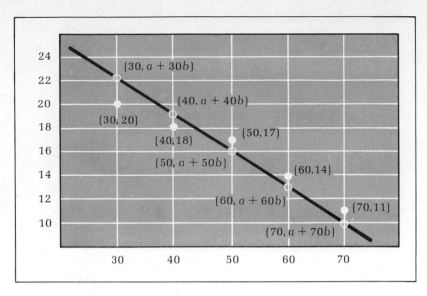

FIGURE 8.6   Vertical Deviations of Data Points from the Line $y = a + bx$

briefly in Chapter 2, there are some situations when mean-absolute-deviation regression lines would be appropriate (see, for example, the articles listed in the bibliography at the end of this chapter); but, by and large, it is more often useful to deal with squared deviations. This is especially true in cases where we have reason to believe that one or both columns of data come from normally distributed populations.

The rule relating the sum of squared deviations to the best-fitting line is shown below.

THE LEAST-SQUARES LINE

A typical set of n data pairs (x_1, y_1), (x_2, y_2), ... (x_n, y_n) to be fitted to a line $y = a + bx$ will have its sum of squared deviations equal to

$$SSD = \Sigma(y - a - bx)^2$$

where SSD is the *sum of squared deviations*.

For any particular set of data points, a line having equation $y = a + bx$ will be the best-fitting line if we use the values of a and b that result in the smallest possible value of SSD. The regression line resulting from this method of reasoning is called the *least-squares line*.

Formulas for a and b

For different sets of data points, we would expect to come up with different numerical values of a and b, because the regression lines would most likely be different in each case. So what we really need are formulas for a and b that we can use to compute their numerical values from the data.

The numbers a and b can be determined from the data using the algebraically derived formulas

$$b = \frac{n\Sigma xy - (\Sigma x)(\Sigma y)}{n\Sigma x^2 - (\Sigma x)^2} \tag{8.1}$$

and

$$a = \frac{\Sigma y - b\Sigma x}{n} \tag{8.2}$$

It should be noted that the numerical value of b must be calculated first, because it is itself used in the calculation of a. Another important observation is that Σx^2 is not the same as $(\Sigma x)^2$. For example, if $x_1 = 2$, $x_2 = 5$, and $x_3 = 1$, we would have

$$\Sigma x^2 = 2^2 + 5^2 + 1^2 = 4 + 25 + 1 = 30$$

but

$$(\Sigma x)^2 = (2 + 5 + 1)^2 = 8^2 = 64.$$

The formulas for a and b involve sums of several quantities, namely the following:

Σx = sum of x values
Σy = sum of y values
Σx^2 = sum of squares of x values
Σxy = sum of products of each x, multiplied by its corresponding y

The simplest way to keep track of all the calculations necessary to obtain the regression line is by way of a table of the sort illustrated in Table 8.2. (In our table, we will also include the calculation of Σy^2, the sum of squares of the y values, which will turn out to be needed in estimating the possible error inherent in using the regression line to predict future occurrences.) Using the sums obtained in Table 8.2, we can complete the calculation of a

TABLE 8.2 Preliminary Calculations in Linear Regression

Speed x (mph)	Efficiency y (mpg)	x^2	y^2	xy
30	20	900	400	600
40	18	1,600	324	720
50	17	2,500	289	850
60	14	3,600	196	840
70	11	4,900	121	770
Sums				
250	80	13,500	1,330	3,780

n = 5

and b by inserting the sums into their proper places in formulas (8.1) and (8.2):

$$b = \frac{n\Sigma xy - (\Sigma x)(\Sigma y)}{n\Sigma x^2 - (\Sigma x)^2} = \frac{5(3,780) - (250)(80)}{5(13,500) - (250)^2}$$

$$= \frac{18,900 - 20,000}{67,500 - 62,500} = \frac{-1,100}{5,000} = -.22$$

$$a = \frac{\Sigma y - b\Sigma x}{n} = \frac{80 - (-.22)(250)}{5} = \frac{80 + 55}{5} = 27.$$

Having calculated a = 27 and b = −.22, we therefore know that the equation of the regression line y = a + bx is in this case y = 27 − .22x.

Superimposing the Regression Line upon the Scattergram

It is often useful, for purposes of visual communication and analysis, to superimpose the graph of the regression line upon the scattergram of the data, as in Fig. 8.7.

We superimpose the graph of the line y = 27 − .22x on the scattergram of Fig. 8.1 by choosing two points on the line and connecting them, as in Fig. 8.5 (see discussion in Section 8.A). If we choose x = 30, then y = 27 − .22x = 27 − (.22)(30) = 20.4, so that (30, 20.4) is on the regression line. Choosing x = 60, the corresponding y = 27 − (.22)(60) = 13.8, so that (60, 13.8) is a second point on the regression line. The two points, (30, 20.4) and (60, 13.8),

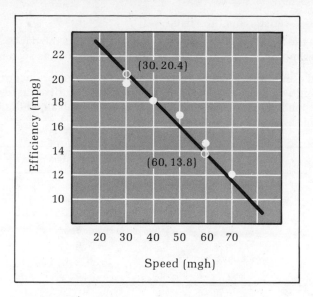

FIGURE 8.7   Superimposing the Regression Line
upon the Scattergram

indicated by open circles, are graphed together with the regression line in
Fig. 8.7.

EXERCISES   8.B

8.7. The manager of an independent supermarket would like to know the relationship, if there is one, between the amount of display space occupied by a local
brand of tuna and the dollar value of weeky sales of that item. The amount of
space occupied was varied over six randomly selected weeks, and the following
data were obtained.

Week Beginning	Display Space (square feet)	Total Sales (hundreds of dollars)
June 13	5	7
Feb 21	2	5
Dec 19	5	7
Aug 8	8	8
Jan 24	2	4
Oct 10	8	9

a. Draw a scattergram of the data.

b. Find the equation of the straight line that best fits the data.

c. Superimpose the graph of the regression line upon the scattergram.

8.8. An agricultural research organization tested a particular chemical fertilizer to try to find whether an increase in the amount of fertilizer used would lead to a corresponding increase in the food supply. They obtained the following data based on seven plots of arable land.

Pounds of Fertilizer	Bushels of Beans
2	4
1	3
3	4
2	3
4	6
5	5
3	5

a. Draw a scattergram of the data.

b. Find the equation of the regression line that would be used to predict the number of bushels of beans obtained from a number of pounds of fertilizer.

c. Superimpose the graph of the regression line upon the scattergram.

8.9. One psychologist suspects that there is a connection between the rate of inflation in the economy and the rate of divorce in the general population. In an attempt to find a way to predict the divorce rate from the inflation rate, she collects the following data from records of the past several years.

Inflation rate, %	2	4	5	7	10	12
Divorce rate per 1,000 of population	3	7	10	15	25	30

a. Draw a scattergram of the data.

b. Find the equation of the regression line.

c. Superimpose the graph of the regression line upon the scattergram.

8.10. A demographer presented the following data to support his theory that high protein diets tend to reduce fertility levels.

Country	Taiwan	Japan	Italy	West Ger-many	United States	Swe-den
Protein in diet, grams/day	5	10	15	40	60	70
Birth rate per 100 population	4	3	3	2	2	1

  a. Draw a scattergram of the data.

  b. Find the equation of the regression line.

  c. Superimpose the graph of the regression line upon the scattergram.

8.11. A sociologist, undertaking a preliminary study of the effects on society of a proposed rescheduling of the 36-hour work week into three 12-hour days, measures the productivity of employees in situations where an experimental 10-hour day is in operation. She collects the following data, which accords with the law of diminishing returns, on one randomly selected employee each hour.

Employee	M	N	P	Q	R	S	T	U	V	W
Hour	1	2	3	4	5	6	7	8	9	10
Accumulated productivity	3	6	8	9	10	11	11	12	12	13

  a. Draw the scattergram of the data.

  b. Calculate the equation of the regression line that would be used to predict accumulated productivity hour-by-hour.

  c. Superimpose the graph of the regression line upon the scattergram.

8.12. Coronado Nautotronics of San Diego, bidding for the contract to produce the radar displays for the Navy's new fleet of patrol hydrofoils, collects the following information in an attempt to determine the cost curve for the radar display.

Quantity produced	10	50	100	160	200	320	630	800
Total cost of production run	7.0	8.5	9.0	9.4	9.5	10.0	10.5	10.8

  a. Draw a scattergram of the data.

  b. Calculate the equation of the regression line for predicting the total cost of the production run.

  c. Superimpose the graph of the regression line upon the scattergram.

8.13. An economist wants to determine the daily demand equation for rolled steel in a small industrial town. She collects the following data relating the price with the quantity of rolled steel that can be sold at that price.

Tons of Rolled Steel That Can Be Sold	Price per Ton (hundreds of dollars)
1.0	50.10
2.0	12.60
2.5	8.00
4.0	3.20
5.0	2.00
6.3	1.25

a. Draw a scattergram of the data.

b. Calculate the regression equation for predicting price based on quantity.

c. Superimpose the graph of the regression line upon the scattergram.

8.14. The following data compare the gross national products (GNP's) of Japan and the United States, measured in billions of dollars at 1973 exchange rates:*

Year	Japan GNP	U.S. GNP
1955	23.9	403.7
1960	43.1	511.4
1965	88.4	696.3
1970	197.2	993.3
1971	225.0	1068.8
1972	299.4	1155.2
1973	415.7	1289.1

a. Draw a scattergram of the data.

b. Calculate the equation that best predicts the U.S. GNP in terms of the Japan GNP.

c. Superimpose the graph of the regression line upon the scattergram.

8.15. A study of effective federal income tax rates, based on an analysis of actual tax returns, yielded the following information:†

1977 Adjusted Gross Income (thousands of dollars)	1977 Effective Federal Tax Rate (percent)
16.2	10.98
21.6	13.15
27.0	15.43
32.4	17.46
37.8	19.23
43.2	20.71
48.6	22.34
54.0	24.03
59.4	24.56
64.8	27.51
70.2	27.72
75.6	28.35
81.0	29.35

a. Draw the scattergram of the data.

*I. Frank and R. Hirono (eds.), How the United States and Japan See Each Other's Economy, New York: Committee for Economic Development, 1974, p. 11.

†Based on a report in Business Week, September 12, 1977, page 83.

b. Determine the equation of the regression line for predicting the effective tax rate for various levels of adjusted gross income.

c. Superimpose the regression line upon the scattergram.

8.16. The following data compare, over a 14-year period in the recent past, expenditures for new homes in the United States with non-real estate mortgage debt:*

Annual Data (billions of dollars, seasonally adjusted)		
Year	Mortgage Debt for Non-Real Estate Purposes	Aggregate Outlays for New Homes
1960	0.1	19.9
1961	3.6	17.4
1962	4.7	19.1
1963	8.4	18.7
1964	10.3	19.1
1965	10.8	19.0
1966	10.5	18.4
1967	11.6	15.8
1968	12.1	20.8
1969	14.4	21.2
1970	14.0	18.1
1971	18.3	26.6
1972	25.4	35.6
1973	28.0	39.9

a. Draw a scattergram of the data.

b. Determine the regression equation that predicts aggregate outlays for new homes from a knowledge of non-real estate mortgage debt.

c. Superimpose the graph of the regression line upon the scattergram.

SECTION 8.C   STRENGTH OF THE LINEAR RELATIONSHIP

Using formulas (8.1) and (8.2) for a and b will give us a regression line for any set of paired data. Unfortunately, therefore, we can find a regression line even for a set of data that is fundamentally nonlinear. (Some examples of nonlinear data are presented in Table 8.3, and their scattergrams are illustrated in Fig. 8.8).

It is often difficult, especially with large amounts of data, to decide from either the table of data or the scattergram whether the relationship that generated the data may be considered linear or not. Even if the underlying rela-

*The Conference Board, *Statistical Bulletin*, December 1974, p. 13.

TABLE 8.3   Some Nonlinear Sets of Data

(a) Parabolic		(b) Logarithmic		(c) Exponential	
x	y	x	y	x	y
8	9	1	1	5	2
1	16	3	4	3	3
6	1	8	8	1	10
3	4	14	9.5	2	5
5	0	20	10	4	2
2	10	10	9	6	1
		5	6		
		2	3		

tionship really were a linear one, it would be unreasonable to expect all the sample data points to fall exactly on a straight line.

To deal with this problem, what we need is a way of measuring the extent to which a set of data can be considered to represent an underlying linear relationship; namely, a way to measure the strength of the relationship between the data and the regression line calculated from the data. We would use such a measure to answer the question, "Does the regression line present a valid pictorial representation of the behavior of the data?"

Let's look again at the typical set of n data pairs (x_1, y_1), (x_2, y_2), ..., (x_n, y_n). In general, y varies; that is to say, the y values $y_1, y_2, \ldots, y_n$ are not all the same. If all the numbers $y_1, y_2, \ldots, y_n$ were the same, they would each be equal to the mean of the group, $\bar{y}$. Therefore, we can consider the sum of squared deviations from the mean,

$$TV = \Sigma(y - \bar{y})^2$$

as a measure of the total variation of y, where TV is the total variation. For example, in Table 8.4 we analyze the variation inherent in the engine efficiency data of Table 8.1. From the fourth column of Table 8.4, we can see that the total variation of engine efficiency is

$$TV = \Sigma(y - \bar{y})^2 = 50.$$

The sum of squared deviations, SSD, discussed in Section 8.B, is actually the variation of the y values away from the regression line, because it is the sum of the squared deviations of the actual y values from their values

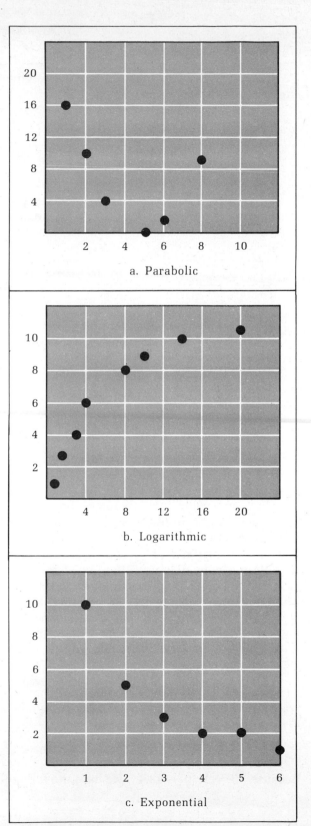

FIGURE 8.8
Scattergrams
of Nonlinear Data

a. Parabolic

b. Logarithmic

c. Exponential

TABLE 8.4   Analysis of the Variation of Engine Efficiency
(regression line: $y = 27 - 0.22x$)

Actual Data Points		Total Variation		Variation from Line		Variation with Line
x	y	$\bar{y}$	$(y - \bar{y})^2$	$a + bx$	$(y-a-bx)^2$	$(a + bx - \bar{y})^2$
30	20	16	16	20.4	0.16	19.36
40	18	16	4	18.2	0.04	4.84
50	17	16	1	16.0	1.00	0.00
60	14	16	4	13.8	0.04	4.84
70	11	16	25	11.6	0.36	19.36
Sums	80		50		1.60	48.40

$a + bx$ predicted by the regression line. In the engine efficiency example,

$$SSD = \Sigma(y - a - bx)^2 = 1.60,$$

as can be seen from the sixth column of Table 8.4.

Well, the total variation in engine efficiency, as recorded in the data, is 50, while the variation away from the regression line is 1.60. The difference $50 - 1.60 = 48.40$ indicates that there is still a substantial portion of the total variation that is not part of the variation away from the regression line. To what can we attribute this remaining variation?

Let's look at the problem from another point of view. Suppose there were no variation at all in y. Then every y value would equal $\bar{y}$, and we would have the total variation $TV = \Sigma(y-\bar{y})^2 = 0$. In fact, however, there is variation in y. Suppose the only variation in y were due to the influence of the regression line $y = a + bx$. Then every y would equal its corresponding $a + bx$. The resulting total variation would then be $TV = \Sigma(y - \bar{y})^2 = \Sigma(a + bx - \bar{y})^2$, since y and $a + bx$ would always be the same. The quantity $\Sigma(a+bx -\bar{y})^2$ is therefore the variation in y that can be attributed to the effect of the regression line. We denote this term as

$$VR = \Sigma(a + bx - \bar{y})^2,$$

where VR means variation due to regression. In the far right column of Table 8.4, we made the calculation that

$$VR = \Sigma(a + bx - \bar{y})^2 = 48.40.$$

Because $50 = 1.60 + 48.40$, we have the following formula:

TOTAL VARIATION OF y

TV = SSD + VR.

In words, we can express this assertion as follows:

total variation of y = variation away from the regression line
+ variation due to the influence of regression

In our example, then, SSD and VR together account for the entire variation TV of y. Since $1.60 + 48.40 = 50$, there is no variation left to be attributed to anything else. We might ask the question, "Does this happen in every possible example or do we have a special case here?" The answer is that TV *always* equals SSD + VR. Because of this fact, TV is usually denoted by SS_{total}, and referred to as the *total sum of squares*; SSD is denoted by SSE and called the *error sum of squares*; and VR is denoted by SSR and named the *regression sum of squares*. It is an algebraic fact that

$$\Sigma(y-\overline{y})^2 = \Sigma(y-a-bx)^2 + \Sigma(a+bx-\overline{y})^2,$$

namely, that SS_{total} = SSE + SSR for *all* sets of paired data.

THE EXPLAINED VARIATION
Because TV is the total variation in y, and VR is the variation in y that can be attributed to the influence of a regression line, the ratio VR/TV represents the proportion of variation in y that can be attributed to the influence of an underlying regression line. VR/TV is often referred to as *the proportion of the variation in y explained by regression*, or, more simply, the *explained variation*.

In our example,

$$\frac{VR}{TV} = \frac{SSR}{SS_{total}} = \frac{48.40}{50} = .968 = 96.8\%,$$

so that 96.8% of the variation in engine efficiency can be explained on the basis of its linear relationship $y = 27 - .22x$ with speed.

The quantity VR/TV is also called the *coefficient of linear determination*, because it measures the extent to which variation in y is determined by its linear relationship with x. When we say "determined" here, we mean determined algebraically and numerically only; we have not proved, nor can we prove conclusively, using statistics alone, the existence of any type of cause-and-effect relationship between x and y.

As it turns out, it is possible to calculate the coefficient of determination directly from a table like Table 8.2 by means of a relatively straightforward formula. First we calculate

$$r = \frac{n\Sigma xy - (\Sigma x)(\Sigma y)}{\sqrt{n\Sigma x^2 - (\Sigma x)^2}\sqrt{n\Sigma y^2 - (\Sigma y)^2}} \tag{8.3}$$

which is called the *coefficient of linear correlation*. (In the next section, we will discuss additional ways of making use of the correlation coefficient.) Then the coefficient of linear determination is given by the following formula:

THE COEFFICIENT OF LINEAR DETERMINATION

$$\frac{VR}{TV} = \frac{SSR}{SS_{total}} = r^2$$

Table 8.2 contains all the basic information needed to compute the correlation coefficient r in our example:

$$r = \frac{5(3,780) - (250)(80)}{\sqrt{5(13,500) - (250)^2}\sqrt{5(1,330) - (80)^2}}$$

$$= \frac{18,900 - 20,000}{\sqrt{67,500 - 62,500}\sqrt{6,650 - 6,400}} = \frac{-1,100}{\sqrt{5,000}\sqrt{250}}$$

$$= \frac{-1,100}{(70.71)(15.81)} = -.984$$

Because $r = -.984$, the coefficient of determination is given by

$$\frac{VR}{TV} = \frac{SSR}{SS_{total}} = r^2 = (-.984)^2 = .968 = 96.8\%.$$

You should observe that we get exactly the same value for VR/TV using the correlation coefficient as we obtained earlier by the direct method of Table 8.4.

Before going on to the next topic, let's calculate coefficients of determination for the three sets of nonlinear data appearing in Table 8.3 and Fig.

8.8. As we will see, they contrast markedly with the 96.8% of the engine efficiency example. These nonlinear examples show that the coefficient of linear determination is *not* a good measure of the strength of a relationship *unless* that relationship is a linear one. We will begin with parabolic data.

Parabolic Data

For the parabolic (that is, parabolic in shape) data of Table 8.3, column a, we first make the preliminary calculations appearing in Table 8.5. We then calculate r:

$$ r = \frac{n\Sigma xy - (\Sigma x)(\Sigma y)}{\sqrt{n\Sigma x^2 - (\Sigma x)^2}\sqrt{n\Sigma y^2 - (\Sigma y)^2}} $$

$$ = \frac{6(126) - (25)(40)}{\sqrt{6(139) - (25)^2}\sqrt{6(454) - (40)^2}} $$

$$ = \frac{-244}{\sqrt{209}\sqrt{1,124}} = \frac{-244}{(14.46)(33.53)} = .503. $$

Therefore, $r^2 = (-.503)^2 = .253 = 25.3\%$ of the variation in y can be attributed to the best possible linear relationship with x. It follows that the regression line fails to acount for almost 75% of the variation in y. These numbers provide a strong hint that x and y are really not linearly related. Therefore, no attempt should be made to find the best-fitting straight line, for even that one will not fit very well. (Note that $r^2 = 25.3\%$ does not represent the strength of the *parabolic* relationship.)

TABLE 8.5   Preliminary Calculations for Parabolic Data

x	y	x^2	y^2	xy
8	9	64	81	72
1	16	1	256	16
6	1	36	1	6
3	4	9	16	12
5	0	25	0	0
2	10	4	100	20
25	40	139	454	126

n = 6

TABLE 8.6.   Preliminary Calculations for Logarithmic Data

x	y	x²	y²	xy
1	1	1	1	1
3	4	9	16	12
8	8	64	64	64
14	9.5	196	90.25	133
20	10	400	100	200
10	9	100	81	90
5	6	25	36	30
2	3	4	9	6
63	50.5	799	397.25	536

n = 8

Logarithmic Data

For the logarithmic data of Table 8.3, column b, the preliminary calculations appear in Table 8.6. We have

$$r = \frac{n\Sigma xy - (\Sigma x)(\Sigma y)}{\sqrt{n\Sigma x^2 - (\Sigma x)^2}\sqrt{n\Sigma y^2 - (\Sigma y)^2}}$$

$$= \frac{8(536) - (63)(50.5)}{\sqrt{8(799) - (63)^2}\sqrt{8(397.25) - (50.5)^2}}$$

$$= \frac{1{,}106.5}{\sqrt{2{,}423}\sqrt{627.75}} = \frac{1{,}106.5}{(49.22)(25.05)} = .897.$$

Therefore the coefficient of determination is

$$r^2 = (.897)^2 = .805 = 80.5\%,$$

indicating that the regression line accounts for all but about 20% of the variation in y. You should observe that the logarithmic data is considerably more linear than the parabolic data having $r^2 = 25.3\%$, but appreciably less linear than the engine efficiency data where $r^2 = 96.8\%$.

Exponential Data

Finally, considering the exponential data of Table 8.3, column c, we make the preliminary calculations in Table 8.7. As usual, we first compute the correlation coefficient r:

$$r = \frac{n\Sigma xy - (\Sigma x)(\Sigma y)}{\sqrt{n\Sigma x^2 - (\Sigma x)^2}\sqrt{n\Sigma y^2 - (\Sigma y)^2}}$$

$$= \frac{6(53) - (21)(23)}{\sqrt{6(91) - (21)^2}\sqrt{6(143) - (23)^2}}$$

$$= \frac{-165}{\sqrt{105}\sqrt{329}} = \frac{-165}{(10.25)(18.14)} = -.888.$$

Therefore, the coefficient of determination is given by $r^2 = (-.888)^2 = .789 = 78.9\%$, again more linear than the parabolic data, but less linear than the linear data.

One more remark: In addition to its use in simplifying the calculation of the coefficient of determination, the correlation coefficient r immediately gives information about the direction of the trend of the data. In particular, if r turns out to be negative, this means that the regression line slopes downward to the right, in the manner of the scattergrams in Figs. 8.7 and 8.8, part c. If, on the other hand, r is positive, the line slopes upward to the right, as in Figs. 8.3, 8.5, and 8.8, part b.

The reason for this association of r with slope is the close algebraic relationship between the formulas for r and b. In fact,

$$r = \frac{n\Sigma xy - (\Sigma x)(\Sigma y)}{\sqrt{n\Sigma x^2 - (\Sigma x)^2}\sqrt{n\Sigma y^2 - (\Sigma y)^2}} = \frac{b\sqrt{n\Sigma x^2 - (\Sigma x)^2}}{\sqrt{n\Sigma y^2 - (\Sigma y)^2}},$$

so that the algebraic sign of r ($+$ or $-$) is always the same as the sign of b. But, for $y = a + bx$, when r (and so b) is negative, y grows smaller as x grows larger so the line slopes downward. Analogously, if r (and so b) is positive, y grows larger as x grows larger, so the line slopes upward. Therefore, taking note of the sign of r and its square, the coefficient of determination, we can form a good mental picture of the regression line and how well the data fit it.

TABLE 8.7 Preliminary Calculations for Exponential Data

x	y	x^2	y^2	xy
5	2	25	4	10
3	3	9	9	9
1	10	1	100	10
2	5	4	25	10
4	2	16	4	8
6	1	36	1	6
21	23	91	143	53

n = 6

422   LINEAR REGRESSION AND CORRELATION

EXERCISES 8.C

8.17. Based on the data presented in Exercise 8.7, what proportion of the variation in total sales can be explained by a linear relationship between total sales and display space?

8.18. From the data of Exercise 8.8, how much of the variation in yield of beans can be attributed to a linear relationship with the amount of fertilizer?

8.19. On the basis of the data given in Exercise 8.9 and the linear relationship described there, how much of the variation in the divorce rate can be ascribed to corresponding variations in the inflation rate?

8.20. Using the demography data of Exercise 8.10, what percentage of the variation in birth rates can be explained in terms of variations in the protein content of national diets and their linear relationship with the national birth rate?

8.21. Use the productivity data of Exercise 8.11 to calculate the proportion of variation in accumulated productivity that can be attributed to the passage of time during the workday.

8.22. Calculate the percentage of variation in the total cost of the production run in Exercise 8.12 that can be explained on the basis of variations in the quantity produced.

8.23. Based on the steel demand data of Exercise 8.13, determine the percentage of variation in the price per ton that can be accounted for by variations in the quantity that can be sold.

8.24. Use the data of Exercise 8.14 to find out how much of the variation in U.S. gross national product during the years studied can be mathematically explained on the basis of variations in Japan's gross national product.

8.25. Calculate, using the information in Exercise 8.15, the proportion of variation in the 1977 effective federal income tax rates that can be attributed to variations in adjusted gross incomes.

8.26. What proportion of the variation in aggregate outlays for new homes can be explained by a linear relationship with non-real estate mortgage debt, based on the data of Exercise 8.16?

8.27. In an attempt to develop an aptitude test measuring an individual's aptitude for pursuing a career in computer programming, a personnel officer first compares mathematical aptitude scores of already employed computer programmers with their job performance ratings. The data follow.

Person	A	B	C	D	E	F	G
Math aptitude score	2	5	0	4	3	1	6
Job performance rating	8	5	1	8	9	5	1

a. Draw a scattergram of the data.

b. Calculate the coefficient of linear determination.

c. What proportion of variation in job performance ratings can be explained on the basis of a linear relationship with mathematical aptitude scores?

8.28. Demand for Napoleon rubies increases as the price goes down, because more people are able to buy them. When the price is high, demand is also high since they are a much sought-after status symbol. At moderate prices, however, not too many are sold, because the price is too high for many people but yet not high enough to result in demand as status symbols. In particular, some recent data relating price and demand of these items went as follows.

Price per carat, hundreds of dollars	0.5	1	3.5	5	7	8
Demand, thousands	160	130	40	50	130	200

a. Construct the scattergram of the data.

b. What percentage of the variation in demand for Napoleon rubies can be attributed to the best possible linear relationship between price and demand?

8.29. In a study of the efficiency of automobile engines with respect to consumption of gasoline, one subcompact car was operated at various speeds, and the fuel consumption (in miles per gallon) was carefully measured. The data are as follows.

Speed, miles/hour	10	20	30	40	50	60
Efficiency, miles/gallon	15	20	30	35	25	20

a. Draw a scattergram of the data.

b. Calculate the proportion of variation in efficiency that can be explained on the basis of a linear relationship between speed and efficiency.

8.30. A study of air passenger traffic on routes across the North Atlantic in June 1977, revealed the following information on the number of passengers carried by randomly selected airlines on their scheduled flights and charter flights:*

Airline	Scheduled	Charter
Irish International	17,162	8,737
British Airways	122,698	17,600
Air India	14,584	0
Trans World Airlines	188,815	36,179
El Al	26,760	0
Air Canada	90,391	23,194
Alitalia	39,844	8,186

a. Construct a scattergram of the data.

*Based on information reported in *Aviation Week & Space Technology*, November 14, 1977, page 34.

    b. What proportion of the variation among airlines in number of charter passengers can be explained on the basis of a linear relationship with the airline's number of passengers on its regularly scheduled flights?

SECTION 8.D   TESTING FOR THE EXISTENCE OF A LINEAR RELATIONSHIP

The correlation coefficient (for historical reasons, sometimes referred to as the *Pearson product-moment correlation coefficient*) can be used to test a statistical hypothesis regarding whether or not a linear relationship exists within the population from which a sample of data points was drawn.

    For the validity of the small-sample *correlation test* of linearity, it is necessary that both the x and y values be drawn from populations that can reasonably be assumed to have normal distributions. Under the required assumption of normality, if any relationship at all exists between x and y, then that relationship must be a linear one. (This fact is a consequence of advanced statistical theory.) The existence of such a linear relationship is expressed by a parameter ρ (pronounced "row"), which is the true coefficient of linear correlation. The correlation coefficient r is merely a sample estimate of ρ based on n data points.

    We want to test the hypothesis

    H: $\rho = 0$ (no correlation—there is no linear relationship),

against the alternative

    A: $\rho \neq 0$ (significant correlation—there is a significant underlying linear relationship).

If ρ were really near 0, r would most likely be small, and so VR/TV = r^2 would represent a very small percentage of the total variation of y. This would indicate that a relationship between x and y does not exist, and that x and y are in fact independent. On the other hand, a significantly large value of ρ, in either the positive or negative direction, would be consistent with a high percentage of the variation of y being accounted for by the regression line. So rejection of H: $\rho = 0$ in favor of A: $\rho \neq 0$ would support the existence of a significant linear relationship in the underlying popuation.

    For simplicity, we can abbreviate our testing problem as:

    H: $\rho = 0$ (no linear relationship)

vs.

    A: $\rho \neq 0$ (a significant relationship which must be linear).

TABLE 8.8   Correlation Coefficients of Sets of Data

Data Set	n	r
Engine efficiency	5	−.984
Parabolic	6	−.503
Logarithmic	8	.897
Exponential	6	−.888

TABLE 8.9   Comparison t values for Correlation Test, $t_{\alpha/2}[n-2]$

n	n−2	$\alpha = .10$	$\alpha = .05$	$\alpha = .01$
5	3	2.353	3.182	5.841
6	4	2.132	2.776	4.604
8	6	1.943	2.447	3.707

To test H against A, we use the test statistic

$$t = \frac{r\sqrt{n-2}}{\sqrt{1-r^2}} \qquad (8.4)$$

which has been shown by advanced theoretical analysis to have the t distribution with $n-2$ degrees of freedom. We would therefore reject H (nonlinearity) in favor of A (linearity) at significance level α if

$$|t| > t_{\alpha/2}[n-2].$$

Let's apply the technique to each of the four sets of data discussed in the previous section. (It is important to observe that r will always fall between −1 and +1 for any set of data. This is due to the fact that r^2 is a proportion and so must always lie between 0 and 1. *A value of r outside the range −1 to +1 signals the presence of a computational error.*) The first step in carrying out the test of linearity is to look up the comparison values $t_{\alpha/2}[n-2]$. We obtain these from Table A.7 in the Appendix, for various levels of α, and compile them in Table 8.9.

It remains only to calculate, from Table 8.8, the numerical value of the test statistic of formula (8.4) and to compare it against the corresponding t value listed in Table 8.9.

Engine Efficiency Data

For the engine efficiency example, we have

$$t = \frac{r\sqrt{n-2}}{\sqrt{1-r^2}} = \frac{(-.984)\sqrt{5-2}}{\sqrt{1-(-.984)^2}} = \frac{(-.984)\sqrt{3}}{\sqrt{1-.968}} = \frac{(-.984)\sqrt{3}}{\sqrt{.0317}}$$

$$= \frac{(-.984)(1.732)}{(.178)} = -9.6$$

We are to reject

H: $\rho = 0$ (no linear relationship)

in favor of

A: $\rho \neq 0$ (a significant linear relationship),

if

$$|t| > t_{\alpha/2}[n-2].$$

Clearly, $|t| = 9.6$ exceeds $t_{\alpha/2}[3]$ for all reasonable levels of α, as can be seen from the top row of numbers in Table 8.9. This means that, whatever your preselected level of α (within the reasonable range), you would be led to reject H in favor of A, concluding that efficiency is related to speed by a linear relationship. (That relationship is, of course, the one given by the regression line $y = 27 - .22x$ of Section 8.B.)

Parabolic Data

If we were to assume that the parabolic data came from a normally distributed population, Table 8.8 indicates that

$$t = \frac{r\sqrt{n-2}}{\sqrt{1-r^2}} = \frac{(-.503)\sqrt{6-2}}{\sqrt{1-(-.503)^2}} = \frac{(-.503)\sqrt{4}}{\sqrt{1-.253}} = \frac{(-.503)(2)}{\sqrt{.747}}$$

$$= \frac{-1.006}{.864} = -1.16.$$

Here $|t| = 1.16$, which fails to exceed $t_{\alpha/2}[4]$, recorded in the middle row of Table 8.9 for any reasonable level of α. Therefore, we would fail to reject H at every reasonable α, concluding that we have no evidence to indicate the presence of a linear or any other relationship in the underlying population.

Logarithmic Data

Making the same assumption for the logarithmic data, we see from Table 8.8 that

$$t = \frac{r\sqrt{n-2}}{\sqrt{1-r^2}} = \frac{(.897)\sqrt{8-2}}{\sqrt{1-(.897)^2}} = 4.97.$$

Here $|t| = 4.97$, which exceeds the values of $t_{\alpha/2}[6]$ recorded in the bottom row of Table 8.9 for every reasonable α. Therefore we would reject H and conclude, at every reasonable level α, that there is significant evidence of a linear relationship between x and y in the underlying normal population. In fact, looking at its scattergram in Fig. 8.8, part b, we see that it is only slightly nonlinear. While a logarithmic curve would describe the data best, our correlation test of linearity shows that the best regression line would not be very bad.

Exponential Data

A somewhat similar situation prevails in the case of the exponential data whose scattergram appears in Fig. 8.8, part c. Here

$$t = \frac{r\sqrt{n-2}}{\sqrt{1-r^2}} = \frac{(-.888)\sqrt{6-2}}{\sqrt{1-(.888)^2}} = -3.86,$$

so that $|t| = 3.86$. The comparison t values in the middle row of Table 8.9 for $n = 6$ show that we should reject H if $\alpha = .10$ or $\alpha = .05$, but we should not reject H if $\alpha = .01$. We therefore have one more example of how the subjective choice of α and the assumption of a normal distribution influences the outcome of a statistical decision process.

   If it is known that the x and y data points come from underlying populations that are normally distributed, the correlation test of linearity should be performed before any calculation of the regression line is made. Such a procedure will protect against the waste of time and effort involved in calculating a regression line that will be essentially useless in case the data is later judged to be nonlinear. (In fact, it is not entirely useless. A comparison of the regression line with the actual data may provide some hints as to what to do next in the analysis. See Chapter 14 for a discussion of this topic.)

   When it is not reasonable to assume normality in the underlying populations, the best we can do is to use the coefficient of linear determination. Of course, in the latter situation, we cannot obtain the degree of precision in significance levels that is available in the case of normality.

EXERCISES 8.D

8.31. At level of significance $\alpha = .05$, does it seem that a (linear) relationship exists between display space and total sales as discussed in Exercises 8.7 and 8.17?

8.32. At significance level $\alpha = .05$, may we assume that there is a linear relationship between amount of fertilizer and bushels of beans studied earlier in Exercises 8.8 and 8.18?

8.33. At level $\alpha = .01$, does it make sense to use a linear equation to describe the relationship between inflation rate and divorce rate based on the data analyzed in Exercises 8.9 and 8.19?

8.34. Using a level of significance of $\alpha = .01$, does a (linear) relationship between protein content of diet and national birth rate, based on the data of Exercises 8.10 and 8.20, appear to exist?

8.35. At level $\alpha = .05$, does the cost curve based on the data of Exercises 8.12 and 8.22 seem to be linear?

8.36. Can the demand curve of Exercises 8.13 and 8.23 be considered linear at level $\alpha = .05$?

8.37. At level of significance $\alpha = .10$, can we validly use a linear equation to describe the relationship between mathematical aptitude and job performance in computer programming, based on the data of Exercise 8.27?

8.38. An analysis of airline revenues received from passenger-oriented and freight-oriented operations uncovered the following data in millions of dollars for seven domestic trunk lines.*

	Source of Revenues	
Airline	Passengers	Freight
Eastern	383.6	12.4
Northwest	159.9	18.9
Braniff	135.9	7.1
Western	139.2	8.4
Continental	137.6	12.9
Delta	409.5	22.9
National	112.7	5.0

a. Draw a scattergram of the data.

b. Assuming that the data come from underlying normal populations, test at level $\alpha = .05$, whether or not the volume of revenues generated by the two sources are linearly related.

*Selected from information presented in *Aviation Week & Space Technology*, September 19, 1977, page 37.

TABLE 8.10   Predicted Efficiency Levels for Various Speeds

Speed (x) (mph)	Computation 27 − .22x	Predicted Efficiency (y) (mpg)
25	27 − (.22)(25)	21.5
55	27 − (.22)(55)	14.9
80	27 − (.22)(80)	9.4

SECTION 8.E   CONFIDENCE INTERVALS FOR PREDICTIONS

To complete our discussion of linear regression and correlation, we now tackle the fourth question raised at the beginning of the chapter, "How can we use the data to predict the efficiency at other speeds, for example 25, 55, or 80, and how accurate will our predictions be?"

We first recall from Section 8.B that the regression line relating speed (x) and engine efficiency (y) has equation $y = 27 - .22x$. Fig. 8.7 represents this relationship graphically, illustrating the linear trend of variation in engine efficiency relative to variations in speed. What this means is that, insofar as we are able to determine from the five data points, when the speed is x miles per hour, our best estimate of the corresponding engine efficiency level will be $27 - .22x$ miles per gallon. In Table 8.10, we use the equation of the regression line to predict the efficiency level for the speeds 25, 55, and 80. We see that at 25 mph, we predict an efficiency of 21.5 mpg; at 55 mph, we predict 14.9 mpg; and at 80 mph, we predict 9.4 mpg. For any desired value x of speed, a predicted value y of engine efficiency can be worked out in the same way.

Now that we have predicted efficiency to be 14.9 mpg when the speed is 55 mph, it is important to know how accurate our prediction can claim to be. This problem is comparable to the one discussed in Chapter 5, where we were trying to estimate the true mean μ of a normally distributed population from a sample of n data points $x_1, x_2, \ldots, x_n$. There we concluded that the sample mean $\bar{x} = (\Sigma x)/n$ is used as an estimate of μ, with the accuracy of the estimate expressed by a confidence interval: we can be $(1-\alpha)100\%$ sure that

$$\bar{x} - t_{\alpha/2}[n-1]\frac{s}{\sqrt{n}} \le \mu \le \bar{x} + t_{\alpha/2}[n-1]\frac{s}{\sqrt{n}},$$

where

$$s = \sqrt{\frac{\Sigma(x - \bar{x})^2}{n-1}}$$

is the sample standard deviation of the n data points.

If in the regression situation, we can reasonably assume for each value of x that the possible y data points are normally distributed about the true value of y lying on the regression line, we can set up confidence intervals. (From Fig. 8.7 we can see that the assumption of normality is not unreasonable; two possible bits of evidence for normality are the facts that: (1) some data points fall above the regression line, while others fall below it; and (2) most points fall close to the line, while relatively few fall far away.) The normality assumption indicates, for x = 55 and its predicted value y = 14.9, that the true engine efficiency for a speed of 55 mph is as likely to be above 14.9 mpg as below it, but more likely to be near 14.9 mpg than far away from it.

The number that plays the role of the sample standard deviation, informing us as to how variable the possible values of y are likely to be from the predicted value, is called the standard error of the estimate (s_e) and is given by the following formula:

STANDARD ERROR OF THE ESTIMATE

$$s_e = \sqrt{\frac{\Sigma(y - a - bx)^2}{n-2}}$$

The numerator $\Sigma(y-a-bx)^2$, what we have called SSD ("sum of squared deviations") or SSE ("error sum of squares"), is analogous to $\Sigma(x-\bar{x})^2$ for a single set of data points. Just as there is a short-cut formula for the standard deviation, there is a more easily usable formula for s_e, which is as follows:

$$s_e = \sqrt{\frac{\Sigma y^2 - a\Sigma y - b\Sigma xy}{n-2}} \tag{8.5}$$

Using the preliminary calculations in Table 8.2 and the facts that $a = 27$ and $b = -.22$, we can compute the standard error of the estimate for the engine efficiency problem. We have

$$s_e = \sqrt{\frac{\Sigma y^2 - a\Sigma y - b\Sigma xy}{n-2}} = \sqrt{\frac{1{,}330 - (27)(80) - (-.22)(3{,}780)}{5-2}}$$

$$= \sqrt{\frac{1{,}330 - 2{,}160 - (-831.6)}{3}} = \sqrt{\frac{1{,}330 - 2{,}160 + 831.6}{3}}$$

$$= \sqrt{\frac{1.6}{3}} = \sqrt{.533} = .73.$$

Because the numerator under the square root sign is really SSD, namely a sum of squares, it can never be negative. *Therefore, a negative number under the square root sign in s_e is a signal of an error in arithmetic or, possibly, too much roundoff in some calculations.*

Assuming, then, that y varies in a normally distributed manner about its true value, we can write down the formula for a $(1-\alpha)100\%$ confidence interval for the true value of y at a given value of x. We denote by x_o the value of x at which we are interested in predicting y. In the engine efficiency example, we should take $x_o = 55$ because we want to predict the engine efficiency at a speed of 55 mph. The predicted engine efficiency is then $a + bx_o = 27 - (.22)(55) = 14.9$. If we denote the true value of y as y_o, when $x = x_o$, we can be $(1-\alpha)100\%$ sure that

$$a + bx_o - E_o \le y_o \le a + bx_o + E_o, \tag{8.6}$$

where the probable bound E_o on the error is given by

$$E_o = t_{\alpha/2}[n-2]\, s_e \sqrt{1 + \frac{1}{n} + \frac{n(x_o - \bar{x})^2}{n\Sigma x^2 - (\Sigma x)^2}}. \tag{8.7}$$

In predicting efficiency when the speed is 55 mph, we need the following components for working out a 90% confidence interval for the efficiency:

$$x_o = 55$$
$$a + bx_o = 27 - (.22)(55) = 14.9$$

$$\left.\begin{array}{l} \alpha = .10 \\ \alpha/2 = .10/2 = .05 \\ n = 5 \end{array}\right\} \quad t_{\alpha/2}[n-2] = t_{.05}[3] = 2.353$$

$$s_e = .73$$
$$\bar{x} = \Sigma x/n = 250/5 = 50$$
$$n(x_o - \bar{x})^2 = 5(55 - 50)^2 = 5(5)^2 = 125$$
$$n\Sigma x^2 - (\Sigma x)^2 = 5(13,500) - (250)^2 = 5,000,$$

using the calculations of Table 8.2 wherever necessary. We then have

$$E_o = t_{\alpha/2}[n-2]\, s_e \sqrt{1 + \frac{1}{n} + \frac{n(x_o - \bar{x})^2}{n\Sigma x^2 - (\Sigma x)^2}}$$

$$= (2.353)(.73) \sqrt{1 + \frac{1}{5} + \frac{125}{5,000}}$$

$$= (1.718) \sqrt{1 + .2 + .025} = (1.718) \sqrt{1.225}$$

$$= (1.718)(1.108) = 1.9.$$

Therefore we obtain our 90% confidence interval as follows:

$$a + bx_0 - E_0 \leq y_0 \leq a + bx_0 + E_0,$$

namely that

$$14.9 - 1.9 \leq y_0 \leq 14.9 + 1.9,$$

or, finally,

$$13.0 \leq y_0 \leq 16.8.$$

In words, we can state our conclusion as follows: If the car under study were to be driven at a speed of 55 miles per hour, the interval 13.0 to 16.8 miles per gallon of gasoline is a 90% confidence interval for the predicted mileage rate.

The key component of the formula for the error E_0 is the term

$$n(x_0 - \bar{x})^2 .$$

This term is the only part of E_0 that involves the number x_0. The quantity $(x_0 - \bar{x})^2$ is the squared distance between x_0 and $\bar{x}$; and so the term $n(x_0 - \bar{x})^2$ is small for values of x_0 that are close to $\bar{x}$, and larger for values of x_0 that are farther away from $\bar{x}$. This means that the farther x_0 is from $\bar{x}$, the larger E_0 itself is. In fact, as we consider values of x_0 farther and farther away from $\bar{x}$, the possible error E_0 grows in size, so our estimate of y_0 correspondingly becomes increasingly unreliable. This situation, which is illustrated graphically in Fig. 8.9, points up the fact that a prediction based on the regression line is most accurate only when it involves numbers well within the range of the original data points, and it loses accuracy as it proceeds beyond that range.

For example, let's consider the accuracy of the three predictions recorded in Table 8.10. For 90% confidence intervals, all components of the formula for E_0 are the same as in the above calculation for $x_0 = 55$, except the various values of x_0 itself. In particular, we have

$$E_0 = (2.353)(.73) \sqrt{1 + \frac{1}{5} + \frac{5(x_0 - 50)^2}{5,000}}$$

for all possible 90% confidence intervals involved in the engine efficiency problem. When $x_0 = 55$, we have already shown above that $E_0 = 1.9$. Other values of x_0 appearing in Table 8.10 are $x_0 = 25$ and $x_0 = 80$. We calculate E_0 for both of these values. For $x_0 = 25$,

$$E_0 = (1.718) \sqrt{1 + .2 + \frac{5(25 - 50)^2}{5,000}} = (1.718) \sqrt{1 + .2 + .625}$$

$$= (1.718) \sqrt{1.825} = (1.718)(1.35) = 2.32,$$

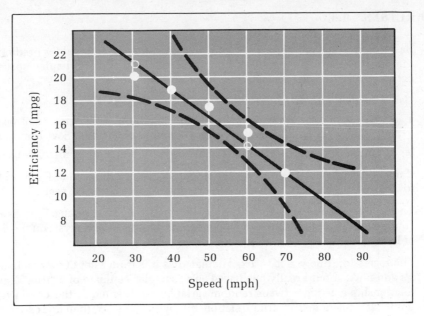

FIGURE 8.9   90% Confidence Intervals for Predicting Efficiency at Various Speeds

and for $x_o = 80$,

$$E_o = (1.718) \sqrt{1 + .2 + \frac{5(80 - 50)^2}{5,000}} = (1.718) \sqrt{1 + .2 + .9}$$

$$= (1.718) \sqrt{2.1} = (1.718)(1.45) = 2.49.$$

We summarize these results in Table 8.11.

The increasing size of the 90% confidence interval, as x_o moves away from the central location of the original data is illustrated in Fig. 8.9. Note that the confidence *band* is at its narrowest for $x_o = \Sigma x/n = 50$, where it extends a length 1.88 in both directions, and as x_o lies farther and farther away from 50, the band widens more and more.

TABLE 8.11   Errors of Prediction for Various Speeds

Speed (x_o)	Predicted Efficiency	Probable Bound on Error (E_o)
25	21.5	2.32
55	14.9	1.90
80	9.4	2.49

EXERCISES 8.E

8.39. Based on the data of Exercise 8.7, find a 90% confidence interval for predicting the weekly sales of the brand of tuna involved if the allotted display space is eventually set at 6 square feet.

8.40. Using the data appearing in Exercise 8.8, find

    a. a 90% confidence interval for predicting the number of bushels of beans harvested from a plot of land on which 6 pounds of fertilizer have been used;

    b. an 80% confidence interval for the harvest of beans if no fertilizer is used.

8.41. If next year's inflation rate looks like it will be 8%, use the data of Exercise 8.9 to calculate a 95% confidence interval for the divorce rate next year.

8.42. The typical protein content of the daily diet in Greece is 18 grams. Find a 90% confidence interval for the birth rate in Greece, basing your analysis on the data of Exercise 8.10.

8.43. The regression line $y = a + bx$, calculated as it is from a set of data points, n x's and n y's, should really be viewed as a "sample" estimate of a "true" linear relationship existing between random variables represented by the x and y data points. If we denote this "true" relationship by $y = \alpha + \beta x$, then α is called the *true intercept* and β is called the *true slope* of the regression line. The numbers a and b are then *estimates* of these true values α and β. The question then arises of how good an estimate of α is a and how good an estimate of β is b. The answers are given by the following confidence intervals.

    We can be $(1-\alpha)100\%$ sure that

$$a - E_a \leqslant \alpha \leqslant a + E_a,$$

where

$$E_a = t_{\alpha/2}[n-2]s_e \sqrt{\frac{1}{n} + \frac{n \cdot \bar{x}^2}{n(\Sigma x^2) - (\Sigma x)^2}},$$

and we can be $(1-\alpha)100\%$ sure that

$$b - E_b \leqslant \beta \leqslant b + E_b,$$

where

$$E_b = t_{\alpha/2}[n-2]s_e \sqrt{\frac{n}{n(\Sigma x^2) - (\Sigma x)^2}}.$$

(Unfortunately, since a knowledge of b is necessary for the calculation of a in formula (8.2), it would not be proper to calculate both of the above confidence intervals using the same set of data. We would need two independently selected sets of data in order to get independent confidence intervals for both α and β using the above formulas.) Based on the data of Exercise 8.7, find a 90% confidence interval for the true intercept α used for predicting weekly tuna sales.

8.44. From the data of Exercise 8.8, find an 80% confidence interval for the true slope β of the regression line used in predicting the size of the bean harvest.

8.45. Use the data of Exercise 8.9 to find a 90% confidence interval for the true slope β for predicting next year's divorce rate.

8.46. Find a 90% confidence interval for the true intercept α for predicting birth rates, using the data of Exercise 8.10.

8.47. The probable bound E_o given by formula (8.7) allows us to calculate confidence intervals for a particular value of y corresponding to our value of x. Suppose we need instead a confidence interval for the mean (or "expected") value of y corresponding to a certain value of x. The appropriate formula for the mean value of y at a certain x level would have us replace E_o of formula (8.7) by

$$E_\mu = t_{\alpha/2}[n-2]s_e\sqrt{\frac{1}{n} + \frac{n(x_o-\overline{x})^2}{n\Sigma x^2 - (\Sigma x)^2}}$$

and then we can be $(1-\alpha)100\%$ sure that

$$a + bx_o - E_\mu \leq y_\mu \leq a + bx_o + E_\mu$$

where y_μ denotes the mean value of y when $x = x_o$. Based on the data of Exercise 8.7, find a 90% confidence interval for the mean volume of tuna sales among those weeks when 6 square feet of display space are allotted.

8.48. Using the data appearing in Exercise 8.8, find a 90% confidence interval for the average number of bushels of beans harvested from a plot of land on which 6 pounds of fertilizer have been used.

8.49. Find a 95% confidence interval for the expected divorce rate in a year in which the inflation rate is 8%. Use the data of Exercise 8.9.

8.50. Find a 90% confidence interval for the average birth rate in countries having a typical protein content of 18 grams in the daily diet. Base your analysis on the data of Exercise 8.10.

SECTION 8.F COMPLETE REGRESSION ANALYSIS OF A BUSINESS PROBLEM

The many topics introduced in this chapter often appear in practice as merely components of a single problem and its comprehensive solution. In this final section of Chapter 8, we present an example of a regression-correlation problem, completely solved in a logical sequence of steps from beginning to end.

EXAMPLE 8.1. Real Estate Values. An investment analyst conducted a study aimed at determining how prices for a certain category of real estate vary with the passage of time. She looked up some data on a random selection of properties that had been sold recently and computed the percentage increase in price over the last time the property had changed hands. The data follow.

Property	A	B	C	D	E	F
Time period (quarters)	3	5	1	7	5	9
Growth rate (percentage increase in price)	10	13	8	13	11	15

The following questions are of interest to the investment analyst:

a. What proportion of the variation in growth rates of prices among the properties can be attributed to the best possible linear relationship with corresponding variations in the time periods?
b. What does the scattergram of the data look like?
c. At significance level $\alpha = .05$, can we consider the growth rate to be linearly related to the length of the time period?
d. If so, what is the equation of regression line that will predict growth rates for various given time periods?
e. Find a 90% confidence interval for the growth rate of the price of a property over a time period of 12 quarters.

SOLUTION.   a. To find the proportion of variation in growth rates of prices that can be explained on the basis of variations in the time period, we have to compute the coefficient of linear determination. We first set up the table of preliminary calculations, Table 8.12.

Here, the growth rate, the variable to be predicted, is denoted by y, while the time period is denoted by x.

The coefficient of determination is r^2, where

$$r = \frac{n\Sigma xy - (\Sigma x)(\Sigma y)}{\sqrt{n\Sigma x^2 - (\Sigma x)^2}\sqrt{n\Sigma y - (\Sigma y)^2}}$$

$$= \frac{6(384) - (30)(70)}{\sqrt{6(190) - (30)^2}\sqrt{6(848) - (70)^2}}$$

$$= \frac{2{,}304 - 2{,}100}{\sqrt{1{,}140 - 900}\sqrt{5{,}088 - 4{,}900}} = \frac{205}{\sqrt{240}\sqrt{188}}$$

$$= \frac{204}{(15.5)(13.7)} = .96 \ .$$

Therefore $r^2 = (.96)^2 = .922 = 92.2\%$ of the variation in growth rate can be attributed to the best linear relationship that exists between growth rate and time period.

b. The scattergram of the real estate data appears in Fig. 8.10.

c. We test the hypothesis

H: $\rho = 0$ (no linear relationship)

vs.

A: $\rho \neq 0$ (significant linear relationship)

at level $\alpha = .05$ by calculating

$$t = \frac{r\sqrt{n-2}}{\sqrt{1-r^2}}$$

TABLE 8.12.   Preliminary Calculations for Real Estate Example

Time Period x (quarters)	Growth Rate y (percentage increase)	x²	y²	xy
3	10	9	100	30
5	13	25	169	65
1	8	1	64	8
7	13	49	169	91
5	11	25	121	55
9	15	81	225	135
Sums 30	70	190	848	384

n = 6

FIGURE 8.10   Scattergram of Real Estate Data

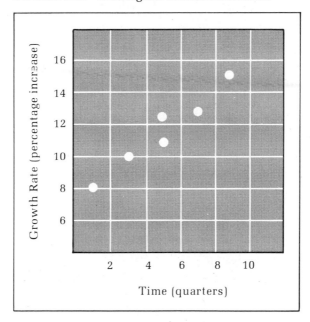

and rejecting H if $|t| > t_{\alpha/2}[n-2] = t_{.025}[4] = 2.776$. We have

$$t = \frac{r\sqrt{n-2}}{\sqrt{1-r^2}} = \frac{(.96)\sqrt{6-2}}{\sqrt{1-(.96)^2}} = \frac{(.96)\sqrt{4}}{\sqrt{1-.9216}} = \frac{1.92}{\sqrt{.0784}}$$

$$= 1.92/.28 = 6.857.$$

Therefore $|t| = 6.857 > 2.776$, so we reject H and conclude at level $\alpha = .05$ that growth rate and length of time period seem to be linearly related.

d. The regression line has equation $y = a + bx$, where y represents growth rate (the variable to be predicted) and x represents length of time period. Here

$$b = \frac{n\Sigma xy - (\Sigma x)(\Sigma y)}{n\Sigma x^2 - (\Sigma x)^2} = \frac{204}{240} = .85$$

where we have obtained the quantities $n\Sigma xy - (\Sigma x)(\Sigma y) = 204$ and $n\Sigma x^2 - (\Sigma x)^2 = 240$ not by computation, but by merely looking their values up among the calculations we made to find the correlation coefficient r. Notice that the numerator in the formula for b is exactly the same as the numerator in the formula for r, while the denominator of b is one of the factors appearing under a square root sign in the expression for r. Now

$$a = \frac{\Sigma y - b\Sigma x}{n} = \frac{70 - (.85)(30)}{6} = \frac{70 - 25.5}{6} = \frac{44.5}{6} = 7.42 .$$

The regression line therefore has equation $y = 7.42 + .85x$.

As mentioned before, for illustrative purposes, as well as to provide a check on our calculations, we often superimpose the graph of the regression line upon the scattergram. To do this, we choose two values of x within the range of the original data, say $x = 4$ and $x = 10$. If $x = 4$, then $y = 7.42 + (.85)(4) = 10.82$, so that $(4, 10.82)$ lies on the regression line, and if $x = 10$, then $y = 7.42 + (.85)(10) = 15.92$, so that $(10, 15.92)$ lies on the regression line. The regression line (formed by connecting these two points) is shown superimposed on the scattergram in Fig. 8.11.

e. To find a 90% confidence interval for the percentage increase y_o of a property that resold after a time period of 12 quarters, we set $\alpha = .10$ and $x_o = 12$ and then proceed to accumulate all the components of the confidence interval formulas $a + bx_o$ and

$$E_o = t_{\alpha/2}[n-2]\, s_e \sqrt{1 + \frac{1}{n} + \frac{n(x_o - \overline{x})^2}{n\Sigma x^2 - (\Sigma x)^2}} .$$

We list the components as follows:

$a + bx_o = 7.42 + (.85)(12) = 17.62$.

$t_{\alpha/2}[n-2] = t_{.05}[4] = 2.132$.

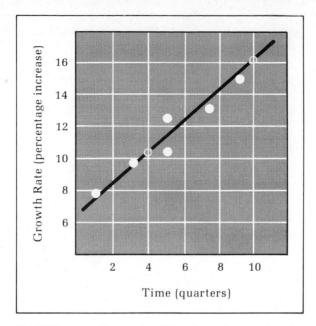

FIGURE 8.11   Regression Line Superimposed on Scattergram

$$s_e = \sqrt{\frac{\Sigma y^2 - a\Sigma y - b\Sigma xy}{n-2}} = \sqrt{\frac{848 - (7.42)(70) - (.85)(384)}{6-2}}$$

$$= \sqrt{\frac{848 - 519.4 - 326.4}{4}} = \sqrt{\frac{2.2}{4}} = \sqrt{.55} = .742$$

$$n(x_o - \bar{x})^2 = 6\left(12 - \frac{30}{6}\right)^2 = 6(12-5)^2 = 6(7)^2 = 6(49) = 294$$

$n\Sigma x^2 - (\Sigma x)^2 = 240$ (from the expression for r) .

Therefore

$$E_o = (2.132)(.742)\sqrt{1 + \frac{1}{6} + \frac{294}{240}} = (1.582)\sqrt{1 + .167 + 1.225}$$

$$= (1.582)\sqrt{2.392} = (1.582)(1.547) = 2.45.$$

The 90% confidence interval is then

$$a + bx_o - E_o \le y_o \le a + bx_o + E_o$$

$$17.62 - 2.45 \le y_o \le 17.62 + 2.45$$

$$15.17 \le y_o \le 20.07$$

TABLE 8.13   BASIC Computer Program for Linear Regression and
Correlation

```
10      REMARK: LINEAR REGRESSION AND CORRELATION
20      DIM X(100),Y(100)
30      PRINT "HOW MANY PAIRS OF DATA ARE THERE";
40      INPUT N
50      X1 = 0
60      X2 = 0
70      Y1 = 0
80      Y2 = 0
90      P = 0
100     FOR K = 1 TO N
110        PRINT "WHAT ARE X,Y FOR PAIR";K;
120        INPUT X(K),Y(K)
130        REMARK: COMPUTE THE COLUMN TOTALS
140        X1 = X1 + X(K)
150        X2 = X2 + X(K)**2
160        Y1 = Y1 + Y(K)
170        Y2 = Y2 + Y(K)**2
180        P = P + X(K)*Y(K)
190        NEXT K
200     PRINT
210     PRINT "SIGMA X =";X1,"        SIGMA Y =";Y1
220     PRINT "SIGMA X2 =";X2,"       SIGMA Y2 =";Y2,"SIGMA XY =";P
230     PRINT
240     D9 = N*X2 - X1**2
250     N9 = N*P - X1*Y1
260     D8 = N*Y2 - Y1**2
270     REMARK: CALCULATE THE CORRELATION COEFFICIENT
280     R = N9/(SQR(D9*D8))
290     PRINT "COEFFICIENT OF LINEAR CORRELATION =";R
300     PRINT "COEFFICIENT OF LINEAR DETERMINATION =";R**2
310     PRINT (R**2)*100;"PERCENT OF VARIATION IN Y MAY BE ATTRIBUTED"
320     PRINT "TO ITS BEST LINEAR RELATIONSHIP WITH X."
330     PRINT
340     REMARK: CALCULATE EQUATION OF REGRESSION LINE
350     B = N9/D9
360     A = (Y1 - B*X1)/N
370     PRINT "EQUATION OF REGRESSION LINE IS Y =";A;"+  ";B;"X"
380     PRINT
390     PRINT "WANT TO FIND SOME POINTS ON LINE (ANSWER Y OR N)";
400     INPUT A$
410     IF A$ = "N" THEN 480
420     PRINT "WHAT IS THE X-VALUE (9999 TO STOP)";
430     INPUT X8
440     IF X8 = 9999 THEN 480
450     PRINT "IF X =";X8;", THEN Y =";A+B*X8;"."
```

We can therefore assert that the interval 15.17 to 20.07 is one of those
that contain the true percentage price increase at which a piece of property sells after 12 quarters.

A computer program written in the BASIC language for undertaking a
complete regression analysis appears in Table 8.13. The solution to Example

TABLE 8.13 (Continued)

```
460     PRINT "WHAT IS THE NEXT X-VALUE (9999 TO STOP)";
470     GO TO 430
480     PRINT
490     PRINT "WANT TO TEST FOR LINEARITY (ANSWER Y OR N)";
500     INPUT A1$
510     IF A1$ = "N" THEN 660
520     PRINT "WHAT IS YOUR LEVEL OF SIGNIFICANCE";
530     INPUT A1
540     PRINT "LOOK UP VALUE OF T";A1/2;"(";N-2;")."
550     PRINT "WHAT IS IT";
560     INPUT T1
570     T = R*SQR(N-2)/SQR(1-R**2)
580     IF ABS(T) > T1 THEN 630
590     PRINT "ABS(T) =";ABS(T);"<";T1;", SO THAT WE CANNOT REJECT H;"
600     PRINT "WE CONCLUDE THAT THERE IS NO SIGNIFICANT LINEAR"
610     PRINT "RELATIONSHIP."
620     GO TO 660
630     PRINT "ABS(T) =";ABS(T);">";T1;", SO THAT WE REJECT H,"
640     PRINT "CONCLUDING THEREBY THAT A SIGNIFICANT LINEAR"
650     PRINT "RELATIONSHIP EXISTS BETWEEN X AND Y."
660     PRINT
670     PRINT"NEED CONFIDENCE INTERVALS FOR PREDICTIONS(ANSWER Y OR N)";
680     INPUT A2$
690     IF A2$ = "N" THEN 920
700     S9 = SQR((Y2 - A*Y1 - B*P)/(N-2))
710     PRINT "STANDARD ERROR OF ESTIMATE IS";S9
720     PRINT "WHAT LEVEL OF CONFIDENCE DO YOU NEED";
730     INPUT C
740     IF C > 1 THEN 770
750     A9 = 1-C
760     GO TO 780
770     A9 = 1-(C/100)
780     PRINT "LOOK UP VALUE OF T";A9/2;"(";N-2;")."
790     PRINT "WHAT IS IT";
800     INPUT T9
810     PRINT "AT WHAT X-VALUE IS PREDICTION NEEDED (9999 TO STOP)";
820     INPUT X0
830     IF X0 = 9999 THEN 920
840     X9 = SQR(1 + (1/N) + (N*(X0-X1/N)**2)/D9)
850     E0 = T9*S9*X9
860     Y0 = A + B*X0
870     PRINT
880     PRINT "THE INTERVAL";Y0-E0;"TO";Y0+E0;
890     PRINT "IS A";100-100*A9;"PERCENT CONFIDENCE"
900     PRINT "INTERVAL FOR THE Y-VALUE WHEN THE X-VALUE IS";X0;"."
910     GO TO 810
920     END
```

8.1, as printed out by the computer program, appears in Table 8.14. Discrepancies between the computer output in Table 8.14 and the solution to the problem as worked out in the text itself are due to the fact that the computer carries many more decimal places of accuracy than can hand calculation.

TABLE 8.14   Computer Printout of Real Estate Example Solution

```
HOW MANY PAIRS OF DATA ARE THERE ? 6
WHAT ARE X,Y FOR PAIR 1    ? 3,10
WHAT ARE X,Y FOR PAIR 2    ? 5,13
WHAT ARE X,Y FOR PAIR 3    ? 1,8
WHAT ARE X,Y FOR PAIR 4    ? 7,13
WHAT ARE X,Y FOR PAIR 5    ? 5,11
WHAT ARE X,Y FOR PAIR 6    ? 9,15

SIGMA X = 30           SIGMA Y = 70
SIGMA X2 = 190         SIGMA Y2 = 848            SIGMA XY = 384

COEFFICIENT OF LINEAR CORRELATION = .960386
COEFFICIENT OF LINEAR DETERMINATION = .92234
 92.234 PERCENT OF VARIATION IN Y MAY BE ATTRIBUTED
TO ITS BEST LINEAR RELATIONSHIP WITH X.

EQUATION OF REGRESSION LINE IS Y = 7.41667 +   .85 X

WANT TO FIND SOME POINTS ON LINE (ANSWER Y OR N) ? y
WHAT IS THE X-VALUE (9999 TO STOP) ? 4
IF X = 4 , THEN Y = 10.8167 .
WHAT IS THE NEXT X-VALUE (9999 TO STOP) ?10
IF X = 10 , THEN Y = 15.9167 .
WHAT IS THE NEXT X-VALUE (9999 TO STOP) ?9999

WANT TO TEST FOR LINEARITY (ANSWER Y OR N) ? y
WHAT IS YOUR LEVEL OF SIGNIFICANCE ? .05
LOOK UP VALUE OF T .025 ( 4 ).
WHAT IS IT ? 2.776
ABS(T) = 6.89252 > 2.776 , SO THAT WE REJECT H,
CONCLUDING THEREBY THAT A SIGNIFICANT LINEAR
RELATIONSHIP EXISTS BETWEEN X AND Y.

NEED CONFIDENCE INTERVALS FOR PREDICTIONS(ANSWER Y OR N) ? y
STANDARD ERROR OF ESTIMATE IS .779957
WHAT LEVEL OF CONFIDENCE DO YOU NEED ? .90
LOOK UP VALUE OF T .05 ( 4 ).
WHAT IS IT ? 2.132
AT WHAT X-VALUE IS PREDICTION NEEDED (9999 TO STOP) ?12

THE INTERVAL 15.045 TO 20.1883 IS A 90. PERCENT CONFIDENCE
INTERVAL FOR THE Y-VALUE WHEN THE X-VALUE IS 12 .
AT WHAT X-VALUE IS PREDICTION NEEDED (9999 TO STOP) ?9999
```

EXERCISES 8.F

8.51. A compensation analyst studying salary patterns among employees of large
corporations accumulated the following data concerning the relationship
between an employee's years of experience and his or her yearly compensation:

Employee	A	B	C	D	E	F	G	H	I	J
Years of experience	10	6	3	2	5	10	10	9	8	5
Compensation (thousands of dollars)	60	15	10	10	5	40	50	40	25	15

a. What proportion of the variation among employees in yearly compensation can be accounted for by a linear relationship between compensation and years of experience?

b. Draw a scattergram of the data.

c. Test, at level $\alpha = .05$, whether or not a linear relationship between years of experience and yearly compensation can reasonably be considered to exist.

d. Determine the regression equation that can be used to predict an employee's yearly compensation from a knowledge of his or her years of experience, and superimpose the graph of the regression line upon the scattergram.

e. Find an 80% confidence interval for the compensation of an employee having seven years of experience.

8.52. The manager of a salmon cannery suspects that the demand for his product is closely related to the disposable income of his target region. To test out his suspicion, he collects the following data for 1981.

Region	Disposable Income (millions of dollars)	Sales Volume (thousands of cases)
A	10	1
B	20	3
C	40	4
D	50	5
E	30	2

a. What percentage of the variation in sales volume among regions can be attributed to a linear relationship with disposable income of the regions?

b. Draw a scattergram of the data.

c. At level $\alpha = .05$, do the data indicate the existence of a linear relationship between disposable income and sales volume?

d. Develop the regression equation that can be used to predict sales volume in a region from a knowledge of the region's disposable income, and superimpose the graph of the regression line upon the scattergram.

e. Find an 80% confidence interval for the sales volume in a region whose disposable income is $25,000,000.

8.53. A Labor Department employment analyst is assigned the task of finding the relationship, if there is any, between the level of an individual's formal education and whether or not that person is currently unemployed. One of her methods of investigation centers on a comparison between the various levels of formal education and the unemployment rate among individuals at that educational level. In the data that follow, the years 0 to 8 are grade school, 9 to 12 are high school, 13 to 16 are college, 17 and 18 are master's degree study, and 19 to 24 are doctoral degree study.

Educational Level, years	Unemployment Rate, %
4	30
18	3
12	6
14	10
8	20
24	6

The analyst would like to use the data to develop a method of predicting a person's chances of being unemployed from a knowledge of that person's educational level.

a. What proportion of the variation in unemployment rates among the various educational levels can be explained by a linear relationship between the two factors?

b. Draw a scattergram of the data.

c. At level $\alpha = .05$, can the employment analyst conclude that a linear relationship appropriately describes the situation under study?

d. Find the equation of the regression line that would be used to predict unemployment rate from a knowledge of the educational level, and superimpose the graph of the regression line upon the scattergram.

e. Find a 90% confidence interval for the unemployment rate among persons having 10 years of formal education.

8.54. In a study of the advertising budgets of small businesses, a consultant to the Small Business Administration collects the following data relating the size of a business's advertising budget with that business's total sales volume.

Advertising budget (hundreds of $)	6	4	10	1	7	5	7
Total sales volume (thousands of $)	50	60	100	30	60	60	40

a. On the basis of the given data, what proportion of variation among total sales volumes can be explained by a linear relationship between the size of the advertising budget and the sales volume?

b. Construct a scattergram of the data.

c. At level $\alpha = .10$, does it seem that a linear relationship is described by the data?

d. Can the relationship be considered linear at level $\alpha = .05$?

e. Determine the equation of the straight line that best predicts sales volume from a knowledge of the advertising budget, and superimpose its graph upon the scattergram.

f. Calculate a 90% confidence interval for the total sales volume of a small business that spends $800 on advertising.

g. Find a 90% confidence interval for the total sales volume of a small business that spends $300 on advertising.

8.55. An analysis of the earnings per share in 1966 and 1976 of a random sample of companies on the *Fortune* 500 list yielded the following data:*

| Corporation | Earnings per Share (dollars) | |
	1966	1976
Eastman Kodak	1.97	4.03
Deere	1.37	4.04
Bristol-Myers	1.56	4.90
Anheuser-Busch	0.75	1.23
Emhart	2.31	5.54
G. D. Searle	0.54	1.18
Sundstrand	1.78	4.42
Harris	2.57	4.42
Questor	0.86	0.70
Tektronix	1.38	3.43
Hoover Ball & Bearing	1.66	3.04
Insilco	1.26	2.01
Green Giant	2.12	1.26
Stanley Works	1.79	3.44
Lear Siegler	0.97	1.75
Polaroid	1.52	2.43
Martin Marietta	1.91	3.32
Kimberly-Clark	1.90	5.21
Honeywell	3.12	5.50
General Foods	1.87	3.02

a. What proportion of the variation in 1976 earnings per share can be accounted for by 1966 earnings per share on the basis of the best linear relationship available?

b. Construct a scattergram of the data.

c. Do the data indicate, at level $\alpha = .01$, that a linear relationship between earnings per share in 1966 and 1976 exists?

d. Determine the equation of the regression line that can be used to predict 1976 earnings per share from a knowledge of the earnings per share 10 years earlier.

e. The Upjohn Company earned $1.28 per share in 1966. Find a 90% confidence interval for its earnings per share in 1976.[†]

8.56. Recent data compared the typical cost of new (i.e., first owner) homes in several metropolitan areas with the typical cost of used homes in those areas. The Federal Home Loan Bank Board would like to use the data to predict the price

*Fortune, May 1977, pp. 369–385.

[†]Actual 1976 earnings per share was $2.62.

of new homes, based on known prices of used homes. A random selection of the data for January 1978, follows:*

	Tens of Thousands of dollars	
Area	Typical Price of Used Homes	Typical Price of New Homes
Tampa–St. Petersburg	4.6	4.6
Kansas City	4.2	5.4
Cleveland	4.9	7.0
Boston	5.4	7.4
Dallas–Fort Worth	4.7	5.6
New York	6.2	8.0

a. What proportion of variation in typical price of new homes can be explained by a linear relationship with typical price of used homes?

b. At level $\alpha = .05$, can we conclude that the actual relationship is a linear one?

c. Draw a scattergram of the data.

d. Determine the equation of the regression line used in predicting the typical price of new homes from a knowledge of the typical price of used homes.

e. Find an 80% confidence interval for the typical price of new homes in Detroit, where a used home typically cost $43,000 in January 1978. (The actual typical price of a new home was $53,500.)

SUMMARY AND DISCUSSION

In developing the techniques introduced in Chapter 8, our goal has been to discover the relationship between two sets of data for the purpose of eventually predicting one of the quantities from a knowledge of the other. The simplest and most fundamental measure of that relationship is the coefficient of linear determination, which indicates the extent to which a straight line having equation $y = a + bx$ describes the data.

A statistic closely tied to the coefficient of linear determination, Pearson's correlation coefficient r, can be used to test hypotheses involving the strength of a linear relationship in cases when the data points come from a normally distributed population. (In fact, under the normality assumption, any relationship that exists must be a linear one.)

If our indicators show that a linear relationship adequately describes the data, the next step would be to specify the regression line, namely the line that most closely approximates all the original data points simultaneously.

*U.S. News & World Report, March 13, 1978, p. 89.

After we calculated the best-fitting (least-squares) line, we showed how to use the line in order to predict one quantity from the other.

A BASIC language computer program for making the required calculations has also been included in our presentation.

Analogous techniques for analyzing sets of data involving relationships among more than two quantities will be introduced in Chapter 9.

CHAPTER 8 BIBLIOGRAPHY

MORE DETAILED TREATMENTS OF LINEAR REGRESSION

CHAPMAN, D. G., and R. A. SCHAUFELE, *Elementary Probability Models and Statistical Inference.* 1970, Waltham, Mass.: Xerox, pp. 208–266.

DIXON, W. I., and F. J. MASSEY, JR., *Introduction to Statistical Analysis,* 3rd ed., 1969, New York: McGraw-Hill, pp. 193–236.

REGRESSION BASED ON ABSOLUTE DEVIATIONS

RAO, M. R., and V. SRINIVASAN, "A Note on Sharpe's Algorithm for Minimizing the Sum of Absolute Deviations in a Simple Regression Problem," *Management Science,* October 1972, pp. 222–225 ("Erratum": July 1973, pp. 1334–1335).

SCHLOSSMACHER, E. J., "An Iterative Technique for Absolute Deviations Curve Fitting," *Journal of the American Statistical Association,* December 1973, pp. 857–859.

SHARPE, W. F., "Mean-Absolute-Deviation Characteristic Lines for Securities and Portfolios," *Management Science,* October 1971, pp. 1–13.

WILSON, H. G., "Least Squares versus Minimum Absolute Deviations Estimation in Linear Models," *Decision Sciences,* April 1978, pp. 322–335.

SPECIAL TOPICS

LIAO, M., "The Effect of Chance Variation on Revenue and Cost Estimations for Break-even Analysis," *The Accounting Review,* October 1976, pp. 922–926.

SEARLE, S. R., "Correlation Between Means of Parts and Wholes," *The American Statistician,* April 1969, pp. 23–24.

TURNER, R. E., and J. C. WIGINTON, "Advertising Expenditure Trajectories: An Empirical Study for Filter Cigarettes 1953–1965," *Decision Sciences,* July 1976, pp. 496–509.

CHAPTER 8 SUPPLEMENTARY EXERCISES

8.57. As part of an analysis of the relationship between smoking and absenteeism, the following data were obtained, relating an individual's number of packs smoked per day with the individual's number of days absent from his or her job.

Individual	A	B	C	D	E	F	G	H
Number of packs smoked per day	0.5	1.5	2.0	0.5	0.0	1.0	3.5	0.0
Number of days absent per year	4	8	15	0	3	10	20	0

a. What proportion of the variation in absenteeism among individuals can be explained on the basis of a linear relationship between absenteeism and amount of smoking?

b. Draw a scattergram of the data.

c. At significance level $\alpha = .05$, is it reasonable to assume, from the data collected, that the relationship between smoking and absenteeism is linear?

d. Calculate the equation of the regression line for predicting an individual's level of absenteeism from that individual's level of smoking, and superimpose the graph of the regression line upon the scattergram.

e. Find an 80% confidence interval for the number of days that an employee who smokes five packs a day will be absent.

8.58. Small businesses are frequently caught in the dilemma of having to locate in a highly-populated area (so as to maintain a needed level of sales volume) and simultaneously to suffer from the high crime rates apparently prevalent in such areas. A sociological study undertaken to investigate the relationship between the population density of a metropolitan area and that area's crime rate yields the following data.

Metropolitan Area	#1	#2	#3	#4	#5	#6
Population density, thousands per square mile	20	15	30	5	10	10
Crimes reported per 10,000 population	10	8	12	1	4	5

a. According to the results of the study, what proportion of variation in the crime rate among various metropolitan areas can be ascribed to variations in population density by way of a linear relationship?

b. Draw a scattergram of the data.

c. Using a level of significance α .05, decide whether or not a linear relationship between population density and crime rate adequately describes the data.

d. Determine the equation of the regression line that best accounts for the relationship between population density and crime rate, and superimpose its graph upon the scattergram.

e. Find a 90% confidence interval for the crime rate in a metropolitan area having population density of 25,000 per square mile.

8.59. Some investors think that prices on the New York Stock Exchange are related to prices on the London Gold Exchange. To test out this theory, one financial analyst compared the Dow-Jones industrial average with the price of gold in London on five randomly selected days. He came up with the following data in an attempt to find out whether he could predict changes in the Dow-Jones average from the behavior of the London gold prices, which close several hours earlier because of time zone differences.

London Gold Prices (dollars)	Dow-Jones Industrial Average
200	560
190	600
160	840
180	700
170	840

a. According to the data, how much of the variation in the Dow-Jones average is attributable to a linear relationship with the price of gold in London?

b. Draw a scattergram of the data.

c. At level $\alpha = .01$, do the data substantiate the existence of a linear relationship between the Dow-Jones average and the price of gold in London?

d. Calculate the regression line for predicting the Dow-Jones average from the price of London gold, and superimpose the graph of the regression line upon the scattergram.

e. Find an 80% confidence interval for the Dow-Jones average if the price of gold in London were to rise to $220.

8.60. A manufacturer of chemicals for use in the pharmaceuticals industry conducts a study of the time it takes after receipt of an order for the chemicals to reach the customer's plant. Because some of the chemicals deteriorate over time, an accurate estimate of the transit time is needed. The following data compare the rail distance from the manufacturer to the customer with the transit time of the most recent shipment to that customer.

Customer	#1	#2	#3	#4	#5	#6
Rail distance, hundreds of miles	5	4	1	3	2	5
Transit time, days	6.0	5.0	3.0	4.5	4.0	5.5

a. Find the proportion of variation in transit times that can be attributed to a linear relationship with the rail distances.

b. Draw a scattergram of the data.

c. At level $\alpha = .05$, would we be justified in using a linear equation to predict the transit time from a knowledge of the rail distance?

    d. Find the equation of the regression line, and superimpose its graph upon the scattergram.

    e. Customer 7 is located 700 miles away from the manufacturer. Find a 90% confidence interval for the time it will take a shipment of chemicals to arrive at his warehouse.

8.61. Some economists believe that as the prime interest rate increases, the value of stocks declines. The following data compare the prime rate with the value of General Nucleonics common stock at five randomly selected times during the past several years:

Prime rate, %	10	6	5	7	12
GN stock prices, dollars per share	60	80	90	70	50

    a. What proportion of the variation in GN stock prices can be accounted for by a linear relationship with the prime interest rate?

    b. Draw a scattergram of the data.

    c. At level $\alpha = .05$, can a relationship be assumed to exist between the prime rate and the price of GN stock?

    d. Determine the equation of the regression line that best predicts GN stock prices from the prime rate, and superimpose the graph of the regression line upon the scattergram.

    e. Find an 80% confidence interval for the value of GN stock if the prime rate were to rise to 15%.

8.62. The following table gives the wholesale price index and the index of industrial production in the United States on July 1 of each of five randomly selected years. Knowledge of the relationship between the two indices would be helpful to an economist attempting to forecast business conditions.

Year No.	Wholesale Prices (Producer Prices)	Industrial Production
1	80	110
2	80	90
3	90	80
4	80	100
5	70	120

    a. What proportion of variation in the index of industrial production seems to be explained by a possible linear relationship with the wholesale price index?

    b. Draw a scattergram of the data.

    c. At level $\alpha = .10$, decide whether or not it is reasonable to consider the two indices to be linearly related.

d. Find the equation of the regression line that can be used to predict the index of industrial production from a knowledge of the wholesale price index, and superimpose the graph of the regression line upon the scattergram.

e. Determine an 80% confidence interval for the index of industrial production when the wholesale price index stands at 100.

8.63. A record of maintenance cost is assembled on six identical metal-stamping machines of different ages in an attempt to find out the relationship, if there is any, between the age of a machine (in years) and the monthly maintenance cost (in dollars) required to keep it in peak operating condition. Such information would be useful in deciding when it is most economical to replace the old machines with new ones. Data on six randomly selected machines follow.

Machine Serial No.	901	887	923	927	891	906
Age, years	2	1	3	4	4	6
Maintenance cost, dollars per month	10	10	30	30	40	50

a. What proportion of the variation in maintenance cost can be attributed, by the way of a linear relationship, to variations in age?

b. Draw a scattergram of the data.

c. At level $\alpha = .05$, can a linear relationship between maintenance cost and machine age be considered to exist?

d. Find the regression equation that predicts monthly maintenance cost in terms of machine age, and superimpose the graph of the regression line upon the scattergram.

e. Find a 90% confidence interval for predicting the monthly maintenance cost of a 7-year-old machine.

8.64. A recent analysis of the use of imported oil in the United States for various years after 1960 (i.e., 1972 is 12 years after 1960, 1967 is 7 years after 1960, etc.), yielded the following information.

Number of Years after 1960	Millions of Barrels of Imported Oil per Day
2	1.5
7	2.0
9	3.0
10	3.5
12	4.5
14	6.5
16	9.0
20	10.0

a. What proportion of variation in volume of imported oil usage can be explained on the basis of a linear relationship with the passage of time?

b. Construct a scattergram of the data.

c. Calculate the equation of the straight line that provides the best linear relationship between the number of years after 1960 and the daily volume of imported oil, and superimpose the graph of the line upon the scattergram.

d. If the trends established during the years covered by the data continue, forecast the daily volume of imported oil for 1985.

8.65. A study of sales trends in the cigarette industry revealed the following information comparing a brand's market share in 1977 with its percentage change in sales from 1976:*

Brand	1977 Market Share (percent)	Percentage Change in Sales Volume from 1976
L & M	1.43	− 10.4
Vantage	2.86	+ 19.4
Marlboro	16.37	+ 5.1
Belair	1.46	− 1.1
Kent	5.09	+ 13.8
True	1.80	+ 28.6
Salem	8.97	+ 2.9
Tareyton	2.19	− 14.8
Virginia Slims	1.61	+ 7.8
Raleigh	2.05	+ 4.2

a. What proportion of variation in percentage change in sales volume can be accounted for by a linear relationship with 1977 market share? (*Warning:* Remember the negative signs!)

b. Draw a scattergram of the data.

c. Find the equation of the regression line for predicting percentage change in sales on the basis of market share, and superimpose the graph of the line upon the scattergram.

d. Use the regression equation to predict the percentage change in sales volume from 1976 for the Camel brand,[†] whose 1977 market share was 4.06%.

e. How much is the prediction in part d worth? Why?

8.66. The *Business Week* survey of corporate performance for the third quarter of 1977 revealed the following information on the sales volume and the profits made by a random sample of 15 corporations during that quarter.[‡]

a. What proportion of the variation in profits among the corporations can be explained on the basis of a linear relationship between sales volume and profits?

*Reported in *Business Week*, Oct. 31, 1977, p. 82.

†Actual percentage change for Camel was − 6.5%.

‡Selected from a massive survey in *Business Week*, November 14, 1977, pp. 106–138.

b. Construct a scattergram of the data.

c. Test, at level of significance $\alpha = .01$, whether a linear relationship between a corporation sales volume and its profits exists.

d. Calculate the regression equation that should be used to predict corporate profits from sales volume, and superimpose the graph of the regression line upon the scattergram.

e. Tandy Corp. had a sales volume of $218,200,000 during the third quarter of 1977. Find a 90% confidence interval for its profits during that quarter.*

Corporation	Sales Volume ($10,000,000)	Profits ($1,000,000)
Allen Group	6.71	1.6
Harris Bankcorp	8.25	7.0
Lone Star Industries	22.46	11.2
Teledyne	55.07	52.8
McGraw-Edison	26.71	13.7
Stokely-Van Camp	11.26	2.3
Hewlett-Packard	34.11	30.5
Lenox	3.89	3.3
Diebold	4.71	1.6
New York Times	12.43	5.8
Dart Industries	39.34	24.4
Baker Industries	4.69	2.4
Armco Steel	93.00	31.7
General Tire & Rubber	51.49	28.3
Neptune International	3.51	1.6

8.67. An analysis of the relationship between corporate earnings and total remuneration paid to its chief executive yielded the following information on 14 randomly selected corporations:†

a. What proportion of the variation in executive remuneration can be explained via a linear relationship with earnings per share?

b. Draw a scattergram of the data.

c. Calculate the equation of the regression line for predicting executive remuneration from a knowledge of earnings per share, and superimpose its graph upon the scattergram.

d. Use the regression equation to predict the 1976 remuneration of the chief executive of Combustion Engineering, which had 1976 earnings of $5.04 per share.‡

* Actual Tandy Corp. profits that quarter were $11,000,000.

† *Forbes*, May 15, 1977, pp. 202–243.

‡ The remuneration was $362,000.

Corporation	1976 Earnings per Share (dollars)	1976 Remuneration of Chief Executive (thousands of $)
American Fletcher	1.69	120
Carborundum	4.14	369
Walt Disney	2.34	167
Fruehauf	4.05	400
Jefferson-Pilot	2.82	167
National Can	2.52	222
Phelps Dodge	2.07	238
Texaco	3.20	661
Winn-Dixie	3.01	325
Union Pacific	4.02	469
St. Regis Paper	3.82	327
Olin Corp.	6.07	406
R. H. Macy	4.37	262
Hershey Foods	3.32	192

8.68. A study of the role of foreign trade in the growth of Brazil's economy yielded the following data on that nation's exports and imports, valued in billions of dollars, for several years in the recent past.*

Year	Billions of dollars	
	Exports	Imports
1972	4.4	5.4
1973	6.8	7.8
1974	8.8	15.0
1975	9.6	14.6
1976	11.0	13.3
1977	13.8	13.8

a. What proportion of the variation in Brazil's imports over the period of years indicated can be explained on the basis of a linear relationship between imports and exports?

b. Construct a scattergram of the data.

c. Determine the equation of the regression line for predicting the nation's imports from a knowledge of its exports, and superimpose the graph of the regression line upon the scattergram.

d. Use the regression equation to forecast Brazil's imports in 1985, when the country's exports are currently projected to be 27.2 billion dollars.

*Reported in Business Week, December 5, 1977, page 73.

8.69. The Social Security bill passed by Congress in the closing days of 1977 contained the following implications regarding the maximum contribution of each employee as the years pass:*

Year	Maximum Annual Pay Taxed ($)	Maximum Tax per Worker ($)
1977	16,500	965.25
1978	17,700	1070.85
1979	22,900	1403.77
1980	25,900	1587.67
1981	29,700	1975.05
1982	31,800	2130.60
1983	33,900	2271.30
1984	36,000	2412.00
1985	38,100	2686.05
1986	40,200	2874.30
1987	42,600	3045.90

a. What proportion of the variation in the maximum tax per worker can be accounted for by a linear relationship with the worker's maximum annual pay taxed?

b. Construct a scattergram of the data.

c. Calculate the equation of the regression line that expresses the maximum tax per worker in terms of the worker's maximum annual pay taxed, and superimpose the graph of the line upon the scattergram.

8.70. A municipal-bond dealer conducted a study of the relationship between a person's federal income tax bracket and the equivalent taxable yield of a 6% tax-exempt bond held by that person. The data follow.

Tax Bracket (%)	26	40	50	54	60	70
Equivalent Taxable Yield of 6% Bond (%)	8	10	12	13	16	21

a. What proportion of the variation in equivalent in taxable yield can be explained by a linear relationship with the tax bracket?

b. Draw a scattergram of the data.

c. Test, at level $\alpha = .01$, whether or not a linear relationship is described by the data.

d. Determine the equation of the best-fitting regression line, and superimpose the graph of the regression line upon the scattergram.

e. One of the dealer's clients is in the 35% federal tax bracket. What would be her equivalent taxable yield on a 6% tax-exempt bond?

* U.S. News & World Report, December 26, 1977/Jan. 2, 1978, p. 89.

8.71. *Aviation Week and Space Technology* listed figures on technical characteristics of U.S.-made business, personal, and utility aircraft.* The following randomly selected and rounded data compare an aircraft's normal fuel capacity (in gallons) with its maximum cruise speed (in miles per hour):

Aircraft	Normal Fuel Capacity (gals.)	Maximum Cruise Speed (mph)
Grumman AA5B Tiger	50	160
Piper PA-38-112 Tomahawk	30	130
Cessna 172 Skyhawk	40	140
Beech 76 Duchess	100	190
Piper PA-28-201 Arrow 3	70	170
Rockwell 114 Commander	70	190
Teal 150 Marlin Amphibian	40	120

a. What proportion of the variation among the planes in maximum cruise speed can be explained on the basis of corresponding variations in fuel capacity?

b. Draw a scattergram of the data.

c. Test, at level of significance $\alpha = .05$, whether a linear relationship between maximum cruise speed and fuel capacity seems to exist.

d. Calculate the equation of the regression line to be used for predicting an aircraft's maximum cruise speed from a knowledge of its normal fuel capacity, and superimpose the graph of the regression line upon the scattergram.

e. The Bellanca 7GCBC Citabria 150S has a normal fuel capacity of 35 gallons. Find an 80% confidence interval for its maximum cruise speed. (Note: Its actual maximum cruise speed is 130 mph, according to the magazine.)

8.72. Purchasers of new cars generally expect a larger discount off list price if they buy a car from a company that is losing money. A recent survey of discounts as related to the company's profit picture came up with the following information on six car purchases.

Company	A	B	C	D	E	F
% Loss last quarter	40	30	10	55	15	50
% Discount on car	18	11	3	21	8	20

a. What proportion of the variation in company-to-company discounts is explainable on the basis of a linear relationship with the company's percentage loss?

b. Draw a scattergram of the data.

c. Use a significance level of $\alpha = .05$ to conduct a test of whether or not the relationship is a linear one.

*Selected from a report in the issue of March 13, 1978, page 119.

d. Calculate the equation of the regression line that predicts the percentage discount, using the company's percentage loss.

e. Superimpose the graph of the regression line upon the scattergram.

f. Company G had a 25% loss last quarter. Find an 80% confidence interval for the percentage discount a customer would get when purchasing a car from that company.

8.73. *Forbes* magazine provided, in its November 10, 1980 issue, page 241, some illustrations of the "bracket creep" phenomenon. Assuming a pay increase equal to the inflation rate of 14%, the magazine compared present salary levels with the additional amount of taxes an individual would pay on his or her 14% pay raise. The data follow:

Present salary ($ thousands)	15	20	25	30	35
Additional taxes ($ hundreds)	4	6	9	13	18

a. What proportion of the variation in additional taxes can be explained on the basis of the best linear relationship with present salary?

b. Use a level of significance of $\alpha = .05$ to test whether or not the relationship between present salary and additional taxes is truly a linear one.

c. Draw a scattergram of the data.

d. Calculate the equation of the regression line for predicting additional taxes in terms of present salary.

e. Superimpose the graph of the regression line upon the scattergram.

f. Find an 80% confidence interval for the additional taxes to be owed by an individual whose present salary is $28,500.

8.74. The issue of *Business Week* of November 24, 1980, page 52, contained some data on how newspapers give substantial discounts to advertisers who run their advertisements in several issues of the paper rather than just once. A sample of the data, in dollars, follows:

Newspaper	Standard Rate, per line	Volume Discounted Rate, per line
Chicago Sun-Times	5.18	4.00
Cleveland Plain Dealer	2.45	2.07
Detroit Free Press	4.47	4.03
Miami Herald News	6.01	5.11
Philadelphia Inquirer	5.94	5.08

a. If we want to predict the discounted rate when we know the standard rate, what percentage of the variation in the discounted rate may be explained on the basis of a linear relationship with the standard rate?

b. Using a significance level of $\alpha = .01$, test the hypothesis that a linear relationship exists between the discounted and standard rates.

c. Draw a scattergram illustrating the data.

d. Calculate the equation of the regression line that best predicts the discounted rate in terms of the standard rate.

e. Superimpose the regression line upon the scattergram.

f. The *Minneapolis Star-Tribune* charged $4.52 per line as its standard rate. Find an 80% confidence interval for its discounted rate.*

*Its actual reported discount rate was $3.98.

9

Multiple Regression and Correlation

During our study of linear regression and correlation, we developed a method of determining the nature and extent of a linear relationship existing between two quantities, so that we could make accurate predictions of an unknown quantity based on firm knowledge of a related quantity. For example, we can forecast a department store's sales volume for the day from a knowledge of its advertising expenditures in that morning's newspaper.

In many situations of interest and importance, however, there are two or even more factors (called *regressors*) that affect the numerical value of the unknown quantity. Sales volume is affected by the morning's weather as well as the extent of advertising. In this chapter, we will study *multiple linear regression*, the technique of predicting an unknown quantity from a knowledge of more than one factor that influences it.

SECTION 9.A THE COEFFICIENT OF MULTIPLE DETERMINATION

In Chapter 8, we expressed the linear relationship between two variables y and x (where y is the variable we want to predict) as an equation of the form

$$y = a + bx,$$

where a and b are numbers calculated from a set of sample data. In this section, we will show how to measure the extent to which the multiple linear regression equation

$$y = a + bx + cw$$

can be used to predict a variable y, based on a knowledge of two other variables, x and w. Here a, b, and c are numbers to be calculated from a set of sample data. (Situations in which y depends on three or more variables are algebraically too complicated to handle directly and are usually developed in terms of computer-oriented matrix methods. Some discussion of standard computer approaches to this problem is presented in Section 9.C.)

EXAMPLE 9.1.   Department Store Sales Volume.   In Portland, Oregon, it rains all day, every day, from October to April. The vice-president in charge of advertising at one small department store claims that the store's daily sales volume can be predicted from a knowledge of the day's advertising costs and the morning's total precipitation. A search of accounting and weather data reveals the following information for six randomly selected days last year:

		Jan 12	Mar 4	Apr 2	Aug 14	Oct 27	Dec 6
Advertising Costs (hundreds of dollars)	(x)	2	3	5	4	6	10
Morning Precipitation (tenths of an inch)	(w)	5	3	1	0	1	5
Sales Volume (thousands of dollars)	(y)	1	9	19	21	25	33

We would like to use the data to answer the following questions. (Unfortunately, with three variables x, w, and y, it is not practical to construct scattergrams or to draw the graph of the regression equation. The graph of the multiple linear regression equation will be a plane in three-dimensional space rather than a line in two dimensions.)

a. How much of the variation in sales volume (y) can be attributed to a linear relationship with advertising costs (x) alone?
b. How much of the variation in sales volume (y) can be attributed to a linear relationship with precipitation (w) alone?
c. How much of the variation in sales volume (y) can be explained by a multiple linear relationship with advertising costs (x) and precipitation (w) together?

TABLE 9.1   Preliminary Calculations for Department Store
Data   (Multiple linear regression)

x	w	y	x^2	w^2	y^2	xw	xy	wy
2	5	1	4	25	1	10	2	5
3	3	9	9	9	81	9	27	27
5	1	19	25	1	361	5	95	19
4	0	21	16	0	441	0	84	0
6	1	25	36	1	625	6	150	25
10	5	33	100	25	1,089	50	330	165
30	15	108	190	61	2,598	80	688	241

$n = 6$

SOLUTION.   To answer questions a, b, and c, we need the calculations appearing in Table 9.1.

a. The answer to this question is the first coefficient of linear determination between x and y. We first calculate the correlation coefficient by formula (8.3) of Chapter 8:

$$r_{xy} = \frac{n\Sigma xy - (\Sigma x)(\Sigma y)}{\sqrt{n\Sigma x^2 - (\Sigma x)^2}\sqrt{n\Sigma y^2 - (\Sigma y)^2}}$$

$$= \frac{6(688) - (30)(108)}{\sqrt{6(190) - (30)^2}\sqrt{6(2,598) - (108)^2}} = \frac{888}{\sqrt{240}\sqrt{3,924}}$$

$$= \frac{888}{(15.49)(62.64)} = .915.$$

Therefore,

$$r_{xy}^2 = (.915)^2 = .837 = 83.7\%$$

of the variation in y can be accounted for by a linear relationship with x alone.

b. To calculate the correlation coefficient between w and y,

$$r_{wy} = \frac{n\Sigma wy - (\Sigma w)(\Sigma y)}{\sqrt{n\Sigma w^2 - (\Sigma w)^2}\sqrt{n\Sigma y^2 - (\Sigma y)^2}}$$

$$= \frac{6(241) - (15)(108)}{\sqrt{6(61) - (15)^2}\sqrt{6(2,598) - (108)^2}} = \frac{-174}{\sqrt{141}\sqrt{3,924}}$$

$$= \frac{-174}{(11.87)(62.64)} = -.234.$$

Therefore, the proportion of the variation in y that can be accounted for by a linear relationship with w alone is

$$r_{wy}^2 = (-.234)^2 = .055 = 5.5\%.$$

c. In order to compute the proportion of variation in y that can be attributed to a multiple linear relationship with x and w together, we calculate the *coefficient of multiple determination*, which has the formula

■ $$R^2 = r_{xy}^2 + (1 - r_{xy}^2)r_{wy:x}^2 \qquad (9.1)$$

or, equivalently,

■ $$R^2 = r_{wy}^2 + (1 - r_{wy}^2)r_{xy:w}^2 \qquad (9.2)$$

where the *coefficients of partial correlation* are given by

■ $$r_{wy:x} = \frac{r_{wy} - r_{xw}r_{xy}}{\sqrt{(1 - r_{xw}^2)(1 - r_{xy}^2)}} \qquad (9.3)$$

and

■ $$r_{xy:w} = \frac{r_{xy} - r_{xw}r_{wy}}{\sqrt{(1 - r_{xw}^2)(1 - r_{wy}^2)}}. \qquad (9.4)$$

Using formula (9.1) to compute the coefficient of multiple determination R^2 requires a prior computation of $r_{wy:x}$ by formula (9.3). We carry this out by first finding r_{xw}.

$$r_{xw} = \frac{n\Sigma xw - (\Sigma x)(\Sigma w)}{\sqrt{n\Sigma x^2 - (\Sigma x)^2}\sqrt{n\Sigma w^2 - (\Sigma w)^2}}$$

$$= \frac{6(80) - (30)(15)}{\sqrt{240}\sqrt{141}} = \frac{30}{(15.49)(11.87)} = .163.$$

Therefore, formula (9.3) yields that

$$r_{wy:x} = \frac{r_{wy} - r_{xw}r_{xy}}{\sqrt{(1 - r_{xw}^2)(1 - r_{xy}^2)}} = \frac{(-.234) - (.163)(.915)}{\sqrt{(1 - [.163]^2)(1 - [.915]^2)}}$$

$$= \frac{-.383}{\sqrt{(1 - .027)(1 - .837)}} = \frac{-.383}{\sqrt{(.973)(.163)}}$$

$$= \frac{-.383}{\sqrt{.159}} = \frac{-.383}{.399} = -.96.$$

It then follows from formula (9.1) that

$$R^2 = r_{xy}^2 + (1 - r_{xy}^2)r_{wy \cdot x}^2$$

$$= .837 + (.163)(-.96)^2$$

$$= .837 + (.163)(.922) = .988$$

$$= 98.8\%.$$

We conclude that x (advertising costs) and w (precipitation) together account for 98.8% of the variation in y (sales volume).

Note that while x by itself accounts for only 83.7% and w by itself accounts for only 5.5%, together they account for more than 83.7% + 5.5%. This shows that the *interaction* between x and w also contributes something to the explanation of variations in y. There may also be situations where x and w together account for *less* than the sum of their individual coefficients of linear determination. Examples of this will appear in the exercises.

Test for Multiple Linearity

By analogy with the correlation test for the existence of a linear relationship, studied in Section 8.D, there is a test for *multiple linearity* based on the coefficient of multiple determination R^2. As with the correlation test, this test is valid only in the normally distributed situation and is really a test for the existence of a dependence relationship among the three variables. Advanced statistical theory then takes over and asserts that, if any relationship exists among these normally distributed variables, it must be a multiple linear one. Instead of the t distribution used in Section 8.D, we require here the F distribution of Table A.10 of the Appendix.

To test the hypothesis

H: There is no significant multiple linear relationship

against the alternative

A: There is a significant multiple linear relationship,

we compute

$$F = \frac{R^2(n-3)}{2(1-R^2)} \qquad\qquad (9.5)$$

and we reject H at level α if $F > F_\alpha[2, n - 3]$. Here n is the number of "triples" of data points. In the example of the department store's sales volume, we see from Table 9.1 that n = 6. To test at level α = .05 for the existence of a

TABLE 9.2 BASIC Computer Program for Solving Multiple Regression and Correlation Problems

```
10       REMARK: MULTIPLE REGRESSION AND CORRELATION
20       DIM X(100),W(100),Y(100),M(3,4)
30       PRINT "HOW MANY SETS OF DATA POINTS ARE THERE";
40       INPUT N
50       X1 = 0
60       X2 = 0
70       W1 = 0
80       W2 = 0
90       Y1 = 0
100      Y2 = 0
110      REMARK: P1 = SIGMA XW, P2 = SIGMA XY, P3 = SIGMA WY
120      P1 = 0
130      P2 = 0
140      P3 = 0
150      FOR K = 1 TO N
160         PRINT "WHAT ARE X,W,Y FOR PAIR";K;
170         INPUT X(K),W(K),Y(K)
180         REMARK: COMPUTE THE COLUMN TOTALS
190         X1 = X1 + X(K)
200         X2 = X2 + X(K)**2
210         W1 = W1 + W(K)
220         W2 = W2 + W(K)**2
230         Y1 = Y1 + Y(K)
240         Y2 = Y2 + Y(K)**2
250         P1 = P1 + X(K)*W(K)
260         P2 = P2 + X(K)*Y(K)
270         P3 = P3 + W(K)*Y(K)
280      NEXT K
290      PRINT
300      PRINT "SIGMA X =";X1;", SIGMA W =";W1;", AND SIGMA Y =";Y1
310      PRINT "SIGMA X2 =";X2;", SIGMA W2 =";W2;", AND SIGMA Y2 =";Y2
320      PRINT "SIGMA XW =";P1;", SIGMA XY =";P2;", AND SIGMA WY =";P3
330      PRINT
340      REMARK: CALCULATE THE CORRELATION COEFFICIENTS
350      D1 = N*X2 - X1**2
360      D2 = N*W2 - W1**2
370      D3 = N*Y2 - Y1**2
380      N1 = N*P1 - X1*W1
390      N2 = N*P2 - X1*Y1
400      N3 = N*P3 - W1*Y1
410      R1 = N1/SQR(D1*D2)
420      R2 = N2/SQR(D1*D3)
430      R3 = N3/SQR(D2*D3)
440      PRINT (R2**2)*100;"PERCENT OF VARIATION IN Y IS "
450      PRINT " ACCOUNTED FOR BY X ALONE."
460      PRINT (R3**2)*100;"PERCENT OF VARIATION IN Y IS"
470      PRINT " ACCOUNTED FOR BY W ALONE."
480      REMARK: CALCULATE THE COEFFICIENT OF MULTIPLE DETERMINATION
490      R4 = (R3-R1*R2)/SQR((1-R1**2)*(1-R2**2))
500      R9 = R2**2 + (1-R2**2)*(R4**2)
510      PRINT R9*100;"PERCENT OF VARIATION IN Y IS"
520      PRINT " ACCOUNTED FOR BY X AND W TOGETHER."
530      REMARK: TEST FOR MULTIPLE LINEARITY
540      PRINT
550      PRINT "WHAT IS YOUR ALPHA LEVEL";
560      INPUT A0
570      PRINT "LOOK UP VALUE OF F";A0;"(2,";N-3;")."
580      PRINT "WHAT IS IT";
590      INPUT F0
600      F = R9*(N-3)/(2*(1-R9))
610      IF F > F0 THEN 650
620      PRINT "BECAUSE F =";F;"<";F0;", WE CANNOT REJECT H;"
```

TABLE 9.2   BASIC Computer Program   (Continued)

```
630     PRINT "THERE IS NO EVIDENCE OF SIGNIFICANT MULTIPLE LINEARITY."
640     STOP
650     PRINT"BECAUSE F =";F;">";F0;", WE REJECT H, CONCLUDING THEREBY"
660     PRINT "THAT Y DEPENDS ON X AND W IN A MULTIPLE LINEAR MANNER."
670     PRINT
680     REMARK: CALCULATE MULTIPLE REGRESSION EQUATION
690     M(1,1) = N
700     M(1,2) = X1
710     M(1,3) = W1
720     M(1,4) = Y1
730     M(2,1) = X1
740     M(2,2) = X2
750     M(2,3) = P1
760     M(2,4) = P2
770     M(3,1) = W1
780     M(3,2) = P1
790     M(3,3) = W2
800     M(3,4) = P3
810     FOR I = 1 TO 3
820        FOR J = 4 TO 1 STEP -1
830           M(I,J) = M(I,J)/M(I,1)
840           NEXT J
850        NEXT I
860     FOR J = 1 TO 4
870        FOR I = 2 TO 3
980           M(I,J) = M(I,J) - M(1,J)
890           NEXT I
900        NEXT J
910     FOR I = 2 TO 3
920        FOR J = 4 TO 2 STEP -1
930           M(I,J) = M(I,J)/M(I,2)
940           NEXT J
950        NEXT I
960     FOR J = 4 TO 2 STEP -1
970        M(1,J) = M(1,J) - M(1,2)*M(2,J)
980        M(3,J) = M(3,J)-M(3,2)*M(2,J)
990        NEXT J
1000    M(3,4) = M(3,4)/M(3,3)
1010    M(3,3) = 1
1020    FOR I = 1 TO 2
1030       FOR J = 4 TO 3 STEP -1
1040          M(I,J) = M(I,J) - M(I,3)*M(3,J)
1050          NEXT J
1060       NEXT I
1070    A = M(1,4)
1080    B = M(2,4)
1090    C = M(3,4)
1100    PRINT "A = ";A;", B =";B;", AND C =";C;", SO THAT THE MULTIPLE"
1110    PRINT "LINEAR REGRESSION EQUATION FOR PREDICTING Y IN TERMS OF"
1120    PRINT "X AND W IS    Y = ";A;" + (";B;")X + (";C;")W."
1130    PRINT
1140    PRINT "DO YOU WANT PREDICTED VALUES OF Y (ANSWER Y OR N)";
1150    INPUT P$
1160    IF P$ = "N" THEN 1260
1170    PRINT "WHAT ARE THE VALUES OF X AND W (9999,9999 TO STOP)";
1180    INPUT X,W
1190    IF X = 9999 THEN 1210
1200    GO TO 1220
1210    IF W = 9999 THEN 1260
1220    Y = A + B*X + C*W
1230    PRINT "FOR X =";X;"AND W =";W;", WE PREDICT Y =";Y;"."
1240    PRINT
1250    GO TO 1170
1260    END
```

TABLE 9.3   Computer Printout of Solution to Example 9.1

```
HOW MANY SETS OF DATA POINTS ARE THERE ? 6
WHAT ARE X,W,Y FOR PAIR 1   ? 2,5,1
WHAT ARE X,W,Y FOR PAIR 2   ? 3,3,9
WHAT ARE X,W,Y FOR PAIR 3   ? 5,1,19
WHAT ARE X,W,Y FOR PAIR 4   ? 4,0,21
WHAT ARE X,W,Y FOR PAIR 5   ? 6,1,25
WHAT ARE X,W,Y FOR PAIR 6   ? 10,5,33

SIGMA X = 30 , SIGMA W = 15 , AND SIGMA Y = 108
SIGMA X2 = 190 , SIGMA W2 = 61 , AND SIGMA Y2 = 2598
SIGMA XW = 80 , SIGMA XY = 688 , AND SIGMA WY = 241

 83.7309 PERCENT OF VARIATION IN Y IS
ACCOUNTED FOR BY X ALONE.
 5.47205 PERCENT OF VARIATION IN Y IS
ACCOUNTED FOR BY W ALONE.
 98.8125 PERCENT OF VARIATION IN Y IS
ACCOUNTED FOR BY X AND W TOGETHER.

WHAT IS YOUR ALPHA LEVEL ? .05
LOOK UP VALUE OF F .05 (2, 3 ).
WHAT IS IT ? 9.55
BECAUSE F = 124.818 > 9.55 , WE REJECT H, CONCLUDING THEREBY
THAT Y DEPENDS ON X AND W IN A MULTIPLE LINEAR MANNER.

A =   3.39344 , B = 3.95956 , AND C =-2.0765 , SO THAT THE MULTIPLE
LINEAR REGRESSION EQUATION FOR PREDICTING Y IN TERMS OF
X AND W IS   Y =   3.39344  + ( 3.95956 )X + (-2.0765 )W.

DO YOU WANT PREDICTED VALUES OF Y (ANSWER Y OR N) ?y
WHAT ARE THE VALUES OF X AND W (9999,9999 TO STOP) ? 7,6
FOR X = 7 AND W = 6 , WE PREDICT Y = 18.6514 .

WHAT ARE THE VALUES OF X AND W (9999,9999 TO STOP) ? 4,2
FOR X = 4 AND W = 2 , WE PREDICT Y = 15.0787 .

WHAT ARE THE VALUES OF X AND W (9999,9999 TO STOP) ? 4,0
FOR X = 4 AND W = 0 , WE PREDICT Y = 19.2317 .

WHAT ARE THE VALUES OF X AND W (9999,9999 TO STOP) ? 9999,9999
```

significant dependence relationship in the data, we would compute F by formula (9.5) and reject H above if

$$F > F_\alpha[2, n - 3] = F_{.05}[2, 3] = 9.55.$$

Recalling that $R^2 = .988$, as we have just calculated, we get

$$F = \frac{R^2(n - 3)}{2(1 - R^2)} = \frac{(.988)(6 - 3)}{2(1 - .988)} = \frac{(.988)(3)}{2(.012)}$$

$$= \frac{2.964}{.024} = 123.5.$$

Because $F = 123.5 > 9.55$, we reject H and conclude, at level $\alpha = .05$, that sales volume data can be adequately described by a multiple linear relationship.

A BASIC computer program that tests for multiple linearity (and also computes the multiple regression equation, if asked to do so) can be found in Table 9.2. The printout of the solution to Example 9.1 appears in Table 9.3.

EXERCISES   9.A

9.1. Two factors that affect the price of sardines in your local grocery store are the availability of sardines in the fishing areas and the costs of transportation. The following data were collected over a recent period.

Year	1961	1967	1970	1974	1977	1979
Estimated size of sardine population of Norway, in trillions (x)	30	30	40	20	20	30
Transportation cost per gross of cans in dollars (w)	6	6	8	10	15	15
Price in cents of 3-oz. can to consumer (y)	28	32	30	40	45	40

a. How much of the variation in price to the consumer can be attributed to a linear relationship with the estimated size of the sardine population?

b. How much of the variation in price can be explained on the basis of a linear relationship with transportation cost?

c. What proportion of the variation in price can be accounted for by a multiple linear relationship between the price and both of the other factors?

d. Using a level of significance of $\alpha = .05$, do the data indicate that a multiple linear relationship is an appropriate method of analyzing the price variation of sardines?

9.2. A psychological study aimed at analyzing the motivational factors that encourage employees to seek advancement in a large corporation uncovered the following data on 10 randomly selected employees:

Employee	A	B	C	D	E	F	G	H	I	J
Number of dependents	3	7	7	0	8	2	4	5	1	1
Years of service in corporation	10	6	3	2	5	10	10	9	8	5
Salary level (thousands of $)	70	25	20	20	15	50	60	50	35	25

a. Calculate the proportion of variation in salary levels that can be explained on the basis of a linear relationship with an employee's number of dependents.

b. Calculate the proportion of variation in salary levels that can be attributed to a linear relationship with an employee's length of service to the corporation.

c. What proportion of the variation in salary levels can be accounted for by a multiple linear relationship with both factors?

9.3. One personnel officer is interested in developing a method of assessing the future performance of applicants for accounting positions in her corporation. She selected a random sample of current and former employees; and compared each one's job performance rating, after one year's experience, with his or her score on the corporation's standard Accounting Applicant Aptitude Test (AAAT) and the number of college courses he or she took in the accounting area. The data follow.

Employee	Number of Courses in Accounting	AAAT Score	One-Year Job Performance Rating
A	1	12	20
B	3	12	40
C	2	10	30
D	10	10	80
E	5	10	50
F	10	3	50
G	12	2	90
H	1	20	30
I	2	20	40
J	1	30	40

a. What proportion of the variation among employees in job performance ratings can be explained on the basis of their variation in AAAT scores?

b. How much of the variation in ratings can be attributed to a linear relationship between the ratings and the number of accounting courses the employee had while in college?

c. What percentage of the variation in ratings is acounted for by a multiple linear relationship with both factors?

d. Use a level of significance of $\alpha = .01$ to test whether or not the relationship in question can reasonably be considered a multiple linear one.

9.4. The I. M. Lyer, Inc., advertising agency conducted a study of television and radio effectiveness by obtaining the following data on six randomly selected shoppers:

Shopper	#1	#2	#3	#4	#5	#6
Average number of hours of TV watched per day	1	2	3	1	1	2
Average number of hours of radio listened to per week	20	10	5	15	10	20
Average number of stores per week at which a purchase is made	25	15	5	20	15	20

a. What percentage of the variation in average number of stores among the shoppers surveyed can be attributed to a linear relationship with number of hours of TV watched?

b. How much of the variation in average number of stores can be explained on the basis of a linear relationship with number of hours of radio listened to?

c. What proportion of the variation is accounted for by a multiple linear relationship with both factors?

9.5. An analysis of the relationship between corporate earnings on the one hand,
and, on the other, sales volume and the total remuneration paid to its chief
executive yielded the following information on 14 randomly selected corporations:*

Corporation	1976 Sales Volume (billions of $)	1976 Executive Remuneration (thousands of $)	1976 Earnings Per Share (dollars)
American Fletcher	0.16	120	1.69
Carborundum	0.61	369	4.14
Walt Disney	0.58	167	2.34
Fruehauf	1.47	400	4.05
Jefferson-Pilot	0.54	167	2.82
National Can	0.92	222	2.52
Phelps Dodge	0.94	238	2.07
Texaco	26.45	661	3.20
Winn-Dixie	3.27	325	3.01
Union Pacific	2.02	469	4.02
St. Regis Paper	1.64	327	3.82
Olin Corp.	1.38	406	6.07
R. H. Macy	1.47	262	4.37
Hershey Foods	0.58	192	3.32

a. What proportion of the variation in earnings per share can be explained on
the basis of a linear relationship with sales volume?

b. What proportion can be explained on the basis of a linear relationship with
executive remuneration?

c. What proportion can be explained on the basis of a multiple linear relationship with both factors?

SECTION 9.B   THE MULTIPLE REGRESSION EQUATION

As we have shown in Section 9.A, the department store sales volume data of
Example 9.1 strongly indicates that sales volume is related in a multiple
linear manner to advertising expenditures and morning precipitation. The
question now is

a. What is the best-fitting multiple linear regression equation of the
form

$$y = a + bx + cw \tag{9.6}$$

*Forbes, May 15, 1977, pp. 202–243.

where a, b, and c are numbers calculated from the data and

y = sales volume (in thousands of dollars)
x = advertising costs (in hundreds of dollars)
w = morning precipitation (in tenths of an inch)

Once we have calculated a multiple regression equation, we usually will want to use it to forecast future daily sales volume. For example, we could answer questions such as the following:

b. If the store spends $700 for next Monday's advertising and the weather service predicts 0.6 inch of precipitation for that morning, what would be the best prediction for Monday's sales volume?

We now proceed to answer these two questions in order.

Determining the Best-Fitting Multiple Linear Regression Equation

We have to compute the numerical values of a, b, and c. The criterion of *least squares*, the same criterion used in Chapter 8, is also the basis of multiple linear regression. Of course, the calculations have to be expanded somewhat to take into account the larger number of variables. It can be shown, using calculus, that a, b, and c are the three unknowns whose values can be determined by solving the following three *simultaneous equations*:

$$na + b\Sigma x + c\Sigma w = \Sigma y$$
$$a\Sigma x + b\Sigma x^2 + c\Sigma xw = \Sigma xy \qquad (9.7)$$
$$a\Sigma w + b\Sigma xw + c\Sigma w^2 = \Sigma wy$$

Inserting the results of the calculations in Table 9.1 into equations (9.7), we obtain

$$6a + 30b + 15c = 108$$

$$30a + 190b + 80c = 688$$

$$15a + 80b + 61c = 241$$

There are many ways of solving three simultaneous equations (as you may have learned in your algebra courses), but none of them is particularly easy. We will present one method here that can be used for hand computation, and your instructor undoubtedly knows of several others. We first set up the *matrix of coefficients* as follows, ignoring the letters a, b, and c, but being sure to keep everything in the proper order:

$$\begin{vmatrix} 6 & 30 & 15 & 108 \\ 30 & 190 & 80 & 688 \\ 15 & 80 & 61 & 241 \end{vmatrix}$$

What we would like to do is to transform this matrix into one of the form

$$\blacksquare \qquad \begin{vmatrix} 1 & 0 & 0 & a \\ 0 & 1 & 0 & b \\ 0 & 0 & 1 & c \end{vmatrix} \qquad\qquad (9.8)$$

If we can transform the first three columns of the matrix into the exact pattern of zeros and ones as shown, then the fourth column will contain the correct values of a, b, and c.

We first divide each row by the number appearing in the first column of that row. Namely, we divide each number in the top row by 6, each number in the middle row by 30, and each number in the bottom row by 15. After this operation, our matrix is as follows:

$$\begin{vmatrix} 1 & 5 & 2.5 & 18 \\ 1 & 6.333 & 2.667 & 22.933 \\ 1 & 5.333 & 4.067 & 16.067 \end{vmatrix}$$

The next step is to keep the top row as it is, but subtract the top row from the middle *and* bottom rows. What we do is take each number in the middle row, for example, and subtract from it the number in the corresponding position of the first row: look at the number in the third column of the middle row. It is 2.667. We subtract from this the number in the third column of the top row, which is 2.5. The new matrix then has $2.667 - 2.5 = .167$ in the third column of its middle row. Doing this with the seven other entries of the middle and bottom rows, we obtain the matrix

$$\begin{vmatrix} 1 & 5 & 2.5 & 18 \\ 0 & 1.333 & .167 & 4.933 \\ 0 & .333 & 1.567 & -1.933 \end{vmatrix}$$

We next divide the middle and bottom rows by their number appearing in the second column; i.e., 1.333 for the middle row and .33 for the bottom row. This step, leaving the top row untouched, yields the matrix:

$$\begin{vmatrix} 1 & 5 & 2.5 & 18 \\ 0 & 1 & .125 & 3.701 \\ 0 & 1 & 4.706 & -5.805 \end{vmatrix}$$

We now subtract the middle row from the bottom row, and leave the top and middle rows untouched. We get:

$$\begin{vmatrix} 1 & 5 & 2.5 & 18 \\ 0 & 1 & .125 & 3.701 \\ 0 & 0 & 4.581 & -9.506 \end{vmatrix}$$

and we then multiply the middle row by the number appearing in the second column of the top row; namely, 5:

$$\begin{vmatrix} 1 & 5 & 2.5 & 18 \\ 0 & 5 & .625 & 18.505 \\ 0 & 0 & 4.581 & -9.506 \end{vmatrix}$$

Next we subtract the middle row from the top row, and leave the middle and bottom rows untouched:

$$\begin{vmatrix} 1 & 0 & 1.875 & -.505 \\ 0 & 5 & .625 & 18.505 \\ 0 & 0 & 4.581 & -9.506 \end{vmatrix}$$

After this, we divide the middle row by the number appearing in its second column; namely, 5:

$$\begin{vmatrix} 1 & 0 & 1.875 & -.505 \\ 0 & 1 & .125 & 3.701 \\ 0 & 0 & 4.581 & -9.506 \end{vmatrix}$$

(Perhaps you can now see the solution taking shape.)

What is the next step? Well, we divide the bottom row by the number appearing in its third column; namely, 4.581:

$$\begin{vmatrix} 1 & 0 & 1.875 & -.505 \\ 0 & 1 & .125 & 3.701 \\ 0 & 0 & 1 & -2.075 \end{vmatrix}$$

We then multiply the bottom row by .125 and subtract it from the middle row. Then we divide the bottom row by .125 to return it to the above form. These three steps are recorded as follows:

$$\begin{vmatrix} 1 & 0 & 1.875 & -.505 \\ 0 & 1 & .125 & 3.701 \\ 0 & 0 & .125 & -.259 \end{vmatrix} \rightarrow \begin{vmatrix} 1 & 0 & 1.875 & -.505 \\ 0 & 1 & 0 & 3.960 \\ 0 & 0 & .125 & -.259 \end{vmatrix}$$

$$\rightarrow \begin{vmatrix} 1 & 0 & 1.875 & -.505 \\ 0 & 1 & 0 & 3.960 \\ 0 & 0 & 1 & -2.072 \end{vmatrix}$$

(Some minor discrepancies in the third decimal place can be expected due to round-off error. The only way to eliminate such discrepancies is to carry more decimal places, as would a computer or pocket calcuator.)

We then multiply the bottom row by 1.875, and subtract it from the top row, obtaining:

$$\begin{vmatrix} 1 & 0 & 1.875 & -.505 \\ 0 & 1 & 0 & 3.960 \\ 0 & 0 & 1.875 & -3.885 \end{vmatrix} \rightarrow \begin{vmatrix} 1 & 0 & 0 & 3.38 \\ 0 & 1 & 0 & 3.960 \\ 0 & 0 & 1.875 & -3.885 \end{vmatrix}$$

We finally divide the bottom row by the number appearing in its third column; namely, 1.875, to get

$$\begin{vmatrix} 1 & 0 & 0 & 3.38 \\ 0 & 1 & 0 & 3.96 \\ 0 & 0 & 1 & -2.072 \end{vmatrix} = \begin{vmatrix} 1 & 1 & 0 & a \\ 0 & 1 & 0 & b \\ 0 & 0 & 1 & c \end{vmatrix}$$

in view of formula (9.8). It follows that in the multiple regression equation (9.6),

$$y = a + bx + cw,$$

we have

$$a = 3.38$$
$$b = 3.96$$
$$c = -2.07$$

so that the multiple regression equation is

$$y = 3.38 + 3.96x - 2.07w.$$

Table 9.3 contains a BASIC computer program for carrying out the above matrix manipulations.

Predicting a Day's Sales Volume

The best prediction for Monday's sales volume, given that $700 will be spent on advertising and 0.6 inches of precipitation will fall that morning, can be obtained by setting $x = 7$ (since advertising costs are listed in hundreds of dollars) and $w = 6$ (since precipitation is listed in tenths of an inch) into the multiple regression equation. We get

$$y = 3.38 + 3.96x - 2.07w$$
$$= 3.38 + (3.96)(7) - (2.07)(6)$$
$$= 3.38 + 27.72 - 12.42$$
$$= 18.68.$$

Because y is expressed in thousands of dollars, we would predict the day's sales volume to be $18,680.

Now that we know that sales volume (y) can be predicted from a knowledge of advertising expenditures (x) and morning precipitation (w), we can estimate any day's sales volume based on various combinations of advertising expenditures and precipitation. For example, if .2 inch of rainfall is predicted

TABLE 9.4   Estimated Department Store Sales Volume

Advertising Expenditures	x	Morning Precipitation (inches)	w	$y = 3.38 + 3.96x - 2.07w$	Estimated Sales Volume
$200	2	.1	1	9.15	$9,150
$200	2	.3	3	5.01	$5,010
$200	2	.5	5	.087	$870
$400	4	.1	1	17.07	$17,070
$400	4	.3	3	12.93	$12,930
$400	4	.5	5	8.79	$8,790
$600	6	.1	1	24.99	$24,990
$600	6	.3	3	20.85	$20,850
$600	6	.5	5	16.71	$16,710

for next Tuesday morning, what would be our predicted sales volume if we spend $400 on advertising? Well, here $x = 4$ and $w = 2$ so that

$$y = 3.38 + 3.96x - 2.07w$$

$$= 3.38 + (3.96)(4) - (2.07)(2)$$

$$= 3.38 + 15.84 - 4.14 = 15.08$$

Therefore, our estimated sales volume for next Tuesday would be $15,080. Finally, suppose that no rainfall is forecast for next Wednesday and we plan to spend $400 on advertising. Then $x = 4$ and $w = 0$, so our estimated sales volume (in thousands of dollars) would be

$$y = 3.38 + 3.96x - 2.07w$$

$$= 3.38 + (3.96)(4) - (2.07)(0)$$

$$= 3.38 + 15.84 - 0 = 19.22,$$

namely $19,220.

Table 9.4 exhibits the estimated sales volume, based on various combinations of advertising expenditures and precipitation levels.

EXERCISES   9.B

9.6. Using the sardine price data of Exercise 9.1:

   a. Calculate the multiple regression equation for predicting price to the consumer, from a knowledge of the sardine population and the transportation cost.

    b. Use the equation to predict the price of a 3-ounce can of sardines in 1985, when the sardine population is estimated to be 25 trillion and transportation costs will be $18 per gross of cans.

9.7. Based on the motivational analysis data of Exercise 9.2:

    a. Determine the multiple regression equation for predicting an employee's salary level, from a knowledge of his or her number of dependents and years of service.

    b. Use the equation to predict the salary level of an employee who has six dependents and four years of service.

9.8. From the performance rating data of Exercise 9.3:

    a. Find the multiple regression equation that should be used to predict an applicant's job performance rating, based on his or her number of accounting courses and AAAT score.

    b. Predict the job performance rating of an applicant who has taken 5 courses in accounting and scored 30 on the AAAT.

9.9. Use the advertising effectiveness data of Exercise 9.4 to:

    a. Calculate the multiple regression equation for predicting the average number of stores at which a shopper makes a purchase each week, based on the shopper's average number of hours of TV per day and average number of hours of radio per week.

    b. Predict the average number of stores per week at which a shopper who watches no television, but listens to about 30 hours of radio per week, makes a purchase.

SECTION 9.C   MULTIPLE REGRESSION ANALYSIS

The numbers a, b, and c in the multiple regression equation

$$y = a + bx + cw \qquad (9.9)$$

that we showed how to calculate in Section 9.B are not really the "true" coefficients in the regression model. If x, w, and y are truly related in a multiple linear manner, then the true multiple linear relationship can be expressed by

$$y = \alpha + \beta x + \gamma w \qquad (9.10)$$

and a, b, and c are merely sample estimates of the true coefficients α, β, and γ, respectively. The F test using equation (9.5) is a hypothesis test of whether or not equation (9.10) is an appropriate way of modeling the data.

    When the number of regressors (here, x and w) is small (say, 2), the formulas presented in Section 9.A for calculating R^2 can be used without too much difficulty. As the number of regressors increases, however, the formulas for R^2 analogous to (9.1) and (9.2) become hopelessly complicated, even for writing a computer program, and so we need an entirely different way of handling the problem. It turns out that the analysis of variance theory introduced in Chapter 7 (in the context of testing for difference between means) can be reinterpreted and applied to the multiple regression problem.

The ANOVA Approach to Multiple Correlation Analysis

If we state that the linear relationship of formula (9.9) expresses the variation of y in terms of variations in x and w, then are we saying that, if we calculate $a + bx + cw$ for a particular pair x and w, that number will equal the y value corresponding to x and w? If that's what we are saying, then we will be making an error, because a glance at Table 9.5 shows that the columns headed y and $a + bx + cw$ are different. In fact, the sum of their squared differences $\Sigma(y-a-bx-cw)^2 = 7.7673$ is called the *error (or residual) sum of squares* and is denoted by SSE. The number SSE (the symbolism comes from the analysis of variance analogue) denotes exactly how erroneous is our statement that y is equal to $a + bx + cw$.

TABLE 9.5   Error Sum of Squares (SSE) in the Department Store Multiple Regression Model $(y = 3.38 + 3.96x - 2.07w)$

x	w	y	$a + bx + cw$	$y - a - bx$ $- cw$	$(y - a - bx$ $- cw)^2$
2	5	1	0.95	0.05	.0025
3	3	9	9.05	−0.05	.0025
5	1	19	21.11	−2.11	4.4521
4	0	21	19.22	1.78	3.1684
6	1	25	25.07	−0.07	.0049
10	5	33	32.63	0.37	.1369
30	15	108	108.00	−0.03	7.7673

SSE $= \Sigma(y - a - bx - cw)^2 = 7.7673$

Big deal. y is not exactly equal to $a + bx + cw$. So what? This is statistics; nothing is ever *exactly* equal to anything else! The *true* relationship could still be the one expressed in formula (9.10). We have to compare the size of SSE with the error we would make in assuming that *there was no* multiple linear relationship. If we were not to assume that a relationship like (9.10) was appropriate, the best we could do in predicting y would be to say that y always equals $\bar{y} = 18$. How much error do we make in using *this* method of prediction? The required computations are given in Table 9.6, and

TABLE 9.6 Total Sum of Squares (SS_{total}) and Regression Sum of Squares (SSR) in the Department Store Multiple Regression Model ($y = 3.38 + 3.96x - 2.07w$)

x	w	y	$\bar{y}$	$y - \bar{y}$	$(y - \bar{y})^2$	$a + bx$ $+ cw - \bar{y}$	$(a + bx$ $+ cw - \bar{y})^2$
2	5	1	18	-17	289	-17.05	290.7025
3	3	9	18	-9	81	-8.95	80.1025
5	1	19	18	1	1	3.11	9.6721
4	0	21	18	3	9	1.22	1.4884
6	1	25	18	7	49	7.07	49.9849
10	5	33	18	15	225	14.63	214.0369
30	15	108	108	0	654	0.03	645.9873

$SS_{total} = \Sigma(y - \bar{y})^2 = 654$

$SSR = \Sigma(a + b + cw - \bar{y})^2 = 645.9873$

Note: $SS_{total} = SSR + SSE$

$654 = 645.9873 + 7.7673$

(discrepancy is due to roundoff in tables)

they show that this error, which is called the *total sum of squares* (and is denoted SS_{total}), equals 654.

To summarize our results so far: if we do not use multiple linear regression, the sum of squares representing our prediction error is $SS_{total} = 654$; while if we do use it, the error is $SSE = 7.7673$. The difference $SS_{total} - SSE$ is called the *regression sum of squares* and is denoted SSR. The larger SSR is, the more likely it is that the regression model is an appropriate way of analyzing the data. In this case

$$SSR = SS_{total} - SSE = 646.2327$$

Note that the last column of Table 9.6 computes $\Sigma(a + bx + cw - \bar{y})^2 = 645.9873$. Actually the formula for SSR is

$$SSR = \Sigma(a + bx + cw - \bar{y})^2$$

but the discrepancy between this and $SSR = SS_{total} - SSE$ is due solely to the many roundoff operations required during the course of the calculations. (A computer solution of the problem would not produce this discrepancy because of the many decimal places carried along by the computer.)

Recall that R^2 is the proportion of variation in y that can be explained on the basis of the multiple linear model. Therefore an alternative formula for R^2 is

$$R^2 = SSR/SS_{total} = 646.2327/654 = 0.988,$$

TABLE 9.7 ANOVA Table for Multiple Linear Regression (*Department store data*)

Source of Variation	Degrees of Freedom	Sum of Squares	Mean Sum of Squares	F
Regression	k	SSR	$MSR = \dfrac{SSR}{k}$	$\dfrac{MSR}{MSE}$
Error	$n-k-1$	SSE	$MSE = \dfrac{SSE}{n-k-1}$	*
Total	n	SS_{total}	*	.*

Source of Variation	Degrees of Freedom	Sum of Squares	Mean Sum of Squares	F
Regression	2	646.2327	323.116	124.8
Error	3	7.7673	2.589	*
Total	5	654	*	*

n = number of "triples" of data points

k = number of regressors

TABLE 9.8 IMSL Multiple Regression Analysis (*For predicting sales volume in terms of advertising costs and morning precipitation*)

```
IMSL MULTIPLE REGRESSION ANALYSIS
(FOR PREDICTING SALES VOLUME IN TERMS OF
ADVERTISING COSTS AND MORNING PRECIPITATION)


COEFFICIENT OF MULTIPLE DETERMINATION

R-SQUARED = 0.9881


ANALYSIS OF VARIANCE

SOURCE     D.F.     SS          MEAN SS    F-VALUE   P-VALUE

REGRESSION  2.     646.24       323.12     124.807   0.001
ERROR       3.       7.77         2.59       **        **
TOTAL       5.     654.00         **         **        **


REGRESSION COEFFICIENT INFERENCES

VARIABLE   REGRESSION   STANDARD      PARTIAL      SIGNIFICANCE
  NAME     COEFFICIENT  ERROR         F-VALUE        P-VALUE

INTERCEPT    3.393      1.564
ADV COST     3.960      0.258         235.790        0.001
M PRECIP    -2.077      0.336          38.098        0.009


(NOTE: REJECT HYPOTHESIS OF SIGNIFICANCE
       IF P-VALUE EXCEEDS YOUR ALPHA LEVEL.)
```

TABLE 9.9 BMDP1R—Multiple Linear Regression, Health Sciences
Computing Facility, University of California, Los Angeles 90024 *(Program
revised November 1979)*

```
BMDP1R - MULTIPLE LINEAR REGRESSION
HEALTH SCIENCES COMPUTING FACILITY
UNIVERSITY OF CALIFORNIA, LOS ANGELES 90024
PROGRAM REVISED NOVEMBER 1979



  PROBLEM TITLE .    .   .
PREDICTING SALES VOLUME IN TERMS OF ADVERTISING COSTS AND MORNING PRECIPITATION


  REGRESSION TITLE. . . . . . . . . . . . . .PREDICTING SALES VOLUME
  DEPENDENT VARIABLE. . . . . . . . . . . . .        3 SALESVOL
  TOLERANCE . . . . . . . . . . . . . . . . .    0.0100
ALL DATA CONSIDERED AS A SINGLE GROUP

MULTIPLE R               0.9940          STD. ERROR OF EST.        1.6090
MULTIPLE R-SQUARE        0.9881

ANALYSIS OF VARIANCE
                         SUM OF SQUARES    DF    MEAN SQUARE    F RATIO    P(TAIL)
          REGRESSION         646.233        2      323.116      124.806    0.00129
          RESIDUAL             7.767        3        2.589

                                                STD. REG
   VARIABLE         COEFFICIENT   STD. ERROR     COEFF       T    P(2 TAIL) TOLERANCE

INTERCEPT              3.39345
ADVCOST      1         3.95956      0.258        0.979    15.355    0.001    0.973405
MPRECIP      2        -2.07650      0.336       -0.394    -6.172    0.009    0.973405



  CORRELATION MATRIX


                   ADVCOST      MPRECIP    SALESVOL
                      1            2          3

ADVCOST      1      1.0000
MPRECIP      2      0.1631       1.0000
SALESVOL     3      0.9150      -0.2339      1.0000
```

which is the same numerical value for R^2 that we got in Section 9.A. (Either
calculated value of SSR can be used.)

The hypothesis test for multiple linearity can be constructed using an
ANOVA table of the sort appearing in Table 7.11 as part of our discussion of
one-way analysis of variance. The ANOVA table appears in Table 9.7 and
shows that $F = 124.8$. Except for rounding discrepancies, this is the same
result for F that we got in Section 9.A. Because $F > F_\alpha[k, n - k - 1] = F_\alpha[2,3]$
$= 9.55$, we reject the hypothesis that there is no multiple linear relationship.

The output of standard "canned" computer programs that are normally
used for solving multiple regression problems is generally expressed as an
ANOVA table of the sort illustrated in Table 9.7.

For examples of how the output of two widely used commercial computer programs can look when applied to the department store example, see
Tables 9.8 and 9.9. It is likely that your college computer center has access
to one or more of such programs.

479

Testing the Significance of the Regressors

One of the major questions in multiple regression with k regressors is, "How significant a role does each regresor play in helping to produce a good prediction of y?" Some regressors may exert a lot of influence on y, while others may be very insignificant.

Everything we say on this topic will apply to the k-regressor situation, which is usually expressed in the form

$$y = \beta_0 + \beta_1 x_1 + \beta_2 x_2 + \ldots + \beta_k x_k \qquad (9.11)$$

where each β_j-value is the coefficient corresponding to its regressor x_j, and β_0 is the constant term called the *intercept*. However, because the significance of each regressor can be tested separately, the procedure is no different in the two-regressor case whose form is given by equation (9.10):

$$y = \alpha + \beta x + \gamma w.$$

Therefore our explanation will be expressed in terms of the easier to understand two-regressor case.

In the two-regressor case, how can we tell how significant a role is played by the regressor x? Well, if $\beta = 0$, the regression equation may as well be written

$$y = \alpha + \gamma w$$

because inclusion of x in the equation contributes nothing beyond that already contributed by w. Similarly if $\gamma = 0$, then the particular value of w is simply irrelevant in making a prediction of y. We can therefore construct a test of the hypothesis

H: regressor x is insignificant if w is already included

against the alternative

A: regressor x is significant even if w is included

by writing the hypothesis and alternative as follows:

H: $\beta = 0$

vs.

A: $\beta \neq 0$.

We reject H at significance level α if $F > F_\alpha[1, n - k - 1]$ where

$$F = \frac{\text{SSE}_\gamma - \text{SSE}}{\text{SSE}/(n - k - 1)} \qquad (9.12)$$

In equation (9.12), the term SSE is the error sum of squares for the regression equation $y = \alpha + \gamma w$, namely the error sum of squares for the regression problem that ignores the regressor x.

The easiest way to find SSE_γ is to calculate R_γ^2 for the regression equation $y = \alpha + \gamma w$. (Because $y = \alpha + \gamma w$ is a simple, rather than a multiple, regression equation, R_γ^2 may be replaced by r_{wy}^2, the coefficient of linear determination. If the original problem has several variables, then R_γ^2 will be an actual coefficient of *multiple* determination.) After we calculate R_γ^2, we recall that

$$R_\gamma^2 = \frac{SSR_\gamma}{SS_{total}} = \frac{SS_{total} - SSE_\gamma}{SS_{total}}.$$

Here

$$SS_{total} = \Sigma(y - \bar{y})^2 = \Sigma y^2 - \frac{(\Sigma y)^2}{n}$$

is the same SS_{total} we used in the earlier portion of the multiple regression problem. If we solve the above algebraic equation for the unknown value of SSE_γ, we get that

$$SSE_\gamma = (1 - R_\gamma^2) SS_{total} . \qquad (9.13)$$

We can then calculate the value of F using the formula

$$F = \frac{(1 - R_\gamma^2)SS_{total} - SSE}{SSE/(n - k - 1)} \qquad (9.14)$$

Let's conduct a test of the hypothesis H: $\beta = 0$ against the alternative A: $\beta \neq 0$ at significance level $\alpha = .05$. For the department store problem we have been working on, $n - k - 1 = 6 - 2 - 1 = 3$ so that we would reject H in favor of A if $F > F_{.05}[1,3] = 10.13$.

We know from the calculations made in Section 9.A that $R_\gamma^2 = r_{wy}^2 = .055$. We have calculated in this section that $SS_{total} = 654$ and $SSE = 7.7673$. Therefore

$$F = \frac{(1 - .055)(654) - 7.7673}{7.7673/3} = \frac{610.263}{2.589} = 235.705 .$$

Because $F = 235.705 > 10.13$, we reject H: $\beta = 0$ and conclude (at level $\alpha = .05$) that the regressor x plays a significant role in predicting y, using a regression equation.

Before we leave this subject, let's conduct a hypothesis test of whether or not the regressor w is significant if x is already included. This test is one of the hypothesis

H: $\gamma = 0$

against the alternative

A: $\gamma \neq 0$.

Because $R_\beta^2 = r_{xy}^2$, we compute

$$F = \frac{(1 - r_{xy}^2)SS_{total} - SSE}{SSE/(n - k - 1)} = \frac{(1 - .837)(654) - 7.7673}{7.7673/3}$$

$$= \frac{98.8347}{2.589} = 38.173$$

and we reject H at level $\alpha = .05$ because $F = 38.173 > F_{.05}[1,3] = 10.13$. Our conclusion is that the regressor w also plays a significant role in basing a prediction on the regression equation (9.10).

In the case of several regressors, as exemplified by equation (9.11), it is even more important to find out which regressors are significant and which are not. To test the significance of the regressor x_j, we set up the hypothesis

H: $\beta_j = 0$

and the alternative

A: $\beta_j \neq 0$

We calculate

$$F_j = \frac{(1 - R_{j*}^2)SS_{total} - SSE}{SSE/(n - k - 1)} \tag{9.15}$$

where R_{j*}^2 is the coefficient of multiple determination of y with respect to all the variables $x_1, x_2, \ldots, x_k$ except x_j. We then reject H at level α if $F > F_\alpha[1, n - k - 1]$.

For more than two regressors, the calculations are too complicated to be done by hand in a reasonable amount of time. Standard computer subroutines (written by professional statisticians and programmers) are included in various statistical packages, some of which are probably available at your educational and commercial institutions. These subroutines print out the coefficient of multiple determination, the ANOVA table, and information required to test for the significance of each of the regressors. Some of them also provide additional material on confidence intervals for predictions and other topics. Examples of such standard printouts appear in Tables 9.8 and 9.9.

The two commercial programs illustrated in Tables 9.8 and 9.9 are more or less typical examples of a number of extremely efficient and useful programs. Table 9.8 illustrates the possible output of the IMSL subroutine RLMUL, one of a large volume of FORTRAN subroutines available from International Mathematical and Statistical Libraries, Inc., of Houston, Texas. The fact that RLMUL is a FORTRAN subroutine is both good news and bad news. The bad news is that, in order to use it, you have to know enough FORTRAN to be able to write your own FORTRAN program to read in the data and print out the information. The subroutine RLMUL is then called somewhere in the interior of your program. (The IMSL subroutine BECOVM is also needed if you are going to try it.) The good news is that, through your control of the FORTRAN program, you can tailor the output to fit any special needs of the project. You can specify the arrangement of the output, the titles of the columns, and any special information to the reader, such as the note at the bottom of Table 9.8.

Table 9.9, on the other hand, contains the output of the UCLA Biomedical program BMDP1R. BMDP1R is a complete program, not simply a FORTRAN subroutine. Again, there is both good news and bad news associated with this. The bad news is that, except for certain valuable options available such as having the pairwise correlations (which we calculated in Section 9.A) printed out as at the bottom of Table 9.9, you have no control over how the output is printed. The BMDP1R program needs some information such as the data and the regressor names; but once given this information, the program always produces a printout exactly as shown in Table 9.9. The good news is that (1) the printout you get contains everything you need anyway, and (2) you don't have to write a FORTRAN program or even know any FORTRAN in order to use the UCLA Biomedical package.

Some, but not all, other generally available "canned" computer programs are those of IBM Corporation's SSP (Scientific Subroutine Package), University of Chicago's SPSS (Statistical Package for the Social Sciences), the SAS Institute's SAS (Statistical Analysis System), and Pennsylvania State University's MINITAB. Of course, you may always write your own program in almost any computer language you know; the BASIC program of Table 9.3 is an example of this. Programmable calculator packages such as those of Hewlett-Packard and Texas Instruments also contain multiple regression programs. When all else fails, the abacus and the slide rule will also work, if used with the formulas given in this chapter.

EXERCISES 9.C

9.10. Construct the ANOVA table for the sardine price data of Exercise 9.1, and:

    a. Test for the validity of the multiple linear regression model at level $\alpha = .05$;

    b. At level $\alpha = .05$, can the transportation cost be considered a significant regressor in predicting the price of a 3-ounce can?

9.11. Construct the ANOVA table based on the motivational analysis data of Exercise 9.2, and:

    a. Use a level of significance of $\alpha = .05$ to test for the validity of the multiple linear regression model;

    b. At level $\alpha = .01$, can we consider an employee's years of service in the corporation to be a significant regressor on his or her salary level?

9.12. Set up the ANOVA table using the performance rating data of Exercise 9.3, and:

    a. Test for the validity of the multiple linear regression model at level $\alpha = .01$;

    b. At level $\alpha = .05$, it is reasonable to consider an employee's number of courses in accounting to be a significant regressor in predicting his or her job performance rating?

9.13. Construct the ANOVA table for the advertising effectiveness data of Exercise 9.4, and:

    a. Test for the appropriateness of the multiple linear regression model, using a level of significance of $\alpha = .05$;

    b. At level $\alpha = .05$, can we consider a shopper's average number of radio hours listened to as a significant regressor in predicting his or her average number of stores at which a purchase is made?

SUMMARY AND DISCUSSION

When a quantity under study depends on more than one factor, the coefficient of multiple determination measures the degree to which a multiple linear regression equation describes a set of data relating to that quantity. As we have seen, the coefficient of multiple determination is based on the intertwining relationships among various coefficients of linear determination. The algebraic technique of setting up and solving three simultaneous equations in three unknowns yielded the multiple regression equation itself.

A BASIC computer program was presented to carry out the calculations required, which were somewhat more complicated than those encountered in earlier chapters of the text. Using an ANOVA table structure, we developed a general method of testing for the validity of the multiple linear regression model and also for the individual significance of each of the regressors.

What we have done in Chapter 9 grew out of the concepts introduced in Chapter 8 and serves to illustrate that the versatility of those concepts transcends the straight-line data they were originally developed to analyze.

CHAPTER 9   BIBLIOGRAPHY

GRAPHICAL AND OTHER INTERPRETATIONS

CROCKER, D. C., "Some Interpretations of the Multiple Correlation Coefficient," The American Statistician, April 1972, pp. 31–33.

FLEISS, J. L., and J. M. TANUR, "A Note on the Partial Correlation Coefficient," The American Statistician, February 1971, pp. 43–45.

MULLET, G. M., "Graphical Illustration of the Simple (Total) and Partial Regression," The American Statistician, December 1972, pp. 25–27.

WEISS, N. S., "A Graphical Representation of the Relationships between Multiple Regression and Multiple Correlation," The American Statistician, April 1970, pp. 25–29.

APPLICATIONS

BRIEF, A. P., M. J. WALLACE, JR., and R. J. ALDAG, "Linear vs. Non-linear Models of the Formation of Affective Reactions: The Case of Job Enlargement," *Decision Sciences*, January 1976, pp. 1–9.

FRANK, R. E., and W. F. MASSY, "The Effect of Retail Promotional Activities on Sales," *Decision Sciences*, October 1971, pp. 405–431.

HEWARD, J. H., and P. M. STEELE, *Business Control Through Multiple Regression Analysis*. 1972, Plymouth, England: Gower Press.

COMPUTER PACKAGES

Health Sciences Computing Facility, University of Caifornia, Los Angeles, *BMDP Biomedical Computer Programs*. 1979, University of California Press.

The IMSL Library: FORTRAN Subroutines in the Areas of Mathematics and Statistics. 1979, International Mathematical & Statistical Libraries, Inc., Sixth Floor, GNB Building, 7500 Bellaire Blvd., Houston, Texas 77036.

CHAPTER 9   SUPPLEMENTARY EXERCISES

9.14. An insurance industry study came up with the following data relating the dollar value of a head of household's life insurance coverage with that individual's formal educational level and the total income of the household, as follows.

Years of formal education	8	12	18	10	6	14
Total household income (thousands of $)	12	20	16	16	8	18
Life insurance coverage (thousands of $)	7	12	16	12	4	16

a. What proportion of the variation in life insurance coverage can be attributed to a linear relationship with years of formal education?

b. How much of the variation is explained by a linear relationship with household income?

c. What percentage of the variation can be accounted for by a multiple linear relationship with both factors?

d. Construct the ANOVA table based on the data, and, at level $\alpha = .05$, test for the existence of a multiple linear relationship in the underlying population.

e. Find the multiple regression equation that the industry can use to predict an individual's life insurance coverage from a knowledge of his or her educational level and household income.

f. At level $\alpha = .01$, can we consider an individual's number of years of formal education to be a significant regressor on his or her life insurance coverage?

g. Estimate the life insurance coverage of a head of household who has completed 16 years of formal education and whose household takes in $20,000 a year.

9.15. An agricultural research organization tested various mixtures of chemical and organic fertilizer to try to find out how variations in the amounts of fertilizer induce variations in food production. The following data were collected from seven virtually identical plots of farmland:

Pounds of Chemical Fertilizer Used	Pounds of Organic Fertilizer Used	Bushels of Beans Harvested
1	8	120
2	6	140
3	5	150
4	4	180
5	3	200
6	2	190
7	2	220

a. Determine the proportions of variation in harvest that can be attributed to separate linear relationships with each of the two fertilizers.

b. Calculate the proportion of variation in harvest that can be explained on the basis of a multiple linear relationship with both fertilizers.

c. What is the best-fitting mutiple linear regression equation for bushels of beans in terms of both kinds of fertilizer?

d. Predict the harvest when 8 pounds of chemical and no pounds of organic fertilizer are used.

e. Predict the harvest when 8 pounds of organic and no pounds of chemical fertilizer are used.

f. Predict the harvest when no fertilizer of either kind is used.

9.16. A psychologist employed by the personnel department of a large corporation developed standardized scales of measurement for the qualities of anxiety, job security, and job managerial responsibility. He would like to be able to predict a manager's anxiety level from a knowledge of the other two factors. The following data were collected on five randomly selected managers employed by the corporation.

Managerial Responsibility Level of Job	Job Security	Anxiety Level
5	2	10
9	4	8
10	5	5
20	9	5
30	10	20

a. What proportion of the variation in anxiety level can be attributed to a linear relationship with managerial responsibility alone?

b. What proportion of the variation in anxiety level can be explained on the basis of a linear relationship with job security alone?

   c. What proportion of the variation in anxiety level can be accounted for by a multiple linear relationship with both factors?

   d. Construct the appropriate ANOVA table, and, at level $\alpha = .05$, test the hypothesis that the relationship can be considered a multiple linear one.

   e. Calculate the multiple regression equation for predicting anxiety level in terms of the other two characteristics.

   f. At level $\alpha = .05$, can we consider the job security index to be a significant regressor for predicting the anxiety level?

   g. Use the regression equation to predict the anxiety level of an employee who has a managerial responsibility score of 15 and a job security score of 8.

9.17. A study of business conditions across the country related the percentage increase in department store sales for August 1977, over those a year earlier in several metropolitan areas, with two other factors.* The data for eight randomly selected metropolitan areas follow.

City	Percentage Rise in Factory Workers' Income over Year	Unemployment Rate	Percentage Rise in Department Store Sales
Fresno	3.0	6.6	10.6
Hartford	7.2	7.2	15.3
Memphis	8.9	5.7	8.3
Columbus	7.5	5.8	10.3
San Diego	3.3	10.0	13.9
Richmond	5.0	4.2	15.0
Grand Rapids	9.9	5.5	18.3
Omaha	8.2	3.9	15.8

   a. Determine the proportion of variation in department store sales rises among the various cities that can be attributed to a linear relationship with the rise in factory workers' incomes.

   b. How much of the variation in sales rises can be explained on the basis of a linear relationship with the unemployment rate?

   c. What proportion of the variation can be accounted for by a multiple linear relationship with both factors?

   d. Construct an ANOVA table based on the data, and test, at level $\alpha = .05$, whether or not the multiple linear model is an appropriate way of modeling this set of data.

   e. At level $\alpha = .05$, can we say that the unemployment rate is a significant regressor in predicting rise in sales?

9.18. One financial analyst feels that a large corporation's net income can be accurately forecast from a knowledge of the corporation's sales volume and stock-

*U.S. News & World Report, October 24, 1977, pp. 73–75.

holders' equity. To check out her theory, she chose a random sample of 12 corporations and found the following data for 1976.*

Corporation	Sales Volume (billions of $)	Stockholders' Equity (billions of $)	Net Income (tens of millions of $)
Gulf Oil	16.45	6.94	81.60
Talley Ind.	0.35	0.09	1.12
Alcoa	2.92	1.69	14.38
Envirotech	0.42	0.10	1.07
Eaton	1.81	0.67	9.09
Cessna Air	0.49	0.18	2.97
Clark Equip	1.26	0.46	6.81
Pitney-Bowes	0.54	0.18	2.08
Schlitz Brew	1.00	0.36	4.99
Interlake	0.71	0.30	3.79
Levi Strauss	1.22	0.36	10.48
Kellwood	0.43	0.06	0.73

a. What proportion of variation in net income among corporations can be explained on the basis of a linear relationship between net income and sales volume?

b. Calculate the proportion of variation in corporate net income that can be accounted for by a linear relationship with stockholders' equity.

c. How much of the variation in net income can be explained on the basis of a multiple linear relationship with both factors?

d. Set up the ANOVA table for the data and, at level $\alpha = .01$, test whether or not the multiple linear relationship is an appropriate model for analyzing the situation under review.

e. Calculate the multiple regression equation for predicting corporate net income, from a knowledge of sales volume and stockholders' equity.

f. At level $\alpha = .05$, can sales volume be considered a significant regressor in predicting net income?

g. Use the regression equation to predict the 1976 net income[†] of Dan River, which in 1976 had sales volume of $500 million and stockholders' equity of $150 million.

*Fortune, May 1977, pp. 366–385.

[†]It was $9,455,000.

10

Index Numbers and Time Series

"WHAT IS PAST IS PROLOGUE," reads the inscription on the National Archives Building in Washington, D.C.* This is the philosophy that underlies the role of time series and trend analysis in business and management forecasting.

By a *time series* we mean a series of data points accumulated over a period of time, usually several years, quarters, or months, as appropriate. The series of data points can represent, for example, monthly sales figures, quarterly business demand, yearly earnings per share, quarterly housing starts, weekly unemployment compensation payments, etc.

Our goal in conducting a statistical analysis of a time series is to determine the nature and extent of the year-to-year, quarter-to-quarter, or month-to-month variation among the data points of the series. We want, in particular, to be able to answer the following questions:

1. What is the overall trend of the data?
2. Is there seasonal variation?
3. Do the numerical values of the data points fluctuate according to some cyclical pattern?

Our answers to these questions will help us develop an understanding of the business relationships involved and will allow us to present more accurate forecasts for use in management decision making.

*The original statement appeared in Shakespeare's *The Tempest*, act II, scene 1.

SECTION 10.A INDEX NUMBERS

Ever since World War II, there has been almost a steady increase in prices in the United States. In addition to causing concern in many sectors of society, this price inflation has created many problems for accountants and financial analysts attempting to make sense out of financial data.

Throughout the 1960s and 1970s, accountants discussed appropriate methods of standardizing financial and economic data to adjust for inflation. Individual corporations also made attempts to adjust their own figures to account for the inflation. (The legal necessity for doing so became a reality in 1976 when the United States Securities and Exchange Commission in *Accounting Series Release, Number 190* set forth rules requiring disclosure of certain replacement cost information.) In many cases, *index numbers* were found to be the most objective, easiest, and least costly method for standard-izing current replacement costs of equipment purchased in pre- or early inflation days.

Basically, an index number is a ratio between a price (or a set of prices) on one date and the price (or set of prices) of equivalent items on another date. (In addition to a price, we can use some other economic factor that can be measured and recorded. See the exercises for some nonprice examples.) One particular date is chosen to serve as a *base time period* or a standard of comparison. For example, if the base year of milk was set at 1975, then the indexes presented for milk prices of other years will be determined by their numerical relationship with the price in 1975. Say a quart of milk cost 30¢ on December 31, 1975 (the base year) and 36¢ on December 31, 1976. Then the 1976 price stands in a ratio of 36¢/30¢ = 1.2 = 120% of the 1975 price. In this case, it is customary to say that the 1976 index of milk is 120.

Published indexes of price changes in various industries and commod-ities are readily available from government as well as industry sources. One of these standard indexes is the Composite Construction Cost Index, which is used in calculating the replacement cost of buildings.

> **EXAMPLE 10.1. Replacement Construction Cost.** An industrial building, perhaps a factory or office structure, was initially purchased for $1.3 million in 1976 when the Composite Construction Cost Index was 150, relative to the base year of 1967. (The index of the base year would always be 100, because the ratio between the 1967 price and the base year (1967) price would always be 1 = 100%.) If 1979 had an index of 170, what would have been the estimated replacement cost of the building?
>
> **SOLUTION.** The Composite Construction Cost Index is formally defined as the ratio
>
> $$I(19XX) = \frac{\text{cost in } 19XX}{\text{cost in } 1967} = \frac{C(19XX)}{C(1967)},$$

because 1967 has been standardized as the base year for this index. It follows algebraically, by cross-multiplying by $C(1967)$, that

$$I(19XX)\,C(1967) = C(19XX),$$

namely that

$$C(19XX) = I(19XX)\,C(1967).$$

Therefore, the ratio between the construction costs in 1979 and 1976 can be expressed as

$$\frac{C(1979)}{C(1976)} = \frac{I(1979)\,C(1967)}{I(1976)C(1967)} = \frac{I(1979)}{I(1976)}$$

because the factor $C(1967)$ has been canceled out of both the numerator and the denominator. Now, multiplying,

$$\frac{C(1979)}{C(1976)} = \frac{I(1979)}{I(1976)}$$

through by $C(1976)$, we obtain that

$$C(1979) = \frac{I(1979)}{I(1976)}C(1976).$$

Upon replacing the symbols in the above formula by their numerical values, namely

$$I(1979) = 170$$

$$I(1976) = 150$$

$$C(1976) = 1.3 \text{ (million dollars)},$$

we find that

$$C(1979) = \frac{170}{150}(1.3) = \frac{221}{150} = 1.47 \text{ (million dollars)}.$$

Therefore, the estimated replacement cost of the building in 1979 is $1.47 million.

The Fundamental Formula of Index Numbers

The main concept that we have used in the above example is the *fundamental formula of index numbers*, which states that

$$\frac{C(19YY)}{C(19XX)} = \frac{I(19YY)}{I(19XX)} \tag{10.1}$$

TABLE 10.1 Consumer Price Index (CPI) for
1960–1970

Year	CPI
1960	88.7
1961	89.6
1962	90.6
1963	91.7
1964	92.9
1965	94.5
1966	97.2
1967	100.0
1968	104.2
1969	109.8
1970	116.3

1967 = base year

Source: Handbook of Labor Statistics, 1978, Table 116, p. 397.

This formula shows the importance of index numbers in estimating projected costs or prices, because it shows that costs or prices (or anything else measured by index numbers) can be analyzed by comparing their index numbers alone, without any knowledge of the base year costs or prices, or even which year was the base year.

Index numbers have also been widely used in property leases since the early 1970s. Again due to the increased pace of inflation, property owners deemed it unwise to be bound to long term leases without a cost-of-living clause. The cost-of-living clause tied rental fees on some properties to one of the widely recognized government indexes. A twenty-year lease, for example, might be tied to one of the Consumer Price Indexes with rent increases based on rises in the index.

The most widely used indexes are the three Consumer Price Indexes and the Producer Price Index. The Consumer Price Indexes CPI and CPI-W measure the average change in prices of all types of consumer goods and services purchased by urban wage earners and clerical workers. The calculations of the indexes are based on studies of actual consumer expenditures on approximately 400 goods and services. The CPI-U reports analogous information on all urban consumers. Before 1978 there was only one Consumer Price Index, the CPI; for 1960–1970, it is listed in Table 10.1. As can be seen from the table, the base year of 1967 has an index of 100 (corresponding to a ratio of 1 = 100%), and prices have been steadily rising during the years covered.

The Producer Price Index measures the average change in prices of all commodities produced or imported for sale in major markets in the United States. Its calculation is determined from price data, collected directly from sellers, on approximately 2700 commodities.

EXERCISES 10.A

10.1. On September 20, 1977, the Bureau of Labor Statistics' "Tuesday index" of the price of industrial raw materials stood at 202.9, compared with the base year average of 100 in 1967. A year earlier, when the Tuesday index was 204.6, a specified amount of industrial raw materials sold for $250,000. Using the index as a guide, estimate the price for the same group of items on September 20, 1977.

10.2. The McGraw-Hill seasonally adjusted index of orders for non-electrical machinery, which uses 1967 as its base year, had a value of 299 in August 1977, up from 268 in August 1976. If a company purchased some machinery in 1977 for $85,000, what would have been its comparable value in 1976?

10.3. The intercity truck tonnage index of the American Truck Associations stood at 114 on September 17, 1977. If a carrier moved 33,000 tons during a comparable period of the base year 1967, what tonnage would it have to have hauled in the 1977 period in order to keep pace with the index?

10.4. The *U.S. News & World Report* index of steel production was at 120.2 on October 8, 1977, up from 111.6 exactly a year earlier. During the 1976 period covered by the index, the O'Hara Steel Mill of suburban Philadelphia produced 8,300 tons. To keep pace with the index, what should have been the company's volume in the 1977 period?

10.5. On October 3, 1977, Standard & Poor's 500 stock index stood at 95.89, whereas a year earlier it reached the 107.83 mark. If a portfolio of the 500 stocks included in the index had a value of $60,000 on October 3, 1976, to what would it have shrunk a year later?

10.6. The index of the price of foodstuffs, compiled by the Bureau of Labor Statistics, reached 204.3 on October 4, 1977, compared to its base year of 1967. Long-range forecasts based on weather and agricultural predictions indicate that the index will stand at 335.2 on October 4, 1983. If the forecast holds up, what would a basket of foodstuffs selling for $42.59 on October 4, 1977, cost six years later?

10.7. A recent article on prices of home appliances in one large metropolitan area revealed the following information.

Item	Units Sold (thousands) 1973	1978	Average Price 1973	1978
Vacuum cleaner	12	14	55	67
Refrigerator	6	5	620	825
Clothes dryer	4	6	220	310
Air conditioner	22	25	550	715
Power lawn mower	8	7	190	270

If we use 1973 as the base year so that 1978 is five years later, standard index number terminology would be as follows:

q_0 = quantity of item sold during base year
q_5 = quantity of item sold five years later
p_0 = base year price of item
p_5 = price of item five years later

We can relabel the table of data as follows:

Item	q_0	q_5	p_0	p_5
Vacuum cleaner	12	14	55	67
Refrigerator	6	5	620	825
Clothes dryer	4	6	220	310
Air conditioner	22	25	550	715
Power lawn mower	8	7	190	270

The data can be used to index the 1978 prices of home appliances as a group with respect to their 1973 prices in a number of ways. In this exercise, you are to calculate the *unweighted index* (also called the *simple aggregative index*), whose formula is shown below.

SIMPLE AGGREGATIVE INDEX FORMULA

$$I_S(1978) = \frac{\Sigma p_5}{\Sigma p_0} \times 100\% .$$

10.8. One defect of the unweighted index calculated in Exercise 10.7 is that, while it claims to represent the overall price change in home appliances, it gives equal influence to clothes dryers and air conditioners even though over five times as many air conditioners as clothes dryers were sold in the base year 1973. This defect can be remedied by using the *weighted aggregative index with base-year weights* (also called the *Laspeyres index*) whose formula is shown below.

LASPEYRES INDEX FORMULA

$$I_L(1978) = \frac{\Sigma q_0 p_5}{\Sigma q_0 p_0} \times 100\% .$$

Calculate the Laspeyres index for the data of Exercise 10.7.

10.9. If quantities of items sold fluctuate rapidly from year to year, it may be more valid, in calculating an index of 1978 prices, to use the 1978 quantities rather than the base-year quantities (which probably are obsolete by 1978). An index that takes account of this situation is the *weighted aggregative index with current-year weights* (also called the *Paasche index*). Its formula is shown below.

PAASCHE INDEX FORMULA
$$I_p(1978) = \frac{\Sigma q_5 p_5}{\Sigma q_5 p_0} \times 100\% \,.$$

Calculate the Paasche index for the data of Exercise 10.7.

10.10. A more stable index of 1978 prices, and those of succeeding years can be obtained by using the *fixed-weight aggregative index*, which uses neither the base-year nor the current-year quantities as weights. This index generally uses as weights the average quantities sold over a period of years. Look at the following table of supplementary data on prices of home appliances.

Item	Units Sold (Thousands) Yearly Average 1970–1978	Average Price			
		1973	1978	1979	1980
	q_A	p_0	p_5	p_6	p_7
Vacuum cleaner	14	55	67	69	72
Refrigerator	6	620	825	840	850
Clothes dryer	5	220	310	315	320
Air conditioner	21	550	715	730	750
Power lawn mower	8	190	270	280	300

a. Calculate the fixed-weight aggregative index using the formula shown below.

FIXED-WEIGHT AGGREGATIVE INDEX FORMULA
$$I_F(1978) = \frac{\Sigma q_A p_5}{\Sigma q_A p_0} \times 100\% \,.$$

---

b. Calculate $I_F(1979)$.

c. Calculate $I_F(1980)$.

10.11. The *ideal index* is an attempt to compromise between the Laspeyres index, having base-year weights, and the Paasche index, having current-year weights.

Using the formula below, calculate the ideal index for the data of Exercise 10.7.

IDEAL INDEX FORMULA

$$I_1(1978) = \sqrt{\frac{\Sigma q_0 p_5}{\Sigma q_0 p_0} \cdot \frac{\Sigma q_5 p_5}{\Sigma q_5 p_0}} \times 100\% \,.$$

10.12. For the data of Exercise 10.10, the *price relative* of an item for 1980 with respect to the base year 1973 is given by the formula below.

PRICE RELATIVE FORMULA

$$\text{P.R.} = \frac{p_7}{p_0} \times 100\%.$$

Using the price relatives, an index which reflects the change in prices from 1973 to 1978 can be constructed. This index is called *the mean of price relatives* and is given by the formula below.

MEAN OF PRICE RELATIVES

$$\text{M.P.R.} = \frac{\Sigma \left(\frac{p_7}{p_0} \times 100 \right)}{n} \%,$$

where n = number of items listed

(n = 5 in the case of the data of Exercise 10.10.)

a. Calculate the mean of price relatives for 1980.

b. Calculate the mean of price relatives for 1979.

c. Calculate the mean of price relatives for 1978.

d. Calculate the mean of price relatives for 1973.

SECTION 10.B THE COMPONENTS OF A TIME SERIES

A major Florida banking corporation reported in its 1978 annual report the quarterly volume of real estate loans it granted during the years 1970 through 1978. This information is presented in Table 10.2. The first point to notice

TABLE 10.2   Quarterly Volume of Real Estate Loans
(1970–1978)   (Millions of dollars)

Year	Quarter	Volume	Year	Quarter	Volume
1970	1	10	1975	1	25
	2	18		2	37
	3	13		3	26
	4	12		4	25
1971	1	11	1976	1	29
	2	18		2	48
	3	12		3	32
	4	11		4	31
1972	1	14	1977	1	34
	2	26		2	52
	3	17		3	35
	4	18		4	34
1973	1	21	1978	1	37
	2	30		2	57
	3	22		3	39
	4	21		4	38
1974	1	26			
	2	43			
	3	28			
	4	30			

from Table 10.2 is that there was considerable variation from quarter to quarter in the dollar amount of real estate loans granted by the bank. In studying time series, our task is to try to find out the nature and sources of this variation.

The so-called *classical time series model* attributes the quarter-to-quarter variation to four different sources, which are referred to as the *components of time series*. These sources are:

1. *Secular trend:* The smooth long-term progress of the data
2. *Seasonal variation:* Short-term variational pattern
3. *Cyclic fluctuation:* Long-term variational pattern
4. *Irregular variation:* Variation not attributable to any of the above three factors.

Secular trend is the component that reflects the long-term behavior of the time series after all the little ups and downs have been smoothed out. Take a look at the graph in Figure 10.1, where you can see that the secular

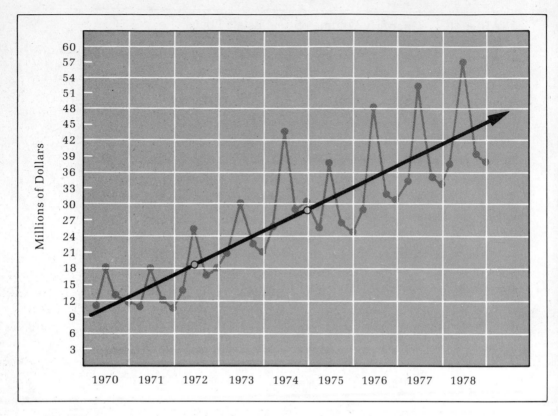

FIGURE 10.1   Time Series of Real Estate Loan Volume with Its Trend Line

trend of real estate loan volume is basically increasing at a moderate rate as the years pass. Secular trend usually reflects general long-term economic growth or decline.

   Seasonal variation includes those ups and downs in the time series that occur on a regular basis over the time period involved and whose occurrence can therefore be reliably predicted once the behavior pattern of the data has been discovered. Seasonal variation is generally attributable to known causes that occur regularly, such as seasonal weather changes (in the athletic equipment business), beginning of the school year in the fall (in the children's clothing business), and the holiday gift-giving season (in the department store business).

   Cyclical fluctuation describes those ups and downs that are known to occur in the economy, but whose occurrence cannot be predicted in advance and does not appear on a regular seasonal basis. The best example of this is the so-called "business cycle" of "boom or bust," those alternating several-year periods of economic growth and decline about whose existence at any particular time economists tend to disagree. Other influences on cyclical fluctuation are changes in political philosophy, which can lead to more or less expenditures on social or military programs.

Irregular variation encompasses all those factors that do not fall into the other three categories and therefore are considered as random unexpected shocks to the time series curve. Some of these are the following: the sudden increase in petroleum prices in the early 1970s; weather problems in Brazil that cause damage to a large portion of the world's coffee crop; sudden appearance of a more efficient technical device such as the cotton gin in the middle 1800s or the microchip in the 1970s; the unsuspected eruption of a volcano that can affect a large region's tourist volume; international events such as wars causing rapid disruption of normal economic processes.

In this text, we shall study the techniques of analyzing time series with *linear secular trend* only. These techniques are based on the methods of linear regression and correlation that were discussed in Chapter 8. Comparable techniques for analyzing curvilinear secular trend are available using analogous methods for curvilinear regression.

To apply the methods of Chapter 8, we set up the table of calculations appearing in Table 10.3. From the sums at the bottom of Table 10.3, we see that the coefficient of linear correlation for the time series is

$$r = \frac{n\Sigma xy - (\Sigma x)(\Sigma y)}{\sqrt{n\Sigma x^2 - (\Sigma x)^2}\sqrt{n\Sigma y^2 - (\Sigma y)^2}}$$

$$= \frac{(36)(21{,}817) - (666)(980)}{\sqrt{(36)(16{,}206) - (666)^2}\sqrt{(36)(31{,}656) - (980)^2}}$$

$$= \frac{785{,}412 - 652{,}680}{\sqrt{583{,}416 - 443{,}556}\sqrt{1{,}139{,}616 - 960{,}400}}$$

$$= \frac{132{,}732}{\sqrt{139{,}860}\sqrt{179{,}216}} = \frac{132{,}732}{(373.98)(423.34)}$$

$$= \frac{132{,}732}{158{,}319.76} = .83838$$

in view of formula (8.3) in Chapter 8. Therefore, the proportion of quarterly variation in total dollar value of real estate loans that can be attributed to linear trend is

$$r^2 = (.83838)^2 = .7029 = 70.29\%,$$

according to the discussion in Section 8.C. This leaves a percentage of

$$100\% - 70.29\% = 29.71\%$$

of the quarterly variation unexplained by the linear trend component.

Our objective in the rest of this chapter is to investigate the possible sources of that remaining 29.71% of the quarterly variation.

TABLE 10.3 Calculation of Linear Trend for Real Estate Loans Time Series (1970–1978)

Year	Quarter	Quarter Number x	Loan Volume y	x^2	y^2	xy
1970	1	1	10	1	100	10
	2	2	18	4	324	36
	3	3	13	9	169	39
	4	4	12	16	144	48
1971	1	5	11	25	121	55
	2	6	18	36	324	108
	3	7	12	49	144	84
	4	8	11	64	121	88
1972	1	9	14	81	196	126
	2	10	26	100	676	260
	3	11	17	121	289	187
	4	12	18	144	324	216
1973	1	13	21	169	441	273
	2	14	30	196	900	420
	3	15	22	225	484	330
	4	16	21	256	441	336
1974	1	17	26	289	676	442
	2	18	43	324	1,849	774
	3	19	28	361	784	532
	4	20	30	400	900	600
1975	1	21	25	441	625	525
	2	22	37	484	1,369	814
	3	23	26	529	676	598
	4	24	25	576	625	600
1976	1	25	29	625	841	725
	2	26	48	676	2,304	1,248
	3	27	32	729	1,024	864
	4	28	31	784	961	868
1977	1	29	34	841	1,156	986
	2	30	52	900	2,704	1,560
	3	31	35	961	1,225	1,085
	4	32	34	1,024	1,156	1,088
1978	1	33	37	1,089	1,369	1,221
	2	34	57	1,156	3,249	1,938
	3	35	39	1,225	1,521	1,365
	4	36	38	1,296	1,444	1,368
Sums		666	980	16,206	31,656	21,817

Before we proceed, however, let's use the method developed in Section 8.B to calculate the equation of the trend line that accounts for 70.29% of the quarterly variation in real estate loan volume. We first compute

$$b = \frac{n\Sigma xy - (\Sigma x)(\Sigma y)}{n\Sigma x^2 - (\Sigma x)^2} = \frac{132{,}732}{139{,}860} = .949,$$

using intermediate results in the calculation of r above. Then

$$a = \frac{\Sigma y - b\Sigma x}{n} = \frac{980 - (.949)(666)}{36}$$

$$= \frac{980 - 632.034}{36} = \frac{347.966}{36} = 9.665.$$

The equation $y = a + bx$ of the trend line is therefore

$$y = 9.665 + .949x,$$

where x is the quarter number, beginning with the first quarter of 1970, and y is the loan volume for that quarter as forecast by the trend line.

Two points on the trend line $y = 9.665 + .949x$ are (10, 19.16) and (20, 28.65) because when x = 10 (second quarter of 1972),

$$y = 9.665 + (.949)(10) = 9.665 + 9.49 = 19.16$$

and when x = 20 (fourth quarter of 1974),

$$y = 9.665 + (.949)(20) = 9.665 + 18.98 = 28.65.$$

The trend line is graphed, together with the time series data, in Fig. 10.1.

EXERCISES   10.B

10.13. Data collected by the National Bureau of Economic Research gives the following yearly figures on per capita gross national product in the United States for the years 1889 through 1953, measured in 1929 dollars.[*]

a. Graph the time series of per capita GNP data.

b. Calculate the proportion of yearly variation in per capita GNP that can be attributed to linear trend.

c. Determine the equation of the trend line of the time series data.

d. Superimpose the graph of the trend line upon the time series graph.

---

[*]As reported in U.S. Department of Commerce, *Long Term Economic Growth 1860–1965: A Statistical Compendium*, U.S. Government Printing Office, 1966, pp. 166–167.

Year	GNP	Year	GNP	Year	GNP
1889	395	1911	621	1933	590
1890	415	1912	640	1934	639
1891	425	1913	653	1935	718
1892	457	1914	592	1936	787
1893	427	1915	601	1937	846
1894	407	1916	675	1938	794
1895	447	1917	651	1939	847
1896	429	1918	711	1940	916
1897	462	1919	710	1941	1040
1898	464	1920	689	1942	1147
1899	497	1921	660	1943	1245
1900	502	1922	689	1944	1327
1901	549	1923	767	1945	1293
1902	543	1924	774	1946	1171
1903	560	1925	782	1947	1139
1904	542	1926	821	1948	1180
1905	571	1927	818	1949	1144
1906	625	1928	817	1950	1236
1907	624	1929	858	1951	1293
1908	561	1930	772	1952	1311
1909	618	1931	721	1953	1341
1910	611	1932	611		

10.14. The following data present, for 1967–1975, the values of the Consumer Buying Plans Index (1969–1970 period is 100), based on bimonthly surveys conducted by National Family Opinion, Inc.* Each year is divided into sixths, each sixth representing two months.

Year	Sixth	Index	Year	Sixth	Index	Year	Sixth	Index
1967	1	97.3	1970	1	99.8	1973	1	102.7
	2	101.9		2	98.0		2	125.2
	3	108.6		3	88.9		3	111.1
	4	107.6		4	93.5		4	111.6
	5	108.8		5	91.3		5	103.2
	6	107.5		6	87.0		6	93.7
1968	1	109.3	1971	1	101.5	1974	1	89.2
	2	98.9		2	112.6		2	87.6
	3	96.3		3	109.3		3	97.5
	4	95.9		4	112.3		4	81.1
	5	101.5		5	109.7		5	93.6
	6	105.2		6	109.3		6	76.8
1969	1	104.8	1972	1	113.3	1975	1	80.0
	2	112.8		2	107.1		2	88.0
	3	114.1		3	106.6		3	100.3
	4	108.2		4	102.5		4	104.6
	5	102.3		5	111.0		5	105.8
	6	99.2		6	111.8		6	108.7

*The Conference Board, *Statistical Bulletin*, Vol. 9, No. 9, September 1976, p. 14.

a. Graph the time series of the Consumer Buying Plans Index.

b. What proportion of bimonthly variation in the index can be explained on the basis of linear trend?

c. Calculate the equation of the trend line of the time series data.

d. Superimpose the graph of the trend line upon the time series graph.

10.15. New plant and equipment expenditures made quarterly by the air transportation industry are listed following, for the years 1967 through 1976.* The data are totals for the nation as a whole and are given in billions of dollars.

Year	Quarter	Total	Year	Quarter	Total
1967	1	.37	1972	1	.50
	2	.72		2	.73
	3	.56		3	.61
	4	.64		4	.63
1968	1	.68	1973	1	.52
	2	.58		2	.72
	3	.64		3	.57
	4	.66		4	.60
1969	1	.68	1974	1	.47
	2	.66		2	.61
	3	.53		3	.43
	4	.64		4	.48
1970	1	.73	1975	1	.44
	2	.80		2	.47
	3	.74		3	.50
	4	.76		4	.43
1971	1	.34	1976	1	.26
	2	.60		2	.42
	3	.39		3	.26
	4	.56		4	.38

a. Graph the time series of new plant and equipment expenditures made by the air transportation industry.

b. How much of the quarterly variation in new plant and equipment expenditures can be accounted for by linear trend?

c. Compute the equation of the trend line.

d. Superimpose the graph of the trend line upon the time series graph.

10.16. The yearly wholesale price index (1953 = 100) in Mexico City is recorded in the following data for the period 1929 through 1971, gathered as part of a study of inflation in Latin America.[†]

---

*U.S. Department of Commerce, Office of Business Economics, *Survey of Current Business*, Vols. 50, 52, 54, 56, p. S-2.

[†]James W. Wilkie, *Statistics and National Policy: UCLA Statistical Abstract of Latin America*, *Supplement 3*, UCLA Latin American Center, 1974, p. 229.

a. Construct a graph that illustrates the behavior of the time series of the wholesale price index in Mexico City.

b. What percentage of the yearly variation in the index can be attributed to linear trend?

c. Determine the equation of the trend line.

d. Superimpose the graph of the trend line upon the graph of the time series itself.

Year	Index	Year	Index	Year	Index
1929	20	1943	35	1957	140
1930	20	1944	45	1958	146
1931	18	1945	50	1959	149
1932	17	1946	58	1960	157
1933	18	1947	59	1961	158
1934	18	1948	63	1962	161
1935	18	1949	67	1963	163
1936	20	1950	74	1964	169
1937	23	1951	89	1965	176
1938	24	1952	98	1966	177
1939	25	1953	100	1967	177
1940	25	1954	108	1968	180
1941	27	1955	124	1969	183
1942	29	1956	132	1970	191
				1971	196

SECTION 10.C   SMOOTHING METHODS AND SEASONAL VARIATION

For the time series of real estate loan volume discussed in Section 10.B, we have shown that linear trend accounts for 70.29% of the quarter-to-quarter variation. In this section our goal is to explain how to determine the influence of seasonal factors on the quarterly variation in loan volume.

Moving Averages

To measure the effect of seasonal variation, we first have to smooth the data, in an attempt to understand the overall flow of information when the individual data points are obscured. This process is analogous to the operation of constructing a frequency distribution (see Section 1.C) to describe the overall pattern of the data in a manner more understandable than the original list of numbers. In our examples, we will use a five-term moving average to smooth the data.

FIVE-TERM MOVING AVERAGE

A five-term moving average is an average over five consecutive quarters, changing with the passage of time. For example, the first average is taken over quarters 1 through 5, the second over quarters 2 through 6, the third over quarters 3 through 7, etc.

Sometimes it may be more appropriate to use moving averages over other lengths of time. We would then speak of a "three-term moving average" or a "twelve-term moving average," for example.

Using the real estate loan data of Table 10.2, we compute the first moving average as follows:

$$MA_1 = \frac{10 + 18 + 13 + 12 + 11}{5} = \frac{64}{5} = 12.8$$

the second as follows:

$$MA_2 = \frac{18 + 13 + 12 + 11 + 18}{5} = \frac{72}{5} = 14.4$$

and the third:

$$MA_3 = \frac{13 + 12 + 11 + 18 + 12}{5} = \frac{66}{5} = 13.2$$

We calculate all the moving averages for the real estate loan data and record them in the fourth column of Table 10.4. We list the moving average for quarter 1 through 5 at the center of the time span, namely at quarter 3. Similarly, the second moving average is listed at quarter 4, and so on.

To illustrate the fact that the calculation of moving averages is a smoothing operation, the original time series is compared with the series of moving averages in Fig. 10.2. A BASIC computer program for smoothing a time series appears in Table 10.5.

Seasonal Variation

The next step in determining the extent of seasonal variation is to compare the actual loan volume achieved during each quarter with the moving average corresponding to that quarter. The ratio LV/MA of the actual quarterly loan

TABLE 10.4   Calculation of Moving Averages and Seasonal Indexes

Year	Quarter	Loan Volume (LV)	Five-Term Moving Average (MA)	Ratio to Moving Average (LV ÷ MA)	Seasonal Index
1970	1	10	–	–	–
	2	18	–	–	–
	3	13	12.8	13 ÷ 12.8 = 1.02	102
	4	12	14.4	12 ÷ 14.4 = .83	83
1971	1	11	13.2	11 ÷ 13.2 = .83	83
	2	18	12.8	18 ÷ 12.8 = 1.41	141
	3	12	13.2	12 ÷ 13.2 = .91	91
	4	11	16.2	11 ÷ 16.2 = .68	68
1972	1	14	16.0	14 ÷ 16.0 = .88	88
	2	26	17.2	26 ÷ 17.2 = 1.51	151
	3	17	19.2	17 ÷ 19.2 = .89	89
	4	18	22.4	18 ÷ 22.4 = .80	80
1973	1	21	21.6	21 ÷ 21.6 = .97	97
	2	30	22.4	30 ÷ 22.4 = 1.34	134
	3	22	24.0	22 ÷ 24.0 = .92	92
	4	21	28.4	21 ÷ 28.4 = .74	74
1974	1	26	28.0	26 ÷ 28.0 = .93	93
	2	43	29.6	43 ÷ 29.6 = 1.45	145
	3	28	30.4	28 ÷ 30.4 = .92	92
	4	30	32.6	30 ÷ 32.6 = .92	92
1975	1	25	29.2	25 ÷ 29.2 = .86	86
	2	37	28.6	37 ÷ 28.6 = 1.29	129
	3	26	28.4	26 ÷ 28.4 = .92	92
	4	25	33.0	25 ÷ 33.0 = .76	76
1976	1	29	32.0	29 ÷ 32.0 = .91	91
	2	48	33.0	48 ÷ 33.0 = 1.45	145
	3	32	34.8	32 ÷ 34.8 = .92	92
	4	31	39.4	31 ÷ 39.4 = .79	79
1977	1	34	36.8	34 ÷ 36.8 = .92	92
	2	52	37.2	52 ÷ 37.2 = 1.40	140
	3	35	38.4	35 ÷ 38.4 = .91	91
	4	34	43.0	34 ÷ 43.0 = .79	79
1978	1	37	40.4	37 ÷ 40.4 = .92	92
	2	57	41.0	57 ÷ 41.0 = 1.39	139
	3	39	–	–	–
	4	38	–	–	–

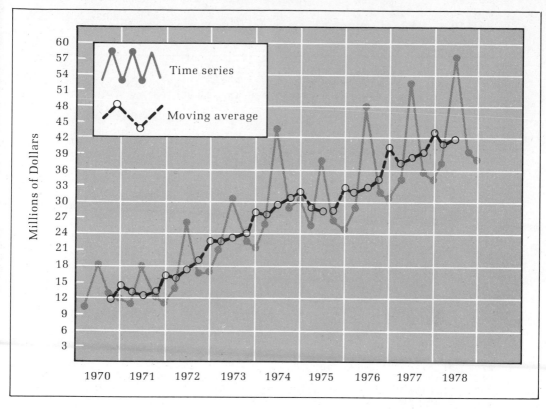

FIGURE 10.2   Comparison of a Time Series with Its Five-Term Moving Average

volume to the moving average appears in the fifth column of Table 10.4. For example, for the third quarter of 1970, the *ratio to moving average* is

$$LV/MA = 13/12.8 = 1.0156 = 1.02 \text{ (to two places)}$$

(It is customary to round these off to two places past the decimal point.) For the fourth quarter of 1970, the ratio to moving average is

$$LV/MA = 12/14.4 = .8333 = .83$$

We multiply the ratio to moving average by 100 in order to determine the *seasonal index*. The seasonal indexes, which are listed in the far right column of Table 10.4, are index numbers of the sort discussed in Section 10.A. They are used to compare the quarterly loan volume with the average loan volume over a five-quarter period.

TABLE 10.5   BASIC Computer Program for Time Series Smoothing
Calculations

```
10      REMARK: SMOOTHING A TIME SERIES
20      DIM T(100),M(5),E(5),A(5,100),S(5,100),R(5,100),M9(5)
30      PRINT "HOW MANY DATA POINTS ARE IN THE SERIES";
40      INPUT N
50      FOR I = 1 TO N
60         READ T(I)
70         NEXT I
80      REMARK: TIME SERIES LISTED (IN ORDER) ON LINES 81 TO 89
90      PRINT "DO YOU WANT MOVING AVERAGES (ANSWER Y OR N)";
100     INPUT M$
110     IF M$ = "Y" THEN 490
120     PRINT "DO YOU WANT EXPONENTIAL SMOOTHING (ANSWER Y OR N)";
130     INPUT E$
140     IF E$ = "Y" THEN 180
150     IF M$ = "Y" THEN 530
160     PRINT "GOODBYE; I CAN'T DO ANYTHING FOR YOU."
170     STOP
180     PRINT "HOW MANY SMOOTHING CONSTANTS DO YOU NEED";
190     INPUT E1
200     IF E1 < 4 THEN 230
210     PRINT "I CANNOT HANDLE MORE THAN 3 CONSTANTS."
220     GO TO 180
230     PRINT "WHAT ARE THEY?"
240     FOR J = 1 TO E1
250        INPUT E(J)
260        NEXT J
270     IF M$ = "Y" THEN 300
275     PRINT
280     PRINT "TIME","EXPONENTIALLY-SMOOTHED SERIES"
290     PRINT "SERIES",
300     FOR J = 1 TO E1
310        IF M$ = "Y" THEN 330
320        PRINT "E =";E(J),
330        S(J,1) = T(1)
340        FOR K = 2 TO N
350           S(J,K) = E(J)*T(K) + (1-E(J))*S(J,K-1)
360           NEXT K
370        NEXT J
380     PRINT
390     IF M$ = "Y" THEN 1080
400     FOR K = 1 TO N
410        PRINT T(K),
415        IF E1 = 1 THEN 460
420        FOR J = 1 TO E1-1
430           PRINT S(J,K),
440           NEXT J
460        PRINT S(E1,K)
470        NEXT K
480     STOP
490     PRINT "WANT RATIO-TO-MOVING-AVERAGE INFO(ANSWER Y OR N)";
500     INPUT R$
510     IF R$ = "Y" THEN 880
520     GO TO 120
530     PRINT "I CAN GIVE YOU UP TO 5 MOVING-AVERAGE SEQUENCES."
540     PRINT "HOW MANY SEQUENCES DO YOU WANT";
550     INPUT M1
560     PRINT "HOW-MANY-TERM MOVING AVERAGES DO YOU NEED";
570     FOR L = 1 TO M1
580        INPUT M(L)
```

TABLE 10.5 *(Continued)*

```
590        M9(L) = INT(M(L)/2)
600        FOR K = M9(L)+1 TO N-M9(L)
610          S = 0
620          FOR H = 1 TO M(L)
630            H1 = K-M9(L)-1
640            S = S + T(H1+H)
650            NEXT H
660          A(L,K) = S/M(L)
670          R(L,K) = T(K)/A(L,K)
680          NEXT K
690        IF E$ = "Y" THEN 1005
700        NEXT L
710      PRINT
720      IF R$ = "Y" THEN 900
730      PRINT "TIME","MOVING AVERAGES OF VARIOUS LENGTHS"
740      PRINT "SERIES","M =";M(1),
750      FOR M2 = 2 TO M1
760        PRINT "M =";M(M2),
770        NEXT M2
780      PRINT
790      FOR K = 1 TO N
800        PRINT T(K),
810        FOR L = 1 TO M1-1
820          PRINT A(L,K),
830          NEXT L
840        IF M1 = 1 THEN 860
850        PRINT A(M1,K)
860        NEXT K
870      STOP
880      M1 = 1
890      GO TO 560
900      PRINT "TIME","MOVING","RATIO TO"
910      PRINT "SERIES","AVERAGE","MOVING AVERAGE"
920      PRINT
970      FOR K = 1 TO N
980        PRINT T(K),A(1,K),R(1,K)
990        NEXT K
1000     STOP
1005     PRINT
1010     PRINT "TIME","MOVING","EXPONENTIALLY-SMOOTHED SERIES"
1020     PRINT "SERIES","AVERAGE",
1030     FOR J = 1 TO E1
1040       PRINT "E =";E(J),
1050       NEXT J
1060     PRINT
1070     GO TO 1100
1080     M1 = 1
1090     GO TO 560
1100     FOR K = 1 TO N
1110       PRINT T(K),A(1,K),
1120       IF E1 = 1 THEN 1160
1130       FOR J = 1 TO E1-1
1140         PRINT S(J,K),
1150         NEXT J
1160       PRINT S(E1,K)
1170       NEXT K
1180     END
```

The BASIC computer program of Table 10.5 yields the printout in Table 10.6 if we request ratio to moving average information.

TABLE 10.6   Computer Printout of Ratio-to-Moving-Average Information

---

```
          81   DATA 10,18,13,12,11,18,12,11,14
          82   DATA 26,17,18,21,30,22,21,26,43
          83   DATA 28,30,25,37,26,25,29,28,32
          84   DATA 31,34,52,35,34,37,57,39,38

     run

     HOW MANY DATA POINTS ARE IN THE SERIES ? 36
     DO YOU WANT MOVING AVERAGES (ANSWER Y OR N) ?y
     WANT RATIO-TO-MOVING-AVERAGE INFO(ANSWER Y OR N) ? y
     HOW-MANY-TERM MOVING AVERAGES DO YOU NEED ?5
```

TIME SERIES	MOVING AVERAGE	RATIO TO MOVING AVERAGE
10	0	0
18	0	0
13	12.8	1.01563
12	14.4	.833333
11	13.2	.833333
18	12.8	1.40625
12	13.2	.909091
11	16.2	.679012
14	16	.875
26	17.2	1.51163
17	19.2	.885417
18	22.4	.803571
21	21.6	.972222
30	22.4	1.33929
22	24	.916667
21	28.4	.739437
26	28	.928571
43	29.6	1.4527
28	30.4	.921053
30	32.6	.920245
25	29.2	.856164
37	28.6	1.29371
26	28.4	.915493
25	29	.862069
29	28	1.03571
28	29	.965517
32	30.8	1.03896
31	35.4	.875706
34	36.8	.923913
52	37.2	1.39785
35	38.4	.911458
34	43	.790698
37	40.4	.915842
57	41	1.39024
39	0	0
38	0	0

TABLE 10.7   Quarterly Pattern of Seasonal Indexes

Year	Quarter 1	2	3	4
1970	–	–	102	83
1971	83	141	91	68
1972	88	151	89	80
1973	97	134	92	74
1974	93	145	92	92
1975	86	129	92	76
1976	91	145	92	79
1977	92	140	91	79
1978	92	139	–	–
Median	91.5	140.5	92.0	79.0

Average Median = (91.5 + 140.5 + 92.0 + 79.0)/4 = 403/4 = 100.75

Quarterly Seasonal Index	90.8	139.5	91.3	78.4
Decimal Equivalent	.908	1.395	.913	.784

To better visualize the seasonal variation inherent in the data, a table of the quarterly behavior patterns can be constructed. This table, which appears in Table 10.7, organizes the seasonal indexes according to quarters. At the bottom of Table 10.7, we record the median seasonal index for each quarter, which may be considered as the average seasonal index for the period of time under study. In Fig. 10.3, we consider the behavior pattern of each quarter as a time series itself, and we graph its variation as time passes.

A general conclusion that can be drawn from the arrangement of seasonal indexes in Table 10.7 is that loan volume in the second quarter is typically 140% of the quarterly average, in the first and third quarters it is about 90%, and in the fourth quarter it is about 80%. These percentages are based on the quarterly medians.

We often use seasonal indexes to "seasonally adjust" data so as to be able to see how the data would look with the effects of seasonal variation removed. This procedure is important in analyzing changes in unemployment rates, for example, to find out if they represent significant trends or merely expected seasonal variation. The first step in the seasonal adjustment of data is to calculate four quarterly seasonal indexes, one for each quarter, by using the formula

$$\text{quarterly seasonal index} = \frac{\text{quarterly median}}{\text{average of quarterly medians}} \times 100. \quad \textbf{(10.2)}$$

At the bottom of Table 10.7, we calculated the average of quarterly medians to be 100.75, so that we get for the first quarter

FIGURE 10.3   Quarterly Pattern of Seasonal Indexes

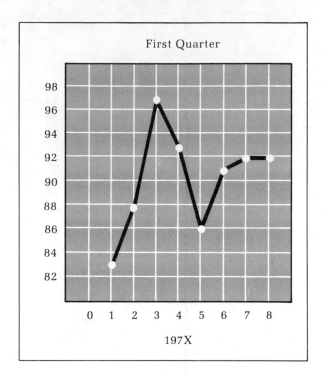

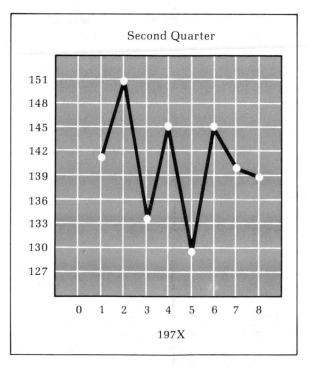

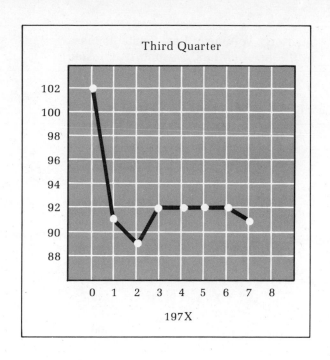

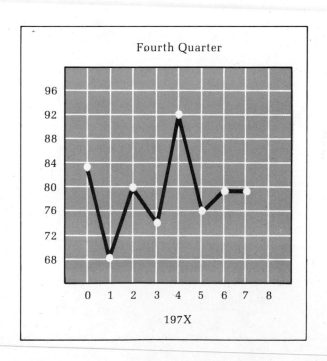

TABLE 10.8   Quarterly Volume of Real Estate Loans
(1970–1978)   *(Millions of dollars, seasonally adjusted)*

Year	Quarter	Actual Loan Volume	÷	Decimal Equivalent	=	Seasonally Adjusted Loan Volume
1970	1	10		.908		11.01
	2	18		1.395		12.90
	3	13		.913		14.24
	4	12		.784		15.31
1971	1	11		.908		12.11
	2	18		1.395		12.90
	3	12		.913		13.14
	4	11		.784		14.03
1972	1	14		.908		15.42
	2	26		1.395		18.64
	3	17		.913		18.62
	4	18		.784		22.96
1973	1	21		.908		23.13
	2	30		1.395		21.51
	3	22		.913		24.10
	4	21		.784		26.79
1974	1	26		.908		28.63
	2	43		1.395		30.82
	3	28		.913		30.67
	4	30		.784		38.27
1975	1	25		.908		27.53
	2	37		1.395		26.52
	3	26		.913		28.48
	4	25		.784		31.89
1976	1	29		.908		31.94
	2	48		1.395		34.41
	3	32		.913		35.05
	4	31		.784		39.54
1977	1	34		.908		37.44
	2	52		1.395		37.28
	3	35		.913		38.34
	4	34		.784		43.37
1978	1	37		.908		40.75
	2	57		1.395		40.86
	3	39		.913		42.72
	4	38		.784		48.47

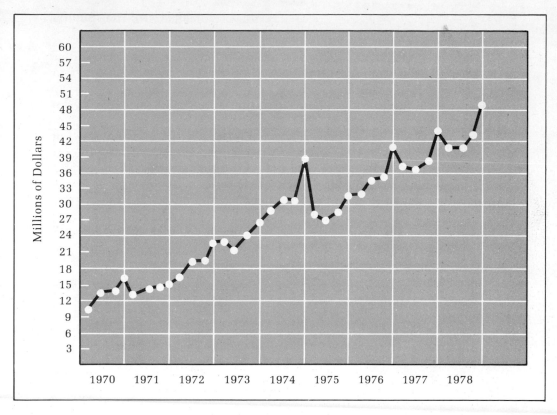

FIGURE 10.4   Time Series of Real Estate Loan Volume   *(Seasonally adjusted)*

$$\text{quarterly seasonal index} = \frac{91.5}{100.75} \times 100 = .908 \times 100 = 90.8$$

The others are calculated similarly and appear in the next-to-last line of Table 10.7.

To seasonally adjust time series data, we now simply divide the actual quarterly data by the decimal equivalent of the proper quarterly seasonal index. The decimal equivalents, obtained by dividing the quarterly seasonal index by 100, appear in the bottom line of Table 10.7. We record the seasonally adjusted quarterly loan volume in Table 10.8. The seasonally adjusted time series is illustrated by the graph of Fig. 10.4.

Exponential Smoothing

If time series were perfectly predictable, that is to say, if the secular trend line were a perfect predictor of time series behavior, there would be no need for smoothing. However, the behavior of a classical time series depends on several factors other than the mere passage of time: seasonal variation, cyclic fluctuation, and irregular variation, to name a few. Therefore, the secular trend line by itself would not be a very good forecasting tool. In order to improve our ability to forecast, it is necessary to smooth the raw data so as to measure the effect of those other factors that influence the behavior of the time series.

The question is, "How smooth is smooth enough?" A one-term moving average would be no smoother than the original time series, while an n-term moving average (n = number of data points in the time series) would yield a single point equal to the average of the whole series—very smooth! A seven-term moving average is a compromise between these two extremes: seven terms of the series, no more and no less, are taken into account in calculating each data point of the smoothed series. No other terms of the series have an effect on the smoothed value at the time period under study, except for the seven data points closest in time.

There is some concern about two characteristics of the moving-average smoothing method: the fact that so little of the series is taken into account at each time period and the fact that the smoothed value depends on future, as well as past, behavior of the time series. A second smoothing method, generally referred to as *exponential smoothing* has been introduced to face these objections.

EXPONENTIAL SMOOTHING

In exponential smoothing, the entire time series up to the time period in question is taken into account in calculating the smoothed value, but no future values are considered.

Exponential smoothing has a further advantage in dealing with short time series for which not many data points are available: a seven-term moving average will eat up so much of the time series that there will be very little left for analysis, but the exponential smoothing method provides a smoothed value of every original data point.

Because everyone agrees that numerical values long past should not have as much of an effect on the smoothed value as more recent values, the precise effect of each point is controlled by the statistician or analyst on the scene, who chooses an appropriate *smoothing constant*. We will denote the smoothing constant by ε (lower-case Greek letter "epsilon") and the kth exponentially smoothed value by ES_k. If the terms of the original time series are denoted by $y_1, y_2, \ldots, y_n$, then the time series of exponentially smoothed values are as follows, where $0 \leq \varepsilon \leq 1$:

$$ES_1 = y_1$$

$$ES_2 = \varepsilon y_2 + (1 - \varepsilon)ES_1$$

$$ES_3 = \varepsilon y_3 + (1 - \varepsilon)ES_2$$

.

.

.

$$ES_k = \varepsilon y_k + (1 - \varepsilon)ES_{k-1} \qquad\qquad (10.3)$$

.

.

.

Formula (10.3) is called the *fundamental formula of exponential smoothing*. The smoothing constant ε always lies between 0 and 1. If ε is close to 1, then the exponentially-smoothed value ES_k is largely determined by y_k, with only a minor role played by the previous smoothed value ES_{k-1}. If ε is close to 0, then ES_k is greatly influenced by ES_{k-1}, and only slightly by y_k. If $\varepsilon = .5$, then ES_k is the average of y_k and ES_{k-1}. In short, ES_k is a weighted mean of the new time series point y_k and the smoothed value ES_{k-1} based on the entire prior series. For the time series of real estate loan volume discussed earlier in this chapter and listed in Table 10.2, we compute exponentially-smoothed time series corresponding to the smoothing constants $\varepsilon = .1$, $\varepsilon = .5$, and $\varepsilon = .9$. The three exponentially-smoothed time series are listed in Table 10.9, and the computer printout of the BASIC program of Table 10.5 yields the information in Table 10.10, when exponential smoothing is requested. Round-off error in Table 10.9 is indicated.

As an example of how Table 10.9 is developed, let's look at quarter number $k = 12$, where the loan volume is $y_{12} = 18$. From formula (10.3), we see that, in the case $\varepsilon = .9$,

$$ES_{12} = .9y_{12} + (1 - .9)ES_{11}.$$

In the previous step of the table, we found that $ES_{11} = 17.78$. Therefore

$$ES_{12} = (.9)(18) + (.1)(17.78) = 17.98.$$

The three different exponentially-smoothed time series are graphed, together, with the original series, in Fig. 10.5. As can be seen from the graph, the $\varepsilon = .1$ series is the smoothest of the three because, at each step, it is composed 90% of the previous smoothed value and only 10% of the new data

TABLE 10.9   Exponentially-Smoothed Time Series

Quarter Number k	Loan Volume y_k	ES_k ($\varepsilon = .1$)	ES_k ($\varepsilon = .5$)	ES_k ($\varepsilon = .9$)
1	10	10.00	10.00	10.00
2	18	10.80	14.00	17.20
3	13	11.02	13.50	13.42
4	12	11.12	12.75	12.14
5	11	11.11	11.88	11.11
6	18	11.80	14.94	17.31
7	12	11.82	13.47	12.53
8	11	11.73	12.23	11.15
9	14	11.96	13.12	13.72
10	26	13.36	19.56	24.77
11	17	13.73	18.28	17.78
12	18	14.16	18.14	17.98
13	21	14.84	19.57	20.70
14	30	16.36	24.78	29.07
15	22	16.92	23.39	22.71
16	21	17.33	22.20	21.17
17	26	18.20	24.10	25.52
18	43	20.68	33.55	41.25
19	28	21.41	30.77	29.33
20	30	22.27	30.39	29.93
21	25	22.54	27.69	25.49
22	37	23.99	32.35	35.85
23	26	24.19	29.17	26.98
24	25	24.27	27.09	25.20
25	29	24.74	28.04	28.62
26	48	27.07	38.02	46.06
27	32	27.56	35.01	33.41
28	31	27.91	33.01	31.24
29	34	28.51	33.50	33.72
30	52	30.86	42.75	50.17
31	35	31.28	38.88	36.52
32	34	31.55	36.44	34.25
33	37	32.09	36.72	36.73
34	57	34.58	46.86	54.97
35	39	35.03	42.93	40.60
36	38	35.32	40.46	38.26

TABLE 10.10   Computer Printout of Exponentially-Smoothed Time Series

```
HOW MANY DATA POINTS ARE IN THE SERIES ? 36
DO YOU WANT MOVING AVERAGES (ANSWER Y OR N) ?y
WANT RATIO-TO-MOVING-AVERAGE INFO(ANSWER Y OR N) ? n
DO YOU WANT EXPONENTIAL SMOOTHING (ANSWER Y OR N) ?y
HOW MANY SMOOTHING_CONSTANTS DO YOU NEED ? 3
WHAT ARE THEY?
?.1
?.5
?.9

HOW-MANY-TERM MOVING AVERAGES DO YOU NEED ?5
```

TIME SERIES	MOVING AVERAGE	EXPONENTIALLY-SMOOTHED SERIES		
		E = .1	E = .5	E = .9
10	0	10	10	10
18	0	10.8	14	17.2
13	12.8	11.02	13.5	13.42
12	14.4	11.118	12.75	12.142
11	13.2	11.1062	11.875	11.1142
18	12.8	11.7956	14.9375	17.3114
12	13.2	11.816	13.4688	12.5311
11	16.2	11.7344	12.2344	11.1531
14	16	11.961	13.1172	13.7153
26	17.2	13.3649	19.5586	24.7715
17	19.2	13.7284	18.2793	17.7772
18	22.4	14.1556	18.1396	17.9777
21	21.6	14.84	19.5698	20.6978
30	22.4	16.356	24.7849	29.0698
22	24	16.9204	23.3925	22.707
21	28.4	17.3284	22.1962	21.1707
26	28	18.1955	24.0981	25.5171
43	29.6	20.676	33.5491	41.2517
28	30.4	21.4084	30.7745	29.3252
30	32.6	22.2675	30.3873	29.9325
25	29.2	22.5408	27.6936	25.4933
37	28.6	23.9867	32.3468	35.8493
26	28.4	24.188	29.1734	26.9849
25	29	24.2692	27.0867	25.1985
29	28	24.7423	28.0434	28.6198
28	29	25.0681	28.0217	28.062
32	30.8	25.7613	30.0108	31.6062
31	35.4	26.2851	30.5054	31.0606
34	36.8	27.0566	32.2527	33.7061
52	37.2	29.551	42.1264	50.1706
35	38.4	30.0959	38.5632	36.5171
34	43	30.4863	36.2816	34.2517
37	40.4	31.1377	36.6408	36.7252
57	41	33.7239	46.8204	54.9725
39	0	34.2515	42.9102	40.5973
38	0	34.6263	40.4551	38.2597

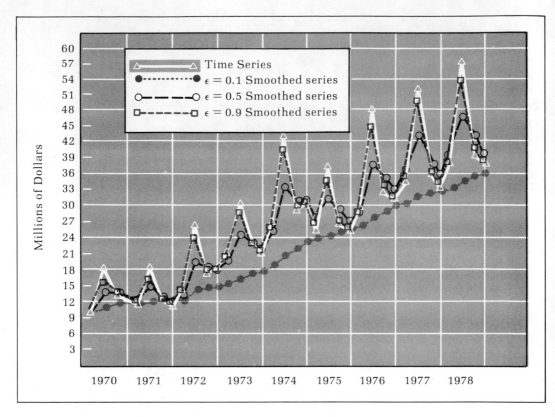

FIGURE 10.5   Comparison of a Time Series with Three of Its Exponentially Smoothed
Versions

point. The ε = .9 series most closely follows the original series because it is
composed 90% of the new data point and only 10% of the previous smoothed
value. The figure illustrates how we can adjust the smoothing constant ε in
order to obtain any desired level of smoothness we may feel is appropriate
to the situation at hand.

   Although we will not go into detail about it here, the exponentially
smoothed time series can be used as a basis for determining seasonal variation, just as the moving-average series can.

EXERCISES 10.C

10.17. For the time series data on the per capita GNP presented in Exercise 10.13:

   a. Calculate the numerical series of five-year moving averages in order to
       smooth the data.

   b. Superimpose the graph of the five-year moving average upon a graph of the original time series.

   c. Considering the period of years to consist of two "seasons," the first being the years ending in the digits 0,1,2,3,4 and the second being the years ending in the digits 5,6,7,8,9, calculate the series of seasonal indexes.

   d. Construct the two graphs that illustrate the pattern of seasonal indexes for the two seasons.

   e. Seasonally adjust the per capita GNP's, and construct the graph of the seasonally adjusted time series.

   f. Exponentially smooth the original time series, using smoothing constant $\varepsilon = .3$.

10.18. Using the time series data on the Consumer Buying Plans Index listed in Exercise 10.14:

   a. Construct the seven-term moving average of the index.

   b. Superimpose the graph of the seven-term moving average upon the graph of the original time series.

   c. Calculate the series of seasonal (bimonthly) indexes.

   d. Draw the six graphs that together portray the pattern of seasonal indexes.

   e. Seasonally adjust the bimonthly values of the Consumer Buying Plans Index, and construct the graph of the seasonally adjusted time series.

   f. Use smoothing constant $\varepsilon = .7$ to smooth the original time series.

10.19. Exercise 10.15 lists quarterly data on new plant and equipment expenditures made by the air transportation industry.

   a. Construct the seven-term moving average of the data.

   b. Superimpose the graph of the seven-term moving average upon the graph of the original time series.

   c. Compute the series of seasonal (quarterly) indexes.

   d. Draw the four graphs that illustrate the behavior pattern of the seasonal indexes.

   e. Seasonally adjust the new plant and equipment expenditure data, and draw the graph of the seasonally adjusted time series.

   f. Calculate the exponentially-smoothed version of the time series, using smoothing constant $\varepsilon = .5$.

10.20. For the time series of Mexico City's wholesale price index, given in Exercise 10.16:

   a. Compute the nine-term moving average of the index.

   b. Superimpose the graph of the nine-term moving average upon the graph of the original time series.

   c. Exponentially smooth the time series, using smoothing constant $\varepsilon = .3$, and superimpose the graph of the exponentially-smoothed version upon that of the original series.

d. Use smoothing constant $\varepsilon = .7$ to smooth the time series, and superimpose the resulting graph of the original series.

10.21. Consider the following illustrative time series in thousands of dollars in sales volume of ski equipment by a year-old sporting goods store:

Month	J	F	M	A	M	J	J	A	S	O	N	D
Sales volume	10	8	6	4	2	1	1	1	3	6	9	12

a. Compute a seven-term moving average of the series, and superimpose its graph on the graph of the original time series.

b. Use smoothing constant $\varepsilon = .4$ to smooth the series, and superimpose the resulting graph on the existing graph from part a.

c. Do you notice anything funny about the moving average?

SECTION 10.D   CYCLICAL FLUCTUATION

From our analysis so far of the real estate loan volume data of Table 10.2, we have found that the trend line of the time series has equation, $y = 9.665 + .949x$, where x is the quarter number (see Table 10.3) and y is the trend line forecasted loan volume. Furthermore, we have obtained in Table 10.8 the seasonally adjusted loan volume, namely the forecasted record of how the data would appear if the effects of seasonal variation were eliminated.

CYCLICAL FLUCTUATION

By the term *cyclical fluctuation* we mean the behavior pattern of the deviations of the seasonally adjusted data from their corresponding trend line forecasts.

Consider quarter number $x = 1$, for example. From Table 10.8, we see that the seasonally adjusted loan volume for that quarter is 11.01 (millions of dollars). On the other hand, the trend line forecast for quarter number $x = 1$ is

$$y = 9.665 + .949x = 9.665 + (.949)(1)$$

$$= 9.665 + .949 = 10.61 \text{ millions of dollars}$$

Therefore, the seasonally adjusted loan volume is $11.01/10.61 = 1.038 = 103.8\%$ "of trend."

The pattern of cyclical fluctuation is determined from the *percent of trend* represented by the seasonally adjusted data.

For another example, look at quarter number x = 14, the second quarter of 1973. Here the trend line forecast is

$$y = 9.665 + .949x = 9.665 + (.949)(14)$$

$$= 9.665 + 13.286 = 22.95 \text{ millions of dollars}$$

A glance at Table 10.8 reveals that the seasonally adjusted loan volume for the second quarter of 1973 is 21.51 millions of dollars. Therefore the

■ $$\text{percent of trend} = \frac{\text{seasonally adjusted data point}}{\text{trend line forecast of data point}} \times 100\%$$ (10.4)

$$= \frac{21.51}{22.95} \times 100\% = .937 \times 100\% = 93.7\% \text{ of trend.}$$

The complete list of cyclical fluctuation measurements, described according to percent of trend, appears in Table 10.11. The cyclical fluctuation pattern is graphed in Fig. 10.5. Those quarters in which the percent of trend was in excess of 100 had a seasonally adjusted loan volume that was greater than the amount forecasted for the period by the trend line. Those quarters having percent of trend below 100 recorded a lower than forecasted seasonally adjusted loan volume.

EXERCISES   10.D

10.22. For the time series data on per capita GNP studied in Exercises 10.13 and 10.17:

    a. Compute the series of trend line forecasts of the per capita GNP, using the equation of the trend line determined in part c of Exercise 10.13.

    b. Combine the trend line forecasts with the series of seasonally adjusted per capita GNP's calculated in part c of Exercise 10.17, in order to determine the series that gives cyclical fluctuation as percent of trend.

    c. Construct the graph that illustrates the pattern of cyclical fluctuation of the time series data.

10.23. Using the time series data on the Consumer Buying Plans Index listed in Exercise 10.14 and further studied in Exercise 10.18:

    a. Calculate the series of trend line forecasts of the Consumer Buying Plans Index, using the trend line equation determined in part c of Exercise 10.14.

    b. Combine the trend line forecasts with the series of seasonally adjusted Consumer Buying Plans Indexes obtained in part e of Exercise 10.18, in order to determine the series of numbers that give the cyclical fluctuation as percent of trend.

    c. Draw the graph indicating the cyclical fluctuation pattern of the time series data.

TABLE 10.11   Cyclical Fluctuation as Percent of Trend   *(Real estate loan volume, 1970–1978)*

Year	Quarter	Seasonally Adjusted Loan Volume (from Table 10.5) (SA)	Quarter Number x	Trend Line $y = 9.665 + .949x$ (TL)	Percent of Trend (SA ÷ TL)
1970	1	11.01	1	10.61	103.8
	2	12.90	2	11.56	111.6
	3	14.24	3	12.51	113.8
	4	15.31	4	13.46	113.7
1971	1	12.11	5	14.41	84.0
	2	12.90	6	15.36	84.0
	3	13.14	7	16.31	80.6
	4	14.03	8	17.26	81.3
1972	1	15.42	9	18.21	84.7
	2	18.64	10	19.16	97.3
	3	18.62	11	20.10	92.6
	4	22.96	12	21.05	109.1
1973	1	23.13	13	22.00	105.1
	2	21.51	14	22.95	93.7
	3	24.10	15	23.90	100.8
	4	26.79	16	24.85	107.8
1974	1	28.63	17	25.80	111.0
	2	30.82	18	26.77	115.1
	3	30.67	19	27.70	110.7
	4	38.27	20	28.65	133.6
1975	1	27.53	21	29.59	93.0
	2	26.52	22	30.54	86.8
	3	28.48	23	31.49	90.4
	4	31.89	24	32.44	98.3
1976	1	31.94	25	33.39	95.7
	2	34.41	26	34.34	100.2
	3	35.05	27	35.29	99.3
	4	39.54	28	36.24	109.1
1977	1	37.44	29	37.19	100.7
	2	37.28	30	38.14	97.7
	3	38.34	31	39.08	98.1
	4	43.37	32	40.03	108.3
1978	1	40.75	33	40.98	99.4
	2	40.86	34	41.93	97.4
	3	42.72	35	42.88	99.6
	4	48.47	36	43.83	110.6

10.24. Exercises 10.15 and 10.19 studied the time series of new plant and equipment expenditures made by the air transportation industry.

   a. Determine the series of trend line forecasts of the expenditures, using the equation of the trend line found in Exercise 10.15, part c.

   b. From the trend line forecasts and the series of seasonally adjusted expenditures determined in Exercise 10.19, part e, derive the pattern of cyclical fluctuation as percent of trend.

   c. Draw the graph that exhibits the behavior pattern of the series of cyclical fluctuations of the time series.

SECTION 10.E   FORECASTING USING CLASSICAL TIME SERIES TECHNIQUES

Our time series analysis of the real estate loan volume data of Table 10.2 can be used to forecast future behavior of the quarterly volume of real estate loans beyond the fourth quarter of 1978. The forecasting procedure operates as follows.

We begin with the equation

$$y = 9.665 + .949x$$

of the trend line that we found in Section 10.B. Here x is the quarter number appearing in the third column (from the left of Table 10.3) and y is the *trend line forecast* of the loan volume (in millions of dollars) for that quarter. As we saw in Section 10.B, however, the trend line forecast accounts for only 70.29% of quarterly variation in loan volume. We can improve the accuracy of our forecast by adjusting the trend line forecast to reflect seasonal variation and cyclical fluctuation, as discussed in Sections 10.C and 10.D. There are, of course, other factors that affect the behavior of the time series, but these are generally more obscure and specialized and we refer to the variation attributable to those factors as *irregular variation*. In this brief introduction, we will not discuss irregular variation any further.

Suppose, for example, we want to forecast loan volume for the third quarter of 1980. If we extend the third column of Table 10.3 into 1980, we see that the quarter number corresponding to the third quarter of 1980 is x = 43. This means that the trend line forecast for that quarter will be

$$y = 9.665 + .949x = 9.665 + (.949)(43)$$

$$= 9.665 + 40.807 = 50.472 \text{ million dollars.}$$

We now try to estimate, for the third quarter of 1980, the number that tells us the cyclical fluctuation as percent of trend. From a glance at Fig. 10.6, it looks as if the cyclical fluctuations have been settling down recently, and we generally have to assume in time series analysis that recent trends will

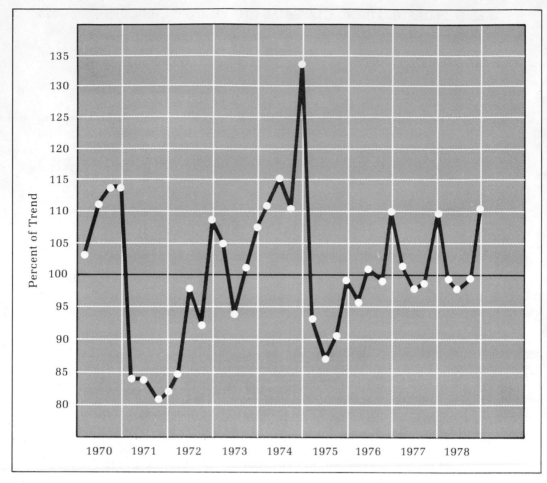

FIGURE 10.6   Cyclical Fluctuations of Real Estate Loan Volume

continue into the not-too-distant future. Table 10.11 contains the percent-of-trend information for all the quarters, but let's take a look at just the third quarter figures, shown in Table 10.12. Based on the material in Table 10.12, for the mean of the various averages for the last three years (although there is considerable flexibility in making the choice of which figure to use), we will view 99.1% as the most appropriate measure of cyclical fluctuation as percent of trend.

The next step is to refer to formula (10.4) of Section 10.D, and to insert the percent-of-trend figure 99.1% = .991 and the trend line forecast 50.472 into it. We get

$$.991 = \frac{\text{seasonally adjusted forecasted point}}{50.472}$$

from which it follows that the seasonally adjusted forecast = (.991)(50.472) = 50.018. The seasonally adjusted forecast of $50.018 million has been

TABLE 10.12 Third-Quarter Percent-of-Trend Data

Year	Quarter	Percent of Trend
1970	3	113.8
1971	3	80.6
1972	3	92.6
1973	3	100.8
1974	3	110.7
1975	3	90.4
1976	3	99.3
1977	3	98.1
1978	3	99.6

Overall median percent of trend = 99.3
Overall mean percent of trend = 98.4
Overall midrange percent of trend = 97.2
Last three years median percent of trend = 99.3
Last three years mean percent of trend = 99.0
Last three years midrange percent of trend = 98.9
Mean of last three years averages = 99.1

obtained from the less accurate trend line forecast by taking account of an appropriate factor reflecting the cyclical fluctuation of the time series.

The final step in sharpening the accuracy of our forecast is to remove the effects of seasonal adjustment for our seasonally adjusted forecast so as to obtain a forecast of the actual loan volume for the third quarter of 1980. According to Table 10.7 the decimal-equivalent seasonal index for the third quarter is .913. The calculations used in constructing Table 10.6 show that

$$\frac{actual\ data\ point}{decimal\text{-}equivalent\ seasonal\ index} = \frac{seasonally\ adjusted}{data\ point}$$

and therefore, crossing over to forecasting,

$$\blacksquare \qquad \frac{actual\ forecast}{decimal\text{-}equivalent\ seasonal\ index} = \frac{seasonally\ adjusted}{forecast} \qquad \textbf{(10.5)}$$

To obtain our actual forecast, we therefore insert the seasonally-adjusted forecast of $50.018 million and the decimal-equivalent seasonal index of .913 into formula (10.5). We have

$$actual\ forecast/.913 = 50.018$$

and so the

$$actual\ forecast = (.913)(50.018)$$

$$= \$45.67\ million.$$

Based on our time series analysis of the data in Table 10.2, we would therefore forecast a loan volume of $45.67 million for the third quarter of 1980.

To summarize the procedure we used in improving the accuracy of a straight trend line forecast, we can write

    F = our forecast
    T = trend line forecast
    C = appropriate measure of cyclical fluctuation as percent of trend
    S = decimal-equivalent seasonal index

Then

$$F = TCS \qquad\qquad (10.6)$$

In the example we just worked out, we had $T = 50.472$, $C = .991$, and $S = .913$, so that

$$F = TCS = (50.472)(.991)(.913)$$

$$= \$45.67 \text{ million.}$$

The only point where individual subjective judgment is required is in the choice of the appropriate measure C of cyclical fluctuation. The numerical values of T and S are found straightforwardly from the equation of the trend line and pattern of seasonal indexes, respectively.

EXERCISES 10.E

10.25. For the time series data on the per capita GNP presentd in Exercise 10.13, forecast the GNP for 1960, using the results of Exercises 10.17 and 10.22.

10.26. Using the time series data on the Consumer Buying Plans Index listed in Exercise 10.14, forecast the value of the index for the third sixth of 1978, applying the results of Exercises 10.18 and 10.23.

10.27. Exercise 10.15 lists quarterly data on new plant and equipment expenditures made by the air transportation industry. Forecast the total expenditure for the fourth quarter of 1979, using the results of Exercises 10.19 and 10.24.

SECTION 10.F   THE AUTOREGRESSIVE FORECASTING MODEL

All the methods introduced so far in our analysis of time series have implicitly assumed that the time period, whether it be measured in quarters, seasons, or any other time unit, is the major factor in establishing the behavior pattern of the series. This was especially evident in Section 10.B, where the secular trend line is a direct expression of the dependence of loan volume on quarter number. Later sections modify the secular trend somewhat, but basically only by assigning each quarter number to a season and then adjusting the secular trend according to seasonal variation parameters.

In this section, the focus will be different. We'll take another look at the time series of real estate loan volume, but we'll virtually ignore the quarter number in making our forecast. What we'll base our analysis upon instead is the relationship between the loan volume figures for two successive quarters, the assumption here being that loan volume in one quarter depends heavily on what the loan volume was in the previous quarter. If we use the notation

$$y_k = \text{loan volume in kth quarter}$$
$$y_{k-1} = \text{loan volume in } (k-1)\text{st quarter}$$

then the *first-order autoregressive forecasting model* looks for coefficients A and B such that

■ $$y_k = A + By_{k-1} \tag{10.7}$$

As you may notice, what we really have here is a linear regression equation of the time series in terms of itself; hence the word "autoregressive." (*Second-order autoregressive forecasting* will be discussed in the exercises.) From formula (10.7) on, the autoregressive forecasting problem is handled just as if it were a linear regression problem (which it really is). The coefficient of linear determination will tell us what proportion of the variation in values of the time series can be explained on the basis of a linear relationship between successive values.

To apply the method of Chapter 8 to the autoregressive forecasting problem, we construct the table of calculations appearing in Table 10.13. Using the sums at the bottom of Table 10.13, we can calculate the coefficient of linear correlation (the *autocorrelation*) for this model:

$$
\begin{aligned}
r_A &= \frac{n\Sigma y_{k-1}y_k - (\Sigma y_{k-1})(\Sigma y_k)}{\sqrt{n\,\Sigma y_{k-1}^2 - (\Sigma y_{k-1})^2}\,\sqrt{n\,\Sigma y_k^2 - (\Sigma y_k)^2}} \\[2mm]
&= \frac{35(29,188) - (942)(970)}{\sqrt{35(30,212) - (942)^2}\,\sqrt{35(31,556) - (970)^2}} \\[2mm]
&= \frac{1,021,580 - 913,740}{\sqrt{170,056}\,\sqrt{163,560}} \\[2mm]
&= \frac{107,840}{(412.378)(404.426)} = .6466.
\end{aligned}
$$

The *coefficient of linear autodetermination* is therefore

$$r_A^2 = (.6466)^2 = .4181 = 41.81\%.$$

TABLE 10.3 Calculations for the First-Order Autoregressive
Forecasting Model

Quarter k	y_{k-1}	y_k	y_{k-1}^2	y_k^2	$y_{k-1}y_k$
(1)	—	—	—	—	—
2	10	18	100	324	180
3	18	13	324	169	234
4	13	12	169	144	156
5	12	11	144	121	132
6	11	18	121	324	198
7	18	12	324	144	216
8	12	11	144	121	132
9	11	14	121	196	154
10	14	26	196	676	364
11	26	17	676	289	442
12	17	18	289	324	306
13	18	21	324	441	378
14	21	30	441	900	630
15	30	22	900	484	660
16	22	21	484	441	462
17	21	26	441	676	546
18	26	43	676	1,849	1,118
19	43	28	1,849	784	1,204
20	28	30	784	900	840
21	30	25	900	625	750
22	25	37	625	1,369	925
23	37	26	1,369	676	962
24	26	25	676	625	650
25	25	29	625	841	725
26	29	48	841	2,304	1,392
27	48	32	2,304	1,024	1,536
28	32	31	1,024	961	992
29	31	34	961	1,156	1,054
30	34	52	1,156	2,704	1,768
31	52	35	2,704	1,225	1,820
32	35	34	1,225	1,156	1,190
33	34	37	1,156	1,369	1,258
34	37	57	1,369	3,249	2,109
35	57	39	3,249	1,521	2,223
36	39	38	1,521	1,444	1,482
(37)	—	—	—	—	—
Sums	942	970	30,212	31,556	29,188

$n = 35$

It follows that 41.81% of the variation in a time series point can be accounted for by the previous point, on the basis of an autoregressive linear model.

The actual autoregressive forecasting formula would then be equation (10.7) with

$$B = \frac{n\Sigma y_{k-1}y_k - (\Sigma y_{k-1})(\Sigma y_k)}{n\Sigma y_{k-1}^2 - (\Sigma y_{k-1})^2}$$

$$= \frac{107{,}840}{170{,}056} = .634$$

$$A = \frac{\Sigma y_k - B\Sigma y_{k-1}}{n} = \frac{970 - (.634)(942)}{35}$$

$$= \frac{372.772}{35} = 10.651$$

Therefore the autoregressive forecasting equation is

$$y_k = 10.651 + .634\, y_{k-1}.$$

In Table 10.14, the autoregressive forecasts are listed together with the actual time series, and the two are graphed together for comparison purposes in Fig. 10.7.

A number of important variants of the linear autoregressive model will be discussed in the exercises. First of all, the autocorrelation coefficient r_A may be higher when computed between y_k and y_{k-2} or between y_k and y_{k-3}, etc., than between y_k and y_{k-1}. This would mean that a better forecasting procedure would be to use the time series values to forecast two, three, or more time periods ahead. This would be done using the equation

$$y_k = A + By_{k-2} \quad \text{or} \quad y_k = A + By_{k-3},$$

etc. The formula (10.7) makes forecasts only one time period ahead.

A second variant would be to use two previous series values to make a forecast; for example,

$$y_k = A + By_{k-1} + Cy_{k-2}.$$

To carry out this procedure, we would use the multiple regression techniques discussed in Chapter 9.

A close examination of Fig. 10.7, in fact, would show that we could substantially improve our ability to forecast the original time series (and this is what we really want to do) if we were to use an autoregressive forecasting model $y_k = A + By_{k-4}$. This would allow us to make our forecasts four time periods ahead, and so would bring into line the two series in Fig. 10.7.

TABLE 10.14   Actual Time Series vs. Autoregressive Forecasts

Quarter k	Actual Value y_k of Time Series	Autoregressive Forecast Using $y_k = 10.651 + .634y_{k-1}$
2	18	16.99
3	13	22.06
4	12	18.89
5	11	18.26
6	18	17.63
7	12	22.06
8	11	18.26
9	14	17.63
10	26	19.53
11	17	27.14
12	18	21.43
13	21	22.06
14	30	23.97
15	22	29.67
16	21	21.40
17	26	23.97
18	43	27.14
19	28	37.91
20	30	28.40
21	25	29.67
22	37	26.50
23	26	34.11
24	25	27.14
25	29	26.50
26	48	29.04
27	32	41.08
28	31	30.94
29	34	30.31
30	52	32.31
31	35	43.62
32	34	32.84
33	37	32.21
34	57	34.11
35	39	46.79
36	38	35.38
(37)	—	(37.74)

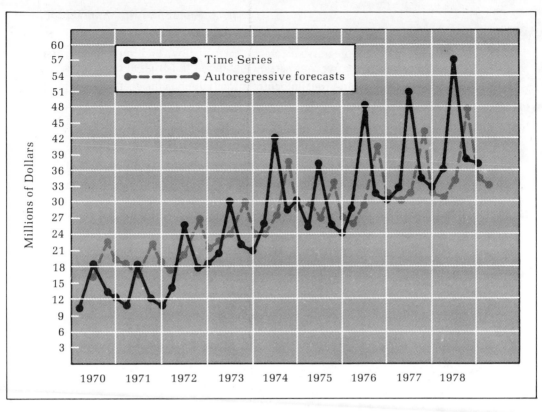

FIGURE 10.7 Comparison of a Time Series with Its Autoregressive Forecasts

EXERCISES 10.F

10.28. Consider the time series data on per capita GNP given in Exercise 10.13.

 a. Calculate the one-time period autocorrelation (between y_k and y_{k-1}) and determine what percentage of variation in series values may be attributed to the autoregressive forecasting equation $y_k = A + By_{k-1}$.

 b. Find the autoregressive forecasting equation, and superimpose the series of autoregressive forecasts on a graph of the original series.

 c. Calculate the two-time period autocorrelation (between y_k and y_{k-2}) and find out what proportion of variation in data points may be explained on the basis of the autoregressive forecasting equation $y_k = A + By_{k-2}$.

 d. Compute the forecasting equation $y_k = A + By_{k-2}$, and superimpose the two-time period autoregressive forecasts on the graph of part b.

10.29. Refer to the time series data on the Consumer Buying Plans Index presented in Exercise 10.14.

 a. Find the one-time-period autocorrelation (between y_k and y_{k-1}) and determine what percentage of variation in individual points of the series can be accounted for by the autoregressive forecasting equation $y_k = A + By_{k-1}$.

b. Calculate the autoregressive forecasting equation, and superimpose the series of autoregressive forecasts on a graph of the original time series.

c. Calcuate the *second-order* autocorrelation (which is the coefficient of multiple linear determination discussed in Section 9.A) to find out what percentage of variation in series values may be expressed in terms of the second-order autoregressive forecasting equation $y_k = A + By_{k-1} + Cy_{k-2}$.

d. Use multiple linear regression techniques, presented in Section 9.B, to compute the second-order forecasting equation $y_k = A + By_{k-1} + Cy_{k-2}$.

10.30. Return to the quarterly data on new plant and equipment expenditures made by the air transportation industry, which are presented in Exercise 10.15.

a. Calculate the four-time-period autocorrelation (between y_k and y_{k-4}) and compute the proportion of variation in series figures that can be attributed to the autoregressive forecasting equation $y_k = A + By_{k-4}$.

b. Find the autoregressive forecasting equation $y_k = A + By_{k-4}$, and superimpose the series of autoregressive forecasts on a graph of the original time series.

c. Compute the second-order autocorrelation, namely the coefficient of multiple determination, to find out what percentage of variation in series values can be attributed to the second-order autoregressive forecasting equation $y_k = A + By_{k-4} + Cy_{k-8}$.

d. Make use of multiple linear regression techniques to calculate the second-order forecasting equation mentioned in part c.

SUMMARY AND DISCUSSION

We have built upon the regression and correlation techniques of Chapter 8 in order to study time series, a special kind of paired data that record the change of business conditions over a long period of time.

We developed methods of analyzing the major components of the classical time series model, namely linear trend, seasonal variation, and cyclical fluctuation. After calculating the coefficient of linear determination to find out what proportion of the variation in the time series data can be attributed to the component of linear trend, we used the methods of linear regression to find the equation of the trend line. Next, we investigated the influence of seasonal variation by using seasonal indexes based on moving averages. Then we used these seasonal indexes to seasonally adjust the data, so as to be able to discover any underlying cyclical fluctuation pattern that may exist in the absence of seasonal variation. All the various components were illustrated and compared in terms of their graphical representations.

We supplemented our study of the classical time series model with an introduction to some modern techniques that have recently attracted great interest among practitioners of time series analysis. Exponential smoothing was presented as an alternative to moving averages, which allows the analyst some more flexibility in deciding how much smoothing he or she wants to exhibit. The autoregressive forecasting model made use of material from

Chapters 8 and 9 in taking advantage of period-to-period links among the points of a time series.

CHAPTER 10   BIBLIOGRAPHY

REFINED TECHNIQUES

BROWN, R. G., "Detection of Turning Points in a Time Series," *Decision Sciences*, October 1971, pp. 383–403.

CHAN, H., and J. HAYYA, "Spectral Analysis in Business Forecasting," *Decision Sciences*, January 1976, pp. 137–151.

VAN NESS, P. H., "Adjusting Polynomial Trend Functions," *Decision Sciences*, October 1973, pp. 563–568.

APPLICATIONS

FOSTER, G., "Quarterly Accounting Data: Time-Series Properties and Predictive-Ability Results," *The Accounting Review*, January 1977, pp. 1–21.

GARDNER, E. S., JR., and D. G. DANNENBRING, "Forecasting with exponential smoothing: some guidelines for model selection," *Decision Sciences*, April 1980, pp. 370–383.

MABERT, V. A., "Statistical Versus Sales Force Executive Opinion Short Range Forecasts: A Time Series Analysis Case Study," *Decision Sciences*, April 1976, pp. 310–318.

MABERT, V. A., "Forecast modification based upon residual analysis: a case study of check volume estimation," *Decision Sciences*, April 1978, pp. 285–296.

MELNICK, E. L., and J. MOUSSOURAKIS, "Seasonal adjustment for the decision maker," *Decision Sciences*, April 1975, pp. 252–258.

CHAPTER 10   SUPPLEMENTARY EXERCISES

10.31. The Federal Reserve Board's index of industrial production uses as its reference point the base year 1967. The following seasonally adjusted values of the index were reported for September 1976 and September 1977, for various industry groupings.*

	1976	1977
Manufacturing	130.5	138.9
Durable goods	122.4	131.9
Nondurable goods	142.3	149.2
Mining	115.5	120.4
Utilities	149.6	157.2

   a. A manufacturing company produced 2.4 million units of its product in September 1976. What would be the equivalent volume of such production

*The Wall Street Journal, October 17, 1977, p. 2.

in September 1977 if this company can be considered typical of those for which the index is computed?

b. A small factory produced 120,000 units of durable goods in September 1977. If it had behaved according to the index of industrial production all these years, how many units would it have produced in September 1967?

c. The U-Sprinklitall Garden Hose Company began production of its nondurable goods in September 1976, with an initial output of 1,500 hoses. To keep up with the index of industrial production, how many should it have produced in September 1977?

d. The Moonglow Mining Company produced 8,500 tons of bauxite in September 1967. It kept pace with the index for the next 14 years. How many tons did it produce in September 1976 and September 1977?

e. Amalgamated Alabama Power Company produced 1.8 million kilowatt-hours of electricity in September 1967. What volume of production was required in September 1977 to maintain its standing relative to the index of industrial production?

10.32. The *U.S. News & World Report* index of business activity, in which the year 1967 is used as the base year, had the following values in August 1976 and August 1977:*

	1976	1977
Consumer prices	171.9	183.3
Wholesale prices	184.8	197.8
Retail store sales	206.9	225.7

a. If a typical family of four spent $600 for a certain combination of goods and services in August 1967, how much would the same combination of items have cost in August 1976 and August 1977?

b. The Rockhard Steel Company spent $280,000 in August 1977 to outfit its new Billings, Montana, mill. It paid the wholesale price for these items. How much would it have cost the company in August 1967 for the same goods?

c. A local chain of department stores sold $900,000 worth of goods in August 1976. To keep pace with the index, what should its sales have been a year later?

10.33. The Labor Department's Bureau of Labor Statistics (BLS) recorded, for the years 1909 through 1965, the number of man-hours spent in the nation's non-agricultural industries. The following data give the yearly index numbers of the original (raw) labor force data, with the 1957–1959 average set at 100:†

*U.S. News & World Report, October 24, 1977, p. 82.

†As reported in U.S. Department of Commerce, Long Term Economic Growth 1860–1965: A Statistical Compendium. 1966, U.S. Government Printing Office, pp. 174–175.

Year	Index	Year	Index	Year	Index
1909	60.9	1928	79.7	1947	91.6
1910	62.9	1929	81.9	1948	93.7
1911	64.2	1930	75.0	1949	90.8
1912	66.7	1931	66.2	1950	93.5
1913	67.5	1932	56.8	1951	96.3
1914	65.1	1933	56.1	1952	96.5
1915	64.8	1934	57.0	1953	98.2
1916	71.4	1935	60.4	1954	94.2
1917	72.7	1936	67.2	1955	98.3
1918	71.7	1937	71.0	1956	101.2
1919	69.0	1938	64.5	1957	100.9
1920	69.8	1939	68.8	1958	98.0
1921	62.3	1940	72.9	1959	101.1
1922	67.9	1941	81.7	1960	102.8
1923	75.0	1942	88.8	1961	102.7
1924	72.4	1943	93.1	1962	104.6
1925	75.4	1944	91.6	1963	106.1
1926	78.8	1945	86.9	1964	108.8
1927	79.2	1946	88.5	1965	112.8

a. Graph the time series of the BLS index.

b. What proportion of yearly variation in the index can be explained on the basis of linear trend?

c. Calculate the equation of the trend line of the time series, and superimpose the graph of the trend line upon the time series graph.

d. Compute the three-term moving average of the time series, and superimpose its graph upon that of the series itself.

e. Construct the series of seasonal index, where the data is assumed to occur in two seasons, the first consisting of those years ending in the digits 9, 0, 1, 2, 3 and the second of those years ending in 4, 5, 6, 7, 8.

f. Draw the two graphs that portray the patterns of each of the two seasons' indexes.

g. Seasonally adjust the time series, and construct the graph of the seasonally adjusted series.

h. Calculate the series of trend line forecasts of the BLS index, using the equation found in part c.

i. Combine the trend line forecasts with the series of seasonally adjusted data found in part g, and determine the cyclical fluctuation of the BLS index as percent of trend.

j. Construct the graph that portrays the pattern of cyclical fluctuation of the time series of the BLS index.

k. Forecast the BLS index for 1975.

l. Exponentially smooth the time series, using smoothing constant $\varepsilon = .7$, and superimpose the graph of the exponentially-smoothed series upon the graph of the original series.

m. Calculate the one-time-period autocorrelation and determine what percentage of variation in series values may be attributed to the autoregressive forecasting equation.

n. Find the autoregressive forecasting equation $y_k = A + By_{k-1}$, and superimpose the series of autoregressive forecasts on a graph of the original time series.

10.34. The following data give the nationwide quarterly totals of expenditures for new plants and equipment made in the durable goods manufacturing industry for the years 1967–1977 (first quarter of 1977 only) in billions of dollars:*

Year	Quarter	Total	Year	Quarter	Total
1967	1	3.14	1972	1	3.29
	2	3.56		2	3.71
	3	3.40		3	3.86
	4	3.96		4	4.77
1968	1	3.06	1973	1	3.92
	2	3.36		2	4.65
	3	3.54		3	4.84
	4	4.16		4	5.84
1969	1	3.36	1974	1	4.74
	2	3.98		2	5.59
	3	4.03		3	5.65
	4	4.49		4	6.64
1970	1	3.59	1975	1	5.10
	2	4.08		2	5.59
	3	3.87		3	5.16
	4	4.26		4	5.99
1971	1	3.11	1976	1	4.78
	2	3.52		2	5.61
	3	3.40		3	6.02
	4	4.12		4	7.19
			1977	1	5.65

a. Graph the time series of the new plant and equipment expenditures data.

b. How much of the quarterly variation in the expenditures can be explained on the basis of linear trend?

c. Determine the equation of the trend line, and superimpose its graph upon the time series graph.

d. Calculate the five-term moving average of the time series, and superimpose its graph upon the time series graph.

*U.S. Department of Commerce, Office of Business Economics, *Survey of Current Business,* Vols. 50, 52, 54, 56, p. S-2.

e. Compute the series of quarterly indexes, and draw the four graphs that illustrate the behavior of the quarterly indexes.

f. Seasonally adjust the time series, and draw the graph of the seasonally adjusted series.

g. Use the trend line equation of part c to find the trend line forecasts of the new plant and equipment expenditures.

h. Based on the trend line forecasts above and the seasonally adjusted data of part f, determine the cyclical fluctuation pattern of the expenditures as percent of trend.

i. Draw a graph illustrating the cyclical fluctuation pattern of the new plant and equipment expenditures.

j. Forecast the total expenditures for the first quarter of 1980.

10.35. New York Stock Exchange data for the first quarter of 1977 showed that General Motors common stock had the following daily closing prices in dollars.*

Week	Day	Price	Week	Day	Price	Week	Day	Price
Jan 3	M	78.000	Jan 31	M	74.625	Feb 28	M	70.750
	T	77.500		T	74.375		T	71.625
	W	76.375		W	74.500		W	70.875
	Th	75.500		Th	74.000		Th	71.250
	F	75.375		F	75.000		F	71.250
Jan 10	M	76.000	Feb 7	M	74.875	Mar 7	M	71.250
	T	75.000		T	73.000		T	71.000
	W	74.875		W	71.000		W	71.000
	Th	75.500		Th	71.375		Th	71.375
	F	75.250		F	69.750		F	71.125
Jan 17	M	74.750	Feb 14	M	70.500	Mar 14	M	71.875
	T	74.250		T	71.250		T	72.125
	W	75.125		W	71.750		W	73.000
	Th	74.625		Th	71.750		Th	73.000
	F	74.750		F	71.375		F	72.375
Jan 24	M	74.625	Feb 21	M	71.063†	Mar 21	M	71.250
	T	75.250		T	70.750		T	71.250
	W	75.375		W	70.250		W	70.250
	Th	75.250		Th	69.875		Th	69.500
	F	75.000		F	70.000		F	68.750

a. Graph the time series of the closing stock prices.

b. Calculate the proportion of daily variation in stock prices that can be attributed to linear trend.

*Reported in Standard & Poor, *Daily Stock Price Record: New York Stock Exchange,* 1977 (first quarter), Part I, p. 160.

†This was a holiday, and the market was closed. To maintain continuity in the data, we have averaged the closing prices of the preceding and following days.

c. Find the equation of the trend line, and superimpose its graph upon that of the time series.

d. Compute the seven-term moving average of the time series, and superimpose its graph on the original graph.

e. Construct the series of five daily indexes, one each for M, T, W, Th, and F, and draw the five graphs that portray the behavior pattern of the daily indexes.

f. Considering the days of the week as seasons, seasonally adjust the data, and construct the graph of the seasonally adjusted time series.

g. Compute the series of trend line forecasts of the stock price, using the equation found in part c.

h. Combine the trend line forecasts with the series of seasonally adjusted prices found in part f in order to come up with a series of numbers that illustrate the cyclical fluctuation of stock prices as percent of trend.

i. Draw the graph that illustrates the cyclical fluctuation pattern of the time series of stock prices.

j. Forecast the stock price for Friday, April 7, 1977.

10.36. Stock dividends and yields of the blocks of stocks that comprise the Dow-Jones Industrial Average had the following dollar values from the first quarter of 1947 to the third quarter of 1977.*

a. Graph the time series of Dow-Jones Industrial Average stock yields.

b. What proportion of the quarterly variation in yield can be explained on the basis of linear trend?

c. Calculate the equation of the trend line, and superimpose its graph upon that of the time series.

d. Compute the nine-term moving average of the time series, and superimpose its graph on that of the original series.

e. Calculate the series of quarterly indexes, and construct the four graphs that illustrate their behavior patterns.

f. Seasonally adjust the data, and draw the graph of the seasonally adjusted time series.

g. Using the trend line equation calculated in part c, compute the series of trend line forecasts of yield.

h. Based on the trend line forecasts and the series of seasonally adjusted yields found in part f, determine the pattern of cyclical fluctuation of the quarterly yields as percent of trend.

i. Construct a graph that illustrates the cyclical fluctuation pattern.

j. Forecast the yield for the second quarter of 1982.

*As reported in *Barron's*, February 13, 1978, p. 38B.

Year	Quar-ter	Yield	Year	Quar-ter	Yield	Year	Quar-ter	Yield
1947	1	1.92	1957	1	5.00	1967	1	7.55
	2	2.22		2	4.79		2	7.36
	3	2.09		3	4.91		3	7.25
	4	2.98		4	6.91		4	8.03
1948	1	2.17	1958	1	4.96	1968	1	7.29
	2	2.59		2	4.62		2	7.73
	3	2.42		3	4.59		3	7.73
	4	4.32		4	5.83		4	8.59
1949	1	2.60	1959	1	4.89	1969	1	9.37
	2	2.85		2	4.59		2	8.08
	3	2.32		3	4.53		3	7.82
	4	5.02		4	6.73		4	8.63
1950	1	3.02	1960	1	5.12	1970	1	7.68
	2	2.98		2	4.83		2	7.80
	3	3.87		3	4.86		3	7.80
	4	6.26		4	6.55		4	8.25
1951	1	3.89	1961	1	5.00	1971	1	7.70
	2	3.48		2	5.05		2	7.80
	3	3.72		3	5.09		3	7.51
	4	5.25		4	7.57		4	7.85
1952	1	3.71	1962	1	5.15	1972	1	7.65
	2	3.55		2	5.23		2	7.87
	3	3.55		3	5.26		3	7.76
	4	4.62		4	7.66		4	8.99
1953	1	3.77	1963	1	5.15	1973	1	8.08
	2	3.95		2	5.52		2	8.27
	3	3.53		3	5.35		3	8.36
	4	4.86		4	7.39		4	10.62
1954	1	4.04	1964	1	7.83	1974	1	8.97
	2	3.92		2	7.16		2	8.87
	3	3.75		3	5.79		3	9.43
	4	5.76		4	10.46		4	10.45
1955	1	4.96	1965	1	6.70	1975	1	9.81
	2	4.24		2	6.79		2	8.97
	3	4.25		3	6.58		3	9.05
	4	8.13		4	8.54		4	9.63
1956	1	5.01	1966	1	7.44	1976	1	9.23
	2	4.98		2	7.26		2	10.19
	3	4.83		3	7.18		3	9.85
	4	8.17		4	10.01		4	12.13
						1977	1	10.46
							2	11.41
							3	10.73

k. Exponentially smooth the time series using, respectively, smoothing constants $\varepsilon = .2$ and $\varepsilon = .8$. Superimpose the graphs of the two smoothed series upon the graph of the original series.

l. Compute the four-time-period autocorrelation, and find what proportion of variation in series values can be explained in terms of the autoregressive forecasting equation $y_k = A + By_{k-4}$.

m. Calculate the above autoregressive forecasting equation, and superimpose the series of autoregressive forecasts upon a graph of the original time series.

V

RECENT TRENDS IN APPROACHING STATISTICAL QUESTIONS

11

Fundamental Bayesian Techniques

In this chapter we return to probability theory by presenting a complete development of Bayes' theorem on conditional probabilities.

Bayes' theorem grows out of the formulas and tree diagrams in Section 3.B and forms the foundation of an extremely useful method of describing the extent of interrelationships among events, because it illuminates the effect of one event upon the probability of another. *Bayesian inference* is an approach to statistical analysis that builds upon the subject matter of Chapter 5 by taking account of the influence of factors other than sample data in making estimates of parameters.

SECTION 11.A   BAYES' THEOREM

From the applied point of view, Bayes' theorem provides an extremely precise way of developing and specifying relationships between events. When properly applied, it brings order out of a chaos of intertwined relationships, and sometimes even gives results so surprising that they would not even have been considered possible by a researcher or manager unacquainted with the procedure. After such a spectacular buildup, you are no doubt wondering what exactly is Bayes' theorem.

Bayes' theorem takes as its point of departure the theory of conditional probability, as developed in Section 3.B. Recall that, if A and B are two events, then

$P(A|B)$ = probability of A, given that B has occurred

while

$P(B|A)$ = probability of B, given that A has occurred.

What is the relationship between these two probabilities? Consider Alcoholic Beverage Marketeers (ABM), Inc., who want to estimate alcoholic beverage consumption at professional baseball games. Suppose A represents the event that an individual regularly consumes alcoholic beverages with food, and B represents the event that an individual is in attendance at a professional baseball game. Then

$P(B|A)$ = probability that an individual is at the baseball game, given that he or she regularly consumes alcoholic beverages with food,

while

$P(A|B)$ = probability that an individual regularly consumes alcoholic beverages with food, given that he or she is at the baseball game.

These probabilities are considerably different from each other; namely, $P(B|A)$ is considerably smaller than $P(A|B)$.

In which probability is ABM more interested? Well, we have to say that they want $P(A|B)$ to be high, since they're in the business of selling alcoholic beverages. In particular, they would like to see a lot of sales at baseball games. Those whose primary interest is selling tickets to baseball games, on the other hand, would like to see $P(B|A)$ kept at a high level, because they want high game attendance among *all* population groups, including consumers of alcoholic beverages.

Even though the two conditional probabilities represent different situations and have different numerical values, we surmise that they ought to be somehow related. In fact, they are, and their relationship is expressed by Bayes' theorem. It goes as follows:

BAYES' THEOREM
If A and B are any two events, then

$$\blacksquare \quad P(B|A) = \frac{P(A|B)P(B)}{P(A|B)P(B) + P(A|B^c)P(B^c)} \qquad (11.1)$$

To justify the validity of Bayes' theorem, we need refer only to formulas (3.4), (3.5), and (3.6), which have already been discussed in Section 3.B. Combining them, we have that

$$P(B|A) = \frac{P(B \cap A)}{P(A)} = \frac{P(A \cap B)}{P(A)} = \frac{P(A|B)P(B)}{P(A|B)P(B) + P(A|B^c)P(B^c)}$$

(We have also used the fact that $B \cap A$ and $A \cap B$ are one and the same, the event consisting of those outcomes that are in both A and B.)

EXAMPLE 11.1. Financial Analysis. Sunset Consulting Group of Santa Monica has developed a financial analysis computer model that can predict imminent collapse in 98% of those companies that really will fail, based on an analysis of the companies' annual reports. Unfortunately, it also predicts collapse in 1% of those companies that are financially solvent. On the average, about one-half of one percent of those companies analyzed actually will fail. If the model asserts that a company in which you own stock will soon collapse, what is the probability it really will?

SOLUTION. Before proceeding with the solution, it would be constructive for you to guess the answer to the question. Check the box below the number that you feel best represents the chances of your company's collapsing, if the model indicates that it will:

99%   98%   95%   90%   80%   50%   33%   20%   15%   10%
☐   ☐   ☐   ☐   ☐   ☐   ☐   ☐   ☐   ☐

Now that you've checked off your guess, let's proceed to the formal analysis of the problem. First we have to specify the events involved. These seem to be:

F = event that your company will *fail*
M = event that the *model* says it will fail.

First of all, we should note that F and M are two distinctly different events, although there are some relationships between them. We would worry more about F than about M; we are not concerned with M in itself, but only because of its relationship with F. In fact, what we want to determine is the strength of this relationship, $P(F|M)$, the conditional probability of your company's failing, given that the computer model says it will.

Now that we have established that $P(F|M)$ is the answer to the question, let's see what information we are given. The statement that, "one-half of one percent of those companies analyzed actually will fail" means that $P(F) = .005$. This is the original probability of F, before the computer model results are known. The fact that imminent collapse is detected by the model in 98% of those that will collapse can be translated mathematically as $P(M|F) = .98$. Finally, $P(M|F^c)$, the probability of the model's thinking your company will fail, given it really won't, is .01.

TABLE 11.1   Information Relating the Financial Analysis and
Financial Solvency

Desired Information	Known Information
$P(F\|M)$	$P(M\|F)$ = .98 $P(M\|F^c)$ = .01 $P(F)$ = .005

In Table 11.1, we summarize the known information and desired information. The primary characteristic apparent from Table 11.1 is that we know the conditional probabilities of M, given some facts about F, but we want to know the conditional probability of F, given M.

For purposes of comparison, let's take another look at what Bayes' theorem, formula (11.1), says. As is apparent from formula (11.1), if we know the conditional probabilities of A, given some facts about B, we can insert them into the formula to obtain the conditional probability of B, given A. Note that Bayes' theorem has the effect of interchanging the events involved in conditional probabilities, and this is exactly what we need to answer our question. If we replace B by F and A by M in Bayes' theorem, we see from the information in Table 11.1 that

$$P(F|M) = \frac{P(M|F)P(F)}{P(M|F)P(F) + P(M|F^c)P(F^c)}$$

$$= \frac{(.98)(.005)}{(.98)(.005) + (.01)(.995)} = \frac{.0049}{.0049 + .00995}$$

$$= \frac{.0049}{.01485} = .33 = 33\%.$$

Here we have used the fact that $P(F^c) = 1 - P(F)$: If one-half of one percent will actually fail, then ninety-nine and one-half percent will not.

The answer to our question, $P(F|M) = .33$, means that, if the computer model asserts that your company will fail, the chances are 33% that it really will. In view of the facts that $P(M|F) = .98$ while $P(M|F^c) = .01$, the value of .33 for $P(F|M)$ might be considered low. (Virtually no one guesses anything near 33% as the answer.) However, a low answer becomes more reasonable when we consider that the .98 refers to 98% of a group that comprises one-half of one percent of the population, while the .01 stands for 1% of the other 99.5%. A modified tree diagram illustrating the fact that $P(F|M) = 33\%$ is presented in Fig. 11.1. The diagram shows that, of 100,000 companies analyzed for imminent finan-

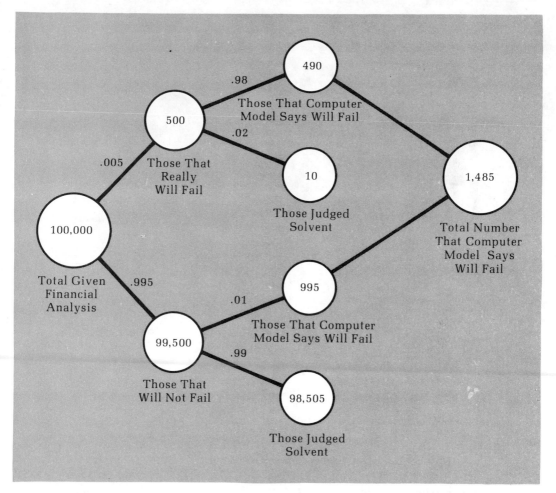

FIGURE 11.1   Status of 100,000 Companies Undergoing Financial Analysis

cial failure, 500 (one-half of one percent) will actually fail, while the other 99,500 will not. Of the 500, 98% or 490 will fail the financial analysis test, while of the 99,500, only 1% or 995 will do so. Therefore, 490 + 995 = 1,485 companies are judged by the computer model to be in danger. Of these 1,485, only 490 or 33% actually will fail.

Tree diagrams of the sort introduced in Chapter 1 can also be applied to this problem. The relevant tree diagram is presented in Fig. 11.2. The interpretation of the tree diagram is as follows: of all companies analyzed, a proportion .01485 are judged to be ripe for collapse. Part of this proportion, an amount .00490, comes from among those that really will

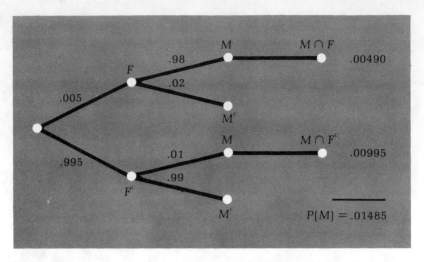

FIGURE 11.2   Tree Diagram of Financial Analysis Example

fail, while the rest, .00995, will not. Of those judged to be about to fail, then, a fraction

$$\frac{.00490}{.01485} = .33 = 33\%$$

actually will. This is, of course, the same result as obtained earlier.

EXAMPLE 11.2.   Department Store Customers.   Willow-Mart, a regional chain of department stores, reaches 45% of all its customers through its downtown urban locations, 40% at suburban malls, and the remaining 15% in rural towns. A marketing questionnaire indicates that 80% of the customers at its downtown urban locations, 10% of those who shop at its suburban mall shops, and 15% of those who shop in its rural stores actually live in the central city. What percentage of those who live in the central city shop at the suburban malls?

SOLUTION.   The first step is to specify the events involved in this problem. We have four basic events:

U = event that customer shops at downtown *urban* location
S = event that customer shops at *suburban* mall
R = event that customer shops in *rural* town
C = event that customer lives in *central* city

We want to know $P(S|C)$, the conditional probability of shopping at a suburban mall, given that the customer lives in the central city. The probabilities available from the data are:

$P(U) = .45$ $P(C|U) = .80$
$P(S) = .40$ $P(C|S) = .10$
$P(R) = .15$ $P(C|R) = .15.$

We apply an expanded version of Bayes' theorem, which is based on formula (3.7) instead of (3.6). The expanded version asserts that

$$\blacksquare \quad P(S|C) = \frac{P(C|S)\,P(S)}{P(C|U)\,P(U) + P(C|S)\,P(S) + P(C|R)\,P(R)} \qquad (11.2)$$

$$= \frac{(.10)(.40)}{(.80)(.45) + (.10)(.40) + (.15)(.15)} = \frac{.04}{.4225} = .095.$$

Therefore 9.5% of those who live in the central city shop at suburban malls.

It would be interesting to also know how many of those who live in the central city shop in rural towns and, for that matter, how many shop at the downtown urban locations. To compute these, we write

$$P(R|C) = \frac{P(C|R)\,P(R)}{P(C|U)\,P(U) + P(C|S)\,P(S) + P(C|R)\,P(R)}$$

$$= \frac{(.15)(.15)}{(.80)(.45) + (.10)(.40) + (.15)(.15)} = \frac{.0225}{.4225} = .053$$

and

$$P(U|C) = \frac{P(C|U)\,P(U)}{P(C|U)\,P(U) + P(C|S)\,P(S) + P(C|R)\,P(R)}$$

$$= \frac{(.80)(.45)}{(.80)(.45) + (.10)(.40) + (.15)(.15)} = \frac{.36}{.4225} = .852.$$

Therefore, we conclude that, of all those customers living in the central city, 85.2% shop at downtown urban locations, 9.5% at suburban malls, and the remaining 5.3% in rural towns.

EXAMPLE 11.3. Weather-Related Merchandise. Long meteorological experience indicates that Syracuse, New York, has rain 20% of the time, snow 10% of the time, cloudiness 30% of the time, and fair weather 40% of the time during November. For inventory planning purposes, Baggs Ninth Avenue (a national chain of widely scattered department stores) must determine the chances of different weather conditions so as to ship the correct volume of umbrellas, snowmobiles, sunglasses, etc. to its various locations.

Baggs' meteorology staff checked on the accuracy of forecasts by the U.S. Weather Service and found out that, for rainy days, the bureau had predicted rain 90% of the time and clouds 10% of the time. For those days on which it snowed, the prediction had been snow 80% of the time, clouds 10% of the time, and fair weather 10% of the time. For those days that turned out to be cloudy, the weather service had forecast clouds 60% of the time, rain 20%, snow 10%, and fair weather 10%. For those days that had fair weather, the prediction was fair weather 70% of the time, rain 10%, clouds 10%, and snow 10%. For tomorrow, a typical November day, the bureau forecasts fair weather. What is the probability that it will snow?

SOLUTION.   The first problem is to define the events involved, and the second is to organize the data. We will be working with the following events:

R = event that it *rains* tomorrow
S = event that it *snows* tomorrow
C = event that it is *cloudy* tomorrow
F = event that it is *fair* tomorrow
PF = event that weather bureau *predicts fair* weather for tomorrow.

We obtain the following probabilities from the data:

$P(R)$ = .20 $P(PF|R)$ =   0
$P(S)$ = .10 $P(PF|S)$ = .10
$P(C)$ = .30 $P(PF|C)$ = .10
$P(F)$ = .40 $P(PF|F)$ = .70.

For example, $P(PF|C)$ is the conditional probability that the weather bureau predicts fair weather, given that the day itself was cloudy; this probability is .10 because, for cloudy days, the bureau had forecast fair weather 10% of the time.

We want to compute $P(S|PF)$, the conditional probability that it will snow tomorrow, given that the weather bureau predicts fair weather. A further extension of the expanded version (11.2) of Bayes' theorem shows that

$$P(S|PF) = \frac{P(PF|S)P(S)}{P(PF|R)P(R) + P(PF|S)P(S) + P(PF|C)P(C) + P(PF|F)P(F)}$$

$$= \frac{(.10)(.10)}{(0)(.20) + (.10)(.10) + (.10)(.30) + (.70)(.40)}$$

$$= .01/.32 = .03125 = 3.125\%.$$

Therefore, if the bureau predicts fair weather, the chances are 3.125% that it will actually snow in Syracuse tomorrow.

EXERCISES 11.A

11.1. Paleontologists have discovered that, in 80% of the cases where wildcat wells have successfully produced substantial amounts of oil, certain prehistoric fossil remains are present in the nearby area. The fossils are present in only 30% of the cases where such wells have not successfully produced oil. On the average, only about 40% of wildcat wells subject to paleontological analysis actually produce oil successfully. If the fossils are present at a site your company is investigating, what is the probability that a well sunk there will successfully produce oil?

11.2. Based on the real estate marketing survey data of Exercise 3.14, what proportion of homeowners are college graduates?

11.3. Use the bond referendum data of Exercise 3.15 to determine what percentage of those who voted against the bond issue do not travel long distances to work or shop.

11.4. Use the psychological test data of Exercise 3.16 to find out the probability that a person who passed the test went on to do satisfactory work that year.

11.5. Based on the underwriting data of Exercise 3.17, calculate the probability that a person to whom a company decides to sell life insurance actually has a serious preexisting disease which, if it had been detected, would have ordinarily disqualified him.

11.6. Using the Census Bureau data of Exercise 3.10, what proportion of households having a color television are headed by persons under 25 years of age?

11.7. An insurance company issues three types of automobile policies: Type G for good risks, Type M for moderate risks, and Type B for bad risks. The company's clients are classified 20% as Type G, 40% as Type M, and 40% as Type B. Accident statistics reveal that a Type G driver has probability .01 of causing an accident in a 12-month period, a Type M driver has probability .02, and a Type B driver has probability .08. A driver who is now applying for insurance has caused an accident within the past 12 months.

    a. What is the probability he is a Type B risk?

    b. What is the probability he is a Type M risk?

    c. What is the probability he is a Type G risk?

11.8. If interest rates are likely to go down, the probability is 70% that the stock market will go up. If interest rates are likely to go up, the chances are 80% that the market will go down. If interest rates are likely to remain stable, the chances are 60% that the market will go up. Financial history indicates that interest rates are likely to go down 30% of the time, up 20% of the time, and remain stable 50% of the time. If the stock market goes up today, what are the prospects for lower interest rates?

11.9. Of used car and major appliance buyers, 80% are good credit risks. Of good credit risks, 70% have charge accounts at department stores, while 40% of bad risks have such charge accounts. If a prospective buyer has a charge account at a department store, what is the probability that she is a bad credit risk anyway?

11.10. Of all persons who eventually go on to hold high management positions in the retailing field, 90% have studied college level accounting or statistics. Of those who never held high management positions, only 30% have studied either of these subjects in college. Only 5% of new hirees will eventually hold high positions in management. Of those new retailing hirees who have studied college level accounting or statistics, what proportion eventually will go on to hold high management positions?

11.11. Soothsayers Associates, Inc., has invented a new-style lie detector that detects lying in 80% of the cases in which the suspect is actually lying. Unfortunately, it also "detects" lying in 40% of the cases in which the suspect is telling the truth. On the average, the suspect is actually lying about 30% of the time. If the lie detector claims that Filbertina is lying, what is the probability she really is?

11.12. Framistan fatigue has a 90% chance of causing premature flexplate failure, while sealant seepage has a 60% chance of doing it. These are the only possible causes of premature flexplate failure. Typically, framistan fatigue occurs in 20% of the units sold, sealant seepage occurs in 30% of the units, and general obsolescence (which does not cause premature flexplate failure) takes the remaining 50% out of service. If your unit experienced premature flexplate failure, what are the chances that the failure was due to sealant seepage?

SECTION 11.B   BAYESIAN INFERENCE

Throughout our study of statistical inference in Chapter 5, we had to face the fundamental question that underlies the application of statistics to real-world problems: "To what extent can a *random sample* be considered a *valid representation* of the *true* existing situation?"

When we computed 95% confidence intervals for parameters, for example, the probability (5%) of our "lack of confidence" was rather substantial. Furthermore, the sample standard deviation, and thus the width of the confidence interval (especially in the case of small samples), might be so large as to make the resulting estimate almost useless. Bayesian inference recognizes these defects in statistical methods based only on sample data, and attempts to improve on them by taking account of additional information that may be available to the decision maker.

What is the nature of statistical information that is not based on sample data? Well, in the management of small businesses especially (but also at certain levels in large ones), decisions are often made by "gut feel." What does this mean? It means that the decision maker has learned from extensive experience and many errors of judgment what the true parameters (e.g., price and demand fluctuations) probably are. The mathematical sophistication of these business people is legendary, but the correctness of their decisions over the long term cannot be denied.*

To illustrate the role of "gut feel" in Bayesian inference, we present the following simplified example, which explicitly points up the application of Bayes' theorem.

> **EXAMPLE 11.4.   Used Appliances.**   A dealer in used and rebuilt appliances has the opportunity to purchase three refrigerators at an exceptionally good price. There is a chance, however, that one or more of the refrigerators might require extensive repairs, which would wipe out anything gained by the low purchase price. In situations like this, experience has shown the dealer that the chances are about 1/24 that none of the three will really be in acceptable condition, about 1/8 that only one will be acceptable, about 1/3 that two will be acceptable, and about 1/2 that all will be acceptable. Because of the limitations of time for and cost of a complete inspection, the dealer decides to pick one at random (with each equally likely to be picked) out of the group of three and to subject it to a thorough inspection. If that one turns out to be acceptable, he agrees to buy all three. If he does decide to buy all three (after testing one), what are the chances that they will all be in acceptable condition?

> **SOLUTION.**   First of all, let's note that, by "gut feel" alone, the probability would be 1/2 that all three are in acceptable condition. We would expect the sampling to increase this probability in case he decides to buy all three. We label the following relevant events:

> N = event that *none* is acceptable
> I = event that *one* is acceptable
> T = event that *two* are acceptable
> A = event that *all* three are acceptable
> B = event that he decides to *buy* them (event that the randomly selected one is acceptable)

> By experience, the dealer feels that

> $P(N) = 1/24 \qquad P(T) = 1/3$
> $P(I) = 1/8 \qquad\ \ P(A) = 1/2$

---

*One story has it that one of these tycoons, being interviewed at his Beverly Hills mansion, was asked how, without an MBA or even a high school education, he was able to amass such a fortune. "Well," he replied, "I buy these things for a nickel and sell 'em for a dime, and that 2% really adds up."

(Probabilities based on gut feel are referred to in Bayesian inference as *subjective probabilities*. This contrasts with *objective probabilities*, which are obtained from formal statistical analysis.) From the dealer's decision rule, we calculate that

$P(B|N)$ = probability he will buy them, given that none is acceptable
= probability that the randomly selected one is acceptable, given that none is acceptable
= 0,
$P(B|I)$ = probability that the randomly selected one is acceptable, given that one is acceptable
= 1/3,
$P(B|T)$ = probability that the randomly selected one is acceptable, given that two are acceptable
= 2/3,
$P(B|A)$ = probability that the randomly selected one is acceptable, given that all are acceptable
= 1.

Applying an extended version of Bayes' theorem, formula (11.2), we obtain the probability that all are acceptable, given that the randomly selected one is:

$$P(A|B) = \frac{P(B|A)P(A)}{P(B|N)P(N) + P(B|I)P(I) + P(B|T)P(T) + P(B|A)P(A)}$$

$$= \frac{(1)(1/2)}{(0)(1/24) + (1/3)(1/8) + (2/3)(1/3) + (1)(1/2)}$$

$$= \frac{1/2}{0 + (1/24) + (2/9) + (1/2)} = \frac{.50}{0 + .0417 + .2222 + .5000}$$

$$= .50/.764 = .654 = 65.4\%.$$

Our conclusion is that, based on an inspection of one randomly selected refrigerator, the dealer was able to increase his chances of buying three good ones to 65.4% from the initial gut feel level of 50%.

Before we move on, let's take a look at what would happen if the dealer were willing to absorb the time and expense of a thorough inspection of two of three refrigerators. We now would use the event

B_2 = event that he decides to buy them
= event that the randomly selected two are acceptable.

Then

$P(B_2|N) = 0$
$P(B_2|I) = 0$
$P(B_2|T) = 1/3$
$P(B_2|A) = 1$

TABLE 11.2   Probabilities of Buying Three
Good Refrigerators   *(Bayesian analysis)*

	Subjective Probabilities Combined with Random Sample of Size:			
	0	1	2	3
Probabilities of Buying	50%	65.4%	81.8%	100%

(Decision rule: Buy only if all refrigerators in the sample are acceptable.)

It follows from Bayes' theorem that

$$P(A|B_2) = \frac{P(B_2|A)P(A)}{P(B_2|N)P(N) + P(B_2|I)P(I) + P(B_2|T)P(T) + P(B_2|A)P(A)}$$

$$= \frac{1/2}{0 + 0 + (1/9) + (1/2)} = \frac{.50}{.611} = .818 = 81.8\%.$$

Therefore, if we add further sampling to the subjective probabilities, we would have an 81.8% chance of getting three good refrigerators when we recommend purchase. A similar analysis shows that, if we were willing to inspect all three, our probability of getting three good ones would be 100% whenever we decided to buy. We summarize these results in Table 11.2.

More detailed statistical anlysis based on the Bayesian principles discussed above allows us to calculate confidence intervals for the mean of a normally distributed population by combining sample data with a business executive's subjective insight.

EXAMPLE 11.5.   Company Cars.   (This example is a Bayesian version of Example 5.3.) A corporation that maintains a large fleet of company cars for the use of its sales staff needs to determine the average number of miles driven monthly per salesperson. The vice president for sales feels that the number of miles driven per month per salesperson is normally distributed with mean 2,500 and standard deviation 300. (The technically correct statement of the vice president's subjective views is that the *prior distribution* of the number of miles is the normal distribution with mean 2,500 and standard deviation 300.) A random sample of 16 monthly car-use records was examined, yielding the information that the 16 salespersons under study had driven a mean of 2,250 miles with a standard deviation of 420 miles. Find a Bayesian 95% confidence interval for the true mean monthly number of miles per salesperson for the entire sales staff.

SOLUTION.   Denoting by μ the true mean number of miles per salesperson, the Bayesian $(1-\alpha)100\%$ confidence interval for μ can be expressed as

■
$$\mu_B - z_{\alpha/2}\, \sigma_B \leq \mu \leq \mu_B + z_{\alpha/2}\, \sigma_B \tag{11.3}$$

(if we are dealing with a normally distributed population), where

■
$$\mu_B = \frac{n\sigma_p^2\, \overline{x} + s^2\, \mu_p}{n\sigma_p^2 + s^2} \tag{11.4}$$

is the Bayesian sample estimate of μ, and

■
$$\sigma_B = \sqrt{\frac{\sigma_p^2 s^2}{n\sigma_p^2 + s^2}} \tag{11.5}$$

is the Bayesian estimated standard deviation of μ. In the above formulas, the vice president's estimates are symbolized by

μ_p = subjective prior normal mean
σ_p = subjective prior normal standard deviation

while, as already known,

n = number of data points in sample
$\overline{x}$ = sample mean
s = sample standard deviation.

Note that the Bayesian estimate μ_B is a *weighted average* of the sample mean $\overline{x}$ and the subjective prior mean μ_p. If there is a lot of data, then $\overline{x}$ is weighted heavily; if not, it is weighted lightly. In the present example, we have

subjective prior estimates	sample data
μ_p = 2,500	n = 16
σ_p = 300	$\overline{x}$ = 2,250
	s = 420

Therefore

$$\mu_B = \frac{n\sigma_p^2\, \overline{x} + s^2\, \mu_p}{n\sigma_p^2 + s^2} = \frac{(16)(300)^2(2,250) + (420)^2(2,500)}{(16)(300)^2 + (420)^2}$$

$$= \frac{(16)(90,000)(2,250) + (176,400)(2,500)}{(16)(90,000) + (176,400)}$$

$$= \frac{3,240,000,000 + 441,000,000}{1,440,000 + 176,400} = \frac{3,681,000,000}{1,616,400}$$

$$= 2,277.28$$

$$\sigma_B = \sqrt{\frac{\sigma_p^2 s^2}{n\sigma_p^2 + s^2}} = \sqrt{\frac{(300)^2(420)^2}{(16)(300)^2 + (420)^2}}$$

$$= \sqrt{\frac{(90,000)(176,400)}{(16)(90,000) + (176,400)}} = \sqrt{\frac{15,876,000,000}{1,616,400}}$$

$$= \sqrt{9,821.83} = 99.11.$$

And, for the 95% confidence interval,

$$z_{\alpha/2} = z_{.025} = 1.96.$$

Therefore, using Bayesian inference, our 95% confidence interval is

$$2,277.28 - (1.96)(99.11) \leq \mu \leq 2,277.28 + (1.96)(99.11)$$

$$2,277.28 - 194.26 \leq \mu \leq 2,277.28 + 194.26$$

$$2,083.0 \ \leq \mu \leq 2,471.5.$$

We can therefore assert that a Bayesian 95% confidence interval for the mean monthly number of miles driven per salesperson for the entire sales staff is 2,083.0 miles to 2,471.5 miles.

Before we leave this subject, let's compare the confidence interval obtained above with the one we obtained in Example 5.3 of Chapter 5. There, our 95% confidence interval was

$$2,026.2 \leq \mu \leq 2,473.8,$$

which is considerably wider than the Bayesian interval

$$2,083.0 \leq \mu \leq 2,471.5.$$

This means that the Bayesian interval provides a tighter estimate of the true mean μ than does the so-called "classical" interval. Why should this be? Well, the Bayesian interval uses all the sample information that the classical interval does, but it also takes account of a "gut feel" subjective estimate. Since the Bayesian interval is therefore based on more information, it logically follows that it would yield a tighter estimate. In fact, looking at formula (11.5), we can see that

$$\frac{\sigma_p^2 s^2}{n\sigma_p^2 + s^2} \leq \frac{\sigma_p^2 s^2}{n\sigma_p^2} = \frac{s^2}{n}$$

so that

$$\sigma_B = \sqrt{\frac{\sigma_p^2 s^2}{n\sigma_p^2 + s^2}} \leqslant \sqrt{\frac{s^2}{n}} = \frac{s}{\sqrt{n}}.$$

Therefore the Bayesian probable error of estimation $E_B = z_{\alpha/2} \sigma_B$ is *always* less than the classical probable error of estimation $E = z_{\alpha/2} (s/\sqrt{n})$. Therefore, the Bayesian interval will always provide a tighter estimate of μ.

EXERCISES 11.B

11.13. An accountant working for a major national CPA firm is assigned the task of verifying the records of a multinational corporation in five European countries. Because of the limitations of time and cost, the accountant has the resources available to study in detail the company's books on only two of the five countries. Based on his experience with similar corporations, the accountant feels that the probabilities are 1/50 that the books on all five countries will be in error, 1/40 that the books on four countries will be in error, 1/20 on three countries, 1/8 on two countries, 1/4 on one country, and 53/100 that the books on every country will be correct.

   a. If he checks the books on two countries carefully and finds both to be correct, what are the chances that the books are correct on all five countries?

   b. If he checks the books on the two countries and finds both to be in error, what are the chances that the books are in error on all five countries?

11.14. A small New York advertising agency would like to recruit two liquor industry experts away from a major San Francisco agency. The New York firm feels that the probability is 2/3 that both people would fit into the New York business community, 1/4 that only one would fit in, and 1/12 that neither would. Unfortunately, their budget can support the expense of inviting only one of the potential recruits to fly in from the West Coast for a personal interview. They decide to select one of the two at random and hire both if the one selected seems to fit in. If he does, what is the probability they both do?

11.15. A municipal bond dealer feels that tax-exempt bonds issued by small cities and rural counties were yielding (as a subjective prior distribution) a normally distributed return with a mean of 5.5% and a standard deviation of .25%. An actual survey of 81 recent issues revealed a mean return of 5.45% with a standard deviation of .36%. Find a Bayesian 98% confidence interval for the true mean yield of all tax-exempt bonds issued by small cities and rural counties. (Compare your interval with that obtained in the solution to Exercise 5.18.)

11.16. A statistician working for the FTC estimates that cigarettes of the Rodeo Rider brand have normally distributed nicotine content (as a subjective prior distribution) with mean 25 mg. and standard deviation 5 mg. A random sample of 25 such cigarettes had 22 mg. of nicotine, on the average, with a standard deviation of 4 mg. Find a Bayesian 80% confidence interval for the true mean nicotine content of Rodeo Rider cigarettes. (Compare your answer with the one to Exercise 5.23.)

SUMMARY AND DISCUSSION

In this chapter on Bayesian analysis, our focus was on Bayesian inference, a method of combining the information yielded by sample data with that based on the decision maker's "gut feel" in order to improve estimates and prediction. The technical foundations of this method lie in Bayes' theorem, a major advance in probability theory that allows us to calculate the probability of an event in the presence of new information that represents either a change in circumstances or a confirmation of a preexisting situation.

CHAPTER 11   BIBLIOGRAPHY

BAYES' THEOREM AND BAYESIAN INFERENCE

AITCHISON, J., "A Geometrical Version of Bayes' Theorem," *The American Statistician*, December 1971, pp. 45–46.
MORGAN, B. W., *An Introduction to Bayesian Statistical Inference*. 1968, Englewood Cliffs, N.J.: Prentice-Hall.

APPLICATIONS

BERHOLD, M., "Procedures to Increase the Validity of Subjective Probability Estimates," *Decision Sciences*, October 1975, pp. 721–730.
FELIX, W. L., JR., "Evidence on Alternative Means of Assessing Prior Probability Distributions for Audit Decision Making," *The Accounting Review*, October 1976, pp. 800–807.
KNOBLETT, J. A., "The Applicability of Bayesian Statistics in Auditing," *Decision Sciences*, July-October 1970, pp. 423–440.
REINMUTH, J. E., "On 'The Application of Bayesian Statistics in Auditing,'" *Decision Sciences*, July 1972, pp. 139–141.
THOMAS, J. and P. CHHABRIA, "Bayesian Models for New Product Pricing," *Decision Sciences*, January 1975, pp. 51–64.
WILLIS, R. E., "A Bayesian Framework for the Reporting of Experimental Results," *Decision Sciences*, October 1972, pp. 1–18.

CHAPTER 11   SUPPLEMENTARY EXERCISES

11.17. A major corporation has several hundred projects in the initial stages of development. These projects are classified into three groups: group I for those having a 90% chance of eventually turning a profit, Group II for those having a 70%

chance of turning a profit, and Group III for those having a 50% chance of turning a profit. Of the corporation's projects, 30% are in Group I, 50% are in Group II, and 20% are in Group III.

a. What percentage of the corporation's projects will eventually turn a profit?

b. Of all those projects that eventually turn a profit, what percentage started out as Group III projects?

11.18. Gulf Fire & Casualty Insurance Company learned that one southern state suffers both coastal hurricanes and inland flooding three years out of every ten. Four years out of every ten it suffers coastal hurricanes, but no inland flooding.

a. What percentage of years does the area suffer coastal hurricanes?

b. For this year, the weather service has forecast coastal hurricanes. What is the probability that inland flooding will also occur?

11.19. Data compiled by the U.S. National Center for Education Statistics on college degrees conferred in 1974 revealed that 14% of the bachelor's degrees were awarded in the field of business and management. In addition, the researchers found that 87% of the business and management degrees were awarded to males, even though males accounted for only 51% of the degrees in other fields.

a. What percentage of college graduates in 1974 were male students receiving their degrees in business and management?

b. What percentage of male college students received their degrees in business and management?

c. What percentage of female graduates received business and management degrees?

11.20. A "stop-smoking" clinic advertised that, in a test region last August, 80% of those who tried and were able to stop smoking had participated in its program, while only 30% of those who tried and failed to stop smoking had participated. Research by the Consumer Fraud Division supported these assertions, but also revealed that, of all smokers in the region who tried to kick the habit last August, 90% had failed, while only 10% succeeded.

a. What percentage of those who tried to stop smoking had participated in the clinic's program?

b. What percentage of the clinic's customers actually were able to stop smoking?

11.21. The noxious oxides of nitrogen comprise 20% of all pollutants in the air, by weight, in a certain metropolitan area. Automobile exhaust accounts for 70% of those noxious oxides, but only 10% of other pollutants in the air.

a. Find out what percentage of pollutants in the air are accounted for by automobile exhaust.

b. Find out how much of the pollution contributed by automobile exhaust can be classified as noxious oxides of nitrogen.

11.22. By her own admission, an applicant taking the CPA exam knew only 50% of the material on which she was being tested. The exam was a multiple-choice

test having four possibilities for the answer to each question, and it covered
in great detail all aspects of the subject matter. On the half of the test covering
material the applicant knew, she naturally got all the answers right; on the
other half, she guessed one of the four options.

a. What proportion of the questions did the applicant answer correctly?

b. If she answered question ;473 correctly, what is the probability she really
knew the answer to it?

11.23. An independent supermarket receives 60% of its fruit from Foster's Fruit
Fields and the other 40% from Gregory's Giant Groves. On the average, about
10% of Foster's shipments arrive overripe, while only 7% of Gregory's do.
Because the farmers issue credits when their fruit arrives overripe, the market
keeps strict records on the subject.

a. What percentage of the market's fruit arrives overripe?

b. What percentage of the overripe fruit comes from Foster, and what percentage from Gregory?

11.24. A real estate sales firm wants to find out whether or not prior experience in
retail sales is good preparation for a career in real estate. Its personnel department notes that only 40% of those it hired over the past five years have developed into productive employees. Of those, 80% had prior retail sales experience, while only 50% of the unproductive employees had such prior
experience. If an applicant for a sales position had no prior retail sales experience, what would the company estimate as his probability of becoming a
productive employee?

11.25. An investor would like to buy four small apartment buildings in the Newport
Beach-Costa Mesa area, but has time to inspect only one before the end of the
taxable year. She feels that the probability is 3/4 that all four are in good
condition, 1/5 that three of the four are, 3/100 that two are, 1/100 that only one
is, and 1/100 that none is.

a. If the one she inspects is in good condition, what is the probability that all
four are in good condition?

b. If the one she inspects is not in good condition, what are the chances that
all four are in bad condition?

c. If the one she inspects is in good condition, what is the probability that the
other three are in bad condition?

11.26. During collective bargaining negotiations, a union official estimated that
candy factory workers were able to produce an average of 15 pounds of chocolate-covered cherries per day, with a standard deviation of 7.0 pounds. Furthermore, their daily production was normally distributed. An analysis
showed that 60 randomly selected employees had a sample mean of 14.8
pounds with a standard deviation of 6.5 pounds. Find a Bayesian 90% confidence interval for the true mean number of pounds produced.

12

Nonparametric Analogues of Parametric Tests

In most of the applied situations studied in the previous chapters using the format of testing statistical hypotheses, there were somewhat stringent requirements on the nature of the data involved.

Many of the small-sample ($n < 30$) tests, for example, could not be applied unless it was reasonable to assume that the underlying populations from which the random samples were drawn were normally distributed. The restriction to normally distributed data was needed because an integral part of the testing procedure made use of either the table of the normal distribution or some other table derived from it, such as that of the t distribution. Prime among the statistical tests falling into this category were the t tests of means in Chapter 6 and the correlation test for a linear relationship in Chapter 8. Our objective in this part is to explain how to handle applied situations in which the prerequisites for using such tests are not met.

All of the statistical tests studied in Chapters 6 and 7 dealt with *parameters* of populations. The concept of a parameter, as you can perhaps recall, was discussed in some detail in Chapter 4 in connection with the binomial and normal distributions. It was pointed out there that both of these distributions possess certain characteristic numbers, called parameters, from which the complete behavior pattern of the distribution can be determined. For the binomial distribution, the parameters were denoted by n and π, while for the normal distribution they were called μ and σ. The tests of Chapters 6 and 7 all involved the study of one or another of these parameters: we had one-, two-, and several-sample tests for population means μ (of normal distributions); one- and two-sample tests for population standard deviations σ (of normal distributions); and one- and two-sample tests for population proportions π (of binomial distributions). Because these tests all involve parameters, they can be referred to as *parametric tests*.

If the underlying populations from which our random samples were selected do not have distributions that behave in accordance with the table of the normal distribution, then these populations cannot be characterized by parameters μ and σ. (A statistical test of whether or not a population can be considered as having the normal distribution will be presented in Section 13.C.) It logically follows that applied problems involving means and standard deviations of nonnormal populations cannot be solved using the usual parametric tests of μ and σ (unless there are enough data points available to satisfy the criteria of the central limit theorem), because such populations do not possess parameters that can be tested in terms of the same statistical properties as can μ and σ. However, because small-sample problems involving nonnormal populations often arise out of applied contexts, it is necessary to develop new statistical techniques for analyzing and solving them. For reasons that should now be obvious, these new techniques are called *nonparametric*.

(Of course, nonparametric techniques can also be applied to normally distributed populations, but it is generally more efficient to use the usual parametric tests in such cases. By "more efficient," we mean that for equal Type I errors, the parametric tests will have smaller Type II errors than the nonparametric ones, if the populations involved are normally distributed.)

The organized study of nonparametric statistics originated in the search for substitute methods for the two-sample t test for the difference of means when the populations involved are nonnormal. Due to the nonparametric nature of the problem, we can no longer test for the difference of means, but only for the difference of other types of averages. The tests we study in this chapter, namely the *sign test* and the *Mann-Whitney test*, are really tests of the difference of medians, rather than the difference of means.

In place of the parametric F test for the difference of standard deviations, we introduce the nonparametric *Mann-Whitney test for equal dispersions*, the notion of dispersion of a set of data away from its average being less precise than the concept of the standard deviation of a set of data from its mean. In the same spirit, we present the *Kruskal-Wallis nonparametric analysis of variance* procedure as a substitute for the normality-based analysis of variance of Chapter 7. All these parametric tests and their nonparametric substitutes are listed formally in Table 12.1.

The correlation test (appearing in Chapter 8) between two sets of data, x's and y's, is actually, you may be surprised to know, a parametric test. The parameter involved is the *true correlation* ρ relating the two normally distributed populations of x's and y's. The correlation ρ is a parameter of the

TABLE 12.1   Parametric Tests and Their Nonparametric Replacements

Section	Parametric Test	Nonparametric Replacement	Section
6.B	One-sample t test	One-sample sign test	12.A
6.D	Two-sample t test	Mann-Whitney test for averages	12.B
6.E	Paired-sample t test	Paired-sample sign test	12.A
6.F	Two-sample F test	Mann-Whitney test for dispersions	12.B
7.B	One-way analysis of variance	Kruskal-Wallis test	12.C
8.D	Correlation test for linearity	Spearman test of rank correlation	12.D

bivariate normal distribution, a two-dimensional probability distribution formed by the intertwining of two normal populations. In the nonnormal case, we use a nonparametric replacement, the *Spearman test of rank correlation*, in order to test for the existence of a close relationship between the x's and y's.

Because nonparametric tests are generally freed from the restrictive conditions that tied the earlier tests to the normal distribution, they are often referred to as *distribution-free* tests. More precisely, distribution-free tests are tests that can be applied to many types of data, regardless of the underlying distribution of the population.

SECTION 12.A   THE SIGN TEST

When we would like to use the one-sample t test or the paired-sample t test on a problem, but are prevented from doing so by the fact that the population from which the data points were selected is probably not normally distributed, the sign test provides a valid way of analyzing the situation.

While the sign test has the valuable advantage that it can be used in cases involving nonnormal data, it suffers from a defect common to all nonparametric tests—a lack of precision regarding exactly how strong a role the data points play in the analysis. For example, as we shall soon see, the one-sample sign test pays no attention at all to the actual numerical values of the data points, but merely whether they are each below or above a number analogous to the hypothesized mean. This means that the one-sample sign test is really a test of whether or not the *true median* is equal to the *hypothesized median*, rather than a test of whether or not the true mean is equal to the hypothesized mean. In fact, the sign test can be viewed as a particular type of test for the binomial proportion $\pi = 1/2$, where the two classifications for the binomial distribution are the categories *above* and *below* the hypothesized median. The paired-sample sign test, used as a replacement for the paired-sample t test, bears a similar relationship to the binomial distribution. Small-sample versions of the sign test base their rejection of the hypothesis on the table of binomial distribution, while large-sample versions use the normal approximation to the binomial.

One-Sample Sign Test

It is instructive to illustrate the *one-sample sign test* by using it to analyze the same situations we faced in the comparable sections of Chapter 6. This method points up very clearly the distinctions between the parallel parametric and nonparametric tests. We begin with the situation of Example 6.2, the setup of which we repeat below:

> **EXAMPLE 12.1.** **Apartment Occupancy Rates.** A real estate investment company with primary holdings in Chicago is considering whether or not to buy and operate properties in Houston. Before it commits itself to investments in the Houston area, it wants to know how occupancy rates for apartment complexes in Houston compare with those in Chicago. Presently, buildings that the company operates in Chicago have an average occupancy rate of 96.1 (per cent). A market research study conducted in Houston recently showed that a random sample of eight comparable apartment buildings there had occupancy rates as follows:
>
> 95.4   95.0   97.5   95.6   93.3   94.2   94.9   97.3
>
> At significance level $\alpha = .05$, do the results of the study support the opinion of one of the company's officers that occupancy rates in Houston are really lower than their counterparts in Chicago?
>
> **SOLUTION.** If we denote by $\tilde{\mu}$ the true median occupancy rate of comparable apartment buildings in Houston, the sign test is capable of testing the hypothesis
>
> H $\tilde{\mu} = 96.1$ (median occupancy rate in Houston is the same as it is in Chicago),
>
> against the alternative
>
> A: $\tilde{\mu} < 96.1$ (median occupancy rate in Houston is below that of Chicago),

---

TABLE 12.2   Calculations for the One-Sample
Sign Test   *(Occupancy rate data)*

x	$\tilde{\mu}_o$	sign of $x - \tilde{\mu}_o$
95.4	96.1	−
95.0	96.1	−
97.5	96.1	+
95.6	96.1	−
93.3	96.1	−
94.2	96.1	−
94.9	96.1	−
97.3	96.1	+
Total number of +'s		2
$\Theta = 2$ $n = 8$ $p = 1/2$		

The number 96.1 here is called the hypothesized median, and by analogy with the hypothesized mean, we use the symbol $\tilde{\mu}_o$ to denote it. The
preliminary computational setup for the one-sample sign test appears
in Table 12.2. Major components of the actual test for the rejection of
H in favor of A are the quantities

Θ = number of data points exceeding hypothesized median $\tilde{\mu}_o$
   = number of plus (+) signs in the list of signs
n = number of data points not equaling $\tilde{\mu}_o$
   = number of pluses, plus number of minuses
π = 1/2

The numbers n and π are used in the small-sample case as the parameters of the appropriate binomial distribution. For a test such as the sign
test, which involves the median, we always use π = 1/2 because, if the
hypothesis H is really true, we should ideally have half pluses and half
minuses in the list of signs.

From the calculations of Table 12.2, we find out that we have to use
the binomial distribution with parameters n = 8 and π = 1/2. The
portion of Table A.2 that is reproduced in Table 12.3 indicates that, if
we use a significance level of α = .0352 instead of α = .05, we should
reject H: $\tilde{\mu}$ = 96.1 in favor of A: $\tilde{\mu}$ < 96.1 if Θ < 2. We derive this
rejection rule from the fact that

$$P(\Theta < 2) = P(\Theta \leqslant 1) = .0352,$$

TABLE 12.3   A Small Portion of Table A.2

n	k	$\pi = .50$
8	0	.0030
	1	.0352
	2	.1445

according to Table 12.3, if Θ is a binomial random variable with parameters $n = 8$ and $\pi = 1/2$.

Therefore, our rejection rule for H against A requires us to reject H if $\Theta < 2$. As it has turned out, however, the calculations in Table 12.2 show that $\Theta = 2$. And if $\Theta = 2$, then it is not true that $\Theta < 2$. Therefore, we cannot reject H at level $\alpha = .0352$. Because .05 is between .0352 and .1445 and we would have to reject H at level $\alpha = .1445$ (for the rejection rule at that level would be $\Theta < 3$), we cannot validly make a decision at level $\alpha = .05$. Our conclusion, then, at level $\alpha = .0352$ (which is the closest we can get to .05), is that the data do not support the assertion that occupancy rates in Houston are really lower, on the average, than their counterparts in Chicago.

Our decision, in regard to the question of Example 12.1, to support the hypothesis that Houston occupancy rates are not lower than those in Chicago was the same decision to which we came in Example 6.2 using the one-sample t test. While it is reassuring that this happened, it is frequently the case that parametric and nonparametric tests, applied to the same set of data, will recommend "opposite" decisions.

In the case of the sign test vs. the t test, this disparity could occur because the true median, for example, is larger than the hypothesized number, while the true mean is not. The decisions are not really opposite because both tests do not really test exactly the same thing, but the appearance of "opposite" is given if we think about the problem in terms of average, rather than the more precise terms, mean and median. Which test is to be believed in the case of "opposite" decisions? Well, if the data come from a normally distributed population, the parametric test takes precedence because it makes use of the data directly, rather than using "shadow" properties of the data, such as signs. (In fact, if the underlying population has the normal distribution, then the mean and median are the same. The parametric test will result in a smaller probability of Type II error.)

On the other hand, if the normality or other conditions for the parametric test are not met, then the nonparametric test dominates because the parametric test simply does not apply, therefore cannot validly be used, and will in fact result in a larger probability of Type II error.

One more comment on the one-sample sign test: if $n > 20$, then the binomial table of the Appendix gives way to the normal approximation to the binomial distribution, by analogy with the large-sample test of proportion discussed in Section 6.G. We then base our decision to reject or accept H on a z test, where

■
$$z = \frac{\theta - \dfrac{n}{2}}{\sqrt{\dfrac{n}{4}}}$$
(12.1)

This formula is a consequence of formula (6.13) of the large-sample test of proportion with $p = \theta/n$ and $\pi_0 = 1/2$.

Paired-Sample Sign Test

To illustrate the small-sample techniques used in the paired-sample sign test, we use the setup from the farm equipment sales experiment of Example 6.7.

> **EXAMPLE 12.2.   Sales of Farm Equipment.**   A regional wholesaler of farm equipment, parts, and supplies wants to investigate whether or not he really needs a staff of sales people to personally present products at the customer's own location. Although he has always had a sales staff, he is wondering whether a change to a mail and phone ordering method would significantly reduce his sales volume. Last year he therefore conducted an experiment to compare sales in Fairbury, where he used his sales staff, and Norfolk, where he operated with mail and phone orders only. These two Nebraska cities were chosen because they were comparable in size and sales volume generated by the wholesaler. A random sample of weekly sales figures for the year, in thousands of dollars, is recorded in Table 12.4. At level $\alpha = .10$, do the results of the experiment indicate that the switch to a mail and phone operation in Norfolk led to a significant reduction in weekly sales volume?

> **SOLUTION.**   To apply the paired-sample sign test, we use the notation

> $\tilde{\mu}_d$ = true median of weekly differences, sales volume in mail-order region minus sales volume in sales staff region

> Then we want to test the hypothesis

> H: $\tilde{\mu}_d = 0$ (switching to mail order *does not* decrease sales)

TABLE 12.4   Farm Equipment Sales Data

| Week Beginning | Thousands of Dollars | |
	Norfolk (mail order)	Fairbury (sales staff)
Aug. 15	1.5	1.4
April 18	1.4	1.4
Jan. 10	2.2	2.6
Nov. 14	2.6	2.5
June 20	3.9	4.8
Feb. 28	4.5	4.3
July 4	3.7	4.0
Dec. 12	3.4	3.5
Jan. 24	4.0	3.9
May 16	2.5	2.6
March 21	1.9	2.6
Oct. 3	1.5	1.4

against the alternative

A: $\tilde{\mu}_d < 0$ (switching to mail order *does* decrease sales).

Table 12.5 contains the calculations necessary to carry out the statistical analysis.

TABLE 12.5   Calculations for the Paired-Sample Sign Test   *(Farm equipment sales data)*

Week Beginning	Mail-Order x_M	Sales Staff x_S	Sign of $d = x_M - x_S$
Aug. 15	1.5	1.4	+
April 18	1.4	1.4	0
Jan. 10	2.2	2.6	−
Nov. 14	2.6	2.5	+
June 20	3.9	4.8	−
Feb. 28	4.5	4.3	+
July 4	3.7	4.0	−
Dec. 12	3.4	3.5	−
Jan. 24	4.0	3.9	+
May 16	2.5	2.6	−
March 21	1.9	1.7	+
Oct. 3	1.5	1.4	+

| Total number of +'s | | | 6 |
| $\Theta = 6$ | $n = 12 - 1 = 11$ | $\pi = 1/2$ | |

TABLE 12.6   A Small Portion of Table A.2

n	k	$\pi = .50$
11	1	.0059
	2	.0327
	3	.1133

The results of the calculations done in Table 12.5 show that we should base our statistical test on the binomial distribution having parameters $n = 11$ and $\pi = 1/2$. From the portion of Table A.2 that is reproduced in Table 12.6, we then see that we should reject H: $\tilde{\mu}_d = 0$ in favor of A: $\tilde{\mu}_d < 0$ at level $\alpha = .1133$ if $\Theta \le 3$ or $\Theta < 4$ and at level $\alpha = .0327$ if $\Theta \le 2$ or $\Theta < 3$. Because $\Theta = 6$, and $.0327 < .10 < .1133$, we therefore cannot reject H at level $\alpha = .10$. Our conclusion, at level $\alpha = .10$, is that switching to mail-order does not seem to significantly decrease sales, according to the given data.

While the decision based on the sign test agreed with our earlier decision in Chapter 6 based on the t test, a comparison of the computations in Tables 6.18 and 12.5 reveals a serious disparity between the two methods. From Table 6.18, we see that the total difference Σd is negative ($= -1.0$), indicating that the mean difference in sales is negative; so that, on the average (in terms of the mean), greater sales were made by the sales staff. From Table 12.5, on the other hand, we see the number $\Theta = 6$, indicating that 6 out of the 11 nonzero signs and consequently the median difference are positive; so that, on the average (in terms of the median), more weeks have greater sales made by mail-order.

The reason for this disparity lies in the major distinction between the mean and the median introduced way back in Section 2.A: the mean takes into account extraordinary large values, while the median does not. The week of June 20, for example, in Table 6.18, has a d value of $-.9$, overcoming the total differences of the weeks of Aug. 15, Nov. 14, Feb. 28, Jan. 24, March 21, and Oct. 3. In Table 12.5, however, June 20 counts as only one minus sign, while those six weeks contribute a total of six plus signs.

In a case of an extreme difference between the mean and the median, it should not be surprising if the t test and the sign test recommend different decisions. (Of course, if the mean and median are different, the underlying population cannot have the normal distribution, and so the result of the t test is not to be believed.)

EXERCISES 12.A

12.1. Based on the U.S. News & World Report unemployment rate data appearing in Exercise 5.17, can we assert that the median nationwide urban unemployment rate was below 6.5% at the time the survey was taken? Use a level of significance of $\alpha = .05$.

12.2. From the data of Exercise 6.35, does the sign test support, at level $\alpha = .05$, the theory that employees have less absenteeism after they stop smoking?

12.3. A company that markets canned carrots advertises that their 15-ounce cans actually have a median weight 15.1 ounces, so that a substantial percentage of cans exceed the stated contents of 15.0 ounces. To test the validity of the company's assertion, a state consumer agency selects a random sample of 11 cans and carefully weighs the contents of each. The 11 weights are as follows:

   15.1   15.0   15.2   14.9   15.0   15.1   15.1   14.9   15.1   15.0   14.6

Can we conclude at level $\alpha = .05$ that:

a. The true median contents are below the 15.1 ounces claimed by the company?

b. The true median contents are below the 15.0 ounces printed on the label?

12.4. *Business Week's* "Investment Outlook Survey: 1978" included the following data on the 1977 percentage change in the market value of outstanding shares of corporation stock.*

Corporation	Percentage Change	Corporation	Percentage Change
Arcata National	11	Brown Group	−6
Columbia Gas	−5	Detroit Edison	14
Firestone Tire	−32	Gillette	−12
IC Industries	19	Laclede Gas	5
Merck	−17	Northrop	−9
Potlatch	−13	Scott & Fetzer	3
TRW	−5	United Technol	8
Xerox	−19	H. F. Ahmanson	23
Avery Intern	−25	Capital Cit. Com.	1
Consolidated Fds	−8	Dravo	26
First Wiscon.	8	Grumman	−5

Based on the above random selection of corporations, can we assert at level $\alpha = .10$ that the median percentage change for 1977 was negative?

12.5. To find out whether or not the on-time performance of air carriers tends to differ from month to month, the following data for August and September of 1977 were collected.[†] At level $\alpha = .10$, do the data indicate that typical on-time performance was better in September than in August?

*Selected from *Business Week*, December 26, 1977, pp. 92–108.

[†]Reported in *Aviation Week & Space Technology*, December 12, 1977, p. 35.

| Airline | Percentage of Flights on Time | |
	August	September
Allegheny	77.66	81.21
Texas Intern.	76.95	79.01
Eastern	78.51	82.38
Hughes Air West	75.97	89.47
National	79.18	80.33
Ozark	70.46	70.39
Continental	70.57	81.26
Braniff	83.20	84.37

12.6. Earnings per share data for a random sample of "Fortune 500" corporations yielded the following information comparing 1975 with 1976.[*]

Corporation	1975 Earnings per Share	1976 Earnings per Share
Caterpillar	4.64	4.45
Republic Steel	4.46	4.07
Campbell Soup	2.61	3.07
Levi Strauss	2.95	4.71
Sunbeam	1.62	2.35
Whittaker	0.10	0.65
Witco Chemical	2.46	4.10
Stokely-Van Camp	3.09	2.54
Natomas	2.35	8.45
Mattel	−1.18	1.52

a. Use the sign test at level $\alpha = .05$ to test whether or not 1976 earnings per share for "Fortune 500" corporations exceeded 1975 earnings, on the average.

b. Use the paired-sample t test at level $\alpha = .05$ to conduct the same test as in part a.

12.7. The following data report the percentage change in construction activity, based on the number of workers employed in construction, in several major cities across the country, from 1976 to 1977.[†] At level of significance $\alpha = .01$, do the data seem to show that construction activity in the cities increased between 1976 and 1977?

---

[*]*Fortune*, May, 1977, pp. 366–385.

[†]*U.S. News & World Report*, October 24, 1977, pp. 73–75.

City	Percentage Change in Construction Activity
Buffalo	− 2.0
Greensboro, N.C.	+ 1.9
Miami	− 1.2
Philadelphia	− 1.5
Rochester	+ 1.6
Chicago	+ 7.5
Dallas-Fort Worth	+ 1.2
Grand Rapids	+ 15.2
Knoxville	+ 20.0
Mobile	+ 11.2
Topeka	+ 5.9
Boise	+ 13.0
Honolulu	− 2.6
Phoenix	+ 21.2
Seattle	+ 28.0

12.8. An insurance industry pamphlet claims that the average yearly cost of a $50,000 5-year renewable term life insurance policy to a 25-year-old male is below $200 per year. A random sample of 10 major insurance companies shows that the costs of their programs are as follows (in dollars):

$$220 \quad 230 \quad 190 \quad 200 \quad 190 \quad 190 \quad 160 \quad 150 \quad 230 \quad 190$$

At level $\alpha = .10$, do the sample data confirm the industry claim that the true median cost falls significantly below $200?

SECTION 12.B   THE MANN-WHITNEY TEST

When a lack of normally distributed data or other conditions required for valid operation of the two-sample t test (conditions listed at the end of Section 6.D) prevents us from using that test, the Mann-Whitney test is sometimes an appropriate nonparametric substitute.

Just as the sign test did not directly use the actual numerical values of the data points, neither does the Mann-Whitney test. However, while the sign test used only facts about which of two numbers is the larger, the Mann-Whitney test uses a more detailed method of comparing the data points— ranking all of the data points in order. Because it manages to squeeze more information out of the data than do the paired comparisons of the sign test, the Mann-Whitney test generally yields more accurate results. By "more accurate" in this context, we mean that, for the same probability α of Type I error, the Mann-Whitney test will have a lower probability of Type II error than will the sign test.

In this chapter, we will study only the Mann-Whitney substitute for the two-sample t test. There exists also a test based on ranks that can be applied to cases involving paired samples, but we will only list references to this test in the bibliography at the end of the chapter. (In Exercise 12.29, you will see an example that illustrates why the Mann-Whitney test we describe here cannot be used for paired samples.)

Mann-Whitney Test for Equal Averages

Following the procedure used in the previous section, we shall reconsider Example 6.4 from Section 6.D, where we discussed the two-sample t test, from the point of view of the Mann-Whitney test.

> **EXAMPLE 12.3. Collection Methods.** The Mainstreet, a regional chain of department stores located primarily in small rural towns, would like to find out which of two suggested methods of collecting accounts receivable is most effective. They first divided a sample of their customers by random assignment into two groups. To one group of six customers, they mailed a series of past due notices at the end of 30 days, 60 days, and 90 days beyond the invoice date. To a second group of nine customers, they made telephone calls at the same time intervals. From the six customers in the mail group, payment was received, respectively, 62, 68, 106, 47, 94, and 43 days after the invoice date. The nine customers in the telephone group paid their bills, respectively, 81, 31, 93, 44, 49, 38, 62, 34, and 36 days past the invoice date.
>
> At level of significance $\alpha = .10$, do the sample results indicate that the telephone method is more effective than the mail method in speeding up collection of accounts receivable?

> **SOLUTION.** To apply the Mann-Whitney test for equal medians, we first introduce the notation
>
> n_M = number of data points involved in *mail* method
> n_T = number of data points involved in *telephone* method.
>
> Because the set of data involved in the mail method is {62, 68, 106, 47, 94, 43}, while the set of data involved in the telephone method is {81, 31, 93, 44, 49, 38, 62, 34, 36}, we have that $n_M = 6$ and $n_T = 9$. It will be important to observe that the mail sample is the smaller sample of the two. We next rank all $n_M + n_T = 6 + 9 = 15$ data points together in order, from smallest to largest, as illustrated in Table 12.7. We test the hypothesis
>
> H: $A_M = A_T$
>
> against the alternative

TABLE 12.7 Ranking of Combined Sets of Data
(Collection methods data)

Rank position	1	2	3	4	5	6	7	8	9	10	11	12	13	14	15
Data points	31	34	36	38	43	44	47	49	62	62	68	81	93	94	106
Mail (M) or telephone (T)	T	T	T	T	M	T	M	T	M	T	M	T	T	M	M

A: $A_M > A_T$

where

A_M = "true median" number of days until payment when *mail* method is used

and

A_T = "true median" number of days until payment when *telephone* method is used.

We next assign each of the 15 data points a *rank*. The *assigned rank* of a point is usually its rank position as shown in Table 12.7, but in the case of *ties* it may be slightly different. Consider, for example, the two 62's, which are tied for the rank positions 9 and 10. What we do is to give each of them an assigned rank of 9.5. (Suppose, in some other problem, we have three data points tied for rank positions 10, 11, and 12. Then we would average the rank positions in question and give each of the three data points an assigned rank of $(10 + 11 + 12)/3 = 11$.) After giving each of the 15 data points its assigned rank, we then remember which of the two samples has the smaller number of data points. (If both samples have the same number of data points, either can be used.) The mail sample is the smaller, so when we compute the number

S = sum of assigned ranks of *smaller* sample,

as worked out in Table 12.8, we obtain that $S = 61.5$.

We then refer to Table A.11 of the Appendix in order to determine the rejection rule at level $\alpha = .10$ for H: $A_M = A_T$ against A: $A_M > A_T$. If it is really true that $A_M > A_T$, we would expect the mail method data to be larger than the telephone method data, on the average. Therefore, the mail method data should occupy the higher ranks. It follows that we should reject H in favor of A if S is appropriately large. In particular, we based our decision on the Mann-Whitney statistic

$$U = n_T n_M + \frac{n_M(n_M + 1)}{2} - S \qquad (12.2)$$

TABLE 12.8   Calculations for the Mann-Whitney Test for Equal
Medians   *(Collection methods data)*

Rank Position	Data Point	Mail (M) or Telephone (T)	Assigned Rank	Of Smaller Sample (M)
1	31	T	1	
2	34	T	2	
3	36	T	3	
4	38	T	4	
5	43	M	5	5
6	44	T	6	
7	47	M	7	7
8	49	T	8	
9	62	M	9.5	9.5
10	62	T	9.5	
11	68	M	11	11
12	81	T	12	
13	93	T	13	
14	94	M	14	14
15	106	M	15	15

Sum $S = 61.5$

TABLE 12.9   A Small Portion of Table A.13
(Values of $U_{.10}[n_L; n_S]$)

n_L	6
9	15

and we use the comparison value $U_\alpha[n_L; n_S] = U_{.10}[9; 6]$ listed in the
table. We reproduce the relevant portion of Table A.11 in Table 12.9.
The rejection rule is as follows: we should reject H: $A_M = A_T$ in favor
of A: $A_M > A_T$ at level $\alpha = .10$ if $U \leqslant U_{.10}[n_M; n_T] = U_{.10}[9; 6] = 15$.
   It turns out that $S = 61.5$ and so

$$U = n_T n_M + \frac{n_M(n_M + 1)}{2} - S = (9)(6) + \frac{(6)(7)}{2} - 61.5$$
$$= 54 + 21 - 61.5 = 13.5.$$

Because $U = 13.5 \leqslant 15$, we should reject H. Therefore, at level $\alpha = .10$, we conclude that the data support the assertion that the telephone method is significantly more effective than the mail method in reducing collection times.

As long as the number of ties is not too large, the procedure used in the above example works satisfactorily. However, if there is an excessive number of ties, an additional *correction factor* enters into the calculations. (In many situations, even as many as half the data points could be involved in ties, and yet the correction factor would not significantly alter the calculations.) The bibliography at the end of the chapter gives references where further details on the correction factor and on the number of ties considered excessive can be found.

EXAMPLE 12.4.   Tea Testing.   It is the job of a tea tester for an English tea-importing company to decide which tea leaves are most suitable for English Breakfast tea. The tea tester has to decide between two lots of leaves, an inexpensive lot and an expensive lot. The tea tester and his laboratory staff come up with the following levels of impurities in eight randomly selected leaves of each lot:

Inexpensive Lot:	3	4	25	9	11	4	12	20
Expensive Lot:	6	5	7	10	6	8	5	4

At significance level $\alpha = .05$, do the data indicate that both lots of leaves have approximately the same median impurity level?

SOLUTION.   We use the Mann-Whitney test to test the hypothesis

H: $A_I = A_E$

vs.

A: $A_I \neq A_E$

where the symbolism should be self-explanatory. Because of the format of the alternative A, we would reject H in favor of A if S turned out to be either too large or too small. After computing U via equation (12.2), we would therefore reject H if either $U \leqslant U_{\alpha/2}[n_I; n_E]$ or $U \geqslant n_I n_E - U_{\alpha/2}[n_I; n_E]$. (Since both sets contain the same number of data points, we can use either sample as the smaller.) The numbers of data points in each sample are $n_I = 8$ and $n_E = 8$. Also $\alpha = .05$. The appropriate portion of Table A.11 is excerpted in Table 12.10. We should therefore reject H: $A_I = A_E$ in favor of A: $A_I \neq A_E$ if either $U \leqslant 13$ or

TABLE 12.10　A Small Portion of Table
A.13　(Values of $U_{.025}[n_L; n_S]$)

	n_S
n_L	8
8	13

TABLE 12.11　Calculations for the Mann-Whitney Test for Equal
Medians　(Tea testing data)

Rank Position	Data Point	Inexpensive (I) or Expensive (E)	Assigned Rank	Of Smaller Sample (I)
1	3	I	1	1
2	4	I	3	3
3	4	I	3	3
4	4	E	3	
5	5	E	5.5	
6	5	E	5.5	
7	6	E	7.5	
8	6	E	7.5	
9	7	E	9	
10	8	E	10	
11	9	I	11	11
12	10	E	12	
13	11	I	13	13
14	12	I	14	14
15	20	I	15	15
16	25	I	16	16
				Sum S = 76

$U \geq 64 - 13 = 51$. (If H is true, the distribution of U is symmetric about its mean $(n_I n_E)/2 = 64/2 = 32$.) It now remains only to compute the numerical value of U.

The calculations in Table 12.11 establish the fact that $S = 76$. Therefore, using equation (12.2), we get that

$$U = (8)(8) + \frac{8(8+1)}{2} - 76 = 64 + 36 - 76 = 24,$$

which is neither ≤ 13 nor ≥ 51. So a numerical value of $U = 24$ does not permit us to reject H. At level $\alpha = .05$, we therefore conclude that the two lots of tea do not differ significantly in impurity levels.

Just as in the case of the sign test, the large-sample version of the Mann-Whitney test is based on the percentage points of the normal distribution. It is a z test using the statistic

$$z = \frac{S - \mu_S}{\sigma_S}$$

where

$$\mu_S = \frac{n_1(n_1 + n_2 + 1)}{2} \tag{12.3}$$

and

$$\sigma_S = \sqrt{\frac{n_1 n_2(n_1 + n_2 + 1)}{12}} \tag{12.4}$$

Here n_1 is the number of data points in the smaller sample, and n_2 is the number of data points in the larger sample. The normal approximation takes effect when n_1 and n_2 are both greater than 10, and the rejection rules are given by Table 6.8.

Mann-Whitney Test for Equal Dispersions

Everything we have learned about making the calculations for the Mann-Whitney test for equal medians holds true for the Mann-Whitney test for equal dispersions as well, except the method of determining the assigned ranks.

To test for equal dispersions, we give the lowest assigned ranks to data points having the extreme, high and low, rank positions. This way a more dispersed sample has a relatively lower sum of ranks than a less dispersed sample, and we are then able to decide whether or not the two samples differ in dispersion.

By the "dispersion" of a set of data, we mean the extent to which they vary from their "average." The standard deviation, the mean absolute deviation, and the range, which we discussed in Section 2.C, are all measures of dispersion, but in the nonparametric framework, we refer to the more general concept of dispersion rather than any specific way to measure it.

Let's take another look, for example, at the tea testing data of Example 12.4. Suppose we wanted to determine whether or not the two lots, inexpensive and expensive, differed in degree of internal variability of impurity level.

In other words, we want to find out if the inexpensive tea leaves have the same degree of uniformity as do the expensive tea leaves. In particular, if we set

D_I = lack of uniformity of impurity level in inexpensive lot, and
D_E = lack of uniformity of impurity level in expensive lot

to be the *true dispersions* of the two samples, we want to test the hypothesis

H: $D_I = D_E$

against the alternative

A: $D_I \neq D_E$.

We still have $n_I = 8$ and $n_E = 8$, and if we use the level of significance $\alpha = .05$, Table 12.10 shows that we again should reject H in favor of A if $U \leq 13$ or $U \geq 51$. The only place where something different occurs is in the calculation of S. We compute S for the test of equal dispersions in Table 12.12.

TABLE 12.12   Calculations for the Mann-Whitney Test for Equal Dispersions   *(Tea testing data)*

Rank Position	Data Point	Inexpensive (I) or Expensive (E)	Adjusted Rank Position	Assigned Rank	Of Smaller Sample (I)
1	3	I	1	1	1
2	4	I	3	5	5
3	4	I	5	5	5
4	4	E	7	5	
5	5	E	9	10	
6	5	E	11	10	
7	6	E	13	14	
8	6	E	15	14	
9	7	E	16	16	
10	8	E	14	14	
11	9	I	12	12	12
12	10	E	10	10	
13	11	I	8	8	8
14	12	I	6	6	6
15	20	I	4	4	4
16	25	I	2	2	2

Sum S = 43

There is an important difference in the treatment of ties between Tables 12.11 and 12.12. In Table 12.11, the data points equal to 4 were tied for the assigned ranks 2, 3, and 4, and were each assigned a rank of

$$\frac{2 + 3 + 4}{3} = 3.$$

In Table 12.12, however, these three data points are tied for the assigned ranks 3, 5, and 7 and are therefore each assigned a rank of

$$\frac{3 + 5 + 7}{3} = 5.$$

From the computations in Table 12.12, we see that $S = 43$. Therefore $U = 64 + 36 - 43 = 100 - 43 = 57$. As we agreed to reject H at level $\alpha = .05$ if either $U \leq 13$ or $U \geq 51$, and in fact $U = 57 \geq 51$, we have to reject H. At level $\alpha = .05$, we conclude on the basis of the given data that the two lots of tea have different dispersions and therefore *do not* have the same internal variability.

Our decision to the effect that the two lots of tea have different dispersions most likely implies that the standard deviations of the two underlying populations are different. As it is one of the requirements for the validity of the two-sample t test that the underlying populations have the same standard deviation, it would be inappropriate for the tea tester to use the t test to test his tea.

One additional point: Technically speaking, the hypothesis H used in the Mann-Whitney test asserts that the two populations being compared have identical distributions. Therefore, in testing for equal averages, we are implicitly assuming that the population dispersions are equal. On the other hand, when testing for equal dispersions, we implicitly assume that the medians are equal. If the medians of the two sets of data are very far apart, the Mann-Whitney test may not be able to discern a difference that really exists between the dispersions. This is due to the fact that the low (nonadjusted) rank positions will all be occupied by the members of one sample, while the high positions will be filled by those of the other, regardless of the degree of internal dispersion within the separate samples. One way to counteract this effect would be to subtract the median of each sample from each data point in that sample. This operation would have no effect at all upon the dispersions, but it would give us two sets of numbers having the same median. It would then be reasonably valid in some situations to run the Mann-Whitney test of equal dispersion. For an example of such a problem, see Exercise 12.30 later in this chapter.

EXERCISES 12.B

12.9. Using the data of Exercise 6.25, decide at level $\alpha = .10$,

    a. whether or not English cheese sells at lower prices in Canada;

    b. whether or not English cheese prices are more variable in the United States.

12.10. Some supermarkets give trading stamps with purchases, while others do not. One consumer organization feels that prices are more variable among super-markets not offering trading stamps, because of their supposed tendency to offer loss-leaders or other price-adjustment gimmicks. To test out its guess, the organization checks out watermelon prices (per pound) at four supermar-kets that give trading stamps and six that do not. The results of the survey follow.

	Watermelon Prices (cents per pound)					
At supermarkets that give stamps	12	18	14	16		
At supermarkets that give no stamps .	10	16	16	12	8	10

At level $\alpha = .05$, do the data support the allegation that prices are more variable at supermarkets not giving trading stamps?

12.11. Personal income showed strong gains throughout the nation in 1977. The following data compare the percentage increases in personal income, for a random sample of states, in the eastern and western regions of the United States for the first three quarters of 1977, as opposed to the analogous period of 1976:*

Eastern States	Percentage Increase	Western States	Percentage Increase
Connecticut	11.2	Arizona	12.1
Massachusetts	10.4	Idaho	13.2
New Jersey	11.2	Nevada	15.3
Pennsylvania	9.3	Alaska	3.9
Florida	10.7	California	12.4
Maryland	9.8	New Mexico	10.9
So. Carolina	12.4	Colorado	10.5
West Virginia	13.4	Washington	11.8
Rhode Island	8.3		
Georgia	12.2		

    a. At level $\alpha = .01$, do the data indicate that state-by-state personal incomes increased to a greater extent in western states than in eastern states?

    b. Can we assert, at level $\alpha = .10$, that the two regions differ significantly in variability of their percentage increases in personal income?

*Selected from *Business Week*, December 19, 1977, p. 95.

12.12. A securities analyst conducted a study of the price/earnings (P/E) ratio of common stocks by various industry groupings. A random sample of several stocks in the building materials and the chemicals industries had the following P/E ratios, based on their stock prices of October 21, 1977, and their earnings for the prior 12-month period.*

Building Materials		Chemicals	
Boise Cascade	7	Union Carbide	7
Weyerhauser	11	Allied Chem.	8
De Soto	7	Stauffer Chem.	6
U.S. Gypsum	9	Dexter	8
General Portland	14	Olin	5
Southwest Forest	4	Emery Indust.	22
Insilco	6	Morton-Norwich	9
Owens-Corning	10	Freeport Min.	12
Kaiser Cement	5	Texas Gulf	15

a. At level $\alpha = .05$, do the results of the study show that there is a significant difference between the two industry groupings in median P/E ratio?

b. Do the data show, at level $\alpha = .05$, that the difference in internal variation between the two groups is significant?

12.13. A study of executive compensation compared the net profits of corporations that paid their chief executives over a half million dollars in 1976 with those of corporations paying less than a half million. Data on randomly selected corporations follow.[†]

Corporations Paying More Than $500,000	Net Profits (Millions)	Corporations Paying Less Than $500,000	Net Profits (Millions)
American Brands	122.0	H. F. Ahmanson	68.3
Borden	112.8	Boeing	102.9
Xerox	358.9	Riggs Nat. Bank	13.2
Studebaker-Worthington	50.2	N L Industries	58.8
Norton Simon	92.4	Merrill Lynch	106.6
Bristol-Meyers	156.8	First Chicago	105.6
LTV	30.7	J. P. Morgan	202.7
Northrop	36.3	Rochester G&E	33.1
Standard Oil Indiana	893.0	Thomas & Betts	16.9
United Technologies	157.4	Hercules	106.8
Gulf & Western	200.2	Fisher Foods	12.8
Fluor	64.9	Louisiana-Pacific	40.4
		Colt Industries	61.6
		Texas Utilities	147.9
		Beatrice Foods	153.1
		Kaiser Steel	43.6

* As reported in *Business Week*, November 14, 1977, pp. 108–109.
[†] *Forbes*, May 15, 1977, pp. 202–243.

Can we assert, at level $\alpha = .05$, that corporations paying their chief executives more than \$500,000 make more profit, on the average, than ones paying less?

12.14. To determine whether or not prices for single-family homes are higher in Southern California than in the Northeastern U.S., a national chain of real estate agencies surveyed a random sample of six recently sold homes in each region and obtained the following information:

Selling Prices (Thousands of Dollars)	
Northeast	Southern California
65	120
50	85
100	60
40	40
50	105
85	70

At level $\alpha = .05$, test whether or not Southern California homes are really more expensive than those in the Northeast.

12.15. Different living costs in various geographic regions often present problems for transferred executives and middle managers. One study determined the family income required in various metropolitan areas around the nation to maintain 319 goods and services considered essential to an average way of life.[*] The results on seven areas in the East and six in the West follow:

East		West	
Area	Income	Area	Income
P	24,420	H	23,530
A	23,560	L	26,530
C	25,890	D	23,330
S	23,020	F	26,900
W	26,740	T	23,000
B	28,070	X	21,650
N	29,000		

At level $\alpha = .05$, do the results of the study indicate that a higher family income is required in the East?

SECTION 12.C   KRUSKAL-WALLIS NONPARAMETRIC ANALYSIS OF VARIANCE

As we emphasized in Chapter 7, the F test on which analysis of variance is based can be used with confidence only when the data points involved come

*Reported in *U.S. News & World Report*, February 20, 1978, p. 69.

from normally distributed populations, each having the same standard deviation. When it is unreasonable to operate under this assumption, the analysis of variance techniques studied in Chapter 7 do not apply. The nonparametric substitute generally used is the Kruskal-Wallis test, also called *nonparametric analysis of variance* or *analysis of variance by ranks*.

Rejection rules for the Kruskal-Wallis test are based on the tables of the chi-square distribution, which we already have had occasion to refer to in several places. There is one basic requirement that must be met in order that the use of the chi-square distribution be valid, namely that each of the samples involved must contain at least 5 data points. In the case of smaller-sized samples, as long as all samples have the same size, we can use a table of rejection values recently worked out by W. V. Gehrlein and E. M. Saniga. Their table appears as Table A.12 of the Appendix.

In this section, we will study a nonparametric replacement for the one-way analysis of variance procedure that we studied in Section 7.B.

EXAMPLE 12.5. Behavioral Management Techniques. E-Fishent Systems, Inc., a rather progressive electronics manufacturing concern, has recently been experimenting with a number of new behavioral approaches to management. To determine which of three leading approaches it should adopt throughout the company, it tests each of them in one of three similar factories. The Swiss approach was used for 6 weeks in the Philadelphia plant, the Australian approach was used for 9 weeks in the Minneapolis plant, and the Japanese approach was used for 11 weeks in the Raleigh plant. The following data record the typical production in units per hour during each week of the experimentation period:

Philadelphia	3	7	9	15	6	2					
Minneapolis	18	14	7	5	11	9	3	11	12		
Raleigh	4	21	11	16	19	23	7	10	17	6	8

At level $\alpha = .05$, can the company conclude that all three approaches lead to the same median production record?

SOLUTION.   To carry out the Kruskal-Wallis test, we use the following notation:

m = number of samples
n_P = number of data points in Philadelphia sample
n_M = number of data points in Minneapolis sample
n_R = number of data points in Raleigh sample
n = $n_P + n_M + n_R$ = total number of data points.

In the case under study, we see that

$m = 3$

$n_P = 6 \qquad n_M = 9 \qquad n_R = 11$

$n = 6 + 9 + 11 = 26.$

The Kruskal-Wallis test is a test of the hypothesis

H: $A_P = A_M = A_R$ (all approaches yield the same median
                      production)

against the alternative

A: the medians are not all the same.

The rejection rule for the Kruskal-Wallis test is as follows: we compute

$$\blacksquare \qquad\qquad H = \frac{12}{n(n+1)} \Sigma \frac{R_i^2}{n_i} - 3(n+1) \qquad\qquad (12.5)$$

and we reject H at level α if

$H > \chi_\alpha^2 [m-1],$

where $m - 1$ is the degrees of freedom for the chi-square statistic. (Referring back to the one-way analysis of variance formula (7.5) in Section
7.B, we can see that the Kruskal-Wallis statistic H is really the treatment
sum of squares based on ranks, rather than on the actual data points.)
In the situation at hand, we are using $\alpha = .05$, and $df = m - 1 = 3$
$- 1 = 2$. It follows that we should reject H in favor of A at level $\alpha =$
.05 if

$H > \chi_{.05}^2 [2] = 5.991.$

Note that we cannot use Table A.12 of the Appendix because the sample
sizes n_P, n_M, and n_R are not all the same. But we don't have to, since they
are all greater than 5.
   It now remains only to compute H. We rank all the $n = 26$ data points
in order (as if we were running a Mann-Whitney test). Then we calculate

R_P = sum of assigned ranks of Philadelphia sample

R_M = sum of assigned ranks of Minneapolis sample

R_R = sum of assigned ranks of Raleigh sample

The Kruskal-Wallis statistic is then

$$H = \frac{12}{n(n+1)} \left[\frac{R_P^2}{n_P} + \frac{R_M^2}{n_M} + \frac{R_R^2}{n_R} \right] - 3(n+1) .$$

TABLE 12.13 Preliminary Calculations for Kruskal-Wallis
Nonparametric Analysis of Variance   *(Behavioral management data)*

Rank Position	Data Point	Sample (P, M, or R)	Assigned Rank	Philadelphia Sample (P)	Minneapolis Sample (M)	Raleigh Sample (R)
1	2	P	1	1		
2	3	P	2.5	2.5		
3	3	M	2.5		2.5	
4	4	R	4			4
5	5	M	5		5	
6	6	P	6.5	6.5		
7	6	R	6.5			6.5
8	7	P	9	9		
9	7	M	9		9	
10	7	R	9			9
11	8	R	11			11
12	9	P	12.5	12.5		
13	9	M	12.5		12.5	
14	10	R	14			14
15	11	M	16		16	
16	11	M	16		16	
17	11	R	16			16
18	12	M	18		18	
19	14	M	19		19	
20	15	P	20	20		
21	16	R	21			21
22	17	R	22			22
23	18	M	23		23	
24	19	R	24			24
25	21	R	25			25
26	23	R	26			26
		Rank Sums		$R_p = 51.5$	$R_M = 121$	$R_R = 178.5$
		Number of Data Points		$n_p = 6$	$n_M = 9$	$n_R = 11$

The preliminary calculations are carried out in Table 12.13.

Upon inserting the results at the bottom of Table 12.13 into the formula for H, we obtain that

$$H = \frac{12}{26(26 + 1)}\left[\frac{(51.5)^2}{6} + \frac{(121)^2}{9} + \frac{(178.5)^2}{11}\right] - 3(26 + 1)$$

$$= \frac{12}{(26)(27)}\left[\frac{2,652.25}{6} + \frac{14,641}{9} + \frac{31,862.25}{11}\right] - 3(27)$$

$$= \frac{12}{702}[442.04 + 1,626.78 + 2,896.57] - 81 = \frac{12}{702}[4,965.39] - 81$$

$$= 84.88 - 81 = 3.88$$

We have agreed to reject the hypothesis if H > 5.991, but H turned out to be 3.88 < 5.991. It follows that we cannot reject the hypothesis that all approaches yield the same median production records. At level α = .05, we therefore assert that, on the basis of the collected data, all three approaches seem to be equally effective, on the average.

EXERCISES 12.C

12.16. Based on the aptitude score data of Exercise 7.2, can the brokerage firm conclude, at level α = .01, that there is a significant difference in aptitude among persons in the six occupational classifications?

12.17. Using the data presented in Exercise 7.5, can we say, at level α = .01, that unemployment rates were virtually the same on the average in cities in each of the three regions?

12.18. Can the stockbroker of Exercise 7.6 assert, at level α = .05, that there are significant differences among the various industrial groupings in median earnings per share?

12.19. Based on the information listed in Exercise 7.14, should the banking officials come to the conclusion, at level α = .01, that various industry groups differ significantly in their net incomes as a percentage of stockholders' equity?

12.20. A study of airline traffic in August 1977 revealed the following information on the percentage change in revenue passenger miles (RPM) flown by various airlines in August 1977 as opposed to August 1976, for various categories of routes.*

Domestic Trunks	International	Local Service	Alaska & Hawaii
5.8	8.9	12.4	5.7
9.2	8.7	7.8	40.1
7.1	14.8	19.4	5.6
6.3	−5.3	9.3	7.4

At level α = .05, do the data indicate that significant differences exist among the various routes?

SECTION 12.D THE SPEARMAN TEST OF RANK CORRELATION

We used the (Pearson) coefficient of linear correlation in Section 8.D to test for the existence of a linear relationship between two sets of corresponding data points. We pointed out at the time the requirement that both sets of data

*Based on an article in *Aviation Week & Space Technology*, October 24, 1977, p. 34.

be normally distributed in order for the t test of a linear dependence relationship to be valid. In cases where it is unreasonable to believe that we are working with normally distributed data, the *Spearman test of rank correlation* allows us to rank each set of data in numerical order and then to compute the correlation coefficient of the ranks.

As we have noted in earlier sections of this chapter, we lose some control over the problem when we convert the original data points into shadow properties such as ranks. The unpleasant consequences of the conversion can best be seen in the fact that we cannot use the Mann-Whitney test to check on the difference of means but rather the difference of medians. A similar situation occurs in connection with the Spearman test of rank correlation. It turns out that we cannot use the Spearman test to check on the existence of a *linear* relationship between two sets of data, but only to check on the existence of a *monotone* dependence relationship.

By a monotone relationship between x and y, we mean a relationship in which y increases as x increases, or y decreases as x increases. Examples of monotone increasing relationships can be seen graphically in Figures 8.3, 8.5, 8.8, part b, and 8.11. Monotone decreasing relationships are illustrated in Figures 8.1, 8.7, and 8.8, part c. All linear relationships are monotone, but, as Figure 8.8, part b and part c, shows, not all monotone relationships are linear. On the other hand, the parabolic data of Figure 8.8, part a, express a relationship that is not monotone, because y starts to decrease as x increases, but after a while turns upward and begins to increase.

Just as the Pearson correlation coefficient measures the degree of linearity, the Spearman coefficient represents the degree of "monotonicity." Technically speaking, however, statistical tests based on either of these coefficients are tests for independence between the variables involved.

Although it is not sharp enough to be able to judge the degree of linearity, the Spearman test has two distinct advantages: (1) its usage is valid in the presence of nonnormal data, and (2) it works excellently in situations where at least one of the two data sets consists entirely of rankings rather than measurements. One widely-known example of data that consists entirely of rankings is the weekly (during the football season) rankings of the top 10 college teams. An interesting rank correlation problem here might be to test the correlation between the national rankings of the top 10 teams (ranked data) and the total weight of their 7 first-string linemen (measurement data). It makes no sense to apply the normal-based Pearson test to this question, but the Spearman test fits quite well.

EXAMPLE 12.6. Theft by Employees.   Some union executives have testified that employee theft is inversely correlated with the wage rates prevailing in the employee's industry. To check on this assertion, an undergraduate business student obtained both wage and theft records for various companies. She ranked the companies by wage rates, lowest to highest, and compared the ranking with the estimated dollar loss due to employee theft in 1980. The data appear in Table 12.14. At level α = .01, do the results of her study support the union executives' claim that, as wage rates increase, employee theft decreases?

TABLE 12.14   Data on Wage Rates vs. Theft by Employees

Wage Rate (rank)	Employee Theft in 1980 (thousands of dollars)
1	105.7
2	104.2
3	103.4
4	100.6
5	102.4
6	101.2
7	100.1
8	99.7
9	98.9
10	98.8
11	99.5
12	98.7
13	98.6

SOLUTION.   It is appropriate to test whether or not wage rates and employee theft are related in a monotone decreasing manner, and so the coefficient of rank correlation will be useful to us. For illustrative purposes, we present the scattergram of employee theft in Fig. 12.1.

We are testing the hypothesis

H: There is no dependence relationship between wage rates and employee theft,

against the alternative

A: relationship is a monotone decreasing one.

To test H against A, we first determine

n = number of pairs of data points.

If n > 30, then a modified version of the central limit theorem allows us to use a z test of H against A, where

$$z = r_s \sqrt{n-1} \qquad (12.6)$$

Here r_s = Spearman coefficient of rank correlation.

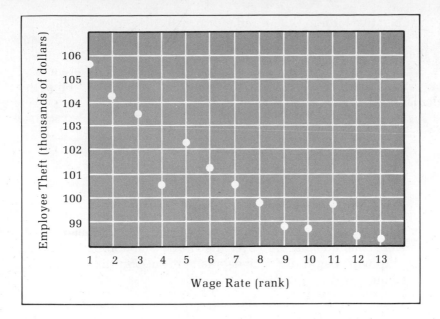

FIGURE 12.1   Scattergram of Employee Theft Data

TABLE 12.15   Rejection Rules in the Presence of Various Alternatives for the Spearman Test of Rank Correlation   (*H: the relationship is not monotone*)

Alternative	Large Samples	Small Samples
	Reject H in favor of A at Significance Level α if:	
A; monotone increasing	$z > z_\alpha$	$r_S > r_S[n; \alpha]$
A: monotone decreasing	$z < -z_\alpha$	$r_S < -r_S[n; \alpha]$
A: monotone	$|z| > z_{\alpha/2}$	$|r_S| > r_S[n; \alpha/2]$

Note: The use of z is valid in the large-sample case when there are more than 30 pairs of data points. For the small-sample case, the percentage points in Table A.13 of the Appendix should be used.

The rejection rule is selected appropriately from Table 12.15. This problem is, however, a small-sample problem, and we reject H in favor of A at level $\alpha = .01$ if $r_S < -r_S[n;\alpha] = -r_S[13;.01] = -.6429$, using the percentage points of Table A.13 of the Appendix. We carry out the calculations of r_S in Table 12.16, using the formula

$$r_S = 1 - \frac{6\Sigma d^2}{n(n^2 - 1)}. \qquad (12.7)$$

The operative term in the formula for r_S is the quantity Σd^2, the sum of the squared differences of the *ranks* of the data points in each pair. As shown in Table 12.16, we rank the data points in each column separately, lowest to highest, we compute the pairwise differences d between the *ranks* of x and y, and we then come up with Σd^2. From the calculations in Table 12.16, we get that

$$r_S = 1 - \frac{6\,\Sigma d^2}{n(n^2 - 1)} = 1 - \frac{6(716)}{13(13^2 - 1)} = 1 - \frac{4{,}296}{(13)(169 - 1)}$$

$$= 1 - \frac{4{,}296}{2{,}184} = 1 - 1.967 = -.967.$$

We have agreed to reject H in favor of A if $r_S < -.6429$. Since it turned out that $r_S = -.967 < -.6429$, we do in fact reject H. At level $\alpha = .01$, we therefore conclude that employee theft bears a monotone decreasing relationship with wage rates. The data tend to support the union executives' claim.

TABLE 12.16   Calculations Leading to the Spearman Coefficient of Rank Correlation   *(Employee theft data)*

Wage Rate (rank) x	Rank r_x	Employee Theft (thousands of dollars) y	Rank r_y	$d = r_x - r_y$	d^2
1	1	105.7	13	-12	144
2	2	104.2	12	-10	100
3	3	103.4	11	-8	64
4	4	100.6	8	-4	16
5	5	102.4	10	-5	25
6	6	101.2	9	-3	9
7	7	100.1	7	0	0
8	8	99.7	6	2	4
9	9	98.9	4	5	25
10	10	98.8	3	7	49
11	11	99.5	5	6	36
12	12	98.7	2	10	100
13	13	98.6	1	12	144
Sums				0	716

$n = 13$

$$r_S = 1 - \frac{6\,\Sigma d^2}{n(n^2 - 1)} = 1 - \frac{6(716)}{13(13^2 - 1)} = 1 - \frac{4{,}296}{(13)(169 - 1)} = 1 - \frac{4{,}296}{(13)(168)}$$

$$= 1 - 1.967 = -.967$$

TABLE 12.17   Interest Rate Data

Date	Discount Rate	Mortgage Rate
A	6.75	8.45
B	4.85	9.50
C	6.50	5.75
D	5.25	7.00
E	5.75	5.25
F	5.05	8.80

EXAMPLE 12.7.   Interest Rates.   Although short-term bank interest rates follow closely behind the Federal Reserve's discount rate, it is known that long-term mortgage rates are less closely related. Over a number of years, Farmers Federal Savings and Loan compared its mortgage rate with the Federal Reserve's discount rate. The data appear in Table 12.17. At level $\alpha = .05$, do the mortgage rate and the discount rate seem to be related in a monotone way?

SOLUTION.   We apply the rank correlation test to test the hypothesis

H: There is no dependence relationship

against

A: the relationship is monotone.

According to the information on rejection rules listed in Table 12.15, we compute r_S and we reject H at level α if $|r_S| > r_S[n; \alpha/2]$.
   Here $n = 6$ and Table A.13 contains the information that $r_S[6; .025] = .8286$. At level $\alpha = .05$, then, we would reject H in favor of A if

$$|r_S| > r_S[6; .025] = .8286.$$

It remains now only to compute r_S for the data of Table 12.17.
   We do the computations in Table 12.18, and they show that

$$r_S = 1 - \frac{6\Sigma d^2}{n(n^2 - 1)} = 1 - \frac{6(56)}{6(6^2 - 1)} = 1 - \frac{336}{(6)(35)}$$

$$= 1 - \frac{336}{210} = 1 - 1.6 = -.6.$$

Therefore $|r_S| = .6$, which does not exceed our rejection level of .8286. Consequently, we cannot reject H at level $\alpha = .05$. Our conclusion is that there is no dependence relationship between the discount rate and

TABLE 12.18   Calculations Leading to the Spearman Coefficient of Rank Correlation   (Interest rate data)

Date	Discount Rate x	Rank r_x	Mortgage Rate y	Rank r_y	$d = r_x - r_y$	d^2
A	6.75	6	8.45	4	2	4
B	4.85	1	9.50	6	−5	25
C	6.50	5	5.75	2	3	9
D	5.25	3	7.00	3	0	0
E	5.75	4	5.25	1	3	9
F	5.05	2	8.80	5	−3	9
Sums					0	56

$n = 6$

$$r_s = 1 - \frac{6\Sigma d^2}{n(n^2 - 1)} = 1 - \frac{6(56)}{(6)(6^2 - 1)} = 1 - \frac{336}{(6)(36 - 1)} = 1 - \frac{336}{(6)(35)}$$

$$= 1 - 1.6 = -.6$$

the mortgage rate. There seems to be no trend in either direction, monotone increasing or monotone decreasing.

One final remark: If there are two or more data points that have the same numerical value and therefore have tied ranks, we divide up the rank positions equally in the same manner as we did for the Mann-Whitney test. For example, if the third, fourth, and fifth ranking data points were all equal, they would each be assigned a rank of $(3+4+5)/3 = 12/3 = 4$. In the presence of tied data, formula (12.7) for r_s is only approximately correct, but unless there is an unusually large number of ties, no adjustment of the formula is necessary.

EXERCISES   12.D

12.21. Referring to the data of Exercise 8.27:

a. At level $\alpha = .05$, does one's job performance rating seem to be related in a monotone way with one's math aptitude score?

b. Draw a scattergram of the data.

12.22. Based on the data of Exercise 8.12:

a. Does the total cost of a production run seem to increase monotonically as the quantity produced increases? Use a level of significance of $\alpha = .01$.

b. Construct a scattergram of the data.

12.23. A psychologist retained by a large national corporation asked the following
two questions of 12 randomly selected employees: (1) What is your hourly
pay? (2) How would you rate your job satisfaction on a scale of 0 (low) to 10
(high)? The results of the job satisfaction survey are presented below:

Respondent	A	B	C	D	E	F	G	H	I	J	K	L
Hourly pay, dollars	8	2	6	4	4	20	10	6	6	4	11	15
Job satisfaction	6	4	5	4	3	8	7	4	1	2	9	5

a. At level α = .05, can we conclude that there is a monotone relationship
between job satisfaction and hourly pay?

b. Draw a scattergram of the data.

12.24. As part of a study of the question of whether or not you should judge a book
by its cover, an advertising executive hired two groups of consultants to judge
each of 10 books according to artistry of the cover and internal literary content.
Each book was given a ranking from 1 to 10, first for its cover and again for its
content. The results follow:

Book No.	1	2	3	4	5	6	7	8	9	10
Cover rank	6	4	10	9	7	8	3	2	5	1
Content rank	5	7	1	2	4	3	8	9	6	10

Do the results of the study indicate at level α = .01 that you should judge a
book by its cover?

12.25. As part of a Civil Aeronautics Board study of the on-time performance of
scheduled air carriers, the following data were collected for various airlines
serving the top 200 city-pair markets in September 1977.*

Airline	Total Number of Flights	Percentage of Flights on Time
Continental	3,517	81.26
National	4,107	80.33
Piedmont	302	83.77
TWA	12,748	83.54
Alaska	328	89.02
Frontier	1,899	86.36
Ozark	1,192	70.39
Texas Intern.	1,839	79.01

At level α = .10, do the results of the study reveal that airlines having fewer
flights produce a better on-time record?

12.26. A study of the value of shares of stock in real estate investment trusts revealed

*Selected from a report in *Aviation Week & Space Technology*, December 12, 1977, p. 35.

the following information comparing their book values with their market values.*

REIT	Book Value per Share	Recent Price per Share
Pitts.–W. Va RR	22.46	6.625
United Realty	17.60	9.125
NW Mutual Life	19.00	11.000
Virginia REIT	9.98	6.500
Fraser Mortg.	16.39	10.750
Continental Ill.	20.58	14.750
Wells Fargo	17.36	13.875
Florida Gulf	15.17	13.250
MONY Mtge.	9.88	10.250
Realty ReFund	18.71	20.750
General Growth	6.34	21.625

At level α = .10, do the data indicate that the market price of a share of REIT stock is related in a monotone way with its book value?

12.27. An analysis of the cigarette business in 1977 yielded the following information relating a brand's market share with its percentage change in sales from 1976.[†]

Brand	Market Share in 1977 (%)	Percentage Change in Sales Volume from 1976
Marlboro	16.37	5.1
Kent	5.09	13.8
Tareyton	2.19	−14.8
True	1.80	28.6
Belair	1.46	−1.1
Salem	8.97	2.9
Vantage	2.86	19.4
Raleigh	2.05	4.2
Virginia Slims	1.61	7.8

Can we conclude from the data, at level α = .05, that the relationship between market share and percentage change in sales volume is a monotone decreasing one?

12.28. The John S. Herold, Inc., petroleum consulting firm conducted a detailed study of the amount of oil and gas reserves held by small oil companies. A random sample of their data follows:[‡]

*Forbes, December 1, 1977, p. 85.

[†]Reported in Business Week, October 31, 1977, p. 82.

[‡]Reported in Forbes, October 15, 1977, p. 78.

Company	Oil Reserves (mil. bbls.)	Gas Reserves (bil. cu. ft.)
Hamilton Bros. Petroleum	33	312
Sabine	29	205
Mitchell Energy & Development	15	423
Amarex	2	100
Universal Resources	4	53
Supron Energy	8	390
Adobe Oil & Gas	13	83
Mesa Petroleum	40	1,700
Crystal Oil	10	62
Crown Central Petroleum	11	172
Noble Affiliates	24	160
Belco Petroleum	108	675

At level $\alpha = .10$, do the data indicate that the relationship between oil and gas reserves held by small oil companies is a monotone one?

SUMMARY AND DISCUSSION

In this chapter, we have tied together a number of loose ends that were left hanging in earlier chapters. Many of the small-sample tests of statistical hypotheses that were studied in Chapters 6, 7, and 8 were based on tables of the t and F distributions, and this meant that they could be validly applied only to sets of data satisfying stringent conditions.

In the present chapter, we have freed ourselves of some of these restrictions by introducing statistical tests based on shadow properties, such as signs and ranks of a set of data, rather than on the numerical values of the data points themselves. While, on the one hand, these tests can be applied to almost all data regardless of the distribution, on the other hand, we get less precise results because we are forced to work with these shadow properties rather than with the actual data points.

Working with a nonparametric test is somewhat analogous to studying the shadow of an object for the purpose of drawing conclusions about the object itself. We can see the general picture, but we lose sight of the sharp distinctions. This lack of clarity has a tendency to increase the probability of a Type II error. For this reason, nonparametric tests should not be used when we are working with normally distributed data. However, when the data do not satisfy the required conditions of the parametric tests, it is in our interest to use nonparametric tests because the parametric tests generally produce a larger probability of Type II error.

The moral of the story is this: When a parametric test is validly applicable, use it; when it is not, make the best of whatever nonparametric test is available.

CHAPTER 12 BIBLIOGRAPHY

DETAILED TREATMENTS INCLUDING CORRECTIONS FOR TIES

EMERSON, J. D. and G. A. SIMON, "Another look at the sign test when ties are present: the problem of confidence intervals," *The American Statistician*, August 1974, pp. 140–142.

MOSTELLER, F. and R. E. K. ROURKE, *Sturdy Statistics*. 1973, Reading, Mass.: Addison-Wesley.

NOETHER, G. E., *Introduction to Statistics: A Nonparametric Approach* (2nd ed.). 1976, Boston: Houghton-Mifflin.

SPECIAL TOPICS

HORRELL, J. F., and V. P. LESSIG, "A note on the multivariate generalization of the Kruskal-Wallis test," *Decision Sciences*, January 1975, pp. 135–141.

NOETHER, G. E., "Distribution-free confidence intervals," *The American Statistician*, February 1972, pp. 39–41.

SCHULMAN, R. S., "A geometric model of rank correlation," *The American Statistician*, May 1979, pp. 77–80.

ZAR, J. H., "Significance testing of the Spearman rank correlation coefficient," *Journal of the American Statistical Association*, September 1972, pp. 578–580.

APPLICATIONS

BAKER, C. R., "An investigation of differences in values: accounting majors vs. non-accounting majors," *The Accounting Review*, October 1976, pp. 886–893.

BASI, B. A., K. J. CAREY, and R. D. TWARK, "A comparison of the accuracy of corporate and security analysts' forecasts of earnings," *The Accounting Review*, April 1976, pp. 244–254.

CHAPTER 12 SUPPLEMENTARY EXERCISES

12.29. A corporate executive who feels the need to lose some weight is considering going on one of the two famous reducing diets, the Drink-More diet and the Eat-Less diet, but only if statistical analysis shows that one or both of these diets are effective in inducing weight loss. As it turns out, five business associates have tried the Drink-More diet, and another five have tried the Eat-Less diet. The weights of the 10 acquaintances, both before beginning and after completing their diet programs, are recorded below:

	Drink-More Diet			Eat-Less Diet	
Associate	Weight Before	Weight After	Associate	Weight Before	Weight After
A	150	150	F	150	141
B	160	160	G	160	150
C	170	170	H	170	160
D	180	179	I	180	170
E	190	141	J	190	179

a. Apply the Mann-Whitney test for equal medians to find at level $\alpha = .05$ whether or not the Drink-More diet is effective.

b. Apply the Mann-Whitney test for equal medians to find at level $\alpha = .05$ whether or not the Eat-Less diet is effective.

c. Compare the results of parts a and b above, and explain why the Mann-Whitney test does not have the ability to distinguish between the two diets.

d. Apply the paired-sample sign test to determine whether or not the hypothesis that the Drink-More diet is not effective can be rejected at level $\alpha = .05$. (At what exact level of significance does the paired-sample sign test reject the hypothesis that the Drink-More diet is not effective?)

e. Apply the paired-sample sign test to determine whether or not the hypothesis that the Eat-Less diet is not effective can be rejected at level $\alpha = .05$.

f. At what exact level of significance does the paired-sample sign test reject the hypothesis that the Eat-Less diet is not effective?

g. Explain how the answers to parts d and f show that the paired-sample sign test has the ability to distinguish between the two diets.

12.30. As part of a comparison of employment patterns in the central city and the suburbs, a researcher accumulates the following data on the monthly incomes of seven randomly selected residents of the central city and five randomly selected residents of the suburbs:

Central city incomes	550	120	401	528	659	896	764
Suburban incomes	1,804	811	960	3,124	1,204		

a. Use the F test of Section 6.F to test at level $\alpha = .05$ whether or not suburban incomes are more variable than incomes in the central city.

b. Use the Mann-Whitney test of equal dispersion at level $\alpha = .05$ to answer the same question as in part a on the basis of the above data.

c. Compare the calculations involved in parts a and b above, and explain why the Mann-Whitney test does not have the ability to distinguish between the two dispersions.

d. Upon observing that the medians of the two sets of data are 550 and 1204, respectively, we can subtract from each data point the median of its own set. We then obtain the following "adjusted data points":

Central city incomes (minus median)	0	-121	-149	-22	109	346	214
Suburban incomes (minus median)	600	-393	-244	1920	0		

Apply the Mann-Whitney test of equal dispersion to the above adjusted data points in order to decide, at level $\alpha = .05$, whether or not suburban incomes are more variable than incomes in the central city.

e. Compare the different results obtained in parts b and d, and explain why the method of part d provides a truer measure of the difference between the two dispersions.

12.31. Several insurance adjusters, concerned about the unusually high repair estimates they seemed to be getting from Fosbert's U-bet Repair Station, brought each of eight damaged cars to Fosbert's and also to the auto repair shop of Nickle's Department Store, a business generally regarded as reliable. They obtained the following estimates, in hundreds of dollars.

Car number	1	2	3	4	5	6	7	8
Fosbert's estimate	2.1	4.5	6.3	3.0	1.2	5.4	7.3	9.3
Nickle's estimate	2.0	3.8	5.9	2.8	1.3	5.0	6.5	8.6

At level $\alpha = .01$, can the adjusters conclude from their survey that Fosbert's estimates are significantly higher than Nickle's?

12.32. The following set of data records the percentage change that 1976 profits represented over those of 1975 for a random sample of American corporations.*

Corporation	Percentage Change	Corporation	Percentage Change
Gulf Oil	16.6	Connecticut Gen	34.8
Carolina Pwr	16.5	Lincoln National	64.9
Amer. Broadcast	319.7	Coastal St. Gas	7.5
Fruehauf	92.3	Becton, Dickin.	20.4
Oscar Mayer	23.8	Interlake	10.3
Crown Cork & Seal	11.0	Columbus & So. Oh.	28.0
Duquesne Light	−7.1	Pacific Pwr	20.0
Texas Eastern	−6.1	AMAX	11.7
Sun Company	61.9	Texaco	4.7

At level $\alpha = .05$, do the data indicate that 1976 profits represented an increase of more than 15%, on the average, over 1975 profits?

12.33. A study of North Atlantic passenger traffic compared the passenger load factors of some randomly selected airlines on the North Atlantic run for June 1976 and June 1977.[†] The data follow.

Airline	Passenger Load Factor	
	June 1976	June 1977
TWA	66.9	63.2
British Airways	66.1	66.1
Air Canada	62.0	68.2
Alitalia	58.5	56.2
Air India	70.2	69.6
Irish Intern.	73.5	69.4
Sabena	66.3	68.2

---

*Forbes, May 15, 1977, pp. 189–201.

[†]Reported in Aviation Week & Space Technology, November 14, 1977, p. 34.

At level α = .10, can we say on the basis of the data that the median passenger load factor on the North Atlantic route changed significantly between June 1976 and June 1977?

12.34. It is the job of a tea tester for an English tea-importing company to decide which tea leaves are most suitable for English Breakfast tea. The tea tester has to decide between three lots of leaves, an inexpensive lot, a moderately-priced lot, and an expensive lot. The tea tester and his laboratory staff discover the following levels of impurities in eight randomly selected leaves of each lot:

Inexpensive lot	3	4	25	9	11	4	12	20
Moderately-priced lot	4	9	16	9	12	6	4	8
Expensive lot	6	5	7	10	6	8	5	4

Do the data indicate, at level α = .01, that the three lots of leaves manifest significant differences in impurity content?

12.35. Applicants for the position of an armed bank security guard are routinely given a psychological test of their ability to remain "cool" under extremely trying conditions. The following set of data classifies some recent test scores according to the age group of the applicant:

21–25	26–30	31–35	36–40	41–45
80	70	41	68	72
69	64	73	84	90
92	80	65	29	44
83	76	88	63	88
94	91	67	87	66
75	59	78	56	34
85		77	86	98
82		95	48	87
77			96	

At level α = .05, do the data provide sufficient evidence to assert that there are significant differences between the various age groups in ability to remain cool?

12.36. An investment counseling firm analyzed data on the Fortune 500 corporations in an attempt to find out whether there is a significant difference between the top 250 corporations and the second 250 corporations in total percentage return to investors. The following data list the yearly average percentage return over the period 1966–1976 for several randomly selected corporations:*

a. Do the data indicate, at level α = .05, that there is a significant difference between the median percentage return to investors of the two groups of corporations?

b. At level α = .05, can we say that the two groups of corporations vary internally to significantly different extents in percentage return to investors?

*Fortune, May, 1977, pp. 366–385.

1976 Sales Rank	Top 250 Corporations	Return	1976 Sales Rank	Next 250 Corporations	Return
8	IBM	8.67	256	Brown Group	3.36
53	Minn. Min. Manu.	5.62	342	Baker Intern.	28.39
112	B. F. Goodrich	0.67	353	Square D	8.32
48	Amerada Hess	6.13	447	Nalco Chem.	10.24
166	Scott Paper	1.16	464	Roper	8.00
95	Uniroyal	−1.90	486	Mattel	8.20
221	Pennzoil	6.66	418	Kellwood	7.04
143	Kimberly-Clark	10.40	383	Texasgulf	0.94
242	Chromalloy	6.10	317	Newmont Mining	9.15
184	Walter Kidde	−2.31	286	Whittaker	−0.22
133	Texas Instruments	8.34	332	Sybron	1.39
77	Pepsico	10.43	374	Revere Copper	−6.18
118	Pfizer	4.41	403	Gardner-Denver	10.19

12.37. A financial analyst conducted a study of the debt position of major corporations in four industry groups (chemicals, electronics, general machinery, and miscellaneous manufacturing) in order to find out whether or not significant differences exist among the financing strategies of the four groups. For a random sample of corporations in each group, the following data list the 1977 long-term debt of the corporation as a percentage of that corporation's total capital investment.*

Long-term Debt as Percent of Total Investment

Chemicals	Electronics	General Machinery	Miscellaneous Manufacturing
35.2	12.2	44.5	44.8
26.0	30.7	25.1	51.2
33.5	20.6	35.0	30.5
30.4	31.9	13.4	21.9
0.0	46.4	26.6	41.0
35.5	1.6	21.6	19.4
30.1	28.1	26.4	9.0
23.8	16.8	42.0	0.6
37.4		21.5	42.8
			28.6
			39.9
			21.8
			30.0

*Selected from information given in *Business Week*, October 17, 1977, pp. 74–86.

Should the data convince the financial analyst that percentage-wise long-term debt differs significantly among the four industry groups? Use $\alpha = .05$.

12.38. *Business Week's* "investment outlook survey" for 1978 compared two estimates for corporate earnings per share for 1978, one based on the corporation's 5-year trend and the other based on the consensus of investment analysts and economists. Data on randomly selected corporations follow.*

Corporation	5-Year Trend Estimate	Consensus Estimate
American Express	3.75	4.30
Chrysler	5.27	3.50
El Paso Electric	1.26	1.35
Honeywell	5.82	7.00
Kellogg	2.29	1.98
Norris Indust.	4.68	4.33
Rochester Teleph.	1.93	2.20
United Brands	1.07	1.25
Wickes	1.60	3.20
Beatrice Foods	2.66	2.68
Cummins Engine	9.20	9.00
Marriott	1.06	1.15
Pepsico	2.46	2.50
J. P. Stevens	3.87	3.38
Avnet	3.16	2.80
Congoleum	1.75	3.53
First Pennsylvania	1.43	2.40
Lockheed	5.71	3.00
Omark Industries	2.61	2.75
Seven-Up	3.03	2.78

At level $\alpha = .05$, do the data indicate that the relationship between the two estimates is a monotone increasing one?

12.39. Official city reports for August 1977, gave the following[†] information on extent of unemployment and percentage change in construction activity for several randomly selected cities across the country. At level $\alpha = .10$, can we assert that, on the basis of the data, there is a monotone relationship between the unemployment rate and the change in construction activity?

* From *Business Week*, December 26, 1977, pp. 92–108.
[†] As reported in *U.S. News & World Report*, October 24, 1977, pp. 73–75.

City	Unemployment Rate (%)	Change in Construction Jobs Compared with Previous Year (%)
Bridgeport	7.7	−2.3
Pittsburgh	6.8	2.1
Savannah	8.3	8.0
Cleveland	5.5	−3.7
Fargo	2.9	4.5
Knoxville	4.2	20.0
Oklahoma City	4.0	38.1
Topeka	4.0	5.9
Casper	2.9	0.0
Phoenix	6.5	21.2
Sacramento	7.7	15.1
New Orleans	8.3	−0.7

12.40. The following data compare a corporation's net profits for 1976 with the total remuneration that year of the chief executive:*

Corporation	Net Profits (millions)	Remuneration (thousands)
Amer. Petrofina	46	170
Bethlehem Steel	168	388
Champion Spark Plug	45	258
Continental Oil	460	525
Duke Power	174	145
First Nat. Boston	43	185
General Mills	101	358
Homestake Mining	22	131
Kimberly-Clark	121	402
Oscar Mayer	33	380
National Tea	−40	118
Pacific Lumber	28	149
Raytheon	85	353
Shell Oil	706	356
Tandy	64	688
United Brands	14	250
Willamette Indust.	42	158

At level $\alpha = .05$, do the data tend to show that the higher-profit companies have higher-paid chief executives?

*Forbes, May 15, 1977, pp. 202–243.

13

Special
Nonparametric
Tests

We continue our study of nonparametric statistics by focusing on three special hypothesis testing problems that involve concepts not at all related to averages and dispersions of sets of data.

The first is a statistical test of whether or not two events are independent, the second gives us a method of deciding whether a population has a binomial, normal, or other probability distribution, and the third is a test of whether or not a set of consecutive data points is behaving randomly or according to some recognizable pattern.

As distinguished from the nonparametric tests mentioned in Chapter 12, these are distribution-free tests that are not involved in any way with parameters or substitutes for parameters, such as averages, dispersions, or correlations. We first encounter the *chi-square test of independence*, which has as its objective a decision as to whether two different events (for example, an individual's investment decision and the format of a corporation's annual report) are independent of each other. Here, independence is not a numerical characteristic, but rather a qualitative one. If the chi-square test indicates that there is a dependence relationship, the extent and direction of the dependence relationship may be measured by the *Goodman-Kruskal index of predictive association*.

In addition to its role in deciding whether two classifications are independent or dependent, the chi-square test can also be used to study the probability distribution of the population from which a random sample has been selected. As we have seen in earlier chapters, it is often important to know whether or not we are dealing with a normally distributed population; if the population is normal, the more accurate t tests for differences and linearity can be used, but if the population is not normal, we often have no choice but to apply a nonparametric test. We therefore study in Section 13.C, the *chi-square test of goodness-of-fit*, a test of whether or not an underlying population can reasonably be considered to have the binomial or normal distribution.

Finally, we discuss a third example of such a distribution-free test, the *Wald-Wolfowitz test of randomness*, the objective of which is to judge whether a sequence of data indicates random behavior in a population (for example, closing prices of a stock on successive days) or whether it instead reveals the presence of trends or rapid fluctuations.

SECTION 13.A   THE CHI-SQUARE TEST OF INDEPENDENCE

One of the major objectives of a researcher or manager's statistical analysis of a set of data is to discover and verify relationships existing among the many components of business situations. The search for and measurement of relationships was our primary goal in our earlier study of z scores, two-sample t and z tests, regression and correlation, and conditional probability. The present section continues this development by focusing on yet another aspect of the problem, a method of measuring whether or not certain events occur independently of each other.

Because the chi-square test of independence attempts to determine from a collection of statistical data whether or not two events are independent, its theoretical foundation lies in the basic property

$$P(C \cap D) = P(C) P(D)$$

of independent events, as expressed in formula (3.8) way back in Chapter 3.

> **EXAMPLE 13.1.   Individual Investment Decisions.**   A financial public relations firm has to know what effect corporate annual reports exert on the investment decisions of individual investors. To find out, they prepared the annual report of one of their clients in two different formats (both of which satisfied SEC disclosure requirements, of course!).
>
> Among those included in a random sample of 1,000 investors who were not already shareholders of their client's stock, some received Format D, which included extensive financial data; others received Format B, which consisted primarily of full-color pictures of company operations; and the rest received no annual report at all. A follow-up survey of the 1,000 potential investors yielded the results presented in the *contingency table* of Table 13.1. The figures in Table 13.1 are interpreted as follows:

TABLE 13.1 Contingency Table of Investment Decision Data

Annual Report Record	Investment Decision		
	Invested in Company	Did Not Invest	Row Totals
Did not receive annual report	44	306	350
Received Format B	19	131	150
Received Format D	37	463	500
Column totals	100	900	1,000

Of the 1,000 people surveyed, 350 received no annual report. Of those 350, 44 eventually invested in the company, while the remaining 306 did not. Those in the sample who received Format B numbered 150, and of these 19 eventually invested in the company, while 131 did not. Finally, 500 persons out of the 1,000 received Format D, with 37 eventually investing in the company and 463 foregoing the opportunity to do so.

The question we now want to answer is, "Does an individual's investment decision depend, to a significant extent, on his or her receipt of annual report information?" Use a significance level of $\alpha = .05$ to test the hypothesis that the investment decision is *independent* of the receipt of annual report information.

SOLUTION. The basic principle underlying the chi-square test goes as follows: We calculate a hypothetical contingency table, usually called the *expected contingency table,* showing how the data would have turned out if investment in a company were really independent of receipt of an annual report. Then we compare this expected contingency table with the actual contingency table that shows how the data really did turn out. If the actual table looks very similar to the expected table, we would conclude that the two characteristics (namely, "investment in company" and "receipt of annual report") are probably independent. If, on the other hand, the actual table looks quite different from the expected table, we would have to say that the two characteristics are probably not independent.

We will introduce the chi-square statistic as a measure of the difference between the two contingency tables, and we will reject the hypothesis of independence if the chi-square value exceeds a corresponding number in the chi-square table, Table A.8 of the Appendix. The chi-square statistic measures the difference between two contingency tables in a manner analogous to the way the two-sample t statistic measures the difference between two sample means.

We now proceed to the calculation of the expected contingency table of the investment decision data of Table 13.1. We are dealing with the following five events:

NA = event that investor received no annual report
FB = event that investor received Format B
FD = event that investor received Format D
IC = event that investor invested in company
NI = event that investor did not invest in company

From Table 13.1, we can find the probabilities of these events:

$P(NA)$ = 350/1,000, because 350 of the 1,000 received no report
$P(FB)$ = 150/1,000, because 150 of the 1,000 received Format B
$P(FD)$ = 500/1,000, because 500 of the 1,000 received Format D
$P(IC)$ = 100/1,000, because 100 of the 1,000 invested in company
$P(NI)$ = 900/1,000, because 900 of the 1,000 did not invest in company

From the actual contingency table on the left side of Table 13.2, we see that the event $NA \cap IC$ describes the behavior of 44 persons. This means that there were exactly 44 persons who both received no annual report and also invested in the company. Let's now compute the expected number of persons described by the event $NA \cap IC$ in the case of independence. If NA and IC are independent events, then from equation (3.8), we know that the proportion of persons described by $NA \cap IC$ is given by

$$P(NA \cap IC) = P(NA)P(IC) = (350/1,000)(100/1,000)$$

$$= 35,000/1,000,000 = 35/1,000.$$

Therefore, if 35/1,000 of the persons involved in the survey are expected to be described by the event $NA \cap IC$, this means that 35/1,000 of the 1,000 persons, namely

$$35/1,000 \times 1,000 = 35$$

persons are expected to be described by $NA \cap IC$. This argument accounts for the number 35 in the upper left corner of the expected contingency table of Table 13.2.

TABLE 13.2   Actual and Expected Contingency Tables for Investment Decision Data

Actual Contingency Table				Expected Contingency Table			
	IC	NI	Row Σ		IC	NI	Row Σ
NA	44	306	350	NA	35	315	350
FB	19	131	150	FB	15	135	150
FD	37	463	500	FD	50	450	500
Column Σ	100	900	1,000	Column Σ	100	900	1,000

We can streamline the above calculation by noting in advance that the number of persons described by $NA \cap IC$ will be $P(NA \cap IC) \times 1{,}000$. Therefore, we can run the entire calculation directly as follows:

$$P(NA \cap IC) \times 1{,}000 = P(NA)P(IC)(1{,}000)$$
$$= \left(\frac{350}{1{,}000}\right)\left(\frac{100}{1{,}000}\right)(1{,}000)$$
$$= \frac{350 \times 100}{1{,}000} = 35$$

due to the cancellation of the last 1,000 with one of those in the denominator.

We now proceed to calculate the expected number of persons described by $FB \cap IC$ under the assumption of independence. We have

$$P(FB \cap IC) \times 1{,}000 = P(FB)P(IC)(1{,}000)$$
$$= \left(\frac{150}{1{,}000}\right)\left(\frac{100}{1{,}000}\right)(1{,}000)$$
$$= \frac{150 \times 100}{1{,}000} = 15.$$

This calculation explains the appearance of the 15 in the expected contingency table of Table 13.2 at the intersection of the row headed "FB" and the column headed "IC."

The remaining cells of the expected contingency table are filled as follows:

$$P(FD \cap IC) \times 1{,}000 = P(FD)P(IC)(1{,}000)$$
$$= \left(\frac{500}{1{,}000}\right)\left(\frac{100}{1{,}000}\right)(1{,}000)$$
$$= \frac{500 \times 100}{1{,}000} = 50$$

$$P(NA \cap NI) \times 1{,}000 = P(NA)P(NI)(1{,}000)$$
$$= \left(\frac{350}{1{,}000}\right)\left(\frac{900}{1{,}000}\right)(1{,}000)$$
$$= \frac{350 \times 900}{1{,}000} = 315$$

$$P(FB \cap NI) \times 1{,}000 = P(FB)P(NI)(1{,}000)$$
$$= \left(\frac{150}{1{,}000}\right)\left(\frac{900}{1{,}000}\right)(1{,}000)$$
$$= \frac{150 \times 900}{1{,}000} = 135$$

and

$$P(FD \cap NI) \times 1,000 = P(FD)P(NI)(1,000)$$

$$= \left(\frac{500}{1,000}\right)\left(\frac{900}{1,000}\right)(1,000)$$

$$= \frac{500 \times 900}{1,000} = 450.$$

For several reasons, including the ability to have a check on the calculation, it is useful to notice that the row and column totals in the expected contingency table are exactly the same as those in the actual contingency table. For example, in Table 13.2, the top rows in each of the contingency tables sum to 350, as $44 + 306 = 350$; and $35 + 315 = 350$. This indicates that the process leading to the expected contingency table merely rearranges the survey results to make them conform with the requirements of independent events.

Because of the fact that the row and column sums must be the same in both contingency tables, it is possible to get away with computing only two of the six *expected frequencies*. This is, in fact, the way the "pros" (professional statisticians, of course) do it. In Table 13.3, for example, only the 35 and the 15 have been calculated using the formula for independent events. The remaining four cells have been filled by subtracting those two numbers from the row and column totals. In particular,

$$350 - 35 = 315$$
$$150 - 15 = 135$$
$$100 - 35 - 15 = 50$$
$$500 - 50 = 450$$

TABLE 13.3   Expected Frequency Computations

	IC	NI	Row Σ
NA	35	315	350
FB	15	135	150
FD	50	450	500
Column Σ	100	900	1,000

$$\frac{350 \times 100}{1,000} = 35 \qquad \frac{150 \times 100}{1,000} = 15$$

The hand calculations of the components of the expected contingency table can be considerably simplified by using this procedure. The number of cells for which the entire computation must be carried out is called the *degrees of freedom* of the contingency table. To find the degrees of freedom, we use the formula

$$df = (r-1)(c-1) \tag{13.1}$$

where

r = number of rows in table
c = number of columns in table

The formula arises from the fact that it would not be necessary to fill directly those cells belonging to the last row or the last column, for these could be obtained indirectly by subtraction. Therefore, the block of cells to be filled by computation consists of one less row and one less column than the original contingency table, and so contains $(r-1)(c-1)$ boxes. The contingency table for the investment decision data has three rows and two columns, so that r = 3 and c = 2. It follows that df = $(3-1)(2-1)$ = $(2)(1)$ = 2, so that it is necessary to directly fill only two of the six boxes. (Not any two boxes chosen at random, but two selected strategically!)

To decide the question of the dependence of the investment decision on the receipt of the annual report, it remains only to measure the difference between the actual and expected contingency tables of Table 13.2. We do this by means of the chi-square statistic. If we denote by f the number (actual frequency) appearing in a cell of the actual contingency table, and by e the number (expected frequency) appearing in the corresponding cell of the expected contingency table, then the chi-square statistic is given by the formula

$$\chi^2 = \Sigma \frac{(f-e)^2}{e} \tag{13.2}$$

where there is one term in the sum for each cell in the table. If the classifications of receipt of annual report and investment decision are truly independent, this fact would be indicated by each e being relatively close to its corresponding f. Therefore, in case of independence, the numerical value of χ^2 would be small. It follows that a large value of χ^2 would tend to deny independence, and so would indicate a significant dependence relationship between receipt of annual report and investment decision. What value of χ^2 would be the cut-off point between large and small, between an indication of dependence and a confirmation of independence? The cut-off point for a significance level of α in testing

H: receipt of annual report and investment decision are
     independent,

vs.

A: they are dependent,

would be χ_α^2 [df], a number to be found in Table A.8 of the Appendix. That is, we would reject H (independence) in favor of A (dependence) at level α if $\chi^2 > \chi_\alpha^2$ [df].

Using a significance level of $\alpha = .05$, and recalling that $df = (r-1)(c-1) = 2$ in this case, we see from Table A.8 that $\chi_{.05}^2$ [2] = 5.991. (The relevant portion of Table A.8 can be found in Table 13.4.) Therefore, we will reject H at level $\alpha = .05$ and conclude that the investment decision depends on receipt of the annual report only if $\chi^2 > 5.991$.

We now proceed to the computation of χ^2. The calculations in Table 13.5 implement the formula for χ^2 above. The columns headed "f" and "e" are filled by the numbers appearing in the proper cells of the contingency tables of Table 13.2. From Table 13.5 it follows that

$$\chi^2 = \Sigma \frac{(f-e)^2}{e}$$

$$= \frac{(44-35)^2}{35} + \frac{(19-15)^2}{15} + \frac{(37-50)^2}{50} + \frac{(305-315)^2}{315}$$

$$+ \frac{(131-135)^2}{135} + \frac{(463-450)^2}{450}$$

$$= 2.3143 + 1.0667 + 3.3800 + .2571 + .1185 + .3756$$

$$= 7.5122.$$

Therefore $\chi^2 = 7.512 > 5.991$, so we reject H (independence) at level $\alpha = .05$. We conclude that, according to the data, there seems to be a

TABLE 13.4   A Small Portion of Table A.8

df	$\chi_{.05}^2$
2	5.991

TABLE 13.5 Calculation of χ^2 for Investment Decision Data

Cell	f	e	$f - e$	$(f - e)^2$	$\dfrac{(f - e)^2}{e}$
$NA \cap IC$	44	35	9	81	2.3143
$FB \cap IC$	19	15	4	16	1.0667
$FD \cap IC$	37	50	-13	169	3.3800
$NA \cap NI$	306	315	-9	81	.2571
$FB \cap NI$	131	135	-4	16	.1185
$FD \cap NI$	463	450	13	169	.3756
Sums	1,000	1,000	0	χ^2	$= 7.5122$

dependence relationship between the receipt of an annual report and the investment decision of the individual.

Further research and analysis would be required to determine the precise nature of this dependence relationship. It seems, however, from Table 13.2, that fewer than expected investors who received a Format D annual report went on to invest in the company. It, therefore, appears, at first glance, that Format D should not be used by a company attempting to attract new investors.

EXAMPLE 13.2. How Good is Dehydrated Hamburger? The Good Dry 'n' Fast Food Company (GDNFF) wants to find out whether or not its experimental powdered hamburger substitute ("just add water") tastes good enough to compete with the real thing at the supermarket checkout stand. The company conducts a market research study in which 500 randomly selected persons are asked to compare the taste of a hamburger the pollster gives them with that of ones they ordinarily eat. A group of 300 of the participants (the control group) are given real hamburgers, while the remaining 200 (the experimental group) are given the rehydrated powdered substitute. All 500 received the impression that they got a real hamburger. Then they are asked whether the hamburger they received tasted better, worse, or about the same as the ones they usually eat. The resulting data are presented in Table 13.6. At level $\alpha = .05$, do the results of the market survey indicate that a person's reaction depends upon whether he or she was given a real hamburger or the powdered substitute?

SOLUTION. We want to test the hypothesis

H: independent (powdered substitute seems to taste as
good as real hamburger)

TABLE 13.6   Contingency Table of GDNFF Market Survey Data

| Type of Hamburger Given | Reaction | | | |
	Tastes as Good or Better	No Opinion	Tastes Worse	Row Totals
Real one	50	240	10	300
Powdered substitute	25	160	15	200
Column totals	75	400	25	500

vs.

A: dependent (powdered substitute tastes different from real hamburger).

Using the contingency table of Table 13.6, the company's statistician observes that the degrees of freedom are

$$df = (r-1)(c-1) = (2-1)(3-1) = (1)(2) = 2$$

because there are two rows and three columns in the table. The chi-square test of independence, then, will have us reject H in favor of A at level $\alpha = .05$ if $\chi^2 > \chi^2_{.05}[2] = 5.991$. That number can be found in Table A.8 of the Appendix. We now proceed to the calculation of χ^2.

   The following events will be involved in the computational aspects of this problem:

RH = event that person received real hamburger
PS = event that person received powdered substitute
TG = event that person thought it tasted as good or better
NO = event that person had no opinion on taste
TW = event that person thought it tasted worse

We calculate the expected contingency table in the quickest way possible and present the results in Table 13.7. We have, under the assumption of independence, that

$$P(RH \cap TG) \times 500 = \frac{300 \times 75}{500} = 45$$

$$P(RH \cap NO) \times 500 = \frac{300 \times 400}{500} = 240$$

and we fill the remaining cells by subtracting these numbers from the row and column totals where called for. On the basis of the information

TABLE 13.7   Actual and Expected Contingency Tables of GDNFF Market Survey Data

	Actual Contingency Table					Expected Contingency Table			
	TG	NO	TW	Row Σ		TG	NO	TW	Row Σ
RH	50	240	10	300	RH	45	240	15	300
PS	25	160	15	200	PS	30	160	10	200
Column Σ	75	400	25	500	Column Σ	75	400	25	500

TABLE 13.8   Calculation of χ^2 for the GDNFF Market Survey Data

Cell	f	e	$f - e$	$(f - e)^2$	$\dfrac{(f - e)^2}{e}$
$RH \cap TG$	50	45	5	25	.5556
$PS \cap TG$	25	30	-5	25	.8333
$RH \cap NO$	240	240	0	0	.0000
$PS \cap NO$	160	160	0	0	.0000
$RH \cap TW$	10	15	-5	25	1.6667
$PS \cap TW$	15	10	5	25	2.5000
Sums	500	500	0	χ^2	$= 5.5556$

presented in Table 13.7, we calculate the numerical value of χ^2, using the procedure outlined in Table 13.8.

Table 13.8 gives us the result that $\chi^2 = 5.556$, and it follows that $\chi^2 = 5.556 < 5.991$, so we cannot reject the hypothesis of independence at level $\alpha = .05$. The results, then, tend to support the company's hope that their new product tastes as good as real hamburger.

At this point in the discussion, a statistician for the FTC, concerned with the possibility of false advertising, challenges the conclusions of the company's analysis. She suggests instead that all persons having no opinion be eliminated from further consideration, and only those persons indicating a definite opinion after tasting the test material be included in the computations. Such a procedure is common, especially in advertising, where results of a survey are presented as being based on "all those responding to the question," or "all those expressing an opinion," or "all those indicating a preference," or some similar criterion for throwing away an often substantial portion of the data.

The FTC's statistician anchors her analysis on the contingency table illustrated in Table 13.9. This contingency table has $r = 2$ rows and $c = 2$ columns, so that the number of degrees of freedom is

$$df = (r-1)(c-1) = (2-1)(2-1) = 1$$

In the FTC's view, then, the hypothesis H (independence) should be rejected at level $\alpha = .05$ in favor of A (dependence) if χ^2 turns out to be larger than $\chi^2_{.05}[1] = 3.841$, as determined from Table A.8 of the Appendix. In Table 13.10, we compare the actual and expected contingency tables based on the FTC's view of the GDNFF market survey data.

The expected contingency table at the right side of Table 13.10 presents an especially clear illustration of the way in which such tables represent the independence of the two classifications, type of hamburger and reaction to it. We observe first that 75 of the 100 persons under study (3 out of every 4) reported that the hamburger received tasted as good or better, rather than worse. If reaction is truly independent of which type was received, this same proportion of persons feeling better (namely, 3 out of every 4) should be observed in each of the hamburger groups. Therefore, of the 60 given a real hamburger, 45 should have said it tasted as good or better, while of the 40 given the powdered substitute, 30 should have said it tasted as good or better. The

TABLE 13.9   The FTC's View of the GDNFF Market Survey Data

	Reaction		
Type Given	Tastes as Good or Better	Tastes Worse	Row Totals
Real one	50	10	60
Powdered substitute	25	15	40
Column totals	75	25	100

TABLE 13.10   Actual and Expected Contingency Tables of GDNFF Market Survey   (Data based on the FTC's view of the statistics)

Actual Contingency Table				Expected Contingency Table			
	TG	TW	Row Σ		TG	TW	Row Σ
RH	50	10	60	RH	45	15	60
PS	25	15	40	PS	30	10	40
Column Σ	75	25	100	Column Σ	75	25	100

TABLE 13.11   The FTC's Calculation of χ^2 for the GDNFF Market Survey Data

Cell	f	e	$f - e$	$(f - e)^2$	$\dfrac{(f - e)^2}{e}$
$RH \cap TG$	50	45	5	25	.5556
$PS \cap TG$	25	30	-5	25	.8333
$RH \cap TW$	10	15	-5	25	1.6667
$PS \cap TW$	15	10	5	25	2.5000
Sums	100	100	0	χ^2	$= 5.5556$

expected contingency table reflects these proportions exactly. The extent to which figures in the actual contingency table differ from them specifies the extent to which type of hamburger and reaction to it are not independent.

In Table 13.11, we calculate the numerical value of χ^2. According to Table 13.11, then, $\chi^2 = 5.556$. Because $\chi^2 = 5.556 > 3.841$, we are led to reject the hypothesis of independence at significance level $\alpha = .05$. Based on the FTC's view of the data, then, we would conclude that reaction does depend on type of hamburger received. The fact that fewer persons (25) actually thought the powdered substitute tasted worse than the number of persons (30) that would be expected to do so in case of independence tends to indicate that the new product does not seem to taste as good as or better than an ordinary hamburger.

What lessons can be drawn from comparing these two competing analyses of the GDNFF market survey data? When all 500 persons participating in the survey were included in the statistical analysis, the χ^2 value of 5.556 did not permit us to reject the hypothesis of independence at level $\alpha = .05$, so we had to conclude that there was insufficient evidence to assert that the powdered substitute tasted worse. However, upon elimination from further analysis of those 400 persons who reported that they had no opinion, the resulting χ^2 value of 5.556 was sufficient to reject independence. What happened to cause these paradoxical results?

On a strictly theoretical level, removing the 400 persons and the two cells $RH \cap NO$ and $PS \cap NO$ from the contingency tables resulted in a lowering of the degrees of freedom from 2 to 1. This caused the rejection value χ^2_α to drop from 5.991 to 3.841. Therefore, as 5.556 does not exceed 5.991 but does exceed 3.841, the decisions are not to reject independence in the first case, but to reject it in the second case.

On the basis of purely practical considerations, the results of the theoretical analysis make good sense. If 400 of the 500 persons, including 240 of the 300 given the new product, a substantial majority of those participating in the survey, reported that they had no opinion, then it cannot be considered likely that the new product tastes really worse than ordinary hamburger. Removing these 400 persons from consideration is equivalent to ignoring the fact that so many of those given the new product had no opinion of it. (Of course, it is possible that the "no opinion" people really had an opinion, but they weren't willing to take the time or make the effort to express it. This is one of the problems encountered all too often in random sampling.) After ignoring such a substantial percentage of those participating in the study, we should not be surprised at any change of conclusion, because whatever result happened to emerge would lack credibility. The procedure suggested by the FTC's statistician, then, is invalid because it excluded 80% of the data from consideration and bases its decision on a selected 20% of the data.

One final technical comment: A requirement for proper usage of the chi-square test is that there be a sufficiently large amount of data to guarantee that each expected frequency (e) be 5 or more. The specification that $e \geqslant 5$ is to be considered as an empirical rule of thumb, although it is supported by theoretical facts. Naturally, the more data that are available, the more reliable our statistical analysis will be. The requirement that $e \geqslant 5$ in every case merely puts some control on the number of data points considered acceptable. (We ran into this argument earlier in our use of the chi-square distribution in connection with the Kruskal-Wallis nonparametric analysis of variance test in Section 12.C.)

EXERCISES 13.A

13.1. In a recent study of whether or not corporate annual reports provide useful information to stockholders, it was suggested that persons with job experience in accounting were better able to make use of the information contained in corporate balance sheets. The following data resulted from a stockholder survey.

Found balance sheet useful?	Had Job Experience in Accounting?	
	Yes	No
Very	100	100
Somewhat	40	60
Little	60	140

At level $\alpha = .025$, do the data indicate that a stockholder's ability to use the balance sheet depends on whether or not he has had job experience in accounting?

13.2. As part of an insurance company's analysis of its policy structure, it conducted an actuarial study of the mortality rates attributable to leading diseases. The following data were gathered to compare different geographical regions of the United States. The numbers in the table indicate the causes of 290 randomly selected cases in five states.

Cause of Death	Alabama	Hawaii	Kansas	New York	Oregon
Heart trouble	32	14	38	40	36
Cancer	16	11	18	20	15
Brain damage	12	5	14	10	9

At level $\alpha = .05$, can we conclude from the data that cause of death depends significantly on geographical region?

13.3. The following data record the positions of 400 members of the various state legislatures on repeal of a certain banking law.

Party affiliation	Position on Repeal	
	For	Against
Republican	120	30
Democratic	110	140

At significance level $\alpha = .01$, does there seem to be a dependence relationship between party affiliation and position on repeal of the law?

13.4. An economist conducted a study of the effect of the presence of local water resources on vegetable prices in 1000 localities west of the Mississippi River. The resulting data presented below relate a locality's availability of water with average retail vegetable prices in that locality.

Water resources	Retail Vegetable Prices		
	Low	Moderate	High
Scarce	20	50	130
Satisfactory	90	300	110
Abundant	190	50	60

Are we justified in asserting at level $\alpha = .05$ that local vegetable prices depend to a significant extent on local water resources?

13.5. The head of marketing analysis for a chain of stop-smoking clinics wants to know whether the fact of being male or female affects one's ability to break the smoking habit. To answer the question, she selects a random sample of 70 men and 30 women from lists of participants in the clinics and obtains the following data:

Status	Male	Female
Still not smoking	45	15
Returned to smoking	25	15

At significance level α = .01, can the market analyst assert that ability to stop smoking is independent of sex?

13.6. As part of a personnel study of desirable characteristics of real estate sales-people, each of 200 salespeople was classified by two characteristics: knowledge of quantitative analysis and dollar volume of sales generated. The results appear in the following table.

Dollar volume of sales generated	Knowledge of Quantitative Analysis		
	Little	Some	Much
Low	32	14	14
Medium	45	27	28
High	23	9	8

Do the results of the study indicate, at significance level α = .05, that dollar volume of sales depends on knowledge of quantitative analysis?

13.7. A television advertising monitoring service conducted a survey to determine if there was a relationship between household income level and the number of television sets in the household. Based upon 1,000 interviews, they obtained the following results.

Number of television sets	Household Income Level				
	Less Than $5000	$5000-$20,000	$20,000-$30,000	$30,000-$50,000	More Than $50,000
0 or 1	71	189	72	47	21
2 or 3	51	146	127	111	65
4 or more	8	15	21	22	34

At level α = .01, do the survey results show that the number of television sets per household is dependent on household income level?

13.8. A recent analysis of automobile accidents revealed to the insurance industry the following information regarding the value of seat belts in preventing fatalities:

Area of occurrence of accident	Number of Fatal Accidents	
	Wearing Seat Belts	Not Wearing Seat Belts
Urban streets	180	320
Suburban freeways	100	300
Rural highways	20	80

At level $\alpha = .05$, do different areas seem to differ in regard to the effect of seat belts on accident fatalities?

13.9. A mail survey of 420 financial analysts across the nation yielded the following information on their views concerning the likelihood of increased inflation in the months ahead:

Employer's business	Views on Future Inflation		
	Less Inflation	About the Same	More Inflation
Bank	16	43	81
Stock brokerage	16	26	28
Real estate agency	13	29	18
Insurance company	13	22	65
Government	12	30	8

At level $\alpha = .01$, do the results of the study indicate that a financial analyst's views on inflation depend to a significant extent on what type of business his or her employer is in?

13.10. Robert's Advertising Agency, Inc., conducted a survey of television viewing with regard to which age groups watch what popular prime-time programs. Contact with a number of randomly-selected viewers revealed the following information:

Program	Age Group		
	16–25	26–40	above 40
Game of the Week	30	70	50
Alfie's Troubles	40	20	60
Everybody Wins!	20	30	90

At level $\alpha = .01$, does a program a person watches seem to depend to a significant extent on that person's age group?

SECTION 13.B   THE INDEX OF PREDICTIVE ASSOCIATION

Our objective in the present section will be to develop a technique of measuring the strength and direction of dependence relationships revealed in a contingency table.

When the hypothesis of independence is rejected by the chi-square test, the question arises of how dependent the classifications are and what manner of dependence exists. To answer this deeper question, we might be tempted to assume that the larger the chi-square value, the more heavily dependent the classifications are. Unfortunately, this argument is not strictly correct.

The chi-square test is really a measure of *how certain we are that the classifications are not independent,* not a measure of *how dependent the classifications themselves are.* Furthermore, the chi-square statistic does not address itself to the problem of establishing the nature and direction of any dependence relationship that may exist.

Many methods of adequately measuring dependence, using data displayed in a contingency table, have been developed. Here we will study one of the most useful, the Goodman-Kruskal index of predictive association. The index of predictive association measures how valuable it is to know a person's category in one classification when we want to predict his category in the other.

EXAMPLE 13.3.   Business' Attitude toward Pollution.   A recent survey of business executives and college students sought their views on whether business in general was acting responsibly in the area of air and water pollution. The data in Table 13.12 record the opinions of 150 randomly selected college students and 250 randomly selected business executives on the subject. If we apply the chi-square test of independence to the contingency table of Table 13.12, we obtain the conclusion, at all reasonable levels of significance, that there is a dependence relationship between status and opinion on the subject. Let's try to find out how strong that relationship is. In particular, we ask the question, "If we know the status of a respondent, how well can we predict his opinion?" (It is to answer this question that we use the index of predictive association.)

SOLUTION.   Suppose we choose one respondent at random out of the 400 and try to predict his or her opinion. Because 240 out of the 400 feel that business is not acting responsibly, while only 160 feel that it is (according to the data of Table 13.12), our best decision would be to say that the selected respondent feels that business is not acting responsibly. Using such a decision process, we would have a probability of $160/400 = 40\%$ of making an error in our decision. In arriving at the decision that the selected respondent feels this way, we have not used any information regarding the status of that respondent.

TABLE 13.12   Survey Results on Business' Attitude toward Pollution

Status	Opinion		
	Business Is Not Acting Responsibly	Business Is Acting Responsibly	Row Totals
College student	140	10	150
Business executive	100	150	250
Column totals	240	160	400

Now, because the chi-square test indicates the presence of a dependence relationship, we might be able to reduce our error probability by making use of information regarding the status of the selected respondent. Suppose, for example, we know that the respondent is a college student. Table 13.12 shows that 140 out of the 150 college students feel that business is not acting responsibly, while the remaining 10 feel otherwise. Our best strategy then would be to assert that the student feels that business is not acting responsibly; we would be correct in 140 of the 150 cases, and we would be in error in 10 of the 150 cases. If, on the other hand, we know that the respondent is a business executive, our best strategy would be to assert that the respondent feels that business is acting responsibly, because Table 13.12 shows that 150 of the 250 executives feel that way, and only the remaining 100 do not. In making such a decision, we would therefore be correct in 150 cases and incorrect in 100. Knowing the status of the respondent, then, our eventual decision would be correct in $140 + 150 = 290$ of the 400 cases, and erroneous in $10 + 100 = 110$ of them. This gives us a probability of error of $110/400 = 27.5\%$. Information about status has therefore permitted us to reduce our probability of error in predicting a respondent's opinion from 40% to 27.5%, a decline of 12.5%. This 12.5% reduction represents a proportion $12.5/40 = .3125$ of the original error probability of 40%. Therefore, knowledge of status has the effect of reducing our chances of an erroneous prediction by a factor of 31.25%.

The index of predictive association, which we will denote by the Greek letter λ (pronounced "lambda"), is the number $.3125 = 31.25\%$. It measures the extent of dependence between status and opinion by signifying that, in predicting a respondent's opinion, we can reduce our probability of making an erroneous prediction by a factor of almost one-third, from 40% down to 27.5%. We present in Table 13.13 the formal structure of the discussion leading to the numerical value of λ for the survey data of Table 13.12.

Before attempting to handle more complex contingency tables, it would be useful to have a computing formula for λ. A model of a typical contingency table having n rows and m columns appears in Table 13.14. For specific examples of actual contingency tables based on this model, you should look at Tables 13.1 (a 3 × 2 table), 13.6 (a 2 × 3 table), 13.9 (a 2 × 2 table), and 13.12 (a 2 × 2 table). We further define the following items:

LR_1 = largest number in row R_1
LR_2 = largest number in row R_2
.
.
.
LR_n = largest number in row R_n
$L(\Sigma C)$ = largest of the column totals

TABLE 13.13   Calculation of Index of Predictive Association for Opinion Survey Data

Without Knowledge of Status

Best Decision on Opinion	Number of Erroneous Cases	Total Number of Cases	Proportion of Erroneous Cases
Not acting responsibly	160	400	160/400 = 40%

With Knowledge of Status

Status	Best Decision on Opinion	Number of Erroneous Cases	Total Number of Cases	Proportion of Erroneous Cases
Student	Not acting responsibly	10	150	
Executive	Acting responsibly	100	250	
Sums		110	400	110/400 = 27.5%

$$\lambda = \frac{40 - 27.5}{40} = \frac{12.5}{40} = .3125 = 31.25\%$$

TABLE 13.14 A Typical $n \times m$ Contingency Table

Rows	Columns				Row Totals
	Column C_1	Column C_2	...	Column C_m	
Row R_1			.		ΣR_1
Row R_2			.		ΣR_2
.		.	.	.	.
.	.		.	.	
.					
Row R_n			.		ΣR_n
Column totals	ΣC_1	ΣC_2	.	ΣC_m	T

For an example of how to determine these items, let's take a look at Table 13.12. Here $LR_1 = 140$ because row R_1 consists of the numbers 140 and 10, $LR_2 = 150$ because R_2 consists of the numbers 100 and 150, and $L(\Sigma C) = 240$ because $\Sigma C_1 = 240$ and $\Sigma C_2 = 160$. We denote by $\Sigma(LR)$ the sum of the largest numbers in each row, i.e.,

$$\Sigma(LR) = LR_1 + LR_2 + \ldots + LR_n.$$

For the data of Table 13.12, we have that

$$\Sigma(LR) = LR_1 + LR_2 = 140 + 150 = 290.$$

We are now ready to present the formula for λ, the Goodman-Kruskal index of predictive association:

■
$$\lambda = \frac{\Sigma(LR) - L(\Sigma C)}{T - L(\Sigma C)} \tag{13.3}$$

In words, λ measures the relative decrease in the error when predicting the column of a member of the population under study, knowing the member's row as opposed to not knowing the member's row. For the data of Table 13.12, we have that

$$\Sigma(LR) = LR_1 + LR_2 = 140 + 150 = 290$$

$$L(\Sigma C) = \text{larger of } \{240, 160\} = 240$$

$$T = 400$$

Therefore

$$\lambda = \frac{290 - 240}{400 - 240} = \frac{50}{160} = .3125 = 31.25\%,$$

TABLE 13.15   Computations Leading to λ for the Opinion Survey Data

Status	Opinion Not Acting Responsibly	Acting Responsibly	Row Totals	Largest of Row
Student	140	10	150	$140 = LR_1$
Executive	100	150	250	$150 = LR_2$
Column totals	240	160	$400 = T$	$290 = \Sigma(LR)$

Largest column total $L(\Sigma C) = 240$

$$\lambda = \frac{\Sigma(LR) - L(\Sigma C)}{T - L(\Sigma C)} = \frac{290 - 240}{400 - 240} = \frac{50}{160} = 31.25\%$$

which is the same answer as obtained earlier. Table 13.15 presents an organized procedure for working out λ from data displayed in a contingency table. It is vital to remember that λ deals only with predicting the column when the row is known. This means that the contingency table must be constructed in such a way that the classification to be predicted is one of the columns, not one of the rows. If necessary the rows and columns will have to be interchanged. Our next example will give an illustration of this situation.

EXAMPLE 13.4.   School Facilities and Housing Prices.   Consider the data of Table 13.16 relating a community's school facilities with average housing prices in that area. The data resulted from an economist's study of the effect of the quality of locally operated schools on housing prices in 1000 localities west of the Mississippi River. The chi-square test of Section 13.A indicates, at all reasonable levels of significance, that there is a dependence relationship existing between the level of housing prices and a community's school facilities. The next step is, therefore, to measure the extent of the existing dependency by calculating the index of predictive association.

TABLE 13.16   Data Relating School Facilities and Housing Prices

Community School Facilities	Level of Housing Prices Low	Moderate	High	Row Totals
Good	10	70	120	200
Satisfactory	40	400	60	500
Mediocre	150	130	20	300
Column totals	200	600	200	1000

TABLE 13.17   Computations Leading to λ for the Data Relating School Facilities and Housing Prices

Community School Facilities	Level of Housing Prices			Row Totals	Largest of Row
	Low	Moderate	High		
Good	10	70	120	200	$120 = LR_1$
Satisfactory	40	400	60	500	$400 = LR_2$
Mediocre	150	130	20	300	$150 = LR_3$
Column totals	200	600	200	$1,000 = T$	$670 = \Sigma(LR)$

Largest column total $L(\Sigma C) = 600$

$$\lambda = \frac{\Sigma(LR) - L(\Sigma C)}{T - L(\Sigma C)} = \frac{670 - 600}{1,000 - 600} = \frac{70}{400} = .175 = 17.5\%$$

SOLUTION.   By analogy with our procedures in connection with the opinion survey data of Table 13.15, we develop the needed calculations in Table 13.17. These calculations are aimed at producing λ, the relative decrease in our error of predicting a locality's level of housing prices based on a knowledge of the locality's school facilities, as compared with the error we make having no knowledge of school facilities. Table 13.17 shows that $\Sigma(LR) = 670$ and $L(\Sigma C) = 600$, so that

$$\lambda = \frac{\Sigma(LR) - L(\Sigma C)}{T - L(\Sigma C)} = \frac{670 - 600}{1,000 - 600} = \frac{70}{400} = .175 = 17.5\%.$$

This means that a knowledge of local school facilities will reduce our error of predicting housing prices by a factor of 17.5%.

Suppose, on the other hand, that we want to use knowledge of a locality's housing prices in order to predict the quality of local school facilities. By what factor will such knowledge decrease our error of prediction? In order to answer this question by applying our formula (13.3) for λ, it is first necessary to interchange the rows and columns of the contingency table of Table 13.16. Table 13.18 contains the computations for the λ value of the interchanged contingency table.

The result of our computations in Table 13.18, to the effect that λ = 34%, means that a knowledge of local housing prices will reduce our error by a factor of 34% (more than one-third) when we try to predict the quality of local school facilities.

Now that we have shown how to calculate and interpret the significance of the index of predictive association, it would be useful to list a few important facts about λ:

TABLE 13.18   Computations Leading to λ for the School Facilities
Housing Prices Data   *(For predicting school facilities from a*
knowledge of housing prices)

Level of Housing Prices	Community School Facilities			Row Totals	Largest of Row
	Good	Satisfac- tory	Medio- cre		
Low	10	40	150	200	150
Moderate	70	400	130	600	400
High	120	60	20	200	120
Column totals	200	500	300	1,000 = T	670 = Σ(LR)

Largest column total $L(\Sigma C) = 500$

$$\lambda = \frac{\Sigma(LR) - L(\Sigma C)}{T - L(\Sigma C)} = \frac{670 - 500}{1,000 - 500} = \frac{170}{500} = .34 = 34\%$$

IMPORTANT FACTS ABOUT
THE INDEX OF PREDICTIVE ASSOCIATION
  1. λ is always between 0 and 1; a negative value for λ or a value
     greater than 1 is a sure indication of a computational error.
  2. λ = 0 only when knowledge of the row of a member of the
     population gives no help at all in predicting the member's
     column.
  3. λ = 1 only when knowledge of a member's row allows us to
     predict the column with certainty.

EXAMPLE 13.5.   Individual Investment Decisions.   Referring to the
investment decision data of Table 13.1, we recall from Section 13.A that
the chi-square test indicated the existence of a dependence relationship
between the receipt of annual reports and individual investment decisions. It would therefore be useful to compute λ, the proportional reduction in our error of predicting an individual's investment decision having knowledge of the annual report he received, as opposed to having
no knowledge of the latter.

SOLUTION.   From Table 13.1, we construct Table 13.19, which sets
up the computation of λ, and we come up with the surprising result that

TABLE 13.19   Computations Leading to λ for the Investment
Decision Data

| | Investment Decision | | | |
	Invested in Company	Did Not Invest	Row Totals	Largest of Row
Did not receive annual report	44	306	350	$306 = LR_1$
Received Format B	19	131	150	$131 = LR_2$
Received Format D	37	463	500	$463 = LR_1$
Column totals	100	900	$1{,}000 = T$	$900 = \Sigma(LR)$

Largest column total $L(\Sigma C) = 900$

$$\lambda = \frac{\Sigma(LR) - L(\Sigma C)}{T - L(\Sigma C)} = \frac{900 - 900}{1{,}000 - 900} = \frac{0}{100} = 0 = 0\%$$

$$\lambda = \frac{\Sigma(LR) - L(\Sigma C)}{T - L(\Sigma C)} = \frac{900 - 900}{1{,}000 - 900} = 0.$$

This seems to say that, despite the chi-square test's confirmation of an
existing dependence relationship, knowledge of a person's annual
report category will not at all improve our ability to accurately predict
that person's investment decisions. Why is this so? Well, for every possible annual report situation (namely None, Format B, and Format D)
our best decision would be to predict that the investor did not invest in
the company because the great majority of persons in each annual report
category did not invest in the company. On the other hand, even if we
had no knowledge of an individual's annual report category, our best
guess would also be to predict that he did not invest in the company
because 900 of the 1,000 persons under study did not invest. It follows
that knowledge of the annual report category will not result in any
change of our prediction and will therefore result in no reduction of our
error of prediction. The factor $\lambda = 0$ is, consequently, an accurate measure of the value of knowledge of the annual report category in predicting the investment decision.

What, then, about a dependence relationship whose existence is indicated by the chi-square test? Well, it would have to be a relationship that did
not involve the ability to improve our prediction of an individual's investment
decision by using knowledge about the annual report category. It is important
to remember that there are many ways of having a dependence relationship:
dependence means only that the two classifications are not independent.

EXERCISES 13.B

13.11. Based on the data of Exercise 13.1, how much can we reduce our error of predicting the extent to which a stockholder found the balance sheet useful, by knowing whether or not he or she has had job experience in accounting?

13.12. Using the data of Exercise 13.2, to what extent can an actuary reduce her error of predicting the cause of death of death if she knows the state in which the death occurred?

13.13. From the data presented in Exercise 13.3, determine the factor by which a banking industry lobbyist can reduce her error of predicting a legislator's vote, if she learns the legislator's party affiliation.

13.14. Based on the data of Exercise 13.4, how much can the economist reduce his error in predicting a locality's retail vegetable price level if he takes cognizance of the locality's water resources?

13.15. Can the marketing analyst of Exercise 13.5 reduce her error of predicting whether or not a participant will return to smoking by knowing that participant's sex? If so, by what factor?

13.16. Can a personnel analyst, using the data of Exercise 13.6, reduce his error of predicting a salesperson's dollar volume by testing that salesperson for knowledge of quantitative analysis? If so, by what factor?

13.17. By how much will knowledge of a household's income level allow the monitoring service of Exercise 13.7 to reduce its error of predicting a household's number of televisions?

13.18. The following data record the positions of 400 mayors with regard to whether or not they would favor oil drilling within their city limits or nearby offshore, if the oil royalties were to go into the city's treasury. The mayors are classified according to region of the country in which their cities are located.

Region of country	Position on Oil Drilling	
	Favor	Oppose
East	50	150
Midwest	75	25
South	40	10
West	35	15

By what factor can our error of predicting a mayor's position on the issue be lessened if we take into account the region of the country in which the mayor's city is located?

13.19. Concerning their views on a proposed tax cut, a survey of 100 persons shows that the persons surveyed can be cross-classified according to income level and view on the issue as follows:

	Oppose Tax Cut	Favor Tax Cut	Don't Know
Lower	100	25	75
Middle	75	400	225
Upper	25	75	0

To what extent can we reduce our error of predicting a person's views on the tax cut issue if we make use of information concerning the person's income level?

13.20. As part of a study of whether or not pension plans are a major factor in collective bargaining discussions, the following data were gathered on the number of workers covered by various collective bargaining agreements.* The workers are classified by industry and by whether or not their agreement mentions pension plans.

Industry	Number of Workers (in Thousands)	
	Plans Mentioned	Not Mentioned
Mining	242	86
Construction	908	164
Manufacturing	6,281	2,857
Utilities & communications	1,042	228
Transportation	898	388
Wholesale trade	340	139
Retail trade	158	282
Finance, insurance, & real estate	78	656
Services	191	118

By how much can we reduce our error of predicting whether or not a worker's collective bargaining agreement mentions a pension plan, if we know the worker's industry, as opposed to not knowing the industry?

SECTION 13.C.   THE CHI-SQUARE TEST OF GOODNESS-OF-FIT

Our objective in this concluding section on the chi-square test is to show how the test may be used to help discover the probability distribution of an underlying population from a random sample of data points from that population.

---

*J. Namias, *Handbook of Selected Sample Surveys in the Government*, 1969, New York: St. Johns University Press, p. 210.

As we discussed in Chapter 4, the probability distribution of a population specifies the general behavior pattern of sets of data that are randomly selected from that population. In particular, we have studied in great detail the normal probability distribution, and we observed in many situations how useful it is to know whether or not a particular population is normally distributed. The method introduced in this section, the *chi-square test of goodness-of-fit*, will allow us to conduct a test of the hypothesis that our random sample comes from a normally distributed population.

Uniform Distribution

We'll begin with the uniform distribution, which we mentioned briefly in Sections 3.A and 4.G.

> **EXAMPLE 13.6. Problem Businesses.** To decide how to distribute its staff of commercial loan officers, specially trained in handling bankruptcies, throughout its four branches, the Bank of the Delta conducted a survey on the number of business failures in those communities where the branches are located. The data appear in Table 13.20. The loan board would like to test, at level $\alpha = .05$, the hypothesis that "problem" businesses are uniformly distributed among the four communities the bank serves.
>
> **SOLUTION.** The board would like to find out whether or not it is reasonable to make its loan officer assignments under the assumption that all four areas have the same number of problem businesses. (This question can even be viewed as one involving independence—whether or not the number of problem businesses is independent of area.) We have to compare the actual table of data, Table 13.20, with the expected table in the case of the uniform distribution. Because the total number of problem businesses in the survey is $28 + 15 + 31 + 26 = 100$, the uniform distribution would give each of the four areas an allotment of 25 problem businesses. From this calculation, we set up the comparison of actual and expected tables appearing in Table 13.21.

TABLE 13.20   Distribution of Problem Businesses

Area	The Hill	Greenbelt	Harbor	Old Town
Number of problem businesses	28	15	31	26

TABLE 13.21 Actual and Expected Data on Problem Businesses

	Actual Table of Data					Expected Table of Data			
T.H.	G.	H.	O.T.	Sum	T.H.	G.	H.	O.T.	Sum
28	15	31	26	100	25	25	25	25	100

We use the chi-square statistic to measure the difference between these tables, a procedure similar to that introduced in Section 13.A for the purpose of testing independence. In particular, we are testing the hypothesis

H: problem businesses are uniformly distributed among the four areas,

vs.

A: they are not uniformly distributed.

Our rejection rule for this type of chi-square test is to reject H at level α if $\chi^2 > \chi_\alpha^2$ [df], where $df = m - 1$. Here m is the number of cells appearing in the table of data, namely Table 13.20.

Noting that $m = 4$, we would reject the hypothesis H of uniformity at level $\alpha = .05$ if $\chi^2 > \chi_{.05}^2$ [3] $= 7.815$. For computing the actual numerical value of χ^2, we set up a table of calculations like that in Tables 13.5, 13.8, and 13.11. The relevant calculations appear in Table 13.22, from which we see that

$$\chi^2 = \Sigma \frac{(f - e)^2}{e} = .3600 + 4.0000 + 1.4400 + .0400$$

$$= 5.8400.$$

TABLE 13.22 Calculation of χ^2 for the Problem Businesses Data

Cell	f	e	$f - e$	$(f - e)^2$	$\dfrac{(f - e)^2}{e}$
The Hill	28	25	3	9	.3600
Greenbelt	15	25	-10	100	4.0000
Harbor	31	25	6	36	1.4400
Old Town	26	25	1	1	.0400
Sums	100	100	0	χ^2	$= 5.8400$

Because we agreed to reject H if χ^2 turned out to be larger than $\chi^2_{.05}$ [3] $= 7.815$, our value of $\chi^2 = 5.8400$ does not permit us to reject H at level $\alpha = .05$. Therefore, the data seem to indicate, at level $\alpha = .05$, that the hypothesis of uniform distribution of problem businesses is a reasonable one. More specifically, the significance level of 5% means that, if problem businesses were truly uniformly distributed among the four areas, the chances exceed 5% of getting a χ^2 value even larger than the 5.8400 that we got, and therefore this assumption of uniformity can be considered somewhat probable. (The chances of getting a χ^2 value larger than 7.815, however, are not larger than 5%, so if we had gotten a $\chi^2 >$ 7.815, we would have been led to reject H and to assume that problem businesses were not uniformly distributed.)

Binomial Distribution

Our next example illustrates a problem that can be solved by applying the chi-square test of goodness-of-fit for the binomial distribution.

> **EXAMPLE 13.7. A Marketing Survey Study.** Some marketing analysts concerned with the overall validity of the marketing survey approach to consumer-preference testing feel that, when testing a new product in terms of a poll of personal preferences, the one listed first has a built-in advantage. In an attempt to measure the extent of this advantage, one marketing analyst set up the following experiment.
>
> She invented eight names of nonexistent products and listed them in pairs, randomly alternating them as brand names in four different product groups. She then conducted a mock marketing poll by asking 1,000 randomly selected consumers to vote their product preferences in each of the four groups. She hypothesized that each consumer would judge the four groups independently, but would have a probability of 60% of casting his or her vote for the brand name listed first in each group. The "ballots" were classified according to how many of the voter's four choices were the products listed first in their groups, and the resulting data appear in Table 13.23. At level $\alpha = .01$, do the results of the mock poll support the assertion that a consumer has a 60% probability of preferring a brand name listed first?
>
> **SOLUTION.** The hypothesis we want to test can be rephrased as follows: a consumer has probability $.60 = 60\%$ of voting for the first-listed brand name in any given group, independently of how that consumer voted in any other group. We are really testing the data of Table 13.23 for the binomial distribution with parameters $n = 4$ and $\pi = .60$. To set up the expected table of data, we therefore base our calculations on

TABLE 13.23   Ballot Results on Listing of Brand Names

Number of times on a ballot the first-listed brand name was preferred	0	1	2	3	4
Number of ballots	35	170	350	330	115

TABLE 13.24   Binomial Proportions for n = 4, π = .60   (F = number of votes on a ballot for first-listed brand name)

k	P(F ⩽ k)
0	.0256
1	.1792
2	.5248
3	.8704
4	1.0000

the portion of Table A.2, the table of the binomial distribution, which appears in Table 13.24.

We can now use the information in Table 13.24 in order to calculate the expected table of data that corresponds to the actual data of Table 13.23. What we need are the theoretical binomial proportions $P(F=0)$, $P(F=1)$, $P(F=2)$, $P(F=3)$, and $P(F=4)$, referring to the number of ballots involved in each of the five cells of Table 13.23. From the proportions listed in Table 13.24, we can calculate that

$$P(F=0) = P(F⩽0) = .0256$$
$$P(F=1) = P(F⩽1) - P(F⩽0) = .1792 - .0256 = .1536$$
$$P(F=2) = P(F⩽2) - P(F⩽1) = .5248 - .1792 = .3456$$
$$P(F=3) = P(F⩽3) - P(F⩽2) = .8704 - .5248 = .3456$$
$$P(F=4) = P(F⩽4) - P(F⩽3) = 1.0000 - .8706 = .1296$$

Because the survey involved the preferences of a total of $35 + 170 + 350 + 330 + 115 = 1,000$ voters, we would multiply each of the above proportions by 1,000 to get the expected frequencies of each cell. This procedure would yield the table of expected frequencies appearing in Table 13.25.

TABLE 13.25   Actual and Expected Data on Listing of Brand
Names

Actual Table of Data					Expected Table of Data				
0	1	2	3	4	0	1	2	3	4
35	170	350	330	115	25.6	153.6	345.6	345.6	129.6

Now that we are ready to calculate the value of χ^2 for the data on
listing of brand names preferred by customers, it would be appropriate
to formalize the hypothesis and alternative under consideration. We are
testing the following hypothesis about a parameter of a binomial distribution:

H: π = .60 (a consumer has probability .60 of voting for
         first-listed brand name)

vs.

A: $\pi \neq .60$.

We compute χ^2, and we reject H at significance level α if $\chi^2 > \chi^2_\alpha$ [df],
where again $df = m - 1$ for m the number of cells in the original table
of data (Table 13.23 in this case). For the listing of brand names data,
m = 5 so that we would reject H at level α = .01 if $\chi^2 > \chi^2_{.01}$ [4] =
13.277. In Table 13.26, we carry out the calculation of χ^2 in the usual
way, and we obtain the result that χ^2 = 7.608. Because χ^2 = 7.608 <
13.277, we cannot reject H at level α = .01, and so we conclude that the
survey data tend to support the assertion that a consumer indeed has
a 60% chance of preferring the first-listed brand name.

TABLE 13.26   Calculation of χ^2 for Market Survey Study Data

Cell	f	e	$f - e$	$(f - e)^2$	$\dfrac{(f - e)^2}{e}$
0	35	25.6	9.4	88.36	3.452
1	170	153.6	16.4	268.96	1.751
2	350	345.6	4.4	19.36	.056
3	350	345.6	− 15.6	243.36	.704
4	115	129.6	− 14.6	213.16	1.645
Sums	1,000	1,000.0	0	χ^2	= 7.608

TABLE 13.27   Table for the Chi-Square Goodness-of-Fit Test for the Binomial Distribution   *(Market survey study data, 1,000 data points, $\pi = .60$)*

Cell	f	k	$P(F=k)$ $n=4, \pi=.60$	$\times$ **1,000** $=$ e	$f - e$	$(f - e)^2$	$\dfrac{(f - e)^2}{e}$
0	35	0	.0256	25.6	9.4	88.36	3.452
1	170	1	.1536	153.6	16.4	268.96	1.751
2	350	2	.3456	345.6	4.4	19.36	.056
3	330	3	.3456	345.6	-15.6	243.36	.704
4	115	4	.1296	129.6	-14.6	213.16	1.645
Sums	**1,000**		1.0000	**1,000.0**	0	χ^2	$= 7.608$

It is possible to streamline the calculations required for this type of problem by combining Tables 13.23, 13.24, 13.25, and 13.26 into one single all-inclusive table. Without further comment, we illustrate the procedure in Table 13.27.

Normal Distribution

As a final example of the use of the chi-square test in testing the probability distribution of a population, we present the chi-square goodness-of-fit test for the normal distribution. The concepts involved here are the same as those involved in the test for the binomial distribution. The expected frequencies are, of course, calculated on the basis of the table of the normal distribution, Table A.4, instead of Table A.2. Because the normal distribution is continuous, rather than discrete, we have to use "intervals" rather than "cells," but this change presents no additional problems in setting up a computational table analogous to Table 13.27.

> **EXAMPLE 13.8.   Usage of Salad Dressing.**   Let's take a look at the salad dressing usage data of Section 1.B (Table 1.3) and test it for normality at significance level $\alpha = .05$.
>
> **SOLUTION.**   Before carrying out the chi-square test of normality, it turns out to be necessary to first compute the mean and standard deviation of the set of data under study. This must be done in order to be able to come up with the z scores we need for making use of Table A.4. As routine (though tedious) calculations will reveal, this set of data has sample mean $\bar{x} = 29.61$ and sample standard deviation $s = 3.493$.

TABLE 13.28   Adjustment of Intervals for Chi-Square Test of
Normality   (Salad dressing usage data)

Original Interval	Adjusted Interval	Upper End Point x	$z = \dfrac{x - \bar{x}}{s}$	z Score Interval
20 to 22	$-\infty$ to 22.5	22.5	-2.04	$-\infty$ to -2.04
23 to 25	22.5 to 25.5	25.5	-1.18	-2.04 to -1.18
26 to 28	25.5 to 28.5	28.5	-0.32	-1.18 to -0.32
29 to 31	28.5 to 31.5	31.5	0.54	-0.32 to 0.54
32 to 34	31.5 to 34.5	34.5	1.40	0.54 to 1.40
35 to 37	34.5 to 37.5	37.5	2.26	1.40 to 2.26
38 to 40	37.5 to ∞	∞	∞	2.26 to ∞

Note: $z = \dfrac{x - \bar{x}}{s}$, where $\bar{x} = 29.61$ and $s = 3.493$.

The next step is to organize the data into a reasonable frequency distribution. For the salad dressing usage data, we have already constructed a frequency distribution, and it appears in Table 1.4. We then modify the intervals to close up all gaps in the manner of Table 1.5, and next we extend the first and last intervals to $-\infty$ and ∞, respectively. Finally, we transform the intervals into z scores. The process of setting up the intervals is presented in detail in Table 13.28. Now we are ready to begin the chi-square test.

For each of the z score intervals in Table 13.28, we calculate the proportion of z scores that would fall in the interval if the data were really normally distributed. We get these proportions from Table A.4, and they are listed in the third column of Table 13.29.

Because there are 100 data points in our survey, we have to multiply each of these proportions by 100 in order to determine the expected number of data points for each interval. These are listed in the fourth column of Table 13.29. We are testing the hypothesis:

H: Daily salad dressing usage is normally distributed

vs.

A: It is not normally distributed.

In the test of the hypothesis of normality, our rejection rule is as follows: we reject H if $\chi^2 > \chi_\alpha^2 [df]$, where $df = m - 3$, for m the number of z score

TABLE 13.29   Table for the Chi-Square Goodness-of-Fit Test for the Normal Distribution   (Salad dressing usage data, 100 data points)

z Score Interval	f	Normal Proportion in Interval	$\times \underline{100} = e$	Combined Intervals f	e	$f - e$	$(f - e)^2$	$\dfrac{(f - e)^2}{e}$
$-\infty$ to -2.04	4	.0207	2.07	12	11.90	.10	.01	.0008
-2.04 to -1.18	8	.0983	9.83					
-1.18 to -0.32	23	.2555	25.55	23	25.55	-2.55	6.5025	.2545
-0.32 to 0.54	34	.3309	33.09	34	33.09	.91	.8281	.0250
0.54 to 1.40	26	.2138	21.38	26	21.38	4.62	21.3444	.9983
1.40 to 2.26	4	.0689	6.89	5	8.08	-3.08	9.4864	1.1741
2.26 to ∞	1	.0119	1.19					
Sums	100	1.0000	100.00	100	100.00	0	$\chi^2 =$	2.4527

Note: The f's are taken from the frequency distribution of Table 1.4.

intervals at the final stage (which is still to come) of the adjustment process.* Let's take $\alpha = .05$, but we will have to wait a while before we know the values of m and df.

A very important aspect of the calculations carried out in connection with Table 13.29 appears in the columns of that table headed "combined intervals." You may recall from the last paragraph of Section 13.A that proper usage of the chi-square test requires that each expected frequency (e) be 5 or larger in magnitude. In the column of Table 13.29 headed "e," all numbers exceed 5, except for 2.07 and 1.19. These two numbers are unacceptable values of e and therefore must be taken care of in some way before the chi-square process can continue. The best way to take care of them is to absorb their intervals into adjacent ones, thereby building the two large intervals $-\infty$ to -1.18 (having $f = 4 + 8 = 12$ and $e = 2.07 + 9.83 = 11.90$) and 1.40 to ∞ (having $f = 4 + 1 = 5$ and $e = 6.89 + 1.19 = 8.08$). This is what is done in the column of Table 13.29 headed "combined intervals." Now all e's exceed 5, and we can proceed with the operation.

It remains to compute df for the quantity $\chi^2_{.05}[df]$, as we have taken $\alpha = .05$. As we have mentioned earlier, $df = m - 3$, where m is the number of z score intervals at the final stage. Now, although Table 13.29 began with 7 intervals, it finished up with only 5, due to the absorption of 2 intervals by adjacent ones. Therefore $m = 5$, and so $df = m - 3 = 5 - 3 = 2$. It follows that $\chi^2_{.05}[df] = \chi^2_{.05}[2] = 5.991$.

We should therefore reject the hypothesis H of normality if $\chi^2 > 5.991$. As the standard chi-square calculations in the right half of Table 13.29 show, $\chi^2 = 2.4527 < 5.991$. Therefore, we cannot reject H. The chi-square test of normality, then, supports the assertion that daily salad dressing usage is normally distributed.

EXERCISES   13.C

13.21. The 35 forecasts of the percentage price increase for 1977, which are presented in Exercise 1.11, have sample mean 5.52 and sample standard deviation 0.5. At significance level $\alpha = .05$, can the magazine conclude that the forecasts are normally distributed?

13.22. In view of the fact that the numbers of attempts in Exercise 1.18 have sample mean 9.35 and sample standard deviation 7.68, is the psychologist justified, at level $\alpha = .01$, in assuming that the times are normally distributed?

---

*The degrees of freedom (df) drop to $m - 3$ in testing for normality due to the fact that we need to first calculate $\bar{x}$ and s. Generally, we subtract from m the number of quantities needed from the data in order to compute the expected frequencies. For normality, the three quantities are $\bar{x}$, s, and the number of data points. If we wanted to test a set of data for the Poisson distribution, we would use $df = m - 2$ because the two quantities would be the number of data points and a sample estimate of the single parameter λ.

13.23. A city housing agency conducted a survey of residential units in order to be better able to forecast future housing needs. An analysis of 300 single-family homes, classified according to number of bedrooms, yielded the following data:

Number of Bedrooms	Homes
1	70
2	75
3	70
4	50
5	35

At level α = .05, does the survey indicate that single-family homes are distributed uniformly throughout the five number-of-bedroom categories?

13.24. A distributor of office equipment has six outlets serving different regions of the country. A random sample of sales records for filing cabinets shows the number of units sold by each outlet, as follows.

Region	North-east	Mid-west	South east	North Central	South Central	West Coast
Number of units sold	85	75	60	65	90	75

Do the records show that, at level α – .01, sales of filing cabinets are uniformly distributed throughout the six outlets?

13.25. The Mendelian theory of genetics implies that, of a certain breed of cow, one-sixteenth (6.25%) will be all brown, one-fourth (25%) will be brown with black spots, three-eighths (37.5%) will be brown with white spots, one-fourth (25%) will be white with brown spots, and one-sixteenth (6.25%) will be all white. Such a distribution is said to have the proportions 1:4:6:4:1. The following data has been taken from a pricing study of a herd of 400 cows of the breed under discussion. (The actual distribution within the herd often affects the price at which it is sold.)

All Brown	Brown with Black Spots	Brown with White Spots	White with Brown Spots	All White
30	105	160	90	15

Do the results of the pricing study accord with predictions of the Mendelian theory, at significance level α = .01?

13.26. A seed company advertises that each of its seeds has an 80% chance of germinating. A detailed analysis of 50 of the company's seed packets, containing four seeds each, gives the following information.

Number of seeds that germinated	0	1	2	3	4
Number of packets	6	4	15	20	5

The company's claim of an 80% germination probability will be substantiated if the number of seeds germinated per packet has a distribution that is binomial with parameters $n = 4$ and $\pi = .80$. Is the company's claim substantiated at level $\alpha = .05$?

13.27. The current values of 500 common stocks that comprise the frequency distribution of Exercise 1.12 have sample mean $\bar{x} = 29.9$ dollars and sample standard deviation $s = 16.45$ dollars. Can we conclude at level $\alpha = .01$ that the stock prices are normally distributed?

13.28. In order to estimate the number of directory-assistance personnel required at its West River facility, Plateau Telephone Company conducted a survey of the arrival rate of calls at the directory-assistance office. In 60 randomly selected minutes throughout one typical week, the following numbers of calls were received each minute:

1	2	4	3	4	3
3	1	4	5	6	7
5	2	0	4	1	5
4	3	3	1	2	3
2	2	4	2	6	5
3	3	2	3	4	1
2	6	2	5	2	4
4	2	4	2	5	2
3	5	5	4	6	7
5	3	6	7	4	6

Recalling that the parameter of a Poisson distribution is its mean and noting that the data have sample mean $\bar{x} = 3.57$, test at level $\alpha = .01$ the hypothesis that the number of calls has a Poisson distribution with parameter $\lambda = 3.6$. (Remember the preceding footnote of this section (below Example 13.8), which asserted that we use $df = m - 2$ in testing for the Poisson distribution.)

13.29. Vacancies in entry-level accounting positions at Smirk & Snile, one of the "Big 8000" accounting firms, occur at the rate of two per week, according to Oliver M. Snile, one of the firm's senior partners. To check on Mr. Snile's assertion, Harry P. Smirk, another senior partner, recorded the following vacancy data over 50 randomly selected weeks in the not-too-distant past:

0	2	1	3	1	3	1
1	0	2	4	3	6	2
4	1	0	2	4	7	
5	3	1	0	1	2	
2	2	2	1	0	4	
3	2	3	1	1	0	
0	8	1	1	2	3	
1	1	3	3	2	2	

At level $\alpha = .05$, do the data support the assertion that the weekly number of vacancies is a Poisson-distributed random variable having parameter $\lambda = 2$?

SECTION 13.D.   THE WALD-WOLFOWITZ TEST OF RANDOMNESS

In many problems of applied interest, the questions do not involve numerical characteristics of a population, such as averages or proportions, but instead are concerned with its more qualitative behavior. We saw in Section 13.A one example of this type of question, where the chi-square test of independence was used to determine whether or not two different ways of classifying elements of a population were independent of each other.

In this section, our objective will be to develop a method of testing whether or not a given sequence of data has been produced by some sort of random process or whether the elements of the sequence are following some identifiable pattern.

> **EXAMPLE 13.9.   London Gold Prices.**   Each day, on the London gold exchange, the price of gold either rises or falls. While some observers believe the price of gold fluctuates randomly, others feel that the prices rise and fall in *trends*; and, more precisely, that there are several consecutive days of falling prices followed by several consecutive days of rising prices. To check on these opposing theories, one financial analyst of a major international banking concern recorded the behavior of price of gold over a recent 40-day period as follows:
>
> FRRRRRFFRRRRRRRFRFRRRFFFRRRRRRRFFFRRRRRRR
>
> where a fall in price from the preceding day's close is indicated by *F* and a rise is indicated by *R*. At significance level $\alpha = .05$, do the data indicate that gold prices rise and fall in trends, as opposed to random fluctuation?
>
> **SOLUTION.**   According to theoretical principles on which the Wald-Wolfowitz test of randomness is based, the mean number of runs occurring in a string of *F*'s and *R*'s is
>
> ■
> $$\mu_W = \frac{2n_F n_R}{n_F + n_R} + 1 \qquad (13.4)$$
>
> where
>
> n_F = number of *F*'s in the sequence
> n_R = number of *R*'s in the sequence.
>
> By a run, we mean a string of consecutive letters that are the same. For example, the sequence of data on the behavior of London gold prices is composed of the following runs:

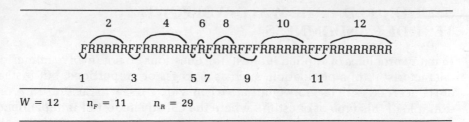

FIGURE 13.1   Marking Off the Runs   *(London gold data)*

1. a run of one F	7. a run of one F
2. a run of five R's	8. a run of three R's
3. a run of two F's	9. a run of three F's
4. a run of six R's	10. a run of six R's
5. a run of one F	11. a run of three F's
6. a run of one R	12. a run of eight R's

We denote the number of runs by the letter W, and for this example, we have $W = 12$. It is convenient to mark off the runs directly on the sequence of data, as was done in Fig. 13.1. It can be clearly seen from the figure that $W = 12$.

Now, how can the number of runs be used to test for trends in the data? Basically the idea is this: if the F's and R's were generated randomly, rather than according to any noticeable pattern, the actual number of runs would be somewhat close to the mean number μ_W calculated by formula (13.4) above. However, if there are noticeable trends in the data, there would be fewer runs than expected, because trend behavior will tend to produce longer runs than usual. The longer runs occur because trend behavior means that several consecutive days of falling prices are followed by several consecutive days of rising prices, etc. If the runs are longer, there would have to be fewer of them in the sequence of 40 data points. Therefore a numerical value of W significantly *below* μ_W would be evidence in favor of trends, as opposed to random behavior. In particular, we test the hypothesis

H: The data sequence exhibits random behavior,

against the alternative

A: The data sequence reveals trends.

If both n_F and n_R exceed 10, a version of the central limit theorem shows that the numbers

$$z = \frac{W - \mu_W}{\sigma_W}$$

(13.5)

have the standard normal distribution, where

$$\blacksquare \qquad \sigma_W = \sqrt{\frac{2n_F n_R(2n_F n_R - n_F - n_R)}{(n_F + n_R)^2(n_F + n_R - 1)}} \qquad (13.6)$$

Because $W < \mu_W$ tends to indicate the presence of trends, it makes sense to reject H in favor of A at level α if $z < -z_\alpha$ because the farther W is below μ_W, the more negative z will be. Therefore, at level $\alpha = .05$, we will agree to reject H in favor of A if $z < -z_{.05} = -1.64$. It remains only to compute z. We already know that $W = 12$, so all we have to do is to calculate μ_W and σ_W. But $n_F = 11$ and $n_R = 29$, as we can find out by counting the F's and R's in the data sequence appearing in Fig. 13.1. (Both of these exceed 10, so the use of the normal distribution is valid.) Therefore

$$\mu_W = \frac{2n_F n_R}{n_F + n_R} + 1 = \frac{2(11)(29)}{11 + 29} + 1 = \frac{638}{40} + 1 = 15.95 + 1 = 16.95$$

and

$$\sigma_W = \sqrt{\frac{2n_F n_R(2n_F n_R - n_F - n_R)}{(n_F + n_R)^2(n_F + n_R - 1)}} = \sqrt{\frac{2(11)(29)\,[2(11)(29) - 11 - 29]}{(11 + 29)^2(11 + 29 - 1)}}$$

$$= \sqrt{\frac{638(638 - 11 - 29)}{(40)^2(39)}} = \sqrt{\frac{(638)(598)}{(1600)(39)}} = \sqrt{6.114} = 2.473\,.$$

Therefore

$$z = \frac{W - \mu_W}{\sigma_W} = \frac{12 - 16.95}{2.473} = \frac{-4.95}{2.473} = -2.00\,.$$

Now $z = -2.00$, and we have agreed to reject H if $z < -1.64$. But in fact $z = -2.00 < -1.64$, so we do reject H. At level $\alpha = .05$, we therefore conclude that the 40-day sequence of data indicates that London gold prices rise and fall according to trends rather than randomly.

In the above example, we have explained how a fewer than average number of runs indicates the presence of trends in the data sequence. Suppose, instead, there is *rapid fluctuation* behavior in the data sequence. We would then expect more than the average number of runs, for rapid fluctuation behavior would seem to indicate that the data sequence fluctuates quite often between the two letters. On the basis of this discussion, we can construct a table of rejection rules for the Wald-Wolfowitz test of randomness, a table analogous to Tables 6.8 and 13.15. This table appears as Table 13.30.

EXAMPLE 13.10.   Employee Absenteeism.   A major corporation has hired a psychological consultant to analyze the phenomenon of employee absenteeism. By its investigation, the corporation hopes to be

TABLE 13.30   Rejection Rules in the Presence of Various Alternatives for the Wald-Wolfowitz Test of Randomness   (H: the data sequence exhibits random behavior)

Alternative	Reject H in favor of A at Significance Level α if:
A: rapid fluctuation	$z > z_\alpha$
A: trends	$z < -z_\alpha$
A: nonrandom behavior	$|z| > z_{\alpha/2}$

Note: The use of z is valid only in the large-sample case when both components of the data sequence are represented by more than 10 data points. For the small-sample case, special tables must be used.

able to cut down on absenteeism, and, better yet, its tendency to reduce productivity. As part of his analysis, the psychologist carefully checks the records of one particular employee who has an unusually large number of absences. On 50 consecutive working days, this employee had the following attendance record, where A indicates "absent" and P indicates "present":

PPPPAPPPAPPPPAPAPPPAPPPAPAPPPPAAPF PPAPPPAPPPAAPPPA

The psychologist decides to test the employee's attendance record for rapid fluctuations, which may indicate that the employee's job is so boring, tiring, or aggravating that he must take a day off after every few days at work. Randomness, on the other hand, would indicate that the employee merely takes a day off once in a while, for no apparent reason. At level α = .05, can the psychologist conclude that the employee's attendance record exhibits rapid fluctuations?

SOLUTION.   We test the hypothesis

H: Absences occur randomly,

against the alternative

A: Absences occur in rapid fluctuation

at level α = .05. In the case of rapid fluctuation, we would expect more than the average number of runs. Therefore W, and consequently z, would have to be large for us to agree to reject H. Referring to Table 13.30, the table of rejection rules, we see that we are to compute

$$z = \frac{W - \mu_W}{\sigma_W}$$

| 2 | 4 | | 6 8 | 10 | | 12 14 | | 16 | | 18 | 20 | 22 | | 24 |

PPPPÂPPPPÂPPPPPÂPÂPPPPÂPPPPÂPÂPPPPPĀAPPPPPÂPPPÂPPPPÂAPPPÂ

| | 1 | | 3 | | 5 | 7 | 9 | | 11 | 13 | 15 | | 17 | | 19 | 21 | | 23 |

$W = 24$ $n_P = 36$ $n_A = 14$

FIGURE 13.2 Marking Off the Runs *(Employee absenteeism data)*

and to reject H in favor of A if $z > z_{.05} = 1.64$. We proceed to the computation of z. We see from Fig. 13.2 that $W = 24$, $n_P = 36$, and $n_A = 14$. It follows that

$$\mu_W = \frac{2n_P n_A}{n_P + n_A} + 1 = \frac{2(36)(14)}{36 + 14} + 1 = \frac{1{,}008}{50} + 1 = 20.16 + 1 = 21.16$$

and

$$\sigma_W = \sqrt{\frac{2n_P n_A(2n_P n_A - n_P - n_A)}{(n_P + n_A)^2(n_P + n_A - 1)}} = \sqrt{\frac{2(36)(14)\,[2(36)(14) - 36 - 14]}{(36 + 14)^2(36 + 14 - 1)}}$$

$$= \sqrt{\frac{1{,}008(1{,}008 - 36 - 14)}{(50)^2(49)}} = \sqrt{\frac{(1{,}008)(958)}{(2{,}500)(49)}} = \sqrt{7.883} = 2.808 \; .$$

Therefore

$$z = \frac{W - \mu_W}{\sigma_W} = \frac{24 - 21.16}{2.808} = \frac{2.84}{2.808} = 1.012 \; .$$

Since we have agreed to reject H if $z > 1.64$, but $z = 1.012 < 1.64$, we cannot reject H. At level $\alpha = .05$, therefore, the psychologist concludes that the employee's record does not show evidence of rapid fluctuation in its absenteeism pattern.

In each of the two examples presented in this section, there have been more than 10 data points of each symbol, namely $n_F > 10$ and $n_R > 10$ in Example 13.9 and $n_P > 10$ and $n_A > 10$ in Example 13.10. Situations where this occurs are said to be large-sample problems, and we can use the z test to solve them.

When this criterion is not satisfied, we are dealing with a small-sample problem and, just as in Chapters 5, 6, and 12, use of the central limit theorem is no longer valid. In such cases, we simply compute W and compare the number we get with entries in a special table for the small-sample Wald-Wolfowitz test. This table and instructions for its use are available in several journals and books, one of which is the book by Dixon and Massey listed in the bibliography at the end of Chapter 8.

EXERCISES 13.D

13.30. A stockbroker observes the behavior of the Dow-Jones Industrial Average on the New York Stock Exchange for a period of 30 market days. When the average goes up, he marks a U for that day, and when the average goes down, he marks a D. The 30-day record of market behavior is as follows:

$$UUUUUDDUUUUDDDDUUUDDDDUUUUUDDUU$$

Can the ups and downs of the stock market during the 30-day period be considered random, at a level of significance of $\alpha = .01$?

13.31. A quality control supervisor wants to know whether or not defective items come off her assembly line randomly or according to some identifiable pattern. She sampled 35 consecutive items and recorded the defective ones as D's and the acceptable ones as A's. The data follow:

$$DDAAAAADAAAAAAADDAAAAAADDDAAADAAADA$$

At level $\alpha = .05$, do the data indicate that the defectives come off the assembly line randomly or nonrandomly?

13.32. One psychologist working for a large nationwide corporation wants to check on a theory to the effect that employees whose last names begin with letters toward the end of the alphabet (namely, the "Z" end) are not promoted as often as other employees. The developer of the theory attributes this situation to various factors, including the fact that such employees often appear at the bottom of lists and are therefore selected less frequently for challenging tasks and special assignments. In any case, the company psychologist conducted a test of the theory by listing 32 randomly selected employees in alphabetical order and recording next to each name an A if the employee has an *above*-the-median promotion record and a B if *below* the median. The 32 data points, in order, turned out as follows:

$$AABBBABAABAAABBBAAABBBABBABAAABB$$

At level $\alpha = .10$, do the results support the theory that records above and below the median occur in some nonrandom pattern, indicating the presence of trends in various sections of the alphabet?

13.33. An oil industry marketing analyst wants to know whether or not gasoline stations tend to be located at random points along main thoroughfares. He studied gas station locations along Mesa Blvd., the main crosstown street, and noted those intersections at which stations were located. Those intersections that have at least one gas station were recorded with a G, while those that didn't were listed as N. The data follow for 40 consecutive intersections:

$$NNGNNNNNGNGNNNNNNNNGGGGGNNNNNNGGNNNNNNNNN$$

At level $\alpha = .05$, can he assert with confidence that stations are located according to some nonrandom distribution?

SUMMARY AND DISCUSSION

With Chapter 13 we have completed our study of nonparametric statistics by presenting three special situations that call for nonparametric tests.

The first involved the chi-square test of independence, applied to contingency tables of cross-classified data, for the purpose of trying to find out whether two descriptive classifications are independent of each other. In those cases where the chi-square test indicated the existence of a dependence relationship, we presented the Goodman-Kruskal index of predictive association as one way of measuring the nature and extent of the dependence.

Next we showed how the chi-square test can be used to check on the probability distribution of a set of data. We illustrated the chi-square goodness-of-fit test, as it is called, for the uniform, binomial, and normal distributions.

Finally, using the Wald-Wolfowitz test of randomness, we asked whether a set of consecutively generated data represented a rapidly fluctuating pattern or one having trends, as opposed to a process exhibiting random behavior.

CHAPTER 13   BIBLIOGRAPHY

TESTING INDEPENDENCE

GRIZZLE, J. E., "Continuity Correction in the χ^2-Test for 2 × 2 Tables," *The American Statistician*, October 1967, pp. 28–32.

MANTEL, N., and S. W. GREENHOUSE, "What is the Continuity Correction?" *The American Statistician*, December 1968, pp. 27–30.

SCOTT, W. E., JR., "The Development of Knowledge in Organizational Behavior and Human Performance," *Decision Sciences*, January 1975, pp. 142–165.

MEASURING DEPENDENCE

BOOK, S. A., "A Sharpened Goodman-Kruskal Statistic and Its Symmetry Property," *Decision Sciences*, October 1975, pp. 605–613.

COSTNER, H. L., "Criteria for Measures of Association," *American Sociological Review*, June 1965, pp. 341–353.

EPSTEIN, M. J., *The Usefulness of Annual Reports to Corporate Shareholders.* 1975, Los Angeles: Bureau of Business and Economics Research, California State University, Los Angeles.

GOODMAN, L. A. and W. H. KRUSKAL, "Measures of Association for Cross Classifications," *Journal of the American Statistical Association*, December 1954, pp. 732–764.

GOODMAN, L. A. and W. H. KRUSKAL, "Measures of Association for Cross Classifications II: Further Discussion and References," *Journal of the American Statistical Association*, March 1959, pp. 123–163.

GOODNESS-OF-FIT PROBLEMS

CARLSON, J. A., "Are Price Expectations Normally Distributed?" *Journal of the American Statistical Association*, December 1975, pp. 749–754.

EATON, P. W., "Yarnold's Criterion and Minimum Sample Size," *The American Statistician*, August 1978, pp. 102–103.

HORA, S. C., "A Screening Test for the Poisson Process," *Decision Sciences*, July 1978, pp. 414–420.

JACOBS, F. and LOREK, K. S., "Distributional Testing of Data from Manufacturing Processes," *Decision Sciences*, April 1980, pp. 259–271.

SPECIAL TOPICS

GUENTHER, W. C., "Some Remarks on the Runs Test and the Use of the Hypergeometric Distribution," *The American Statistician*, May 1978, pp. 71–73.

MARSHALL, C. W., "A Simple Derivation of the Mean and Variance of the Number of Runs in an Ordered Sample," *The American Statistician*, October 1970, pp. 27–28.

SUMMERS, G. W., "The Coefficient of Concordance and the Q-Sort Technique," *Decision Sciences*, January 1975, pp. 37–41.

CHAPTER 13   SUPPLEMENTARY EXERCISES

13.34. One financial analyst believes that the stock market's Dow-Jones industrial average (DJIA) declines the week following an increase in the prime rate of interest at major banks. He feels, on the other hand, that the DJIA rises the week after the prime rate goes down. To substantiate his opinion, he studied the data for the last 1000 weeks and obtained the following information relating the prime rate behavior one week with the DJIA behavior the next:

	Dow-Jones Industrial Average a Week Later		
Prime rate	Down	Same	Up
Down	100	50	250
Same	70	100	30
Up	330	50	20

a. At level $\alpha = .01$, can the financial analyst conclude that changes in the DJIA really depend on fluctuations in the prime rate of interest?

b. By what factor can the error of predicting changes in the DJIA be reduced if we take into consideration the behavior of the prime rate a week earlier?

13.35. In a recent study of the use of corporate annual reports by prospective and current stockholders, it was hypothesized that college-educated investors were better able to understand a company's balance sheet than were those investors who had not had a college education. To check on the hypothesis, 400 randomly selected stockholders were interviewed and classified according to whether or not they had a college education and whether or not they had difficulty understanding the balance sheet. The results of the survey follow.

	Had Difficulty Understanding the Company Balance Sheet	
Had college education	Yes	No
Yes	145	55
No	155	45

a. At level $\alpha = .05$, do the results of the survey indicate that college-educated investors are better able to understand the balance sheet?

b. By what percentage can the error of predicting whether or not an investor had difficulty with the balance sheet be reduced by the knowledge of whether or not that investor was college-educated?

13.36. A market survey, aimed at analyzing the effect of fluoride toothpaste on the number of cavities in children's teeth, yielded the following data on the change in the number of cavities 200 children had compared with the number at their last checkup.

Type of toothpaste used	Number of Cavities Compared with Last Checkup		
	More	Same	Fewer
Fluoride	5	5	40
Regular	20	20	60
None	20	20	10

a. Decide, at level of significance $\alpha = .05$, whether or not the survey showed that the change in number of cavities depends on the type of toothpaste used.

b. To what extent can our error of predicting the change in number of cavities be reduced by making use of information concerning the type of toothpaste used?

13.37. A chain of fast-food restaurants sells three major items, a hamburger, a fish sandwich, and fried chicken. It has four restaurants in the suburban area, and sales went as follows to 600 randomly selected customers:

	Restaurant Location			
Type of Dinner	West Hills	South Valley	North Lake	Central Mall
Hamburger	45	100	20	135
Fish sandwich	5	30	60	5
Fried chicken	50	70	20	60

a. Test at level $\alpha = .01$ whether or not type of dinner sold tends to depend on location of restaurant.

b. By what factor can the error of predicting the type of dinner a customer will order (so as to be able to stock correct amounts of each) be reduced if the management takes into account the location of the restaurant?

13.38. A recent study of "redlining" in one northeastern metropolitan area was based on an analysis of 700 randomly selected mortgage loans approved between September 1973 and December 1976. The data revealed the following information on the types of loans obtained on central city and suburban property.

	Type of Loan		
Area of property	Conventional	FHA	VA
Central city	213	101	136
Suburban	147	39	64

a. Can the type of loan be considered independent of the property area at level of significance $\alpha = .05$?

b. Test for independence at level $\alpha = .01$.

c. Would the mortgage-lending industry tend to prefer $\alpha = .05$ or $\alpha = .01$?

d. By what factor can the error of predicting the type of loan a property will receive be reduced by taking into account the area in which the property is located?

13.39. The value of the dollar on international currency markets goes up (U) some days and down (D) other days. For a recent 40-day period, the ups and downs went as follows:

DDDDUDDDDDDDDUUDDDUUUUUUDDDDDDDUDDDDDDDDD.

At level $\alpha = .05$, can we consider the price fluctuations of the dollar to be random?

13.40. One over-the-counter issue handled by SSI (Stock Supermarkets, Inc.), a discount brokerage house, had the following price fluctuation pattern on its last 60 trades:

UUUDDUDDDDUUDDDDUUUDUUUUDDDUUUUUDDUUDDDDUUUUDDU
DUDUUUDDDDUUU

At level $\alpha = .01$, do the data indicate that the price of the stock is subject to rapid fluctuation?

13.41. Recent data on population trends show that the average age (in years) of persons moving to Alaska to live is 26.8 with a standard deviation of 19.08, based on a random sample of 50 "immigrants." The age distribution of those 50 persons was as follows:

Age (in years)	Number of Persons
0 to 9	5
10 to 19	15
20 to 29	20
30 to 59	6
60 or older	4

At level of significance $\alpha = .05$, do the data indicate that the ages are normally distributed?

13.42. When one large western city recently contracted with the local Floriffic Fluoridation Franchise to fluoridate its water supply, it insisted that the company make certain to distribute the chemical uniformly throughout the entire city and to file a statistical report afterwards verifying this. (When a city's water supply is fluoridated, there is some question as to whether the chemical is uniformly distributed throughout the city or whether certain neighborhoods receive greater, and possibly dangerous, concentrations than others.) A survey of five neighborhoods yielded the following data on units of fluoride in a gallon of water.

Neighborhood No.	1	2	3	4	5
Units of fluoride	72	85	68	77	73

At level $\alpha = .05$ is it reasonable to assume that the chemical is uniformly distributed throughout the water supply?

13.43. Health Protection Services, Inc., a major regional prepaid health maintenance plan, wants to check on the theory that a live birth is equally likely to be male or female. If this is the case, the number of girls in a family of six children should logically be a binomial random variable with parameters $n = 6$ and $p = 1/2$. A demographic survey of 300 families with 6 children each revealed the following data:

Number of girls in family	0	1	2	3	4	5	6
Number of families in category	5	30	70	90	65	30	10

Can the company conclude from the data, at level $\alpha = .01$, that the "equally likely" theory is valid?

13.44. A financial analyst recorded the following data on the price per earnings ratio of the common stock of 70 randomly selected corporations, as part of a study of the probability distribution of P/E ratios.*

*Business Week, December 26, 1977, pp. 92–108.

8	14	11	8	8	8	7	10	14
44	9	9	8	9	29	9	6	12
10	6	6	17	13	6	4	6	11
6	4	8	8	10	6	8	16	6
7	8	8	17	8	14	8	10	11
11	7	10	11	9	7	8	7	11
7	7	8	9	7	8	4	7	14
7	9	8	6	8	5	7		

At level α = .05, do the data indicate that P/E ratios tend to be normally distributed?

13.45. Piedmont Auto Insurance Company of the South, Inc., has been conducting a long-term study of the rate at which accident claims arrive at its disbursing office. Preliminary indications seem to show that the number of claims arriving each hour is a Poisson-distributed random variable having parameter λ = 3.2. Are these indications substantiated, at level α = .01, by the following data recorded for 100 randomly selected hours in the recent past?

Number of Claims per Hour	Number of Hours in Category
0	4
1	12
2	25
3	22
4	18
5	10
6	4
7	4
8	0
9	1

14

Introduction
to
Data Analysis

It is fair to say that the bulk of this text has been concerned with estimating parameters of probability distributions, testing hypotheses involving averages, dispersions, and shapes of statistical populations, and testing hypotheses about relationships among two or more statistical populations. It may have occurred to you more than once to ask how we manage to get the idea of *which* hypotheses to test. How do we know that we should conduct a test of whether or not the relationship between x and y is a linear one, as opposed to a logarithmic, exponential, parabolic or some other? If we want to find out what the probability distribution of a population is, how do we decide to conduct the chi-square test of *normality*, rather than a test for some other distribution?

In many situations, of course, we know what we want to ask; that is, we have a specific question in mind, so we set it up as a hypothesis and test it. That is what has been dealt with in the course up to now. In many other situations (some say in *most* situations), we don't have any particular question in mind. Our only question may be, "What's going on here?" We want to find out what information a set of data contains, so that we can learn as much as possible about the problem. The organized procedure of going about this open-minded investigation is called "data analysis."

Data analysis in one form or another has been around for several hundred years, especially in questions relating to census-taking, public-health statistics, and actuarial tables. Up until the last couple of decades, it was a procedure conducted mainly in a haphazard fashion, with everyone

657

doing what seemed right at the time. During the 1960s and 1970s, an organized theory of data analysis as a statistical method began to take shape in
a number of papers and books, many of them under the authorship or co-
authorship of John Tukey.

The formal techniques of data analysis, as expounded by Tukey in his
influential book, *Exploratory Data Analysis* (Addison-Wesley, 1977), involve
a considerable amount of specialized vocabulary and analytic procedures.
Our goal in this concluding chapter is not to present the formal details and
jargon, but merely to provide a brief introduction to the spirit and objectives
of data analysis. Accordingly, we aim only to instill in the reader an appreciation of the thought processes involved. To achieve our goal we will illustrate the role of data analysis methods in handling what are probably the two
most fundamental questions in the subject: (1) "What is the probability distribution illustrated by a set of data?" and (2) "What, if any, is the relationship
between two quantities, x and y?"

SECTION 14.A.   STUDYING THE SHAPE
OF A PROBABILITY DISTRIBUTION

Throughout our entire study of statistical inference, the probability distribution of the underlying population exerted a strong effect on our choice of
methods to use in analyzing a problem. Often the only important question
was whether or not the data came from a normally distributed population.
In such cases, the normal distribution provided a reference standard against
which we could compare the data. We have even presented in Section 13.C
a statistical test of whether or not the data may be reasonably considered to
have come from a normal distribution. As we noted in Chapter 4, however,
there is a large number of other common distributions that arise in business
and management decision problems. How can we recognize what distribution
we are working with?

In Table 5.14 of Chapter 5, we have a set of data on various financial
aspects of the top 160 stock-market performers over the years 1973 through
1977, as ranked by *Forbes* magazine. The far right-hand column contains
figures on the current yield (%) of the stock. In order to get some idea of the
probability distribution of current yields, we first construct a "cumulative
frequency distribution" in Table 14.1. (See Chapter 1, Exercises 1.19 to 1.22,
where this concept was introduced.)

An extremely simple method of comparing a set of data against the
normal distribution reference standard is to locate the *upper edges* of the
intervals (the "yield range" column of Table 14.1) on normal distribution
graph paper, often called *probability paper*. We have illustrated this method
in Fig. 14.1. Notice that the upper edges of the distribution are marked off
along the horizontal axis of the graph, while the cumulative fractions are
marked off along the vertical axis.

TABLE 14.1   Cumulative Frequency Distribution of Current Yields   (Forbes' *stock market data*)

Yield Range (%)	Fre-quency	Cumulative Frequency	÷ Sum =	Cumulative Fraction	Upper Edge	Square Root Upper Edge	Squared Upper Edge
0.0 or less	14	14		.088	0.0	0.00	0.00
0.1 to 0.5	0	14		.088	0.5	0.71	0.25
0.6 to 1.0	2	16		.100	1.0	1.00	1.00
1.1 to 1.5	7	23		.144	1.5	1.22	2.25
1.6 to 2.0	7	30		.188	2.0	1.41	4.00
2.1 to 2.5	9	39		.244	2.5	1.58	6.25
2.6 to 3.0	17	56		.350	3.0	1.73	9.00
3.1 to 3.5	17	73		.456	3.5	1.87	12.25
3.6 to 4.0	16	89		.556	4.0	2.00	16.00
4.1 to 4.5	17	106		.663	4.5	2.12	20.25
4.6 to 5.0	18	124		.775	5.0	2.24	25.00
5.1 to 5.5	12	136		.850	5.5	2.35	30.25
5.6 to 6.0	17	153		.956	6.0	2.45	36.00
6.1 to 6.5	5	158		.988	6.5	2.55	42.25
6.6 to 7.0	1	159		.9938	7.0	2.65	49.00
7.1 to 7.5	1	100		1.0000	7.5	2.74	56.25
Sum	160						

The upper edges are graphed against their respective cumulative fractions by the solid dots in Fig. 14.1. If the solid dots had followed a more or less straight line on the probability paper, we would have good reason to believe that the set of data has a normal distribution. Normally distributed data graphs into a straight line on probability paper, and it therefore is fairly easy to recognize when a data set is normal.

Unfortunately, however, we can see in Fig. 14.1 that the line followed by the solid dots is not particularly straight in some places. It would therefore be hard to believe that the set of 160 stock yields was normally distributed throughout the entire range of values. Don't worry, though; here's where the data analysis comes in! Even though the *raw* data itself may not have the normal distribution, it is quite possible that some algebraic transformation applied to the data will produce a straight line on probability paper.

As illustrations, let's apply the *square-root transformation* and the *square transformation* to the data. Both of these algebraic operations, along with a number of others, have proved to be extremely valuable in carrying out a data analysis. To apply the square-root transformation, we simply use

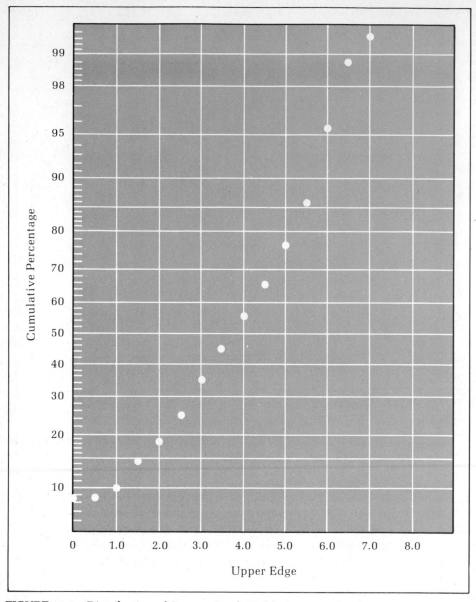

FIGURE 14.1   Distribution of Current Stock Yields Compared with Normal Reference Standard on Probability Paper

the square roots of the upper edges, instead of the upper edges themselves, in marking off the horizontal axis. The resulting points on the graph are the *open circles* in Fig. 14.2. Notice that the *square-root transformation* tends to bend the data *downward and to the right* on probability paper.

It seems that the square-root transformation has given us a curve, which is even farther away from being a straight line than was the original data.

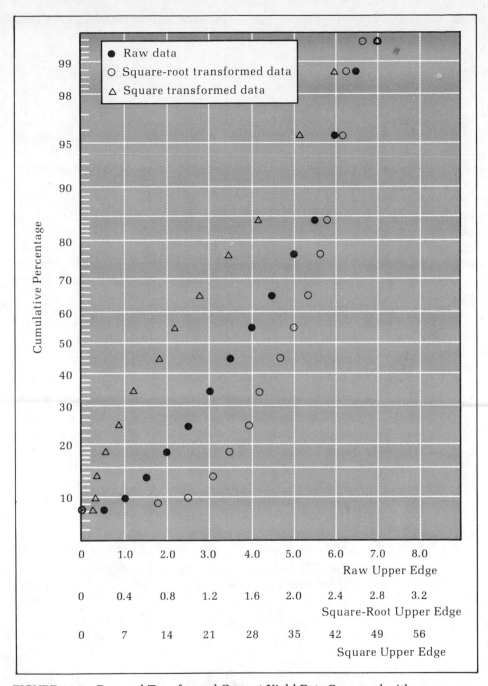

FIGURE 14.2 Raw and Transformed Current Yield Data Compared with
Normal Reference Standard on Probability Paper

Clearly, we are headed in the wrong direction. What we need is a transformation that bends a probability paper image *upward and to the left*. The *square transformation* does this particular job. We square each upper edge in Table 14.1 and mark the square off on the horizontal axis. The square transformation results in the x's on the graph in Fig. 14.2.

It certainly appears that the square transformation has produced an almost-straight line on the probability paper, thereby giving us a strong hint that the squares of the original data points may be reasonably considered to be normally distributed.

The knowledge that we have gained from a data anaysis of the *Forbes* current yield data, namely that the squares of the current yields may be normally distributed, provides a framework for conducting statistical inference on the data. We can proceed to a statistical test of normality of the squared current yields and perhaps even do *t* tests and *F* tests using the squared data points. One goal of data analysis is to bring us from the raw data to this point, where valid statistical inference may begin.

EXERCISES 14.A

14.1. Refer to the *Forbes* stock market data of Table 5.14. Conduct a data analysis of the distribution of the number of shares outstanding of the top 160 companies, by studying the frequency distribution of the data.

    a. Plot the cumulative frequency distribution of the raw data on probability paper, and assess the chances that the raw data have a normal distribution.

    b. Apply the square-root transformation to the raw data, and discuss the likelihood that the square roots of the raw data form a normally distributed set of numbers.

    c. Apply the square transformation, graph the results on probability paper, and decide which of the three choices is closer to being normally distributed.

14.2. Carry out the data analysis described in Exercise 14.1 to study the probability distribution of the price/earnings ratios of the 160 companies in the *Forbes* survey.

14.3. Apply the data analysis procedure outlined in Exercise 14.1 to investigate the probability distribution of the total market values of the stocks of the top 160 companies.

14.4. Conduct a data analysis of the sort discussed in Exercise 14.1, relating the "Latest 12-Month Earnings per Share" data to the normal probability distribution.

14.5. Use normal probability paper to analyze the distribution of salad dressing sales figures, as recorded in the data of Table 1.3 of Chapter 1.

SECTION 14.B.   SEEKING AN APPROPRIATE REGRESSION MODEL

Our major efforts in Chapters 8 and 10 were devoted to finding and using the equation of a straight line to make predictions and forecasts of future behavior. In Chapter 8 we routinely tested for the validity of the linear relationship and we calculated linear regression equations, while in Chapter 10 our analysis of time series centered on the so-called *trend line* of the data. Only briefly in both of these chapters did we allude to the possibility that such efforts would be wasted if the trend of the data did not really follow a straight line. How can we find out what type of equation is most appropriate for modeling a set of data representing a relationship between x and y?

Let's refer again to the *Forbes* magazine stock market data of Table 5.14 and ask about the relationship between a company's current yield and the price/earnings ratio of its stock. A scattergram with the price/earnings ratio on the horizontal axis and the current yield on the vertical axis is the first step in our attempt to determine the relationship.

The scattergram constructed in Fig. 14.3 exhibits a definite trend in the data from upper left to lower right. However, the presence of the points at the lower right of the scattergram seems to upset the linearity indicated on the left side of the picture. What the scattergram seems to be telling us is that the relationship is not really linear. It can be better described as a sloping curve that falls rapidly from 7 to 3 and then flattens out and drops much more slowly from 3 to 1. Rather than the linear equation $y = a + bx$, therefore, we should try to fit the data to the *exponential curve*, whose equation is

$$y = a \cdot b^x$$

(x = price/earnings ratio, while y = current yield).

Those of you who know logarithms can take the logarithms of both sides of the exponential curve's equation and get

$$\log y = \log a + (\log b) \cdot x .$$

Therefore, if we abbreviate $Y = \log y$, $A = \log a$, and $B = \log b$, the exponential curve's equation can be written as

$$Y = A + Bx,$$

which is a linear relationship between x and $Y = \log y$, rather than between x and y. So if we construct a scattergram like that in Fig. 14.3, but use the logarithms of the current yields (instead of the actual current yield values themselves), we should get a straighter line than the one appearing in Fig. 14.3.

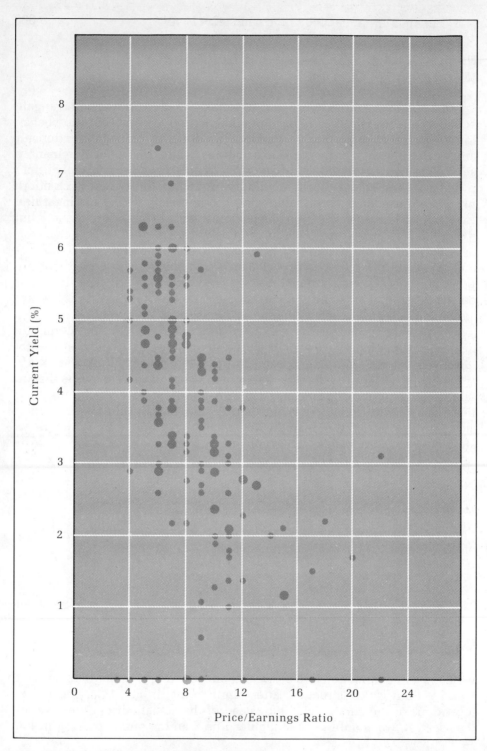

FIGURE 14.3 Scattergram of *Forbes'* Stock Data:
Current Yield vs. Price/Earnings Ratio

Fortunately, however, we don't have to take the logarithms, nor do you have to know anything about logarithms in order to check for the exponential curve relationship. It turns out that special graph paper (called *semi-log paper*) is available to automatically convert y values into log y values, with no effort at all required on the part of the analyst. We illustrate the scattergram as it looks on semi-log paper in Fig. 14.4. One minor problem, however, is that we must ignore those few y values that equal 0 because the logarithm of 0 is not algebraically definable. The remaining points can all be graphed on the logarithmic scale of Fig. 14.4.

A glance at Fig. 14.4 reveals that the semi-log scattergram is considerably more linear than is the "raw" scattergram of Fig. 14.3. This indicates that the exponential curve

$$y = a \cdot b^x$$

provides a better description of the relationship between current yield (y) and price/earnings ratio (x) than does the straight line $y = a + bx$.

In this particular example, it is unlikely that further adjustments of the data, such as the taking of logarithms, square roots, etc., would result in a more linear relationship than that in Fig. 14.4. Sometimes, though, other kinds of graph paper may be needed (and are available) for pursuing a more thorough data analysis, should that seem desirable. Various "curvilinear," rather than linear, relationships may exist between two columns of data. There are, for example, parabolic relationships and sinusoidal (periodic fluctuations) relationships, as well as exponential and logarithmic.

How do we know which relationship to use in conducting statistical inference on the data? How do we know which hypotheses to test? How do we find out which statistical procedures are appropriate? To restate the rationale of data analysis in the words of John Tukey, "The graph paper . . . [is] there . . . as a recognition that the picture-examining eye is the best finder we have of the wholly unanticipated."[*] Earlier, a zealous Tukey had explained somewhat enthusiastically, "In data analysis, a plot of y against x may help us when we know nothing about the logical connection from x to y—even when we do not know whether or not there is one—even when we know that such a connection is impossible."[†] The cryptic last clause here indicates that a graphical framework may be useful in describing a situation, even in cases where a real relationship does not exist.

EXERCISES 14.B

14.6. Try to determine the nature of the relationship, if there is one, between a company's number of shares outstanding and the price/earnings ratio of its stock. Use the *Forbes* stock market data of Table 5.14.

[*]*The American Statistician*, Vol. 34, No. 1 (February, 1980), page 24.

[†]*Exploratory Data Analysis*, p. 131.

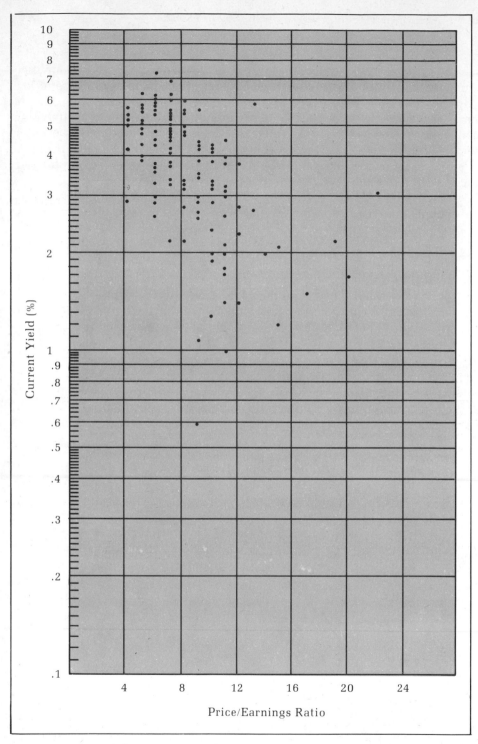

FIGURE 14.4   Semi-log Scattergram of *Forbes'* Stock Data:
Current Yield vs. Price/Earnings Ratio

      a. Draw the scattergram on ordinary graph paper and find out if there is a noticeable bend in the shape.

      b. Construct the scattergram on semi-log paper to find out if you get a straighter curve.

14.7. Find out if there is a relationship between the figures on "Latest 12-Month Earnings per Share" and "Indicated Annual Dividend" appearing in the *Forbes* data.

      a. Draw a scattergram on ordinary graph paper and check for a bend in the data.

      b. See if you get a better linear fit on semi-log paper.

14.8. Carry out a data analysis of the relationship between the recent price of a stock and its current yield, according to the *Forbes* data.

      a. Draw the scattergram on ordinary graph paper and seek a bend in the curve.

      b. Try to get more linearity on semi-log paper.

SUMMARY AND DISCUSSION

In this closing chapter we have provided a brief glance at some objectives and graphical tools of data analysis, which is incidentally both the oldest and the newest mode of statistical thinking.

One goal of data analysis is to study the nature of a set of data and so to find characteristics of it that may be subject to statistical inference. With this aim in mind, we have presented an introduction to the usage of various kinds of graph paper in the search for the probability distribution of a set of data and for a relationship between two columns of data. In particular, we have noted that, if a cumulative frequency distribution graphs as a straight line on probability paper, one is led to believe that the set of data has the normal distribution. A relationship which graphs as a straight line on semi-log paper is revealed as an exponential relationship between two sets, x and y, of data points.

CHAPTER 14   BIBLIOGRAPHY

DATA ANALYSIS IN GENERAL

J. W. Tukey and M. B. Wilk, "Data Analysis and Statistics: An Expository Overview," *Proceedings of the Fall Joint Computer Conference*, American Federation of Information Processing Societies (AFIPS), Vol. 29 (1966), pp. 695–709.

J. W. Tukey, *Exploratory Data Analysis*. 1977, Reading, Mass.: Addison-Wesley.

J. W. Tukey, "We Need Both Exploratory and Confirmatory," *The American Statistician*, Vol. 34 (February 1980), pp. 23–25.

SPECIAL TOPICS

D. C. HOAGLIN, "A Poissonness Plot," *The American Statistician*, Vol. 34 (August 1980), pp. 146–149.

C. L. MALLOWS, "Robust Methods—Some Examples of Their Use," *The American Statistician*, Vol. 33 (November 1979), pp. 179–184.

CHAPTER 14   SUPPLEMENTARY EXERCISES

14.9. Turn to the business and financial section of a daily newspaper serving your region and look at the previous day's New York Stock Exchange quotations. For the first 200 (or as many as you feel you can handle) listed stocks, observe the price change from the day earlier. (The stock either increased in price, decreased, or remained the same.)

   a. Construct a cumulative frequency distribution of the price changes.

   b. Graph the cumulative frequency distribution on probability paper, and note whether the resulting curve is or is not a straight line.

   c. Try to find an algebraic transformation that makes the probability paper curve straighter.

14.10. Obtain in your library or elsewhere, a copy of *Fortune* magazine's first issue in May of any year. This is the issue that contains the *Fortune* Directory of the 500 largest industrial corporations (the so-called "*Fortune* 500"). One of the items listed for each corporation is its number of employees.

   a. Put together a cumulative frequency distribution of the number of employees of the *Fortune* 500 corporations.

   b. Plot the cumulative distribution on probability paper and note that the resulting curve is not very close to being a straight line.

   c. Try to transform the upper edges in such a way that the probability paper curve becomes straighter.

14.11. Return to the Stock Exchange listing you used in Exercise 14.9. Find the column of numbers that contains the closing price of each stock on the day earlier. Call these numbers the x values. As y values use the price changes studied in Exercise 14.9.

   a. Draw a scattergram of the relationship between x and y on ordinary graph paper, and note whether or not the indicated relationship is a straight line.

   b. Does the scattergram follow a straight line more closely if the relationship is graphed on semi-log paper?

14.12. The *Fortune* 500 data includes a ranking of each corporation according to sales volume in dollars and according to number of employees. Taking the sales volume rank as x and the number of employees rank as y, conduct a data analysis of the relationship between x and y.

   a. Use ordinary graph paper to construct a scattergram of the relationship, in order to find out if a straight line is appropriate.

   b. Draw the scattergram on semi-log paper to determine if an exponential curve is more descriptive than a straight line.

Appendix:
Statistical Tables

TABLE A.1.   Square Roots

n	$\sqrt{n}$	$\sqrt{10n}$	n	$\sqrt{n}$	$\sqrt{10n}$	n	$\sqrt{n}$	$\sqrt{10n}$	n	$\sqrt{n}$	$\sqrt{10n}$
1.00	1.000	3.162	1.50	1.225	3.873	2.00	1.414	4.472	2.50	1.581	5.000
1.01	1.005	3.178	1.51	1.229	3.886	2.01	1.418	4.483	2.51	1.584	5.010
1.02	1.010	3.194	1.52	1.233	3.899	2.02	1.421	4.494	2.52	1.587	5.020
1.03	1.015	3.209	1.53	1.237	3.912	2.03	1.425	4.506	2.53	1.591	5.030
1.04	1.020	3.225	1.54	1.241	3.924	2.04	1.428	4.517	2.54	1.594	5.040
1.05	1.025	3.240	1.55	1.245	3.937	2.05	1.432	4.528	2.55	1.597	5.050
1.06	1.030	3.256	1.56	1.249	3.950	2.06	1.435	4.539	2.56	1.600	5.060
1.07	1.034	3.271	1.57	1.253	3.962	2.07	1.439	4.550	2.57	1.603	5.070
1.08	1.039	3.286	1.58	1.257	3.975	2.08	1.442	4.561	2.58	1.606	5.079
1.09	1.044	3.302	1.59	1.261	3.987	2.09	1.446	4.572	2.59	1.609	5.089
1.10	1.049	3.317	1.60	1.265	4.000	2.10	1.449	4.583	2.60	1.612	5.099
1.11	1.054	3.332	1.61	1.269	4.012	2.11	1.453	4.593	2.61	1.616	5.109
1.12	1.058	3.347	1.62	1.273	4.025	2.12	1.456	4.604	2.62	1.619	5.119
1.13	1.063	3.362	1.63	1.277	4.037	2.13	1.459	4.615	2.63	1.622	5.128
1.14	1.068	3.376	1.64	1.281	4.050	2.14	1.463	4.626	2.64	1.625	5.138
1.15	1.072	3.391	1.65	1.285	4.062	2.15	1.466	4.637	2.65	1.628	5.148
1.16	1.077	3.406	1.66	1.288	4.074	2.16	1.470	4.648	2.66	1.631	5.158
1.17	1.082	3.421	1.67	1.292	4.087	2.17	1.473	4.658	2.67	1.634	5.167
1.18	1.086	3.435	1.68	1.296	4.099	2.18	1.476	4.669	2.68	1.637	5.177
1.19	1.091	3.450	1.69	1.300	4.111	2.19	1.480	4.680	2.69	1.640	5.187
1.20	1.095	3.464	1.70	1.304	4.123	2.20	1.483	4.690	2.70	1.643	5.196
1.21	1.100	3.478	1.71	1.308	4.135	2.21	1.487	4.701	2.71	1.646	5.206
1.22	1.105	3.493	1.72	1.311	4.147	2.22	1.490	4.712	2.72	1.649	5.215
1.23	1.109	3.507	1.73	1.315	4.159	2.23	1.493	4.722	2.73	1.652	5.225
1.24	1.114	3.521	1.74	1.319	4.171	2.24	1.497	4.733	2.74	1.655	5.234
1.25	1.118	3.536	1.75	1.323	4.183	2.25	1.500	4.743	2.75	1.658	5.244
1.26	1.122	3.550	1.76	1.327	4.195	2.26	1.503	4.754	2.76	1.661	5.254
1.27	1.127	3.564	1.77	1.330	4.207	2.27	1.507	4.764	2.77	1.664	5.263
1.28	1.131	3.578	1.78	1.334	4.219	2.28	1.510	4.775	2.78	1.667	5.273
1.29	1.136	3.592	1.79	1.338	4.231	2.29	1.513	4.785	2.79	1.670	5.282
1.30	1.140	3.606	1.80	1.342	4.243	2.30	1.517	4.796	2.80	1.673	5.291
1.31	1.145	3.619	1.81	1.345	4.254	2.31	1.520	4.806	2.81	1.676	5.301
1.32	1.149	3.633	1.82	1.349	4.266	2.32	1.523	4.817	2.82	1.679	5.310
1.33	1.153	3.647	1.83	1.353	4.278	2.33	1.526	4.827	2.83	1.682	5.320
1.34	1.158	3.661	1.84	1.356	4.289	2.34	1.530	4.837	2.84	1.685	5.329
1.35	1.162	3.674	1.85	1.360	4.301	2.35	1.533	4.848	2.85	1.688	5.339
1.36	1.166	3.688	1.86	1.364	4.313	2.36	1.536	4.858	2.86	1.691	5.348
1.37	1.170	3.701	1.87	1.367	4.324	2.37	1.539	4.868	2.87	1.694	5.357
1.38	1.175	3.715	1.88	1.371	4.336	2.38	1.543	4.878	2.88	1.697	5.367
1.39	1.179	3.728	1.89	1.375	4.347	2.39	1.546	4.889	2.89	1.700	5.376
1.40	1.183	3.742	1.90	1.378	4.359	2.40	1.549	4.899	2.90	1.703	5.385
1.41	1.187	3.755	1.91	1.382	4.370	2.41	1.552	4.909	2.91	1.706	5.394
1.42	1.192	3.768	1.92	1.386	4.382	2.42	1.556	4.919	2.92	1.709	5.404
1.43	1.196	3.781	1.93	1.389	4.393	2.43	1.559	4.929	2.93	1.712	5.413
1.44	1.200	3.795	1.94	1.393	4.405	2.44	1.562	4.940	2.94	1.715	5.422
1.45	1.204	3.808	1.95	1.396	4.416	2.45	1.565	4.950	2.95	1.718	5.431
1.46	1.208	3.821	1.96	1.400	4.427	2.46	1.568	4.960	2.96	1.720	5.441
1.47	1.212	3.834	1.97	1.404	4.438	2.47	1.572	4.970	2.97	1.723	5.450
1.48	1.217	3.847	1.98	1.407	4.450	2.48	1.575	4.980	2.98	1.726	5.459
1.49	1.221	3.860	1.99	1.411	4.461	2.49	1.578	4.990	2.99	1.729	5.468

Note on usage of this table: For numbers greater than 1, the rule is as follows: Use the column headed $\sqrt{n}$ if there is an odd number of digits to the *left* of the decimal point, and the column headed $\sqrt{10n}$ if there is an even number of digits to the *left* of the decimal point. For numbers less than 1, count the number of zeros to the *right* of the decimal point before you reach the first nonzero digit: if the number of zeros is odd, then use the column headed $\sqrt{n}$, while if the number of zeros is 0 or even, use the column headed $\sqrt{10n}$. Further explanation with examples may be found throughout the text, beginning in Section 2.C.

n	$\sqrt{n}$	$\sqrt{10n}$	n	$\sqrt{n}$	$\sqrt{10n}$	n	$\sqrt{n}$	$\sqrt{10n}$	n	$\sqrt{n}$	$\sqrt{10n}$
3.00	1.732	5.477	3.50	1.871	5.916	4.00	2.000	6.325	4.50	2.121	6.708
3.01	1.735	5.486	3.51	1.873	5.925	4.01	2.002	6.332	4.51	2.124	6.716
3.02	1.738	5.495	3.52	1.876	5.933	4.02	2.005	6.340	4.52	2.126	6.723
3.03	1.741	5.505	3.53	1.879	5.941	4.03	2.007	6.348	4.53	2.128	6.731
3.04	1.744	5.514	3.54	1.881	5.950	4.04	2.010	6.356	4.54	2.131	6.738
3.05	1.746	5.523	3.55	1.884	5.958	4.05	2.012	6.364	4.55	2.133	6.745
3.06	1.749	5.532	3.56	1.887	5.967	4.06	2.015	6.372	4.56	2.135	6.753
3.07	1.752	5.541	3.57	1.889	5.975	4.07	2.017	6.380	4.57	2.138	6.760
3.08	1.755	5.550	3.58	1.892	5.983	4.08	2.020	6.387	4.58	2.140	6.768
3.09	1.758	5.559	3.59	1.895	5.992	4.09	2.022	6.395	4.59	2.142	6.775
3.10	1.761	5.568	3.60	1.897	6.000	4.10	2.025	6.403	4.60	2.145	6.782
3.11	1.764	5.577	3.61	1.900	6.008	4.11	2.027	6.411	4.61	2.147	6.790
3.12	1.766	5.586	3.62	1.903	6.017	4.12	2.030	6.419	4.62	2.149	6.797
3.13	1.769	5.595	3.63	1.905	6.025	4.13	2.032	6.426	4.63	2.152	6.804
3.14	1.772	5.604	3.64	1.908	6.033	4.14	2.035	6.434	4.64	2.154	6.812
3.15	1.775	5.612	3.65	1.910	6.042	4.15	2.037	6.442	4.65	2.156	6.819
3.16	1.778	5.621	3.66	1.913	6.050	4.16	2.040	6.450	4.66	2.159	6.826
3.17	1.780	5.630	3.67	1.916	6.058	4.17	2.042	6.458	4.67	2.161	6.834
3.18	1.783	5.639	3.68	1.918	6.066	4.18	2.045	6.465	4.68	2.163	6.841
3.19	1.786	5.648	3.69	1.921	6.075	4.19	2.047	6.473	4.69	2.166	6.848
3.20	1.789	5.657	3.70	1.924	6.083	4.20	2.049	6.481	4.70	2.168	6.856
3.21	1.792	5.666	3.71	1.926	6.091	4.21	2.052	6.488	4.71	2.170	6.863
3.22	1.794	5.674	3.72	1.929	6.099	4.22	2.054	6.496	4.72	2.173	6.870
3.23	1.797	5.683	3.73	1.931	6.107	4.23	2.057	6.504	4.73	2.175	6.877
3.24	1.800	5.692	3.74	1.934	6.116	4.24	2.059	6.512	4.74	2.177	6.885
3.25	1.803	5.701	3.75	1.936	6.124	4.25	2.062	6.519	4.75	2.179	6.892
3.26	1.806	5.710	3.76	1.939	6.132	4.26	2.064	6.527	4.76	2.182	6.899
3.27	1.808	5.718	3.77	1.942	6.140	4.27	2.066	6.535	4.77	2.184	6.907
3.28	1.811	5.727	3.78	1.944	6.148	4.28	2.069	6.542	4.78	2.186	6.914
3.29	1.814	5.736	3.79	1.947	6.156	4.29	2.071	6.550	4.79	2.189	6.921
3.30	1.817	5.745	3.80	1.949	6.164	4.30	2.074	6.557	4.80	2.191	6.928
3.31	1.819	5.753	3.81	1.952	6.173	4.31	2.076	6.565	4.81	2.193	6.935
3.32	1.822	5.762	3.82	1.954	6.181	4.32	2.078	6.573	4.82	2.195	6.943
3.33	1.825	5.771	3.83	1.957	6.189	4.33	2.081	6.580	4.83	2.198	6.950
3.34	1.828	5.779	3.84	1.960	6.197	4.34	2.083	6.588	4.84	2.200	6.957
3.35	1.830	5.788	3.85	1.962	6.205	4.35	2.086	6.595	4.85	2.202	6.964
3.36	1.833	5.797	3.86	1.965	6.213	4.36	2.088	6.603	4.86	2.205	6.971
3.37	1.836	5.805	3.87	1.967	6.221	4.37	2.090	6.611	4.87	2.207	6.979
3.38	1.838	5.814	3.88	1.970	6.229	4.38	2.093	6.618	4.88	2.209	6.986
3.39	1.841	5.822	3.89	1.972	6.237	4.39	2.095	6.626	4.89	2.211	6.993
3.40	1.844	5.831	3.90	1.975	6.245	4.40	2.098	6.633	4.90	2.214	7.000
3.41	1.847	5.839	3.91	1.977	6.253	4.41	2.100	6.641	4.91	2.216	7.007
3.42	1.849	5.848	3.92	1.980	6.261	4.42	2.102	6.648	4.92	2.218	7.014
3.43	1.852	5.857	3.93	1.982	6.269	4.43	2.105	6.656	4.93	2.220	7.021
3.44	1.855	5.865	3.94	1.985	6.277	4.44	2.107	6.663	4.94	2.223	7.028
3.45	1.857	5.874	3.95	1.987	6.285	4.45	2.109	6.671	4.95	2.225	7.036
3.46	1.860	5.882	3.96	1.990	6.293	4.46	2.112	6.678	4.96	2.227	7.043
3.47	1.863	5.891	3.97	1.992	6.301	4.47	2.114	6.686	4.97	2.229	7.050
3.48	1.865	5.899	3.98	1.995	6.309	4.48	2.117	6.693	4.98	2.232	7.057
3.49	1.868	5.908	3.99	1.997	6.317	4.49	2.119	6.701	4.99	2.234	7.064

(Continued)

TABLE A.1.   Square Roots (continued)

n	$\sqrt{n}$	$\sqrt{10n}$	n	$\sqrt{n}$	$\sqrt{10n}$	n	$\sqrt{n}$	$\sqrt{10n}$	n	$\sqrt{n}$	$\sqrt{10n}$
5.00	2.236	7.071	5.50	2.345	7.416	6.00	2.449	7.746	6.50	2.550	8.062
5.01	2.238	7.078	5.51	2.347	7.423	6.01	2.452	7.752	6.51	2.551	8.068
5.02	2.241	7.085	5.52	2.349	7.430	6.02	2.454	7.759	6.52	2.553	8.075
5.03	2.243	7.092	5.53	2.352	7.436	6.03	2.456	7.765	6.53	2.555	8.081
5.04	2.245	7.099	5.54	2.354	7.443	6.04	2.458	7.772	6.54	2.557	8.087
5.05	2.247	7.106	5.55	2.356	7.450	6.05	2.460	7.778	6.55	2.559	8.093
5.06	2.249	7.113	5.56	2.358	7.457	6.06	2.462	7.785	6.56	2.561	8.099
5.07	2.252	7.120	5.57	2.360	7.463	6.07	2.464	7.791	6.57	2.563	8.106
5.08	2.254	7.127	5.58	2.362	7.470	6.08	2.466	7.797	6.58	2.565	8.112
5.09	2.256	7.134	5.59	2.364	7.477	6.09	2.468	7.804	6.59	2.567	8.118
5.10	2.258	7.141	5.60	2.366	7.483	6.10	2.470	7.810	6.60	2.569	8.124
5.11	2.261	7.148	5.61	2.369	7.490	6.11	2.472	7.817	6.61	2.571	8.130
5.12	2.263	7.155	5.62	2.371	7.497	6.12	2.474	7.823	6.62	2.573	8.136
5.13	2.265	7.162	5.63	2.373	7.503	6.13	2.476	7.829	6.63	2.575	8.142
5.14	2.267	7.169	5.64	2.375	7.510	6.14	2.478	7.836	6.64	2.577	8.149
5.15	2.269	7.176	5.65	2.377	7.517	6.15	2.480	7.842	6.65	2.579	8.155
5.16	2.272	7.183	5.66	2.379	7.523	6.16	2.482	7.849	6.66	2.581	8.161
5.17	2.274	7.190	5.67	2.381	7.530	6.17	2.484	7.855	6.67	2.583	8.167
5.18	2.276	7.197	5.68	2.383	7.537	6.18	2.486	7.861	6.68	2.585	8.173
5.19	2.278	7.204	5.69	2.385	7.543	6.19	2.488	7.868	6.69	2.587	8.179
5.20	2.280	7.211	5.70	2.387	7.550	6.20	2.490	7.874	6.70	2.588	8.185
5.21	2.283	7.218	5.71	2.390	7.556	6.21	2.492	7.880	6.71	2.590	8.191
5.22	2.285	7.225	5.72	2.392	7.563	6.22	2.494	7.887	6.72	2.592	8.198
5.23	2.287	7.232	5.73	2.394	7.570	6.23	2.496	7.893	6.73	2.594	8.204
5.24	2.289	7.239	5.74	2.396	7.576	6.24	2.498	7.899	6.74	2.596	8.210
5.25	2.291	7.246	5.75	2.398	7.583	6.25	2.500	7.906	6.75	2.598	8.216
5.26	2.293	7.253	5.76	2.400	7.589	6.26	2.502	7.912	6.76	2.600	8.222
5.27	2.296	7.259	5.77	2.402	7.596	6.27	2.504	7.918	6.77	2.602	8.228
5.28	2.298	7.266	5.78	2.404	7.603	6.28	2.506	7.925	6.78	2.604	8.234
5.29	2.300	7.273	5.79	2.406	7.609	6.29	2.508	7.931	6.79	2.606	8.240
5.30	2.302	7.280	5.80	2.408	7.616	6.30	2.510	7.937	6.80	2.608	8.246
5.31	2.304	7.287	5.81	2.410	7.622	6.31	2.512	7.944	6.81	2.610	8.252
5.32	2.307	7.294	5.82	2.412	7.629	6.32	2.514	7.950	6.82	2.612	8.258
5.33	2.309	7.301	5.83	2.415	7.635	6.33	2.516	7.956	6.83	2.613	8.264
5.34	2.311	7.308	5.84	2.417	7.642	6.34	2.518	7.962	6.84	2.615	8.270
5.35	2.313	7.314	5.85	2.419	7.649	6.35	2.520	7.969	6.85	2.617	8.276
5.36	2.315	7.321	5.86	2.421	7.655	6.36	2.522	7.975	6.86	2.619	8.282
5.37	2.317	7.328	5.87	2.423	7.662	6.37	2.524	7.981	6.87	2.621	8.289
5.38	2.319	7.335	5.88	2.425	7.668	6.38	2.526	7.987	6.88	2.623	8.295
5.39	2.322	7.342	5.89	2.427	7.675	6.39	2.528	7.994	6.89	2.625	8.301
5.40	2.324	7.348	5.90	2.429	7.681	6.40	2.530	8.000	6.90	2.627	8.307
5.41	2.326	7.355	5.91	2.431	7.688	6.41	2.532	8.006	6.91	2.629	8.313
5.42	2.328	7.362	5.92	2.433	7.694	6.42	2.534	8.012	6.92	2.631	8.319
5.43	2.330	7.369	5.93	2.435	7.701	6.43	2.536	8.019	6.93	2.632	8.325
5.44	2.332	7.376	5.94	2.437	7.707	6.44	2.538	8.025	6.94	2.634	8.331
5.45	2.335	7.382	5.95	2.439	7.714	6.45	2.540	8.031	6.95	2.636	8.337
5.46	2.337	7.389	5.96	2.441	7.720	6.46	2.542	8.037	6.96	2.638	8.343
5.47	2.339	7.396	5.97	2.443	7.727	6.47	2.544	8.044	6.97	2.640	8.349
5.48	2.341	7.403	5.98	2.445	7.733	6.48	2.546	8.050	6.98	2.642	8.355
5.49	2.343	7.409	5.99	2.447	7.739	6.49	2.548	8.056	6.99	2.644	8.361

n	$\sqrt{n}$	$\sqrt{10n}$	n	$\sqrt{n}$	$\sqrt{10n}$	n	$\sqrt{n}$	$\sqrt{10n}$	n	$\sqrt{n}$	$\sqrt{10n}$
7.00	2.646	8.367	7.50	2.739	8.660	8.00	2.828	8.944	8.50	2.915	9.220
7.01	2.648	8.373	7.51	2.740	8.666	8.01	2.830	8.950	8.51	2.917	9.225
7.02	2.650	8.379	7.52	2.742	8.672	8.02	2.832	8.955	8.52	2.919	9.230
7.03	2.651	8.385	7.53	2.744	8.678	8.03	2.834	8.961	8.53	2.921	9.236
7.04	2.653	8.390	7.54	2.746	8.683	8.04	2.835	8.967	8.54	2.922	9.241
7.05	2.655	8.396	7.55	2.748	8.689	8.05	2.837	8.972	8.55	2.924	9.247
7.06	2.657	8.402	7.56	2.750	8.695	8.06	2.839	8.978	8.56	2.926	9.252
7.07	2.659	8.408	7.57	2.751	8.701	8.07	2.841	8.983	8.57	2.927	9.257
7.08	2.661	8.414	7.58	2.753	8.706	8.08	2.843	8.989	8.58	2.929	9.263
7.09	2.663	8.420	7.59	2.755	8.712	8.09	2.844	8.994	8.59	2.931	9.268
7.10	2.665	8.426	7.60	2.757	8.718	8.10	2.846	9.000	8.60	2.933	9.274
7.11	2.666	8.432	7.61	2.759	8.724	8.11	2.848	9.006	8.61	2.934	9.279
7.12	2.668	8.438	7.62	2.760	8.729	8.12	2.850	9.011	8.62	2.936	9.284
7.13	2.670	8.444	7.63	2.762	8.735	8.13	2.851	9.017	8.63	2.938	9.290
7.14	2.672	8.450	7.64	2.764	8.741	8.14	2.853	9.022	8.64	2.939	9.295
7.15	2.674	8.456	7.65	2.766	8.746	8.15	2.855	9.028	8.65	2.941	9.301
7.16	2.676	8.462	7.66	2.768	8.752	8.16	2.857	9.033	8.66	2.943	9.306
7.17	2.678	8.468	7.67	2.769	8.758	8.17	2.858	9.039	8.67	2.944	9.311
7.18	2.680	8.473	7.68	2.771	8.764	8.18	2.860	9.044	8.68	2.946	9.317
7.19	2.681	8.479	7.69	2.773	8.769	8.19	2.862	9.050	8.69	2.948	9.322
7.20	2.683	8.485	7.70	2.775	8.775	8.20	2.864	9.055	8.70	2.950	9.327
7.21	2.685	8.491	7.71	2.777	8.781	8.21	2.865	9.061	8.71	2.951	9.333
7.22	2.687	8.497	7.72	2.778	8.786	8.22	2.867	9.066	8.72	2.953	9.338
7.23	2.689	8.503	7.73	2.780	8.792	8.23	2.869	9.072	8.73	2.955	9.343
7.24	2.691	8.509	7.74	2.782	8.798	8.24	2.871	9.077	8.74	2.956	9.349
7.25	2.693	8.515	7.75	2.784	8.803	8.25	2.872	9.083	8.75	2.958	9.354
7.26	2.694	8.521	7.76	2.786	8.809	8.26	2.874	9.088	8.76	2.960	9.359
7.27	2.696	8.526	7.77	2.787	8.815	8.27	2.876	9.094	8.77	2.961	9.365
7.28	2.698	8.532	7.78	2.789	8.820	8.28	2.877	9.099	8.78	2.963	9.370
7.29	2.700	8.538	7.79	2.791	8.826	8.29	2.879	9.105	8.79	2.965	9.375
7.30	2.702	8.544	7.80	2.793	8.832	8.30	2.881	9.110	8.80	2.966	9.381
7.31	2.704	8.550	7.81	2.795	8.837	8.31	2.883	9.116	8.81	2.968	9.386
7.32	2.706	8.556	7.82	2.796	8.843	8.32	2.884	9.121	8.82	2.970	9.391
7.33	2.707	8.562	7.83	2.798	8.849	8.33	2.886	9.127	8.83	2.972	9.397
7.34	2.709	8.567	7.84	2.800	8.854	8.34	2.888	9.132	8.84	2.973	9.402
7.35	2.711	8.573	7.85	2.802	8.860	8.35	2.890	9.138	8.85	2.975	9.407
7.36	2.713	8.579	7.86	2.804	8.866	8.36	2.891	9.143	8.86	2.977	9.413
7.37	2.715	8.585	7.87	2.805	8.871	8.37	2.893	9.149	8.87	2.978	9.418
7.38	2.717	8.591	7.88	2.807	8.877	8.38	2.895	9.154	8.88	2.980	9.423
7.39	2.718	8.596	7.89	2.809	8.883	8.39	2.897	9.160	8.89	2.982	9.429
7.40	2.720	8.602	7.90	2.811	8.888	8.40	2.898	9.165	8.90	2.983	9.434
7.41	2.722	8.608	7.91	2.812	8.894	8.41	2.900	9.171	8.91	2.985	9.439
7.42	2.724	8.614	7.92	2.814	8.899	8.42	2.902	9.176	8.92	2.987	9.445
7.43	2.726	8.620	7.93	2.816	8.905	8.43	2.903	9.181	8.93	2.988	9.450
7.44	2.728	8.626	7.94	2.818	8.911	8.44	2.905	9.187	8.94	2.990	9.455
7.45	2.729	8.631	7.95	2.820	8.916	8.45	2.907	9.192	8.95	2.992	9.460
7.46	2.731	8.637	7.96	2.821	8.922	8.46	2.909	9.198	8.96	2.993	9.466
7.47	2.733	8.643	7.97	2.823	8.927	8.47	2.910	9.203	8.97	2.995	9.471
7.48	2.735	8.649	7.98	2.825	8.933	8.48	2.912	9.209	8.98	2.997	9.476
7.49	2.737	8.654	7.99	2.827	8.939	8.49	2.914	9.214	8.99	2.998	9.482

(Continued)

TABLE A.1.   Square Roots (continued)

n	$\sqrt{n}$	$\sqrt{10n}$	n	$\sqrt{n}$	$\sqrt{10n}$	n	$\sqrt{n}$	$\sqrt{10n}$	n	$\sqrt{n}$	$\sqrt{10n}$
9.00	3.000	9.487	9.25	3.041	9.618	9.50	3.082	9.747	9.75	3.122	9.874
9.01	3.002	9.492	9.26	3.043	9.623	9.51	3.084	9.752	9.76	3.124	9.879
9.02	3.003	9.497	9.27	3.045	9.628	9.52	3.085	9.757	9.77	3.126	9.884
9.03	3.005	9.503	9.28	3.046	9.633	9.53	3.087	9.762	9.78	3.127	9.889
9.04	3.007	9.508	9.29	3.048	9.638	9.54	3.089	9.767	9.79	3.129	9.894
9.05	3.008	9.513	9.30	3.050	9.644	9.55	3.090	9.772	9.80	3.130	9.899
9.06	3.010	9.518	9.31	3.051	9.649	9.56	3.092	9.778	9.81	3.132	9.905
9.07	3.012	9.524	9.32	3.053	9.654	9.57	3.094	9.783	9.82	3.134	9.910
9.08	3.013	9.529	9.33	3.055	9.659	9.58	3.095	9.788	9.83	3.135	9.915
9.09	3.015	9.534	9.34	3.056	9.664	9.59	3.097	9.793	9.84	3.137	9.920
9.10	3.017	9.539	9.35	3.058	9.670	9.60	3.098	9.798	9.85	3.138	9.925
9.11	3.018	9.545	9.36	3.059	9.675	9.61	3.100	9.803	9.86	3.140	9.930
9.12	3.020	9.550	9.37	3.061	9.680	9.62	3.102	9.808	9.87	3.142	9.935
9.13	3.022	9.555	9.38	3.063	9.685	9.63	3.103	9.813	9.88	3.143	9.940
9.14	3.023	9.560	9.39	3.064	9.690	9.64	3.105	9.818	9.89	3.145	9.945
9.15	3.025	9.566	9.40	3.066	9.695	9.65	3.106	9.823	9.90	3.146	9.950
9.16	3.027	9.571	9.41	3.068	9.701	9.66	3.108	9.829	9.91	3.148	9.955
9.17	3.028	9.576	9.42	3.069	9.706	9.67	3.110	9.834	9.92	3.150	9.960
9.18	3.030	9.581	9.43	3.071	9.711	9.68	3.111	9.839	9.93	3.151	9.965
9.19	3.031	9.586	9.44	3.072	9.716	9.69	3.113	9.844	9.94	3.153	9.970
9.20	3.033	9.592	9.45	3.074	9.721	9.70	3.114	9.849	9.95	3.154	9.975
9.21	3.035	9.597	9.46	3.076	9.726	9.71	3.116	9.854	9.96	3.156	9.980
9.22	3.036	9.602	9.47	3.077	9.731	9.72	3.118	9.859	9.97	3.158	9.985
9.23	3.038	9.607	9.48	3.079	9.737	9.73	3.119	9.864	9.98	3.159	9.990
9.24	3.040	9.612	9.49	3.081	9.742	9.74	3.121	9.869	9.99	3.161	9.995

TABLE A.2   The Binomial Distribution

This table lists the cumulative binomial probabilities $P(B \leq k)$ for a random variable B having the binomial distribution with parameters n and π. The parameter values covered are as follows: n ranges from 1 to 20 by whole numbers, and π ranges from .05 to .95 in units of .05. For each value of n, k ranges from 0 to n. As an example of how to use the table, consider the problem of finding $P(B \leq 6)$ for $n = 12$ and $\pi = .55$. We look in the row corresponding to $n = 12$ and $k = 6$ and in the column headed $\pi = .55$, and there we find the number .4731. Therefore, we know that $P(B \leq 6) = .4731$.

Probabilities of the form $P(B > k)$, $P(B = k)$, and $P(B < k)$ may be found from the entries in this table by using the relationships:

$$P(B > k) = 1 - P(B \leq k)$$
$$P(B = k) = P(B \leq k) - P(B \leq k-1)$$
$$P(B < k) = P(B \leq k) - P(B = k)$$

For values of the parameter n higher than 20, the cumulative binomial formula

$$P(B \leq k) = \sum_{j=0}^{k} C(n,j)\, \pi^{j} (1 - \pi)^{n-j}$$

must be used.

(Continued)

Source: W. J. Conover, *Practical Nonparametric Statistics*, John Wiley & Sons, Inc., 1971, pp. 368–379. Reprinted by permission of John Wiley & Sons, Inc. We would like to thank Professor Conover for calling our attention to some misprints in the original version of the table.

TABLE A.2   The Binomial Distribution (continued)

n	k	π = .05	.10	.15	.20	.25	.30	.35	.40	.45
1	0	.9500	.9000	.8500	.8000	.7500	.7000	.6500	.6000	.5500
	1	1.0000	1.0000	1.0000	1.0000	1.0000	1.0000	1.0000	1.0000	1.0000
2	0	.9025	.8100	.7225	.6400	.5625	.4900	.4225	.3600	.3025
	1	.9975	.9900	.9775	.9600	.9375	.9100	.8775	.8400	.7975
	2	1.0000	1.0000	1.0000	1.0000	1.0000	1.0000	1.0000	1.0000	1.0000
3	0	.8574	.7290	.6141	.5129	.4219	.3439	.2746	.2160	.1664
	1	.9928	.9720	.9392	.8960	.8438	.7840	.7182	.6480	.5748
	2	.9999	.9990	.9966	.9929	.9844	.9730	.9571	.9360	.9089
	3	1.0000	1.0000	1.0000	1.0000	1.0000	1.0000	1.0000	1.0000	1.0000
4	0	.8145	.6561	.5200	.4096	.3164	.2401	.1785	.1296	.0915
	1	.9860	.9477	.8905	.8192	.7383	.6517	.5630	.4752	.3910
	2	.9995	.9963	.9880	.9728	.9492	.9163	.8735	.8208	.7585
	3	1.0000	.9999	.9995	.9984	.9961	.9919	.9850	.9743	.9590
	4	1.0000	1.0000	1.0000	1.0000	1.0000	1.0000	1.0000	1.0000	1.0000
5	0	.7738	.5905	.4437	.3277	.2373	.1681	.1160	.0778	.0503
	1	.9774	.9185	.8352	.7373	.6328	.5282	.4284	.3370	.2562
	2	.9988	.9924	.9734	.9421	.8965	.8369	.7648	.6826	.5931
	3	1.0000	.9995	.9978	.9933	.9844	.9692	.9460	.9130	.8688
	4	1.0000	1.0000	.9999	.9997	.9990	.9976	.9947	.9898	.9815
	5	1.0000	1.0000	1.0000	1.0000	1.0000	1.0000	1.0000	1.0000	1.0000
6	0	.7351	.5314	.3771	.2621	.1780	.1176	.0754	.0467	.0277
	1	.9672	.8857	.7765	.6534	.5339	.4202	.3191	.2333	.1636
	2	.9978	.9842	.9527	.9011	.8306	.7443	.6471	.5443	.4415
	3	.9999	.9987	.9941	.9830	.9624	.9295	.8826	.8208	.7447
	4	1.0000	.9999	.9996	.9984	.9954	.9891	.9777	.9590	.9308
	5	1.0000	1.0000	1.0000	.9999	.9998	.9993	.9982	.9959	.9917
	6	1.0000	1.0000	1.0000	1.0000	1.0000	1.0000	1.0000	1.0000	1.0000
7	0	.6983	.4783	.3206	.2097	.1335	.0824	.0490	.0280	.0152
	1	.9556	.8503	.7166	.5767	.4449	.3294	.2338	.1586	.1024
	2	.9962	.9743	.9262	.8520	.7564	.6471	.5323	.4199	.3164
	3	.9998	.9973	.9879	.9667	.9294	.8740	.8002	.7102	.6083
	4	1.0000	.9998	.9988	.9953	.9871	.9712	.9444	.9037	.8643
	5	1.0000	1.0000	.9999	.9996	.9987	.9962	.9910	.9812	.9643
	6	1.0000	1.0000	1.0000	1.0000	.9999	.9998	.9994	.9984	.9963
	7	1.0000	1.0000	1.0000	1.0000	1.0000	1.0000	1.0000	1.0000	1.0000

n	k	$\pi = .50$	.55	.60	.65	.70	.75	.80	.85	.90	.95
1	0	.5000	.4500	.4000	.3500	.3000	.2500	.2000	.1500	.1000	.0500
	1	1.0000	1.0000	1.0000	1.0000	1.0000	1.0000	1.0000	1.0000	1.0000	1.0000
2	0	.2500	.2025	.1600	.1225	.0900	.0625	.0400	.0225	.0100	.0025
	1	.7500	.6975	.6400	.5775	.5100	.4373	.3600	.2775	.1900	.0975
	2	1.0000	1.0000	1.0000	1.0000	1.0000	1.0000	1.0000	1.0000	1.0000	1.0000
3	0	.1250	.0911	.0640	.0429	.0270	.0156	.0080	.0034	.0010	.0001
	1	.5000	.4252	.3520	.2818	.2160	.1562	.1040	.0608	.0280	.0072
	2	.8750	.8336	.7840	.7254	.6570	.5781	.4880	.3959	.2710	.1426
	3	1.0000	1.0000	1.0000	1.0000	1.0000	1.0000	1.0000	1.0000	1.0000	1.0000
4	0	.0625	.0410	.0256	.0150	.0081	.0039	.0016	.0005	.0001	.0000
	1	.3125	.2415	.1792	.1265	.0837	.0508	.0272	.0120	.0037	.0005
	2	.6875	.6090	.5248	.4370	.3483	.2617	.1808	.1095	.0523	.0140
	3	.9375	.9085	.8704	.8215	.7599	.6836	.5904	.4780	.3439	.1855
	4	1.0000	1.0000	1.0000	1.0000	1.0000	1.0000	1.0000	1.0000	1.0000	1.0000
5	0	.0312	.0185	.0102	.0053	.0024	.0010	.0003	.0001	.0000	.0000
	1	.1875	.1312	.0870	.0540	.0308	.0156	.0067	.0022	.0005	.0000
	2	.5000	.4069	.3174	.2352	.1631	.1035	.0579	.0266	.0086	.0012
	3	.8125	.7438	.6630	.5716	.4718	.3672	.2627	.1648	.0815	.0226
	4	.9688	.9497	.9222	.8840	.8319	.7627	.6723	.5563	.4095	.2262
	5	1.0000	1.0000	1.0000	1.0000	1.0000	1.0000	1.0000	1.0000	1.0000	1.0000
6	0	.0156	.0083	.0041	.0018	.0007	.0002	.0001	.0000	.0000	.0000
	1	.1094	.0692	.0410	.0223	.0109	.0046	.0016	.0004	.0001	.0000
	2	.3438	.2553	.1792	.1174	.0705	.0376	.0170	.0059	.0013	.0001
	3	.6562	.5585	.4557	.3529	.2557	.1694	.0989	.0473	.0158	.0022
	4	.8906	.8364	.7667	.6809	.5798	.4661	.3446	.2235	.1143	.0328
	5	.9844	.9723	.9533	.9246	.8824	.8220	.7379	.6229	.4686	.2649
	6	1.0000	1.0000	1.0000	1.0000	1.0000	1.0000	1.0000	1.0000	1.0000	1.0000
7	0	.0078	.0037	.0016	.0006	.0002	.0001	.0000	.0000	.0000	.0000
	1	.0625	.0357	.0188	.0090	.0038	.0013	.0004	.0001	.0000	.0000
	2	.2266	.1529	.0963	.0556	.0288	.0129	.0047	.0012	.0002	.0000
	3	.5000	.3917	.2898	.1998	.1260	.0706	.0333	.0121	.0027	.0002
	4	.7734	.6836	.5801	.4677	.3529	.2436	.1480	.0738	.0257	.0038
	5	.9375	.8976	.8414	.7662	.6706	.5551	.4233	.2834	.1497	.0444
	6	.9922	.9848	.9720	.9510	.9176	.8665	.7903	.6794	.5217	.3917
	7	1.0000	1.0000	1.0000	1.0000	1.0000	1.0000	1.0000	1.0000	1.0000	1.0000

(Continued)

TABLE A.2   The Binomial Distribution (continued)

n	k	$\pi = .05$	.10	.15	.20	.25	.30	.35	.40	.45
8	0	.6634	.4305	.2725	.1678	.1001	.0576	.0319	.0168	.0084
	1	.9428	.8131	.6572	.5033	.3671	.2553	.1691	.1064	.0632
	2	.9942	.9619	.8948	.7969	.6785	.5518	.4278	.3154	.2201
	3	.9996	.9950	.9786	.9437	.8862	.8059	.7064	.5941	.4770
	4	1.0000	.9996	.9971	.9896	.9727	.9420	.8939	.8263	.7396
	5	1.0000	1.0000	.9998	.9988	.9958	.9887	.9747	.9502	.9115
	6	1.0000	1.0000	1.0000	.9999	.9996	.9987	.9964	.9915	.9819
	7	1.0000	1.0000	1.0000	1.0000	1.0000	.9999	.9998	.9993	.9983
	8	1.0000	1.0000	1.0000	1.0000	1.0000	1.0000	1.0000	1.0000	1.0000
9	0	.6302	.3874	.2316	.1342	.0751	.0404	.0207	.0101	.0046
	1	.9288	.7748	.5995	.4362	.3003	.1960	.1211	.0705	.0385
	2	.9916	.9470	.8591	.7382	.6007	.4628	.3373	.2318	.1495
	3	.9994	.9917	.9661	.9144	.8343	.7297	.6089	.4826	.3614
	4	1.0000	.9991	.9944	.9804	.9511	.9012	.8283	.7334	.6214
	5	1.0000	.9999	.9994	.9969	.9900	.9747	.9464	.9006	.9342
	6	1.0000	1.0000	1.0000	.9997	.9987	.9957	.9888	.9750	.9502
	7	1.0000	1.0000	1.0000	1.0000	.9999	.9996	.9986	.9962	.9909
	8	1.0000	1.0000	1.0000	1.0000	1.0000	1.0000	.9999	.9997	.9992
	9	1.0000	1.0000	1.0000	1.0000	1.0000	1.0000	1.0000	1.0000	1.0000
10	0	.5987	.3487	.1969	.1074	.0563	.0282	.0135	.0060	.0025
	1	.9139	.7361	.5443	.3758	.2440	.1493	.0860	.0464	.0233
	2	.9885	.9298	.8202	.6778	.5256	.3828	.2616	.1673	.0996
	3	.9990	.9872	.9500	.8791	.7759	.6496	.5138	.3823	.2660
	4	.9999	.9984	.9901	.9672	.9219	.8497	.7515	.6331	.5044
	5	1.0000	.9999	.9986	.9936	.9803	.9527	.9051	.8338	.7384
	6	1.0000	1.0000	.9999	.9991	.9965	.9894	.9740	.9452	.8980
	7	1.0000	1.0000	1.0000	.9999	.9996	.9984	.9952	.9877	.9726
	8	1.0000	1.0000	1.0000	1.0000	1.0000	.9999	.9995	.9983	.9955
	9	1.0000	1.0000	1.0000	1.0000	1.0000	1.0000	1.0000	.9999	.9997
	10	1.0000	1.0000	1.0000	1.0000	1.0000	1.0000	1.0000	1.0000	1.0000
11	0	.5688	.3138	.1673	.0859	.0422	.0198	.0088	.0036	.0014
	1	.8981	.6974	.4922	.3221	.1971	.1130	.0606	.0302	.0139
	2	.9848	.9104	.7788	.6174	.4552	.3127	.2001	.1189	.0652
	3	.9984	.9815	.9306	.8389	.7133	.5696	.4256	.2963	.1911
	4	.9999	.9972	.9841	.9496	.8854	.7897	.6683	.5328	.3971
	5	1.0000	.9997	.9973	.9883	.9657	.9218	.8513	.7535	.6331
	6	1.0000	1.0000	.9997	.9980	.9924	.9784	.9499	.9006	.8262
	7	1.0000	1.0000	1.0000	.9998	.9988	.9957	.9878	.9707	.9390
	8	1.0000	1.0000	1.0000	1.0000	.9999	.9994	.9980	.9941	.9852
	9	1.0000	1.0000	1.0000	1.0000	1.0000	1.0000	.9998	.9993	.9978
	10	1.0000	1.0000	1.0000	1.0000	1.0000	1.0000	1.0000	1.0000	.9998
	11	1.0000	1.0000	1.0000	1.0000	1.0000	1.0000	1.0000	1.0000	1.0000

n	k	π = .50	.55	.60	.65	.70	.75	.80	.85	.90	.95
8	0	.0030	.0017	.0007	.0002	.0001	.0000	.0000	.0000	.0000	.0000
	1	.0352	.0181	.0085	.0036	.0013	.0004	.0001	.0000	.0000	.0000
	2	.1445	.0885	.0498	.0253	.0113	.0042	.0012	.0002	.0000	.0000
	3	.3633	.2604	.1737	.1061	.0580	.0273	.0104	.0029	.0004	.0000
	4	.6367	.5230	.4059	.2936	.1941	.1138	.0563	.0214	.0050	.0004
	5	.8555	.7799	.6846	.5722	.4482	.3215	.2031	.1052	.0381	.0058
	6	.9648	.9368	.8936	.8309	.7447	.6329	.4967	.3428	.1869	.0572
	7	.9961	.9916	.9832	.9681	.9424	.8999	.8322	.7275	.5695	.3366
	8	1.0000	1.0000	1.0000	1.0000	1.0000	1.0000	1.0000	1.0000	1.0000	1.0000
9	0	.0020	.0008	.0003	.0001	.0000	.0000	.0000	.0000	.0000	.0000
	1	.0195	.0091	.0038	.0014	.0004	.0001	.0000	.0000	.0000	.0000
	2	.0898	.0498	.0250	.0112	.0043	.0013	.0003	.0000	.0000	.0000
	3	.2539	.1658	.0994	.0536	.0253	.0100	.0031	.0006	.0001	.0000
	4	.5000	.3786	.2666	.1717	.0988	.0489	.0196	.0056	.0009	.0000
	5	.7461	.6383	.5174	.3911	.2703	.1657	.0856	.0339	.0083	.0006
	6	.9102	.8505	.7682	.6627	.5372	.3993	.2618	.1409	.0530	.0084
	7	.9805	.9615	.9295	.8789	.8040	.6997	.5638	.4005	.2252	.0712
	8	.9980	.9954	.9899	.9793	.9596	.9249	.8658	.7684	.6126	.3698
	9	1.0000	1.0000	1.0000	1.0000	1.0000	1.0000	1.0000	1.0000	1.0000	1.0000
10	0	.0010	.0003	.0001	.0000	.0000	.0000	.0000	.0000	.0000	.0000
	1	.0107	.0045	.0017	.0005	.0001	.0000	.0000	.0000	.0000	.0000
	2	.0547	.0274	.0123	.0048	.0016	.0004	.0001	.0000	.0000	.0000
	3	.1719	.1020	.0548	.0260	.0106	.0035	.0009	.0001	.0000	.0000
	4	.3770	.2616	.1662	.0949	.0473	.0197	.0064	.0014	.0001	.0000
	5	.6230	.4956	.3669	.2485	.1503	.0781	.0328	.0099	.0016	.0001
	6	.8281	.7340	.6177	.4862	.3504	.2241	.1209	.0500	.0128	.0010
	7	.9453	.9004	.8327	.7184	.6172	.4744	.3222	.1798	.0702	.0115
	8	.9893	.9767	.9536	.9140	.8507	.7560	.6242	.4557	.2639	.0861
	9	.9990	.9975	.9940	.9865	.9718	.9437	.8926	.8031	.6513	.4013
	10	1.0000	1.0000	1.0000	1.0000	1.0000	1.0000	1.0000	1.0000	1.0000	1.0000
11	0	.0005	.0002	.0000	.0000	.0000	.0000	.0000	.0000	.0000	.0000
	1	.0059	.0022	.0007	.0002	.0000	.0000	.0000	.0000	.0000	.0000
	2	.0327	.0148	.0059	.0020	.0006	.0001	.0000	.0000	.0000	.0000
	3	.1133	.0610	.0293	.0122	.0043	.0012	.0002	.0000	.0000	.0000
	4	.2744	.1738	.0994	.0501	.0216	.0076	.0020	.0003	.0000	.0000
	5	.5000	.3669	.2465	.1487	.0782	.0343	.0117	.0027	.0003	.0000
	6	.7256	.6029	.4672	.3317	.2103	.1146	.0504	.0159	.0028	.0001
	7	.8867	.8089	.7037	.5744	.4304	.2867	.1611	.0694	.0185	.0016
	8	.9673	.9348	.8811	.7999	.6873	.5448	.3826	.2212	.0896	.0152
	9	.9941	.9861	.9698	.9394	.8870	.8029	.6779	.5078	.3026	.1019
	10	.9995	.9986	.9964	.9912	.9802	.9578	.9141	.8327	.6862	.4312
	11	1.0000	1.0000	1.0000	1.0000	1.0000	1.0000	1.0000	1.0000	1.0000	1.0000

(Continued)

TABLE A.2   The Binomial Distribution (continued)

n	k	π = .05	.10	.15	.20	.25	.30	.35	.40	.45
12	0	.5404	.2824	.1422	.0687	.0317	.0138	.0057	.0022	.0008
	1	.8816	.6590	.4435	.2749	.1584	.0850	.0424	.0196	.0083
	2	.9804	.8891	.7358	.5583	.3907	.2528	.1513	.0834	.0421
	3	.9978	.9744	.9078	.7946	.6488	.4925	.3467	.2253	.1345
	4	.9998	.9956	.9761	.9274	.8424	.7237	.5833	.4382	.3044
	5	1.0000	.9995	.9954	.9806	.9456	.8822	.7873	.6652	.5269
	6	1.0000	.9999	.9993	.9961	.9857	.9614	.9154	.8418	.7393
	7	1.0000	1.0000	.9999	.9994	.9972	.9905	.9745	.9427	.8883
	8	1.0000	1.0000	1.0000	.9999	.9996	.9983	.9944	.9847	.9644
	9	1.0000	1.0000	1.0000	1.0000	1.0000	.9998	.9992	.9972	.9921
	10	1.0000	1.0000	1.0000	1.0000	1.0000	1.0000	.9999	.9997	.9989
	11	1.0000	1.0000	1.0000	1.0000	1.0000	1.0000	1.0000	1.0000	.9999
	12	1.0000	1.0000	1.0000	1.0000	1.0000	1.0000	1.0000	1.0000	1.0000
13	0	.5133	.2542	.1209	.0550	.0238	.0097	.0037	.0013	.0004
	1	.8646	.6213	.3983	.2336	.1267	.0637	.0296	.0126	.0049
	2	.9755	.8661	.7296	.5017	.3326	.2025	.1132	.0579	.0269
	3	.9969	.9658	.9033	.7473	.5843	.4206	.2783	.1686	.0929
	4	.9997	.9935	.9740	.9009	.7940	.6543	.5005	.3530	.2279
	5	1.0000	.9991	.9947	.9700	.9198	.8346	.7159	.5744	.4268
	6	1.0000	.9999	.9987	.9930	.9757	.9376	.8705	.7712	.6437
	7	1.0000	1.0000	.9998	.9988	.9944	.9818	.9538	.9023	.8212
	8	1.0000	1.0000	1.0000	.9998	.9990	.9960	.9874	.9679	.9302
	9	1.0000	1.0000	1.0000	1.0000	.9999	.9993	.9975	.9922	.9797
	10	1.0000	1.0000	1.0000	1.0000	1.0000	.9999	.9997	.9987	.9959
	11	1.0000	1.0000	1.0000	1.0000	1.0000	1.0000	1.0000	.9999	.9995
	12	1.0000	1.0000	1.0000	1.0000	1.0000	1.0000	1.0000	1.0000	1.0000
	13	1.0000	1.0000	1.0000	1.0000	1.0000	1.0000	1.0000	1.0000	1.0000
14	0	.4877	.2288	.1028	.0440	.0178	.0068	.0024	.0008	.0002
	1	.8470	.5846	.3567	.1979	.1010	.0475	.0205	.0081	.0029
	2	.9699	.8416	.6479	.4481	.2811	.1608	.0839	.0398	.0170
	3	.9958	.9559	.8535	.6982	.5213	.3552	.2205	.1243	.0632
	4	.9996	.9908	.9533	.8702	.7415	.5842	.4227	.2793	.1672
	5	1.0000	.9985	.9885	.9561	.8883	.7805	.6405	.4859	.3373
	6	1.0000	.9998	.9978	.9884	.9617	.9067	.8164	.6925	.5461
	7	1.0000	1.0000	.9997	.9976	.9897	.9685	.9247	.8499	.7414
	8	1.0000	1.0000	1.0000	.9996	.9978	.9917	.9757	.9417	.8811
	9	1.0000	1.0000	1.0000	1.0000	.9997	.9983	.9940	.9825	.9574
	10	1.0000	1.0000	1.0000	1.0000	1.0000	.9998	.9989	.9961	.9886
	11	1.0000	1.0000	1.0000	1.0000	1.0000	1.0000	.9999	.9994	.9978
	12	1.0000	1.0000	1.0000	1.0000	1.0000	1.0000	1.0000	.9999	.9997
	13	1.0000	1.0000	1.0000	1.0000	1.0000	1.0000	1.0000	1.0000	1.0000
	14	1.0000	1.0000	1.0000	1.0000	1.0000	1.0000	1.0000	1.0000	1.0000

n	k	$\pi = .50$	.55	.60	.65	.70	.75	.80	.85	.90	.95
12	0	.0002	.0001	.0000	.0000	.0000	.0000	.0000	.0000	.0000	.0000
	1	.0032	.0011	.0003	.0001	.0000	.0000	.0000	.0000	.0000	.0000
	2	.0193	.0079	.0028	.0008	.0002	.0000	.0000	.0000	.0000	.0000
	3	.0730	.0356	.0153	.0056	.0017	.0004	.0001	.0000	.0000	.0000
	4	.1938	.1117	.0573	.0255	.0095	.0028	.0006	.0001	.0000	.0000
	5	.3872	.2607	.1582	.0846	.0386	.0143	.0039	.0007	.0001	.0000
	6	.6128	.4731	.3348	.2127	.1178	.0544	.0194	.0046	.0005	.0000
	7	.8062	.6956	.5618	.4167	.2763	.1576	.0726	.0239	.0043	.0002
	8	.9270	.8655	.7747	.6533	.5075	.3512	.2054	.0922	.0256	.0022
	9	.9807	.9579	.9166	.8487	.7472	.6093	.4417	.2642	.1109	.0196
	10	.9968	.9917	.9804	.9576	.9150	.8416	.7251	.5565	.3410	.1184
	11	.9998	.9992	.9978	.9943	.9862	.9683	.9313	.8578	.7176	.4596
	12	1.0000	1.0000	1.0000	1.0000	1.0000	1.0000	1.0000	1.0000	1.0000	1.0000
13	0	.0001	.0000	.0000	.0000	.0000	.0000	.0000	.0000	.0000	.0000
	1	.0017	.0005	.0001	.0000	.0000	.0000	.0000	.0000	.0000	.0000
	2	.0112	.0041	.0013	.0003	.0001	.0000	.0000	.0000	.0000	.0000
	3	.0461	.0203	.0078	.0025	.0007	.0001	.0000	.0000	.0000	.0000
	4	.1334	.0698	.0321	.0126	.0040	.0010	.0002	.0000	.0000	.0000
	5	.2905	.1788	.0977	.0462	.0182	.0056	.0012	.0002	.0000	.0000
	6	.5000	.3563	.2288	.1295	.0624	.0243	.0070	.0013	.0001	.0000
	7	.7095	.5732	.4256	.2841	.1654	.0802	.0300	.0053	.0009	.0000
	8	.8666	.7721	.6470	.4995	.3457	.2060	.0991	.0260	.0065	.0003
	9	.9539	.9071	.8314	.7217	.5794	.4157	.2527	.0967	.0342	.0031
	10	.9888	.9731	.9421	.8868	.7975	.6674	.4983	.2704	.1339	.0245
	11	.9983	.9951	.9874	.9704	.9363	.8733	.7664	.6017	.3787	.1354
	12	.9999	.9996	.9987	.9963	.9903	.9762	.9450	.8791	.7458	.4867
	13	1.0000	1.0000	1.0000	1.0000	1.0000	1.0000	1.0000	1.0000	1.0000	1.0000
14	0	.0000	.0000	.0000	.0000	.0000	.0000	.0000	.0000	.0000	.0000
	1	.0009	.0003	.0001	.0000	.0000	.0000	.0000	.0000	.0000	.0000
	2	.0065	.0022	.0006	.0001	.0000	.0000	.0000	.0000	.0000	.0000
	3	.0287	.0114	.0039	.0011	.0002	.0000	.0000	.0000	.0000	.0000
	4	.0898	.0462	.0175	.0060	.0017	.0003	.0000	.0000	.0000	.0000
	5	.2120	.1189	.0583	.0243	.0083	.0022	.0004	.0000	.0000	.0000
	6	.3953	.2586	.1501	.0753	.0315	.0108	.0024	.0003	.0000	.0000
	7	.6047	.4539	.3075	.1836	.0933	.0383	.0116	.0022	.0002	.0000
	8	.7880	.6627	.5141	.3595	.2195	.1117	.0439	.0115	.0015	.0000
	9	.9102	.8328	.7207	.5773	.4158	.2585	.1298	.0467	.0092	.0004
	10	.9713	.9368	.8757	.7795	.6448	.4787	.3018	.1465	.0441	.0042
	11	.9935	.9830	.9602	.9161	.8392	.7189	.5519	.3521	.1584	.0301
	12	.9991	.9971	.9919	.9795	.9525	.8990	.8021	.6433	.4154	.1530
	13	.9999	.9998	.9992	.9976	.9932	.9822	.9560	.8972	.7712	.5123
	14	1.0000	1.0000	1.0000	1.0000	1.0000	1.0000	1.0000	1.0000	1.0000	1.0000

(Continued)

TABLE A.2 The Binomial Distribution (continued)

n	k	π = .05	.10	.15	.20	.25	.30	.35	.40	.45
15	0	.4633	.2059	.0874	.0352	.0134	.0047	.0016	.0005	.0001
	1	.8290	.5490	.3186	.1671	.0802	.0353	.0142	.0052	.0017
	2	.9638	.8159	.6042	.3980	.2361	.1268	.0617	.0271	.0107
	3	.9945	.9444	.8227	.6482	.4613	.2969	.1727	.0905	.0424
	4	.9994	.9873	.9383	.8358	.6865	.5155	.3519	.2173	.1204
	5	.9999	.9978	.9832	.9389	.8516	.7216	.5643	.4032	.2608
	6	1.0000	.9997	.9964	.9819	.9434	.8689	.7548	.6098	.4522
	7	1.0000	1.0000	.9994	.9958	.9827	.9500	.8868	.7869	.6535
	8	1.0000	1.0000	.9999	.9992	.9958	.9848	.9578	.9050	.8121
	9	1.0000	1.0000	1.0000	.9999	.9992	.9963	.9876	.9662	.9231
	10	1.0000	1.0000	1.0000	1.0000	.9999	.9993	.9972	.9907	.9745
	11	1.0000	1.0000	1.0000	1.0000	1.0000	.9999	.9995	.9981	.9937
	12	1.0000	1.0000	1.0000	1.0000	1.0000	1.0000	.9999	.9997	.9989
	13	1.0000	1.0000	1.0000	1.0000	1.0000	1.0000	1.0000	1.0000	.9999
	14	1.0000	1.0000	1.0000	1.0000	1.0000	1.0000	1.0000	1.0000	1.0000
	15	1.0000	1.0000	1.0000	1.0000	1.0000	1.0000	1.0000	1.0000	1.0000
16	0	.4401	.1853	.0743	.0281	.0100	.0033	.0010	.0003	.0001
	1	.8108	.5147	.2839	.1407	.0635	.0261	.0098	.0088	.0010
	2	.9571	.7892	.5614	.3518	.1971	.0904	.0451	.0183	.0066
	3	.9930	.9316	.7899	.5981	.4050	.2459	.1339	.0651	.0281
	4	.9991	.9830	.9209	.7982	.6302	.4499	.2892	.1666	.0853
	5	.9999	.9967	.9765	.9183	.8103	.6598	.4900	.3288	.1970
	6	1.0000	.9995	.9944	.9733	.9204	.8247	.6881	.5272	.3660
	7	1.0000	.9999	.9989	.9930	.9729	.9256	.8406	.7161	.5629
	8	1.0000	1.0000	.9998	.9985	.9925	.9743	.9329	.8577	.7441
	9	1.0000	1.0000	1.0000	.9998	.9984	.9938	.9809	.9514	.8759
	10	1.0000	1.0000	1.0000	1.0000	.9997	.9984	.9938	.9809	.9514
	11	1.0000	1.0000	1.0000	1.0000	1.0000	.9997	.9987	.9951	.9851
	12	1.0000	1.0000	1.0000	1.0000	1.0000	1.0000	.9998	.9991	.9965
	13	1.0000	1.0000	1.0000	1.0000	1.0000	1.0000	1.0000	.9999	.9994
	14	1.0000	1.0000	1.0000	1.0000	1.0000	1.0000	1.0000	1.0000	.9999
	15	1.0000	1.0000	1.0000	1.0000	1.0000	1.0000	1.0000	1.0000	1.0000
	16	1.0000	1.0000	1.0000	1.0000	1.0000	1.0000	1.0000	1.0000	1.0000

n	k	$\pi = .50$	.55	.60	.65	.70	.75	.80	.85	.90	.95
15	0	.0000	.0000	.0000	.0000	.0000	.0000	.0000	.0000	.0000	.0000
	1	.0005	.0001	.0000	.0000	.0000	.0000	.0000	.0000	.0000	.0000
	2	.0037	.0011	.0003	.0001	.0000	.0000	.0000	.0000	.0000	.0000
	3	.0176	.0063	.0019	.0005	.0001	.0000	.0000	.0000	.0000	.0000
	4	.0592	.0255	.0093	.0028	.0007	.0001	.0000	.0000	.0000	.0000
	5	.1509	.0769	.0228	.0124	.0037	.0008	.0001	.0000	.0000	.0000
	6	.3036	.1818	.0950	.0422	.0152	.0042	.0008	.0001	.0000	.0000
	7	.5000	.3465	.2131	.1132	.0500	.0173	.0042	.0006	.0000	.0000
	8	.6964	.5478	.3902	.2452	.1311	.0566	.0181	.0036	.0003	.0000
	9	.8491	.7392	.5968	.4357	.2784	.1484	.0611	.0168	.0022	.0001
	10	.9408	.8796	.7827	.6481	.4845	.3135	.1642	.0617	.0127	.0006
	11	.9824	.9576	.9095	.8273	.7031	.5387	.3518	.1773	.0556	.0055
	12	.9963	.9893	.9729	.9383	.8732	.7639	.6020	.3958	.1841	.0362
	13	.9995	.9983	.9948	.9858	.9647	.9198	.8329	.6814	.4510	.1710
	14	1.0000	.9999	.9995	.9984	.9953	.9866	.9648	.9126	.7941	.5367
	15	1.0000	1.0000	1.0000	1.0000	1.0000	1.0000	1.0000	1.0000	1.0000	1.0000
16	0	.0000	.0000	.0000	.0000	.0000	.0000	.0000	.0000	.0000	.0000
	1	.0003	.0001	.0000	.0000	.0000	.0000	.0000	.0000	.0000	.0000
	2	.0021	.0006	.0001	.0000	.0000	.0000	.0000	.0000	.0000	.0000
	3	.0106	.0035	.0009	.0002	.0000	.0000	.0000	.0000	.0000	.0000
	4	.0384	.0149	.0049	.0013	.0003	.0000	.0000	.0000	.0000	.0000
	5	.1051	.0486	.0191	.0062	.0016	.0003	.0000	.0000	.0000	.0000
	6	.2272	.1241	.0583	.0229	.0071	.0016	.0002	.0000	.0000	.0000
	7	.4018	.2559	.1423	.0671	.0257	.0075	.0015	.0002	.0000	.0000
	8	.5982	.4371	.2839	.1594	.0744	.0271	.0070	.0011	.0001	.0000
	9	.7728	.6340	.4728	.3119	.1753	.0795	.0267	.0056	.0005	.0000
	10	.8949	.8024	.6712	.5100	.3402	.1897	.0817	.0235	.0033	.0001
	11	.9616	.9147	.8334	.7108	.5501	.3698	.2018	.0791	.0170	.0009
	12	.9894	.9719	.9349	.8661	.7541	.5950	.4019	.2101	.0684	.0070
	13	.9979	.9934	.9817	.9549	.9006	.8729	.6482	.4386	.2108	.0429
	14	.9997	.9990	.9967	.9902	.9739	.9365	.8593	.7176	.4853	.1892
	15	1.0000	.9999	.9997	.9990	.9967	.9900	.9719	.9257	.8147	.5599
	16	1.0000	1.0000	1.0000	1.0000	1.0000	1.0000	1.0000	1.0000	1.0000	1.0000

(Continued)

TABLE A.2   The Binomial Distribution (continued)

n	k	$\pi = .05$	.10	.15	.20	.25	.30	.35	.40	.45
17	0	.4181	.1668	.0631	.0225	.0075	.0023	.0007	.0002	.0000
	1	.7922	.4818	.2525	.1182	.0501	.0193	.0067	.0021	.0006
	2	.9497	.7618	.5198	.3096	.1637	.0774	.0327	.0123	.0041
	3	.9912	.9174	.7556	.5489	.3530	.2019	.1028	.0464	.0184
	4	.9988	.9779	.9013	.7582	.5739	.3887	.2348	.1260	.0596
	5	.9999	.9953	.9681	.8943	.7653	.5968	.4197	.2639	.1471
	6	1.0000	.9992	.9917	.9623	.8929	.7752	.6188	.4478	.2902
	7	1.0000	.9999	.9983	.9891	.9598	.8954	.7872	.6405	.4743
	8	1.0000	1.0000	.9997	.9974	.9876	.9597	.9006	.8011	.6626
	9	1.0000	1.0000	1.0000	.9995	.9969	.9873	.9611	.9081	.8166
	10	1.0000	1.0000	1.0000	.9999	.9994	.9968	.9880	.9652	.9174
	11	1.0000	1.0000	1.0000	1.0000	.9999	.9993	.9970	.9894	.9699
	12	1.0000	1.0000	1.0000	1.0000	1.0000	.9999	.9994	.9975	.9914
	13	1.0000	1.0000	1.0000	1.0000	1.0000	1.0000	.9999	.9995	.9981
	14	1.0000	1.0000	1.0000	1.0000	1.0000	1.0000	1.0000	.9999	.9997
	15	1.0000	1.0000	1.0000	1.0000	1.0000	1.0000	1.0000	1.0000	1.0000
	16	1.0000	1.0000	1.0000	1.0000	1.0000	1.0000	1.0000	1.0000	1.0000
	17	1.0000	1.0000	1.0000	1.0000	1.0000	1.0000	1.0000	1.0000	1.0000
18	0	.3972	.1501	.0536	.0180	.0056	.0016	.0004	.0001	.0000
	1	.7735	.4503	.2241	.0991	.0395	.0142	.0046	.0013	.0003
	2	.9419	.7338	.4797	.2713	.1353	.0600	.0236	.0082	.0025
	3	.9891	.9018	.7202	.5010	.3057	.1646	.0783	.0328	.0120
	4	.9985	.9718	.8794	.7164	.5187	.3327	.1886	.0942	.0411
	5	.9998	.9936	.9581	.8671	.7175	.5344	.3550	.2088	.1077
	6	1.0000	.9988	.9882	.9487	.8610	.7217	.5491	.3543	.2258
	7	1.0000	.9998	.9973	.9837	.9431	.8593	.7283	.5634	.3915
	8	1.0000	1.0000	.9995	.9957	.9807	.9404	.8609	.7368	.5778
	9	1.0000	1.0000	.9999	.9991	.9946	.9790	.9403	.8653	.7473
	10	1.0000	1.0000	1.0000	.9998	.9988	.9939	.9788	.9424	.8720
	11	1.0000	1.0000	1.0000	1.0000	.9998	.9985	.9938	.9797	.9463
	12	1.0000	1.0000	1.0000	1.0000	1.0000	.9997	.9986	.9942	.9817
	13	1.0000	1.0000	1.0000	1.0000	1.0000	1.0000	.9997	.9987	.9951
	14	1.0000	1.0000	1.0000	1.0000	1.0000	1.0000	1.0000	.9998	.9990
	15	1.0000	1.0000	1.0000	1.0000	1.0000	1.0000	1.0000	1.0000	.9999
	16	1.0000	1.0000	1.0000	1.0000	1.0000	1.0000	1.0000	1.0000	1.0000
	17	1.0000	1.0000	1.0000	1.0000	1.0000	1.0000	1.0000	1.0000	1.0000
	18	1.0000	1.0000	1.0000	1.0000	1.0000	1.0000	1.0000	1.0000	1.0000

n	k	π = .50	.55	.60	.65	.70	.75	.80	.85	.90	.95
17	0	.0000	.0000	.0000	.0000	.0000	.0000	.0000	.0000	.0000	.0000
	1	.0001	.0000	.0000	.0000	.0000	.0000	.0000	.0000	.0000	.0000
	2	.0012	.0003	.0001	.0000	.0000	.0000	.0000	.0000	.0000	.0000
	3	.0064	.0019	.0005	.0001	.0000	.0000	.0000	.0000	.0000	.0000
	4	.0245	.0086	.0025	.0006	.0001	.0000	.0000	.0000	.0000	.0000
	5	.0717	.0301	.0106	.0030	.0007	.0001	.0000	.0000	.0000	.0000
	6	.1662	.0826	.0348	.0120	.0032	.0006	.0001	.0000	.0000	.0000
	7	.3145	.1834	.0919	.0383	.0127	.0031	.0005	.0000	.0000	.0000
	8	.5000	.3374	.1989	.0994	.0403	.0124	.0026	.0003	.0000	.0000
	9	.6855	.5257	.3595	.2128	.1046	.0402	.0109	.0017	.0001	.0000
	10	.8338	.7098	.5522	.3812	.2248	.1071	.0377	.0083	.0008	.0000
	11	.9283	.8529	.7361	.5803	.4032	.2347	.1057	.0319	.0047	.0001
	12	.9755	.9404	.8740	.7652	.6113	.4261	.2418	.0987	.0221	.0012
	13	.9936	.9816	.9536	.8972	.7981	.6470	.4511	.2444	.0826	.0088
	14	.9988	.9959	.9877	.9673	.9226	.8363	.6904	.4802	.2382	.0503
	15	.9999	.9994	.9979	.9933	.9807	.9499	.8818	.7475	.5182	.2078
	16	1.0000	1.0000	.9998	.9993	.9977	.9925	.9775	.9369	.8332	.5819
	17	1.0000	1.0000	1.0000	1.0000	1.0000	1.0000	1.0000	1.0000	1.0000	1.0000
18	0	.0000	.0000	.0000	.0000	.0000	.0000	.0000	.0000	.0000	.0000
	1	.0001	.0000	.0000	.0000	.0000	.0000	.0000	.0000	.0000	.0000
	2	.0007	.0001	.0000	.0000	.0000	.0000	.0000	.0000	.0000	.0000
	3	.0038	.0010	.0002	.0000	.0000	.0000	.0000	.0000	.0000	.0000
	4	.0154	.0049	.0013	.0003	.0000	.0000	.0000	.0000	.0000	.0000
	5	.0481	.0183	.0058	.0014	.0003	.0000	.0000	.0000	.0000	.0000
	6	.1189	.0537	.0203	.0062	.0014	.0002	.0000	.0000	.0000	.0000
	7	.2403	.1280	.0576	.0212	.0061	.0012	.0002	.0000	.0000	.0000
	8	.4073	.2527	.1347	.0597	.0210	.0054	.0009	.0001	.0000	.0000
	9	.5927	.4222	.2632	.1391	.0694	.0193	.0043	.0005	.0000	.0000
	10	.7597	.6085	.4366	.2717	.1407	.0569	.0163	.0027	.0002	.0000
	11	.8811	.7742	.6457	.4509	.2783	.1390	.0513	.0118	.0012	.0000
	12	.9519	.8923	.7912	.6450	.4656	.2825	.1329	.0419	.0064	.0002
	13	.9846	.9589	.9058	.8114	.6673	.4813	.2836	.1206	.0282	.0015
	14	.9962	.9880	.9672	.9217	.8354	.6943	.4990	.2798	.0982	.0109
	15	.9993	.9975	.9918	.9764	.9400	.8647	.7287	.5203	.2662	.0581
	16	.9999	.9997	.9987	.9954	.9858	.9605	.9009	.7759	.5497	.2265
	17	1.0000	1.0000	.9999	.9996	.9984	.9944	.9820	.9464	.8499	.6028
	18	1.0000	1.0000	1.0000	1.0000	1.0000	1.0000	1.0000	1.0000	1.0000	1.0000

(Continued)

TABLE A.2   The Binomial Distribution (continued)

n	k	π = .05	.10	.15	.20	.25	.30	.35	.40	.45
19	0	.3774	.1351	.0456	.0144	.0042	.0011	.0003	.0001	.0000
	1	.7547	.4203	.1985	.0829	.0310	.0104	.0031	.0008	.0002
	2	.9335	.7054	.4413	.2369	.1113	.0462	.0170	.0055	.0015
	3	.9869	.8850	.6841	.4551	.2631	.1332	.0591	.0230	.0077
	4	.9981	.9648	.8555	.6733	.4654	.2823	.1500	.0697	.0280
	5	.9998	.9914	.9463	.8369	.6678	.4739	.2968	.1629	.0777
	6	1.0000	.9983	.9837	.9324	.8251	.6655	.4912	.3081	.1727
	7	1.0000	.9997	.9959	.9767	.9225	.8180	.6656	.4878	.3169
	8	1.0000	1.0000	.9992	.9933	.9713	.9161	.8145	.6675	.4940
	9	1.0000	1.0000	.9999	.9984	.9911	.9674	.9125	.8139	.6710
	10	1.0000	1.0000	1.0000	.9997	.9977	.9895	.9653	.9115	.8159
	11	1.0000	1.0000	1.0000	1.0000	.9995	.9972	.9886	.9648	.9129
	12	1.0000	1.0000	1.0000	1.0000	.9999	.9994	.9969	.9884	.9658
	13	1.0000	1.0000	1.0000	1.0000	1.0000	.9999	.9993	.9969	.9891
	14	1.0000	1.0000	1.0000	1.0000	1.0000	1.0000	.9999	.9994	.9972
	15	1.0000	1.0000	1.0000	1.0000	1.0000	1.0000	1.0000	.9999	.9995
	16	1.0000	1.0000	1.0000	1.0000	1.0000	1.0000	1.0000	1.0000	.9999
	17	1.0000	1.0000	1.0000	1.0000	1.0000	1.0000	1.0000	1.0000	1.0000
	18	1.0000	1.0000	1.0000	1.0000	1.0000	1.0000	1.0000	1.0000	1.0000
	19	1.0000	1.0000	1.0000	1.0000	1.0000	1.0000	1.0000	1.0000	1.0000
20	0	.3585	.1261	.0388	.0115	.0032	.0008	.0000	.0000	.0000
	1	.7358	.3917	.1756	.0692	.0243	.0076	.0021	.0005	.0001
	2	.9245	.6769	.4049	.2061	.0913	.0355	.0121	.0036	.0009
	3	.9841	.8670	.6477	.4114	.2252	.1071	.0444	.0160	.0049
	4	.9974	.9568	.8298	.6296	.4148	.2375	.1182	.0510	.0189
	5	.9997	.9887	.9327	.8042	.6172	.4164	.2454	.1256	.0553
	6	1.0000	.9976	.9781	.9133	.7858	.6080	.4166	.2500	.1299
	7	1.0000	.9996	.9941	.9679	.8982	.7723	.6010	.4159	.2520
	8	1.0000	.9999	.9987	.9900	.9591	.8867	.7624	.5956	.4143
	9	1.0000	1.0000	.9998	.9974	.9861	.9520	.8782	.7553	.5914
	10	1.0000	1.0000	1.0000	.9994	.9961	.9829	.9468	.8725	.7507
	11	1.0000	1.0000	1.0000	.9999	.9991	.9949	.9804	.9435	.8692
	12	1.0000	1.0000	1.0000	1.0000	.9998	.9987	.9940	.9790	.9420
	13	1.0000	1.0000	1.0000	1.0000	1.0000	.9997	.9985	.9935	.9786
	14	1.0000	1.0000	1.0000	1.0000	1.0000	1.0000	.9997	.9984	.9936
	15	1.0000	1.0000	1.0000	1.0000	1.0000	1.0000	1.0000	.9997	.9985
	16	1.0000	1.0000	1.0000	1.0000	1.0000	1.0000	1.0000	1.0000	.9997
	17	1.0000	1.0000	1.0000	1.0000	1.0000	1.0000	1.0000	1.0000	1.0000
	18	1.0000	1.0000	1.0000	1.0000	1.0000	1.0000	1.0000	1.0000	1.0000
	19	1.0000	1.0000	1.0000	1.0000	1.0000	1.0000	1.0000	1.0000	1.0000
	20	1.0000	1.0000	1.0000	1.0000	1.0000	1.0000	1.0000	1.0000	1.0000

n	k	$\pi = .50$	.55	.60	.65	.70	.75	.80	.85	.90	.95
19	0	.0000	.0000	.0000	.0000	.0000	.0000	.0000	.0000	.0000	.0000
	1	.0000	.0000	.0000	.0000	.0000	.0000	.0000	.0000	.0000	.0000
	2	.0004	.0001	.0000	.0000	.0000	.0000	.0000	.0000	.0000	.0000
	3	.0022	.0005	.0001	.0000	.0000	.0000	.0000	.0000	.0000	.0000
	4	.0096	.0028	.0006	.0001	.0000	.0000	.0000	.0000	.0000	.0000
	5	.0318	.0109	.0031	.0007	.0001	.0000	.0000	.0000	.0000	.0000
	6	.0835	.0342	.0116	.0031	.0006	.0001	.0000	.0000	.0000	.0000
	7	.1796	.0871	.0352	.0114	.0028	.0005	.0000	.0000	.0000	.0000
	8	.3238	.1841	.0885	.0347	.0105	.0023	.0003	.0000	.0000	.0000
	9	.5000	.3290	.1861	.0875	.0326	.0089	.0016	.0001	.0000	.0000
	10	.6762	.5060	.3325	.1855	.0839	.0287	.0067	.0008	.0000	.0000
	11	.8204	.6831	.5122	.3344	.1820	.0775	.0233	.0041	.0008	.0000
	12	.9165	.8273	.6919	.5188	.3345	.1749	.0676	.0163	.0017	.0000
	13	.9682	.9223	.8371	.7032	.5261	.3322	.1631	.0537	.0086	.0002
	14	.9904	.9720	.9304	.8500	.7178	.5346	.3267	.1444	.0352	.0020
	15	.9978	.9923	.9770	.9409	.8668	.7369	.5449	.3159	.1150	.0132
	16	.9996	.9985	.9945	.9830	.9538	.8887	.7631	.5587	.2946	.0665
	17	1.0000	.9998	.9992	.9969	.9896	.9690	.9171	.8015	.5797	.2453
	18	1.0000	1.0000	.9999	.9997	.9989	.9958	.9856	.9544	.8649	.6226
	19	1.0000	1.0000	1.0000	1.0000	1 0000	1.0000	1.0000	1.0000	1.0000	1.0000
20	0	.0000	.0000	.0000	.0000	.0000	.0000	.0000	.0000	.0000	.0000
	1	.0000	.0000	.0000	.0000	.0000	.0000	.0000	.0000	.0000	.0000
	2	.0002	.0000	.0000	.0000	.0000	.0000	.0000	.0000	.0000	.0000
	3	.0013	.0003	.0000	.0000	.0000	.0000	.0000	.0000	.0000	.0000
	4	.0059	.0015	.0003	.0000	.0000	.0000	.0000	.0000	.0000	.0000
	5	.0207	.0064	.0016	.0003	.0000	.0000	.0000	.0000	.0000	.0000
	6	.0577	.0214	.0065	.0015	.0003	.0000	.0000	.0000	.0000	.0000
	7	.1316	.0580	.0210	.0060	.0013	.0002	.0000	.0000	.0000	.0000
	8	.2517	.1308	.0565	.0196	.0051	.0009	.0001	.0000	.0000	.0000
	9	.4119	.2493	.1275	.0532	.0171	.0039	.0006	.0000	.0000	.0000
	10	.5881	.4086	.2447	.1218	.0480	.0139	.0026	.0002	.0000	.0000
	11	.7483	.5857	.4044	.2376	.1133	.0409	.0100	.0013	.0001	.0000
	12	.8684	.7480	.5841	.3990	.2277	.1018	.0321	.0059	.0004	.0000
	13	.9423	.8701	.7500	.5834	.3920	.2142	.0867	.0219	.0024	.0000
	14	.9793	.9447	.8744	.7546	.5836	.3828	.1958	.0673	.0113	.0003
	15	.9941	.9811	.9490	.8818	.7625	.5852	.3704	.1702	.0432	.0026
	16	.9987	.9951	.9840	.9556	.8929	.7748	.5886	.3523	.1330	.0159
	17	.9998	.9991	.9964	.9879	.9645	.9087	.7939	.5951	.3231	.0755
	18	1.0000	.9999	.9995	.9979	.9924	.9757	.9308	.8244	.6083	.2642
	19	1.0000	1.0000	1.0000	.9998	.9992	.9968	.9885	.9612	.8784	.6415
	20	1.0000	1.0000	1.0000	1.0000	1.0000	1.0000	1.0000	1.0000	1.0000	1.0000

TABLE A.3   Binomial Coefficients

n	C(n,0)	C(n,1)	C(n,2)	C(n,3)	C(n,4)	C(n,5)	C(n,6)	C(n,7)	C(n,8)	C(n,9)	C(n,10)
0	1										
1	1	1									
2	1	2	1								
3	1	3	3	1							
4	1	4	6	4	1						
5	1	5	10	10	5	1					
6	1	6	15	20	15	6	1				
7	1	7	21	35	35	21	7	1			
8	1	8	28	56	70	56	28	8	1		
9	1	9	36	84	126	126	84	36	9	1	
10	1	10	45	120	210	252	210	120	45	10	1
11	1	11	55	165	330	462	462	330	165	55	11
12	1	12	66	220	495	792	924	792	495	220	66
13	1	13	78	286	715	1,287	1,716	1,716	1,287	715	286
14	1	14	91	364	1,001	2,002	3,003	3,432	3,003	2,002	1,001
15	1	15	105	455	1,365	3,003	5,005	6,435	6,435	5,005	3,003
16	1	16	120	560	1,820	4,368	8,008	11,440	12,870	11,440	8,008
17	1	17	136	680	2,380	6,188	12,376	19,448	24,310	24,310	19,448
18	1	18	153	816	3,060	8,568	18,564	31,824	43,758	48,620	43,758
19	1	19	171	969	3,876	11,628	27,132	50,388	75,582	92,378	92,378
20	1	20	190	1,140	4,845	15,504	38,760	77,520	125,970	167,960	184,756

Note on the usage of this table: Remaining elements of this table beyond $C(n, 10)$ can be determined as follows: $C(n, n-k) = C(n, k)$. For example, $C(19, 14) = C(19, 19-5) = C(19,5) = 11,628$. For values of n larger than 20, Pascal's triangle identity can be used: $C(n, k) = C(n-1, k-1) + C(n-1, k)$. For example, $C(21, 9) = C(20, 8) + C(20, 9) = 125,970 + 167,960 = 293,930$.

TABLE A.4 The Standard Normal Distribution

For a set of data having the standard normal distribution (that is, the normal distribution with mean 0 and standard deviation 1), Table A.4 lists for various values of z the proportion of data falling to the left of z, as in the diagram below. The proportion of data points in the shaded region of the diagram is the number in the column headed "proportion," corresponding to the z value in the column headed "z."

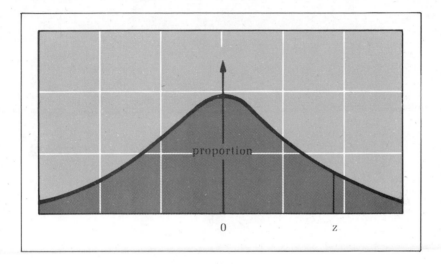

Note on usage of this table: If Z is a random variable having the standard normal distribution, then the number in the "proportion" column of Table A.4 that corresponds to the number z (in the "z" column) is the probability $P(Z \leq z)$, which represents the proportion of the data to the left of z. To find the proportion of data to the right of z, we use the fact that $P(Z > z) = 1 - P(Z \leq z)$. To find the proportion of the data *between* two numbers z_1 and z_2, we observe that $(Pz_1 < Z \leq z_2) = P(Z \leq z_2) - P(Z \leq z_1)$. Therefore, all proportions involving the standard normal distribution can be found by using Table A.4.

Note on the generation of this table: This table was generated by a FORTRAN program, the essential element of which was a 13-term exponential-trigonometric series approximating the standard normal distribution function to nine decimal places whenever $|z| < 7$. Details may be found in the article by P.A.P. Moran, "Calculation of the Normal Distribution Function," *Biometrika*, Vol. 67, No. 3 (1980), pp. 675–676.

TABLE A.4   The Standard Normal Distribution (continued)

Z	PROPORTION	Z	PROPORTION	Z	PROPORTION	Z	PROPORTION	Z	PROPORTION
-4.00	.00003	-3.50	.00023	-3.00	.0013	-2.50	.0062	-2.00	.0228
-3.99	.00003	-3.49	.00024	-2.99	.0014	-2.49	.0064	-1.99	.0233
-3.98	.00003	-3.48	.00025	-2.98	.0014	-2.48	.0066	-1.98	.0239
-3.97	.00004	-3.47	.00026	-2.97	.0015	-2.47	.0068	-1.97	.0244
-3.96	.00004	-3.46	.00027	-2.96	.0015	-2.46	.0069	-1.96	.0250
-3.95	.00004	-3.45	.00028	-2.95	.0016	-2.45	.0071	-1.95	.0256
-3.94	.00004	-3.44	.00029	-2.94	.0016	-2.44	.0073	-1.94	.0262
-3.93	.00004	-3.43	.00030	-2.93	.0017	-2.43	.0075	-1.93	.0268
-3.92	.00004	-3.42	.00031	-2.92	.0017	-2.42	.0078	-1.92	.0274
-3.91	.00005	-3.41	.00032	-2.91	.0018	-2.41	.0080	-1.91	.0281
-3.90	.00005	-3.40	.00034	-2.90	.0019	-2.40	.0082	-1.90	.0287
-3.89	.00005	-3.39	.00035	-2.89	.0019	-2.39	.0084	-1.89	.0294
-3.88	.00005	-3.38	.00036	-2.88	.0020	-2.38	.0087	-1.88	.0301
-3.87	.00005	-3.37	.00038	-2.87	.0021	-2.37	.0089	-1.87	.0307
-3.86	.00006	-3.36	.00039	-2.86	.0021	-2.36	.0091	-1.86	.0314
-3.85	.00006	-3.35	.00040	-2.85	.0022	-2.35	.0094	-1.85	.0322
-3.84	.00006	-3.34	.00042	-2.84	.0023	-2.34	.0096	-1.84	.0329
-3.83	.00006	-3.33	.00043	-2.83	.0023	-2.33	.0099	-1.83	.0336
-3.82	.00007	-3.32	.00045	-2.82	.0024	-2.32	.0102	-1.82	.0344
-3.81	.00007	-3.31	.00047	-2.81	.0025	-2.31	.0104	-1.81	.0351
-3.80	.00007	-3.30	.00048	-2.80	.0026	-2.30	.0107	-1.80	.0359
-3.79	.00007	-3.29	.00050	-2.79	.0026	-2.29	.0110	-1.79	.0367
-3.78	.00008	-3.28	.00052	-2.78	.0027	-2.28	.0113	-1.78	.0375
-3.77	.00008	-3.27	.00054	-2.77	.0028	-2.27	.0116	-1.77	.0384
-3.76	.00008	-3.26	.00056	-2.76	.0029	-2.26	.0119	-1.76	.0392
-3.75	.00009	-3.25	.00058	-2.75	.0030	-2.25	.0122	-1.75	.0401
-3.74	.00009	-3.24	.00060	-2.74	.0031	-2.24	.0125	-1.74	.0409
-3.73	.00010	-3.23	.00062	-2.73	.0032	-2.23	.0129	-1.73	.0418
-3.72	.00010	-3.22	.00064	-2.72	.0033	-2.22	.0132	-1.72	.0427
-3.71	.00010	-3.21	.00066	-2.71	.0034	-2.21	.0136	-1.71	.0436
-3.70	.00011	-3.20	.00069	-2.70	.0035	-2.20	.0139	-1.70	.0446
-3.69	.00011	-3.19	.00071	-2.69	.0036	-2.19	.0143	-1.69	.0455
-3.68	.00012	-3.18	.00074	-2.68	.0037	-2.18	.0146	-1.68	.0465
-3.67	.00012	-3.17	.00076	-2.67	.0038	-2.17	.0150	-1.67	.0475
-3.66	.00013	-3.16	.00079	-2.66	.0039	-2.16	.0154	-1.66	.0485
-3.65	.00013	-3.15	.00082	-2.65	.0040	-2.15	.0158	-1.65	.0495
-3.64	.00014	-3.14	.00084	-2.64	.0041	-2.14	.0162	-1.64	.0505
-3.63	.00014	-3.13	.00087	-2.63	.0043	-2.13	.0166	-1.63	.0516
-3.62	.00015	-3.12	.00090	-2.62	.0044	-2.12	.0170	-1.62	.0526
-3.61	.00015	-3.11	.00094	-2.61	.0045	-2.11	.0174	-1.61	.0537
-3.60	.00016	-3.10	.00097	-2.60	.0047	-2.10	.0179	-1.60	.0548
-3.59	.00017	-3.09	.00100	-2.59	.0048	-2.09	.0183	-1.59	.0559
-3.58	.00017	-3.08	.00103	-2.58	.0049	-2.08	.0188	-1.58	.0571
-3.57	.00018	-3.07	.00107	-2.57	.0051	-2.07	.0192	-1.57	.0582
-3.56	.00019	-3.06	.00111	-2.56	.0052	-2.06	.0197	-1.56	.0594
-3.55	.00019	-3.05	.00114	-2.55	.0054	-2.05	.0202	-1.55	.0606
-3.54	.00020	-3.04	.00118	-2.54	.0055	-2.04	.0207	-1.54	.0618
-3.53	.00021	-3.03	.00122	-2.53	.0057	-2.03	.0212	-1.53	.0630
-3.52	.00022	-3.02	.00126	-2.52	.0059	-2.02	.0217	-1.52	.0643
-3.51	.00022	-3.01	.00131	-2.51	.0060	-2.01	.0222	-1.51	.0655

Z	PROPORTION	Z	PROPORTION	Z	PROPORTION	Z	PROPORTION	Z	PROPORTION
-1.50	.0668	-1.00	.1587	-0.50	.3085	0.0	.5000	0.50	.6915
-1.49	.0681	-0.99	.1611	-0.49	.3121	0.01	.5040	0.51	.6950
-1.48	.0694	-0.98	.1635	-0.48	.3156	0.02	.5080	0.52	.6985
-1.47	.0708	-0.97	.1660	-0.47	.3192	0.03	.5120	0.53	.7019
-1.46	.0721	-0.96	.1685	-0.46	.3228	0.04	.5160	0.54	.7054
-1.45	.0735	-0.95	.1711	-0.45	.3264	0.05	.5199	0.55	.7088
-1.44	.0749	-0.94	.1736	-0.44	.3300	0.06	.5239	0.56	.7123
-1.43	.0764	-0.93	.1762	-0.43	.3336	0.07	.5279	0.57	.7157
-1.42	.0778	-0.92	.1788	-0.42	.3372	0.08	.5319	0.58	.7190
-1.41	.0793	-0.91	.1814	-0.41	.3409	0.09	.5359	0.59	.7224
-1.40	.0808	-0.90	.1841	-0.40	.3446	0.10	.5398	0.60	.7257
-1.39	.0823	-0.89	.1867	-0.39	.3483	0.11	.5438	0.61	.7291
-1.38	.0838	-0.88	.1894	-0.38	.3520	0.12	.5478	0.62	.7324
-1.37	.0853	-0.87	.1922	-0.37	.3557	0.13	.5517	0.63	.7357
-1.36	.0869	-0.86	.1949	-0.36	.3594	0.14	.5557	0.64	.7389
-1.35	.0885	-0.85	.1977	-0.35	.3632	0.15	.5596	0.65	.7422
-1.34	.0901	-0.84	.2005	-0.34	.3669	0.16	.5636	0.66	.7454
-1.33	.0918	-0.83	.2033	-0.33	.3707	0.17	.5675	0.67	.7486
-1.32	.0934	-0.82	.2061	-0.32	.3745	0.18	.5714	0.68	.7517
-1.31	.0951	-0.81	.2090	-0.31	.3783	0.19	.5753	0.69	.7549
-1.30	.0968	-0.80	.2119	-0.30	.3821	0.20	.5793	0.70	.7580
-1.29	.0985	-0.79	.2148	-0.29	.3859	0.21	.5832	0.71	.7611
-1.28	.1003	-0.78	.2177	-0.28	.3897	0.22	.5871	0.72	.7642
-1.27	.1020	-0.77	.2207	-0.27	.3936	0.23	.5910	0.73	.7673
-1.26	.1038	-0.76	.2236	-0.26	.3974	0.24	.5948	0.74	.7704
-1.25	.1057	-0.75	.2266	-0.25	.4013	0.25	.5987	0.75	.7734
-1.24	.1075	-0.74	.2297	-0.24	.4052	0.26	.6026	0.76	.7764
-1.23	.1093	-0.73	.2327	-0.23	.4090	0.27	.6064	0.77	.7794
-1.22	.1112	-0.72	.2358	-0.22	.4129	0.28	.6103	0.78	.7823
-1.21	.1131	-0.71	.2389	-0.21	.4168	0.29	.6141	0.79	.7852
-1.20	.1151	-0.70	.2420	-0.20	.4207	0.30	.6179	0.80	.7881
-1.19	.1170	-0.69	.2451	-0.19	.4247	0.31	.6217	0.81	.7910
-1.18	.1190	-0.68	.2483	-0.18	.4286	0.32	.6255	0.82	.7939
-1.17	.1210	-0.67	.2514	-0.17	.4325	0.33	.6293	0.83	.7967
-1.16	.1230	-0.66	.2546	-0.16	.4364	0.34	.6331	0.84	.7995
-1.15	.1251	-0.65	.2578	-0.15	.4404	0.35	.6368	0.85	.8023
-1.14	.1271	-0.64	.2611	-0.14	.4443	0.36	.6406	0.86	.8051
-1.13	.1292	-0.63	.2643	-0.13	.4483	0.37	.6443	0.87	.8079
-1.12	.1314	-0.62	.2676	-0.12	.4522	0.38	.6480	0.88	.8106
-1.11	.1335	-0.61	.2709	-0.11	.4562	0.39	.6517	0.89	.8133
-1.10	.1357	-0.60	.2743	-0.10	.4602	0.40	.6554	0.90	.8159
-1.09	.1379	-0.59	.2776	-0.09	.4641	0.41	.6591	0.91	.8186
-1.08	.1401	-0.58	.2810	-0.08	.4681	0.42	.6628	0.92	.8212
-1.07	.1423	-0.57	.2843	-0.07	.4721	0.43	.6664	0.93	.8238
-1.06	.1446	-0.56	.2877	-0.06	.4761	0.44	.6700	0.94	.8264
-1.05	.1469	-0.55	.2912	-0.05	.4801	0.45	.6736	0.95	.8289
-1.04	.1492	-0.54	.2946	-0.04	.4841	0.46	.6772	0.96	.8315
-1.03	.1515	-0.53	.2981	-0.03	.4880	0.47	.6808	0.97	.8340
-1.02	.1539	-0.52	.3015	-0.02	.4920	0.48	.6844	0.98	.8365
-1.01	.1563	-0.51	.3050	-0.01	.4960	0.49	.6879	0.99	.8389

(Continued)

TABLE A.4 The Standard Normal Distribution (continued)

Z	PROPORTION	Z	PROPORTION	Z	PROPORTION	Z	PROPORTION	Z	PROPORTION
1.00	.8413	1.50	.9332	2.00	.9772	2.50	.9938	3.00	.99865
1.01	.8438	1.51	.9345	2.01	.9778	2.51	.9940	3.01	.99869
1.02	.8461	1.52	.9357	2.02	.9783	2.52	.9941	3.02	.99873
1.03	.8485	1.53	.9370	2.03	.9788	2.53	.9943	3.03	.99878
1.04	.8508	1.54	.9382	2.04	.9793	2.54	.9945	3.04	.99882
1.05	.8531	1.55	.9394	2.05	.9798	2.55	.9946	3.05	.99885
1.06	.8554	1.56	.9406	2.06	.9803	2.56	.9948	3.06	.99889
1.07	.8577	1.57	.9418	2.07	.9808	2.57	.9949	3.07	.99893
1.08	.8599	1.58	.9429	2.08	.9812	2.58	.9951	3.08	.99896
1.09	.8621	1.59	.9441	2.09	.9817	2.59	.9952	3.09	.99900
1.10	.8643	1.60	.9452	2.10	.9821	2.60	.9953	3.10	.99903
1.11	.8665	1.61	.9463	2.11	.9826	2.61	.9955	3.11	.99906
1.12	.8686	1.62	.9474	2.12	.9830	2.62	.9956	3.12	.99909
1.13	.8708	1.63	.9484	2.13	.9834	2.63	.9957	3.13	.99912
1.14	.8729	1.64	.9495	2.14	.9838	2.64	.9959	3.14	.99915
1.15	.8749	1.65	.9505	2.15	.9842	2.65	.9960	3.15	.99918
1.16	.8770	1.66	.9515	2.16	.9846	2.66	.9961	3.16	.99921
1.17	.8790	1.67	.9525	2.17	.9850	2.67	.9962	3.17	.99924
1.18	.8810	1.68	.9535	2.18	.9854	2.68	.9963	3.18	.99926
1.19	.8830	1.69	.9545	2.19	.9857	2.69	.9964	3.19	.99929
1.20	.8849	1.70	.9554	2.20	.9861	2.70	.9965	3.20	.99931
1.21	.8869	1.71	.9564	2.21	.9864	2.71	.9966	3.21	.99934
1.22	.8888	1.72	.9573	2.22	.9868	2.72	.9967	3.22	.99936
1.23	.8906	1.73	.9582	2.23	.9871	2.73	.9968	3.23	.99938
1.24	.8925	1.74	.9591	2.24	.9875	2.74	.9969	3.24	.99940
1.25	.8943	1.75	.9599	2.25	.9878	2.75	.9970	3.25	.99942
1.26	.8962	1.76	.9608	2.26	.9881	2.76	.9971	3.26	.99944
1.27	.8980	1.77	.9616	2.27	.9884	2.77	.9972	3.27	.99946
1.28	.8997	1.78	.9625	2.28	.9887	2.78	.9973	3.28	.99948
1.29	.9015	1.79	.9633	2.29	.9890	2.79	.9974	3.29	.99950
1.30	.9032	1.80	.9641	2.30	.9893	2.80	.9974	3.30	.99952
1.31	.9049	1.81	.9649	2.31	.9896	2.81	.9975	3.31	.99953
1.32	.9066	1.82	.9656	2.32	.9898	2.82	.9976	3.32	.99955
1.33	.9082	1.83	.9664	2.33	.9901	2.83	.9977	3.33	.99956
1.34	.9099	1.84	.9671	2.34	.9904	2.84	.9977	3.34	.99958
1.35	.9115	1.85	.9678	2.35	.9906	2.85	.9978	3.35	.99960
1.36	.9131	1.86	.9686	2.36	.9909	2.86	.9979	3.36	.99961
1.37	.9147	1.87	.9693	2.37	.9911	2.87	.9979	3.37	.99962
1.38	.9162	1.88	.9699	2.38	.9913	2.88	.9980	3.38	.99964
1.39	.9177	1.89	.9706	2.39	.9916	2.89	.9981	3.39	.99965
1.40	.9192	1.90	.9713	2.40	.9918	2.90	.9981	3.40	.99966
1.41	.9207	1.91	.9719	2.41	.9920	2.91	.9982	3.41	.99967
1.42	.9222	1.92	.9726	2.42	.9922	2.92	.9982	3.42	.99969
1.43	.9236	1.93	.9732	2.43	.9924	2.93	.9983	3.43	.99970
1.44	.9251	1.94	.9738	2.44	.9927	2.94	.9984	3.44	.99971
1.45	.9265	1.95	.9744	2.45	.9929	2.95	.9984	3.45	.99972
1.46	.9279	1.96	.9750	2.46	.9931	2.96	.9985	3.46	.99973
1.47	.9292	1.97	.9756	2.47	.9932	2.97	.9985	3.47	.99974
1.48	.9306	1.98	.9761	2.48	.9934	2.98	.9986	3.48	.99975
1.49	.9319	1.99	.9767	2.49	.9936	2.99	.9986	3.49	.99976

Z	PROPORTION	Z	PROPORTION	Z	PROPORTION	Z	PROPORTION	Z	PROPORTION
3.50	.99977	3.60	.99984	3.70	.99989	3.80	.99993	3.90	.99995
3.51	.99978	3.61	.99985	3.71	.99990	3.81	.99993	3.91	.99995
3.52	.99978	3.62	.99985	3.72	.99990	3.82	.99993	3.92	.99995
3.53	.99979	3.63	.99986	3.73	.99990	3.83	.99994	3.93	.99996
3.54	.99980	3.64	.99986	3.74	.99991	3.84	.99994	3.94	.99996
3.55	.99981	3.65	.99987	3.75	.99991	3.85	.99994	3.95	.99996
3.56	.99981	3.66	.99987	3.76	.99991	3.86	.99994	3.96	.99996
3.57	.99982	3.67	.99988	3.77	.99992	3.87	.99994	3.97	.99996
3.58	.99983	3.68	.99988	3.78	.99992	3.88	.99995	3.98	.99996
3.59	.99983	3.69	.99989	3.79	.99992	3.89	.99995	3.99	.99997

TABLE A.5   The Poisson Distribution

λ	0	1	2	3	4	5	N 6	7	8	9	10	11	12
.05	.9512	.9988	1.0000	—	—	—	—	—	—	—	—	—	—
.10	.9048	.9953	.9998	1.0000	—	—	—	—	—	—	—	—	—
.15	.8607	.9898	.9995	1.0000	—	—	—	—	—	—	—	—	—
.20	.8187	.9825	.9989	.9999	1.0000	—	—	—	—	—	—	—	—
.25	.7788	.9735	.9978	.9999	1.0000	—	—	—	—	—	—	—	—
.30	.7408	.9631	.9964	.9997	1.0000	—	—	—	—	—	—	—	—
.35	.7047	.9513	.9945	.9995	1.0000	—	—	—	—	—	—	—	—
.40	.6703	.9384	.9921	.9992	.9999	1.0000	—	—	—	—	—	—	—
.45	.6376	.9246	.9891	.9988	.9999	1.0000	—	—	—	—	—	—	—
.50	.6065	.9098	.9856	.9982	.9998	1.0000	—	—	—	—	—	—	—
.55	.5769	.8943	.9815	.9975	.9997	1.0000	—	—	—	—	—	—	—
.60	.5488	.8781	.9769	.9966	.9996	1.0000	—	—	—	—	—	—	—
.65	.5220	.8614	.9717	.9956	.9994	.9999	1.0000	—	—	—	—	—	—
.70	.4966	.8442	.9659	.9942	.9992	.9999	1.0000	—	—	—	—	—	—
.75	.4724	.8266	.9595	.9927	.9989	.9999	1.0000	—	—	—	—	—	—
.80	.4493	.8088	.9526	.9909	.9986	.9998	1.0000	—	—	—	—	—	—
.85	.4274	.7907	.9451	.9889	.9982	.9997	1.0000	—	—	—	—	—	—
.90	.4066	.7725	.9371	.9865	.9977	.9997	1.0000	—	—	—	—	—	—
.95	.3867	.7541	.9287	.9839	.9971	.9995	.9999	1.0000	—	—	—	—	—
1.00	.3679	.7358	.9197	.9810	.9963	.9994	.9999	1.0000	—	—	—	—	—
1.10	.3329	.6990	.9004	.9743	.9946	.9990	.9999	1.0000	—	—	—	—	—
1.20	.3012	.6626	.8795	.9662	.9923	.9985	.9997	1.0000	—	—	—	—	—
1.30	.2725	.6268	.8571	.9569	.9893	.9978	.9996	.9999	1.0000	—	—	—	—
1.40	.2466	.5918	.8335	.9463	.9857	.9968	.9994	.9999	1.0000	—	—	—	—
1.50	.2231	.5578	.8088	.9344	.9814	.9955	.9991	.9998	1.0000	—	—	—	—
1.60	.2019	.5249	.7834	.9212	.9763	.9940	.9987	.9997	1.0000	—	—	—	—
1.70	.1827	.4932	.7572	.9068	.9704	.9920	.9981	.9996	.9999	1.0000	—	—	—
1.80	.1653	.4628	.7306	.8913	.9636	.9896	.9974	.9994	.9999	1.0000	—	—	—
1.90	.1496	.4337	.7037	.8747	.9559	.9868	.9966	.9992	.9998	1.0000	—	—	—
2.00	.1353	.4060	.6767	.8571	.9473	.9834	.9955	.9989	.9998	1.0000	—	—	—
2.20	.1108	.3546	.6227	.8194	.9275	.9751	.9925	.9980	.9995	.9999	1.0000	—	—
2.40	.0907	.3084	.5697	.7787	.9041	.9643	.9884	.9967	.9991	.9998	1.0000	—	—
2.60	.0743	.2674	.5184	.7360	.8774	.9510	.9828	.9947	.9985	.9996	.9999	1.0000	—
2.80	.0608	.2311	.4695	.6919	.8477	.9349	.9756	.9919	.9976	.9993	.9998	1.0000	—
3.00	.0498	.1991	.4232	.6472	.8153	.9161	.9665	.9881	.9962	.9989	.9997	.9999	1.0000
3.20	.0408	.1712	.3799	.6025	.7806	.8946	.9554	.9832	.9943	.9982	.9995	.9999	1.0000
3.40	.0334	.1468	.3397	.5584	.7442	.8705	.9421	.9769	.9917	.9973	.9992	.9998	.9999
3.60	.0273	.1257	.3027	.5152	.7064	.8441	.9267	.9692	.9883	.9960	.9987	.9996	.9999
3.80	.0224	.1074	.2689	.4735	.6678	.8156	.9091	.9599	.9840	.9942	.9981	.9994	.9998
4.00	.0183	.0916	.2381	.4335	.6288	.7851	.8893	.9489	.9786	.9919	.9972	.9991	.9997

Note on the usage of this table: If X is a Poisson random variable with a parameter λ, then the entry in the table corresponding to λ and N is the cumulative Poisson probability $P\{X \leq N\}$.

TABLE A.6   Table of Exponentials

x	e^{-x}	x	e^{-x}	x	e^{-x}	x	e^{-x}	x	e^{-x}
0.00	1.0000	0.40	.6703	0.80	.4493	1.20	.3012	1.60	.2019
0.01	.9900	0.41	.6637	0.81	.4449	1.21	.2982	1.61	.1999
0.02	.9802	0.42	.6570	0.82	.4404	1.22	.2952	1.62	.1979
0.03	.9704	0.43	.6505	0.83	.4360	1.23	.2923	1.63	.1959
0.04	.9608	0.44	.6440	0.84	.4317	1.24	.2894	1.64	.1940
0.05	.9512	0.45	.6376	0.85	.4274	1.25	.2865	1.65	.1921
0.06	.9418	0.46	.6313	0.86	.4232	1.26	.2837	1.66	.1901
0.07	.9324	0.47	.6250	0.87	.4190	1.27	.2808	1.67	.1882
0.08	.9231	0.48	.6188	0.88	.4148	1.28	.2780	1.68	.1864
0.09	.9139	0.49	.6126	0.89	.4107	1.29	.2753	1.69	.1845
0.10	.9048	0.50	.6065	0.90	.4066	1.30	.2725	1.70	.1827
0.11	.8958	0.51	.6005	0.91	.4025	1.31	.2698	1.71	.1809
0.12	.8869	0.52	.5945	0.92	.3985	1.32	.2671	1.72	.1791
0.13	.8781	0.53	.5886	0.93	.3946	1.33	.2645	1.73	.1773
0.14	.8694	0.54	.5827	0.94	.3906	1.34	.2618	1.74	.1755
0.15	.8607	0.55	.5769	0.95	.3867	1.35	.2592	1.75	.1738
0.16	.8521	0.56	.5712	0.96	.3829	1.36	.2567	1.76	.1720
0.17	.8437	0.57	.5655	0.97	.3791	1.37	.2541	1.77	.1703
0.18	.8353	0.58	.5599	0.98	.3753	1.38	.2516	1.78	.1686
0.19	.8270	0.59	.5543	0.99	.3716	1.39	.2491	1.79	.1670
0.20	.8187	0.60	.5488	1.00	.3679	1.40	.2466	1.80	.1653
0.21	.8106	0.61	.5434	1.01	.3642	1.41	.2441	1.81	.1637
0.22	.8025	0.62	.5379	1.02	.3606	1.42	.2417	1.82	.1620
0.23	.7945	0.63	.5326	1.03	.3570	1.43	.2393	1.83	.1604
0.24	.7866	0.64	.5273	1.04	.3535	1.44	.2369	1.84	.1588
0.25	.7788	0.65	.5220	1.05	.3499	1.45	.2346	1.85	.1572
0.26	.7711	0.66	.5169	1.06	.3465	1.46	.2322	1.86	.1557
0.27	.7634	0.67	.5117	1.07	.3430	1.47	.2299	1.87	.1541
0.28	.7558	0.68	.5066	1.08	.3396	1.48	.2276	1.88	.1526
0.29	.7483	0.69	.5016	1.09	.3362	1.49	.2254	1.89	.1511
0.30	.7408	0.70	.4966	1.10	.3329	1.50	.2231	1.90	.1496
0.31	.7334	0.71	.4916	1.11	.3296	1.51	.2209	1.91	.1481
0.32	.7261	0.72	.4868	1.12	.3263	1.52	.2187	1.92	.1466
0.33	.7189	0.73	.4819	1.13	.3230	1.53	.2165	1.93	.1451
0.34	.7118	0.74	.4771	1.14	.3198	1.54	.2144	1.94	.1437
0.35	.7047	0.75	.4724	1.15	.3166	1.55	.2122	1.95	.1423
0.36	.6977	0.76	.4677	1.16	.3135	1.56	.2101	1.96	.1409
0.37	.6907	0.77	.4630	1.17	.3104	1.57	.2080	1.97	.1395
0.38	.6839	0.78	.4584	1.18	.3073	1.58	.2060	1.98	.1381
0.39	.6771	0.79	.4538	1.19	.3042	1.59	.2039	1.99	.1367

(Continued)

TABLE A.6   Table of Exponentials (continued)

x	e^{-x}	x	e^{-x}	x	e^{-x}	x	e^{-x}	x	e^{-x}
2.00	.1353	2.40	.0907	2.80	.0608	3.20	.0408	4.00	.01832
2.01	.1340	2.41	.0898	2.81	.0602	3.21	.0404	4.05	.01742
2.02	.1327	2.42	.0889	2.82	.0596	3.22	.0400	4.10	.01657
2.03	.1313	2.43	.0880	2.83	.0590	3.23	.0396	4.15	.01576
2.04	.1300	2.44	.0872	2.84	.0584	3.24	.0392	4.20	.01500
2.05	.1287	2.45	.0863	2.85	.0578	3.25	.0388	4.25	.01426
2.06	.1275	2.46	.0854	2.86	.0573	3.26	.0384	4.30	.01357
2.07	.1262	2.47	.0846	2.87	.0567	3.27	.0380	4.35	.01291
2.08	.1249	2.48	.0837	2.88	.0561	3.28	.0376	4.40	.01228
2.09	.1237	2.49	.0829	2.89	.0556	3.29	.0373	4.45	.01168
2.10	.1225	2.50	.0821	2.90	.0550	3.30	.0369	4.50	.01111
2.11	.1212	2.51	.0813	2.91	.0545	3.31	.0365	4.55	.01057
2.12	.1200	2.52	.0805	2.92	.0539	3.32	.0362	4.60	.01005
2.13	.1188	2.53	.0797	2.93	.0534	3.33	.0358	4.65	.00956
2.14	.1177	2.54	.0789	2.94	.0529	3.34	.0354	4.70	.00910
2.15	.1165	2.55	.0781	2.95	.0523	3.35	.0351	4.75	.00865
2.16	.1153	2.56	.0773	2.96	.0518	3.36	.0347	4.80	.00823
2.17	.1142	2.57	.0765	2.97	.0513	3.37	.0344	4.85	.00783
2.18	.1130	2.58	.0758	2.98	.0508	3.38	.0340	4.90	.00745
2.19	.1119	2.59	.0750	2.99	.0503	3.39	.0337	4.95	.00708
2.20	.1108	2.60	.0743	3.00	.0498	3.40	.0334	5.00	.00674
2.21	.1097	2.61	.0735	3.01	.0493	3.41	.0330	5.10	.00610
2.22	.1086	2.62	.0728	3.02	.0488	3.42	.0327	5.20	.00552
2.23	.1075	2.63	.0721	3.03	.0483	3.43	.0324	5.30	.00499
2.24	.1065	2.64	.0714	3.04	.0478	3.44	.0321	5.40	.00452
2.25	.1054	2.65	.0707	3.05	.0474	3.45	.0317	5.50	.00409
2.26	.1044	2.66	.0699	3.06	.0469	3.46	.0314	5.60	.00370
2.27	.1033	2.67	.0693	3.07	.0464	3.47	.0311	5.70	.00335
2.28	.1023	2.68	.0686	3.08	.0460	3.48	.0308	5.80	.00303
2.29	.1013	2.69	.0679	3.09	.0455	3.49	.0305	5.90	.00274
2.30	.1003	2.70	.0672	3.10	.0451	3.50	.0302	6.00	.00248
2.31	.0993	2.71	.0665	3.11	.0446	3.55	.0287	6.25	.00193
2.32	.0983	2.72	.0659	3.12	.0442	3.60	.0273	6.50	.00150
2.33	.0973	2.73	.0652	3.13	.0437	3.65	.0260	6.75	.00117
2.34	.0963	2.74	.0646	3.14	.0433	3.70	.0247	7.00	.00091
2.35	.0954	2.75	.0639	3.15	.0429	3.75	.0235	7.50	.00055
2.36	.0944	2.76	.0633	3.16	.0424	3.80	.0224	8.00	.00034
2.37	.0935	2.77	.0627	3.17	.0420	3.85	.0213	8.50	.00020
2.38	.0926	2.78	.0620	3.18	.0416	3.90	.0202	9.00	.00012
2.39	.0916	2.79	.0614	3.19	.0412	3.95	.0193	10.00	.00005

TABLE A.7 The t Distribution (values of t_α)

df	$t_{.25}$	$t_{.10}$	$t_{.05}$	$t_{.025}$	$t_{.01}$	$t_{.005}$	$t_{.0025}$	$t_{.001}$
1	1.000	3.078	6.314	12.706	31.821	63.657	127.321	318.309
2	.816	1.886	2.920	4.303	6.965	9.925	14.089	22.327
3	.765	1.638	2.353	3.182	4.541	5.841	7.453	10.214
4	.741	1.533	2.132	2.776	3.747	4.604	5.598	7.173
5	.727	1.476	2.015	2.571	3.365	4.032	4.773	5.893
6	.718	1.440	1.943	2.447	3.143	3.707	4.317	5.208
7	.711	1.415	1.895	2.365	2.998	3.499	4.029	4.785
8	.706	1.397	1.860	2.306	2.896	3.355	3.833	4.501
9	.703	1.383	1.833	2.262	2.821	3.250	3.690	4.297
10	.700	1.372	1.812	2.228	2.764	3.169	3.581	4.144
11	.697	1.363	1.796	2.201	2.718	3.106	3.497	4.025
12	.695	1.356	1.782	2.179	2.681	3.055	3.428	3.930
13	.694	1.350	1.771	2.160	2.650	3.012	3.372	3.852
14	.692	1.345	1.761	2.145	2.624	2.977	3.326	3.787
15	.691	1.341	1.753	2.131	2.602	2.947	3.286	3.733
16	.690	1.337	1.746	2.120	2.583	2.921	3.252	3.686
17	.689	1.333	1.740	2.110	2.567	2.898	3.223	3.646
18	.688	1.330	1.734	2.101	2.552	2.878	3.197	3.610
19	.688	1.328	1.729	2.093	2.539	2.861	3.174	3.579
20	.687	1.325	1.725	2.086	2.528	2.845	3.153	3.552
21	.686	1.323	1.721	2.080	2.518	2.831	3.135	3.527
22	.686	1.321	1.717	2.074	2.508	2.819	3.119	3.505
23	.685	1.319	1.714	2.069	2.500	2.807	3.104	3.485
24	.685	1.318	1.711	2.064	2.492	2.797	3.090	3.467
25	.684	1.316	1.708	2.060	2.485	2.787	3.078	3.450
26	.684	1.315	1.706	2.056	2.479	2.779	3.067	3.435
27	.684	1.314	1.703	2.052	2.473	2.771	3.057	3.421
28	.683	1.313	1.701	2.048	2.467	2.763	3.047	3.408
29	.683	1.311	1.699	2.045	2.462	2.756	3.038	3.396
30	.683	1.310	1.697	2.042	2.457	2.750	3.030	3.385
35	.682	1.306	1.690	2.030	2.438	2.724	2.996	3.340
40	.681	1.303	1.684	2.021	2.423	2.704	2.971	3.307
45	.680	1.301	1.679	2.014	2.412	2.690	2.952	3.281
50	.679	1.299	1.676	2.009	2.403	2.678	2.937	3.261
55	.679	1.297	1.673	2.004	2.396	2.668	2.925	3.245
60	.679	1.296	1.671	2.000	2.390	2.660	2.915	3.232
70	.678	1.294	1.667	1.994	2.381	2.648	2.899	3.211
80	.678	1.292	1.664	1.990	2.374	2.639	2.887	3.195
90	.677	1.291	1.662	1.987	2.368	2.632	2.878	3.183
100	.677	1.290	1.660	1.984	2.364	2.626	2.871	3.174
200	.676	1.286	1.652	1.972	2.345	2.601	2.838	3.131
500	.675	1.283	1.648	1.965	2.334	2.586	2.820	3.107
1,000	.675	1.282	1.646	1.962	2.330	2.581	2.813	3.098
2,000	.675	1.282	1.645	1.961	2.328	2.578	2.810	3.094
10,000	.675	1.282	1.645	1.960	2.327	2.576	2.808	3.091
∞	.674	1.282	1.645	1.960	2.326	2.576	2.807	3.090

Source: Enrico T. Federighi, "Extended tables of the percentage points of Student's t distribution," *Journal of the American Statistical Association*, Vol. 54 (1959), page 684. Reproduced by permission of the American Statistical Association.

TABLE A.8   The Chi-Square Distribution (values of χ_α^2)*

df	$\chi_{.995}^2$	$\chi_{.99}^2$	$\chi_{.975}^2$	$\chi_{.95}^2$	$\chi_{.05}^2$	$\chi_{.025}^2$	$\chi_{.01}^2$	$\chi_{.005}^2$	df
1	.0000393	.000157	.000982	.00393	3.841	5.024	6.635	7.879	1
2	.0100	.0201	.0506	.103	5.991	7.378	9.210	10.597	2
3	.0717	.115	.216	.352	7.815	9.348	11.345	12.838	3
4	.207	.297	.484	.711	9.488	11.143	13.277	14.860	4
5	.412	.554	.831	1.145	11.070	12.832	15.086	16.750	5
6	.676	.872	1.237	1.635	12.592	14.449	16.812	18.548	6
7	.989	1.239	1.690	2.167	14.067	16.013	18.475	20.278	7
8	1.344	1.646	2.180	2.733	15.507	17.535	20.090	21.955	8
9	1.735	2.088	2.700	3.325	16.919	19.023	21.666	23.589	9
10	2.156	2.558	3.247	3.940	18.307	20.483	23.209	25.188	10
11	2.603	3.053	3.816	4.575	19.675	21.920	24.725	26.757	11
12	3.074	3.571	4.404	5.226	21.026	23.337	26.217	28.300	12
13	3.565	4.107	5.009	5.892	22.362	24.736	27.688	29.819	13
14	4.075	4.660	5.629	6.571	23.685	26.119	29.141	31.319	14
15	4.601	5.229	6.262	7.261	24.996	27.488	30.578	32.801	15
16	5.142	5.812	6.908	7.962	26.296	28.845	32.000	34.267	16
17	5.697	6.408	7.564	8.672	27.587	30.191	33.409	35.718	17
18	6.265	7.015	8.231	9.390	28.869	31.526	34.805	37.156	18
19	6.844	7.633	8.907	10.117	30.144	32.852	36.191	38.582	19
20	7.434	8.260	9.591	10.851	31.410	34.170	37.566	39.997	20
21	8.034	8.897	10.283	11.591	32.671	35.479	38.932	41.401	21
22	8.643	9.542	10.982	12.338	33.924	36.781	40.289	42.796	22
23	9.260	10.196	11.689	13.091	35.172	38.076	41.638	44.181	23
24	9.886	10.856	12.401	13.848	36.415	39.364	42.980	45.558	24
25	10.520	11.524	13.120	14.611	37.652	40.646	44.314	46.928	25
26	11.160	12.198	13.844	15.379	38.885	41.923	45.642	48.290	26
27	11.808	12.879	14.573	16.151	40.113	43.194	46.963	49.645	27
28	12.461	13.565	15.308	16.928	41.337	44.461	48.278	50.993	28
29	13.121	14.256	16.047	17.708	42.557	45.722	49.588	52.336	29
30	13.787	14.953	16.791	18.493	43.773	46.979	50.892	53.672	30

TABLE A.9 | Table of Random Digits

26488 65184	35213 08911	41771 78761	34881 71011	38769 24817
03377 88678	27022 84903	84513 47052	14385 22169	12337 20836
67360 11209	04703 39197	93794 87287	50308 34204	64143 29808
07859 05673	77437 56001	96183 50414	77605 50494	86175 30721
21174 79881	31784 29177	80907 52946	67446 20201	60409 14530
14916 68132	65010 74111	93003 60176	43652 38637	24846 17161
81615 80233	87860 88405	03328 09140	30501 39434	67801 36037
94308 19026	51141 52030	91280 38243	06521 09759	37885 79255
18737 84431	10210 19135	62392 33368	65096 21635	52768 00335
91808 11434	84510 09724	44804 75493	23848 07525	51220 32981
98109 97606	32651 03797	75607 71252	53208 11220	12464 56543
57438 59987	03073 88112	13637 38195	83731 35312	90660 36780
39419 40518	60169 67593	77874 81275	71598 17925	79731 59571
45324 63563	99292 39986	24099 98128	46527 40652	16618 04084
41069 15749	28541 31100	25983 21706	09643 07666	01573 52145
03825 02399	75383 34402	35331 47832	39017 53635	74501 68789
82887 42543	92642 70000	63614 14826	49424 89663	85509 14280
28795 19978	07330 87676	27559 19564	29702 24204	97089 69210
59117 31864	20413 01258	48115 99390	66464 10103	13691 82780
81841 85493	06282 32567	42696 58381	46118 83007	08915 36396
26454 28487	32450 58558	16169 18995	77039 12866	40426 18284
07573 05228	92017 16315	85887 84969	37262 95862	40855 07788
26070 07301	70134 60577	49443 92914	51603 39632	09308 18023
78789 17386	65408 33061	61336 76264	31684 07421	74411 94873
89502 63120	42146 28239	80403 14009	58786 99347	38137 71018
49900 89267	81039 37192	86746 13303	33858 41254	05114 18748
04596 00469	84147 41439	86139 46286	85496 98215	14644 54949
03032 90178	97586 58485	14777 86467	30520 61690	34076 13101
60407 73065	52537 22210	07620 55251	26638 73767	84644 13066
20049 85142	71961 44114	43840 67648	06484 11383	42522 86489
81870 83709	97610 79950	42214 31597	87211 89188	87472 62569
63161 59799	41696 37290	02170 57803	12988 91659	20840 41839
99417 04688	84622 36662	69190 58303	10249 46573	41177 77616
56280 67821	02461 86316	53174 01469	18511 48170	54274 35223
49640 27628	66188 83630	42092 55752	37156 21654	73948 39454
40985 49558	10525 84926	12748 94011	26487 94169	94770 79757
83866 16648	90314 32706	31814 08857	86393 14702	96954 81180
99746 49460	24504 08069	64188 23469	15887 35938	57377 50222
25531 37403	28661 31842	83494 64740	75668 91909	48839 71560
58525 46229	46661 77378	62943 84154	55650 85352	05221 74645
59168 77022	38849 48418	37059 81142	43724 69633	48123 83852
45556 78930	17822 64085	97000 58207	03992 00524	33338 63320
95941 09941	26646 79930	08941 71644	28300 07758	88565 99968
69974 07396	30168 81184	35167 14183	93215 20689	95890 85617
27790 47571	99301 08247	98616 45998	62406 07600	31083 04375
32065 65133	03253 34611	90918 60893	85595 52459	21400 40246
50824 88029	37647 93690	97465 39804	11207 28485	48862 11813
13251 27156	23257 95377	30224 88312	51181 98961	30460 40066
19281 67639	61205 85952	06230 58813	79229 03874	00138 55781
22161 31143	48783 53936	65274 50813	06125 59984	42227 94356

Source: The Rand Corporation, *A Million Random Digits with 100,000 Normal Deviates*, The Free Press, 1955, p. 293. Reprinted with permission of the Rand Corporation.

TABLE A.10 The F Distribution (values of $F_{.05}$)*

DEGREES OF FREEDOM FOR NUMERATOR

	1	2	3	4	5	6	7	8	9	10	12	15	20	24	30	40	60	120	∞
1	161	200	216	225	230	234	237	239	241	242	244	246	248	249	250	251	252	253	254
2	18.5	19.0	19.2	19.2	19.3	19.3	19.4	19.4	19.4	19.4	19.4	19.4	19.4	19.5	19.5	19.5	19.5	19.5	19.5
3	10.1	9.55	9.28	9.12	9.01	8.94	8.89	8.85	8.81	8.79	8.74	8.70	8.66	8.64	8.62	8.59	8.57	8.55	8.53
4	7.71	6.94	6.59	6.39	6.26	6.16	6.09	6.04	6.00	5.96	5.91	5.86	5.80	5.77	5.75	5.72	5.69	5.66	5.63
5	6.61	5.79	5.41	5.19	5.05	4.95	4.88	4.82	4.77	4.74	4.68	4.62	4.56	4.53	4.50	4.46	4.43	4.40	4.37
6	5.99	5.14	4.76	4.53	4.39	4.28	4.21	4.15	4.10	4.06	4.00	3.94	3.87	3.84	3.81	3.77	3.74	3.70	3.67
7	5.59	4.74	4.35	4.12	3.97	3.87	3.79	3.73	3.68	3.64	3.57	3.51	3.44	3.41	3.38	3.34	3.30	3.27	3.23
8	5.32	4.46	4.07	3.84	3.69	3.58	3.50	3.44	3.39	3.35	3.28	3.22	3.15	3.12	3.08	3.04	3.01	2.97	2.93
9	5.12	4.26	3.86	3.63	3.48	3.37	3.29	3.23	3.18	3.14	3.07	3.01	2.94	2.90	2.86	2.83	2.79	2.75	2.71
10	4.96	4.10	3.71	3.48	3.33	3.22	3.14	3.07	3.02	2.98	2.91	2.85	2.77	2.74	2.70	2.66	2.62	2.58	2.54
11	4.84	3.98	3.59	3.36	3.20	3.09	3.01	2.95	2.90	2.85	2.79	2.72	2.65	2.61	2.57	2.53	2.49	2.45	2.40
12	4.75	3.89	3.49	3.26	3.11	3.00	2.91	2.85	2.80	2.75	2.69	2.62	2.54	2.51	2.47	2.43	2.38	2.34	2.30
13	4.67	3.81	3.41	3.18	3.03	2.92	2.83	2.77	2.71	2.67	2.60	2.53	2.46	2.42	2.38	2.34	2.30	2.25	2.21
14	4.60	3.74	3.34	3.11	2.96	2.85	2.76	2.70	2.65	2.60	2.53	2.46	2.39	2.35	2.31	2.27	2.22	2.18	2.13
15	4.54	3.68	3.29	3.06	2.90	2.79	2.71	2.64	2.59	2.54	2.48	2.40	2.33	2.29	2.25	2.20	2.16	2.11	2.07
16	4.49	3.63	3.24	3.01	2.85	2.74	2.66	2.59	2.54	2.49	2.42	2.35	2.28	2.24	2.19	2.15	2.11	2.06	2.01
17	4.45	3.59	3.20	2.96	2.81	2.70	2.61	2.55	2.49	2.45	2.38	2.31	2.23	2.19	2.15	2.10	2.06	2.01	1.96
18	4.41	3.55	3.16	2.93	2.77	2.66	2.58	2.51	2.46	2.41	2.34	2.27	2.19	2.15	2.11	2.06	2.02	1.97	1.92
19	4.38	3.52	3.13	2.90	2.74	2.63	2.54	2.48	2.42	2.38	2.31	2.23	2.16	2.11	2.07	2.03	1.98	1.93	1.88
20	4.35	3.49	3.10	2.87	2.71	2.60	2.51	2.45	2.39	2.35	2.28	2.20	2.12	2.08	2.04	1.99	1.95	1.90	1.84
21	4.32	3.47	3.07	2.84	2.68	2.57	2.49	2.42	2.37	2.32	2.25	2.18	2.10	2.05	2.01	1.96	1.92	1.87	1.81
22	4.30	3.44	3.05	2.82	2.66	2.55	2.46	2.40	2.34	2.30	2.23	2.15	2.07	2.03	1.98	1.94	1.89	1.84	1.78
23	4.28	3.42	3.03	2.80	2.64	2.53	2.44	2.37	2.32	2.27	2.20	2.13	2.05	2.01	1.96	1.91	1.86	1.81	1.76
24	4.26	3.40	3.01	2.78	2.62	2.51	2.42	2.36	2.30	2.25	2.18	2.11	2.03	1.98	1.94	1.89	1.84	1.79	1.73
25	4.24	3.39	2.99	2.76	2.60	2.49	2.40	2.34	2.28	2.24	2.16	2.09	2.01	1.96	1.92	1.87	1.82	1.77	1.71
30	4.17	3.32	2.92	2.69	2.53	2.42	2.33	2.27	2.21	2.16	2.09	2.01	1.93	1.89	1.84	1.79	1.74	1.68	1.62
40	4.08	3.23	2.84	2.61	2.45	2.34	2.25	2.18	2.12	2.08	2.00	1.92	1.84	1.79	1.74	1.69	1.64	1.58	1.51
60	4.00	3.15	2.76	2.53	2.37	2.25	2.17	2.10	2.04	1.99	1.92	1.84	1.75	1.70	1.65	1.59	1.53	1.47	1.39
120	3.92	3.07	2.68	2.45	2.29	2.18	2.09	2.02	1.96	1.91	1.83	1.75	1.66	1.61	1.55	1.50	1.43	1.35	1.25
∞	3.84	3.00	2.60	2.37	2.21	2.10	2.01	1.94	1.88	1.83	1.75	1.67	1.57	1.52	1.46	1.39	1.32	1.22	1.00

TABLE A.10 The F Distribution (values of $F_{.01}$)* (continued)

									DEGREES OF FREEDOM FOR NUMERATOR										
	1	2	3	4	5	6	7	8	9	10	12	15	20	24	30	40	60	120	∞
1	4,052	5,000	5,403	5,625	5,764	5,859	5,928	5,982	6,023	6,056	6,106	6,157	6,209	6,235	6,261	6,287	6,313	6,339	6,366
2	98.5	99.0	99.2	99.2	99.3	99.3	99.4	99.4	99.4	99.4	99.4	99.4	99.4	99.5	99.5	99.5	99.5	99.5	99.5
3	34.1	30.8	29.5	28.7	28.2	27.9	27.7	27.5	27.3	27.2	27.1	26.9	26.7	26.6	26.5	26.4	26.3	26.2	26.1
4	21.2	18.0	16.7	16.0	15.5	15.2	15.0	14.8	14.7	14.5	14.4	14.2	14.0	13.9	13.8	13.7	13.7	13.6	13.5
5	16.3	13.3	12.1	11.4	11.0	10.7	10.5	10.3	10.2	10.1	9.89	9.72	9.55	9.47	9.38	9.29	9.20	9.11	9.02
6	13.7	10.9	9.78	9.15	8.75	8.47	8.26	8.10	7.98	7.87	7.72	7.56	7.40	7.31	7.23	7.14	7.06	6.97	6.88
7	12.2	9.55	8.45	7.85	7.46	7.19	6.99	6.84	6.72	6.62	6.47	6.31	6.16	6.07	5.99	5.91	5.82	5.74	5.65
8	11.3	8.65	7.59	7.01	6.63	6.37	6.18	6.03	5.91	5.81	5.67	5.52	5.36	5.28	5.20	5.12	5.03	4.95	4.86
9	10.6	8.02	6.99	6.42	6.06	5.80	5.61	5.47	5.35	5.26	5.11	4.96	4.81	4.73	4.65	4.57	4.48	4.40	4.31
10	10.0	7.56	6.55	5.99	5.64	5.39	5.20	5.06	4.94	4.85	4.71	4.56	4.41	4.33	4.25	4.17	4.08	4.00	3.91
11	9.65	7.21	6.22	5.67	5.32	5.07	4.89	4.74	4.63	4.54	4.40	4.25	4.10	4.02	3.94	3.86	3.78	3.69	3.60
12	9.33	6.93	5.95	5.41	5.06	4.82	4.64	4.50	4.39	4.30	4.16	4.01	3.86	3.78	3.70	3.62	3.54	3.45	3.36
13	9.07	6.70	5.74	5.21	4.86	4.62	4.44	4.30	4.19	4.10	3.96	3.82	3.66	3.59	3.51	3.43	3.34	3.25	3.17
14	8.86	6.51	5.56	5.04	4.70	4.46	4.28	4.14	4.03	3.94	3.80	3.66	3.51	3.43	3.35	3.27	3.18	3.09	3.00
15	8.68	6.36	5.42	4.89	4.56	4.32	4.14	4.00	3.89	3.80	3.67	3.52	3.37	3.29	3.21	3.13	3.05	2.96	2.87
16	8.53	6.23	5.29	4.77	4.44	4.20	4.03	3.89	3.78	3.69	3.55	3.41	3.26	3.18	3.10	3.02	2.93	2.84	2.75
17	8.40	6.11	5.19	4.67	4.34	4.10	3.93	3.79	3.68	3.59	3.46	3.31	3.16	3.08	3.00	2.92	2.83	2.75	2.65
18	8.29	6.01	5.09	4.58	4.25	4.01	3.84	3.71	3.60	3.51	3.37	3.23	3.08	3.00	2.92	2.84	2.75	2.66	2.57
19	8.19	5.93	5.01	4.50	4.17	3.94	3.77	3.63	3.52	3.43	3.30	3.15	3.00	2.92	2.84	2.76	2.67	2.58	2.49
20	8.10	5.85	4.94	4.43	4.10	3.87	3.70	3.56	3.46	3.37	3.23	3.09	2.94	2.86	2.78	2.69	2.61	2.52	2.42
21	8.02	5.78	4.87	4.37	4.04	3.81	3.64	3.51	3.40	3.31	3.17	3.03	2.88	2.80	2.72	2.64	2.55	2.46	2.36
22	7.95	5.72	4.82	4.31	3.99	3.76	3.59	3.45	3.35	3.26	3.12	2.98	2.83	2.75	2.67	2.58	2.50	2.40	2.31
23	7.88	5.66	4.76	4.26	3.94	3.71	3.54	3.41	3.30	3.21	3.07	2.93	2.78	2.70	2.62	2.54	2.45	2.35	2.26
24	7.82	5.61	4.72	4.22	3.90	3.67	3.50	3.36	3.26	3.17	3.03	2.89	2.74	2.66	2.58	2.49	2.40	2.31	2.21
25	7.77	5.57	4.68	4.18	3.86	3.63	3.46	3.32	3.22	3.13	2.99	2.85	2.70	2.62	2.53	2.45	2.36	2.27	2.17
30	7.56	5.39	4.51	4.02	3.70	3.47	3.30	3.17	3.07	2.98	2.84	2.70	2.55	2.47	2.39	2.30	2.21	2.11	2.01
40	7.31	5.18	4.31	3.83	3.51	3.29	3.12	2.99	2.89	2.80	2.66	2.52	2.37	2.29	2.20	2.11	2.02	1.92	1.80
60	7.08	4.98	4.13	3.65	3.34	3.12	2.95	2.82	2.72	2.63	2.50	2.35	2.20	2.12	2.03	1.94	1.84	1.73	1.60
120	6.85	4.79	3.95	3.48	3.17	2.96	2.79	2.66	2.56	2.47	2.34	2.19	2.03	1.95	1.86	1.76	1.66	1.53	1.38
∞	6.63	4.61	3.78	3.32	3.02	2.80	2.64	2.51	2.41	2.32	2.18	2.04	1.88	1.79	1.70	1.59	1.47	1.32	1.00

Source: This table is abridged from E. S. Pearson and H. O. Hartley (eds.), Biometrika Tables for Statisticians, vol. 1, table 18, pp. 159 and 161, Cambridge University Press, 1962, and is reproduced with permission of the "Biometrika" trustees.

TABLE A.11   The Mann-Whitney Test

The tables on the next five pages are used to determine rejection rules for the Mann-Whitney two-sample test when the smaller sample contains 20 or fewer data points and the larger sample contains 40 or fewer data points. Here n_s is the number of data points in the smaller sample and n_L is the number of data points in the larger sample. If T is the sum of assigned ranks of the smaller sample, then the Mann-Whitney U-statistic is the number

$$U = n_L n_S + \frac{n_S(n_S+1)}{2} - T.$$

The table lists, for values of α = .005, .01, .025, .05, and .01, the largest number k for which $P(U \leqslant k) = P(U \geqslant n_L n_S - k) \leqslant \alpha$. This largest number k is denoted $U_A [n_L; n_S]$. The rejection rules are as follows:

1. Reject H: $A_S = A_L$ in favor of A: $A_S > A_L$ at significance level α if the calculated value of U is less than or equal to $U_\alpha [n_L; n_S]$. (Here a_S is the true median of the smaller sample, and A_L is the true median of the larger sample.)

2. Reject H: $A_S = A_L$ in favor of A: $A_S < A_L$ at significance level α if the calculated value of U is greater than or equal to $n_L n_S - U_\alpha [n_L; n_S]$.

3. Reject H: $A_S = A_L$ in favor of A: $A_S \neq A_L$ at significance level α if the calculated value of U is either less than or equal to $U_{\alpha/2} [n_L; n_S]$, or it is greater than or equal to $n_L n_S - U_{\alpha/2} [n_L n_S]$.

Note: T is the sum of assigned ranks of the smaller sample when the data points are ranked in order from smallest to largest.

Source: Roy C. Milton, "An extended table of critical values for the Mann-Whitney (Wilcoxon) two-sample statistic," *Journal of the American Statistical Association*, Vol. 59 (1964), pages 927–933. Reproduced by permission of the American Statistical Association.

										Values of $U_{.005}[N_L; n_S]$										
										n_S										
n_L	1	2	3	4	5	6	7	8	9	10	11	12	13	14	15	16	17	18	19	20
1	—																			
2	—	—																		
3	—	—	—																	
4	—	—	—	—																
5	—	—	—	—	0															
6	—	—	—	0	1	2														
7	—	—	—	0	1	3	4													
8	—	—	—	1	2	4	6	7												
9	—	—	0	1	3	5	7	9	11											
10	—	—	0	2	4	6	9	11	13	16										
11	—	—	0	2	5	7	10	13	16	18	21									
12	—	—	1	3	6	9	12	15	18	21	24	27								
13	—	—	1	3	7	10	13	17	20	24	27	31	34							
14	—	—	1	4	7	11	15	18	22	26	30	34	38	42						
15	—	—	2	5	8	12	16	20	24	29	33	37	42	46	51					
16	—	—	2	5	9	13	18	22	27	31	36	41	45	50	55	60				
17	—	—	2	6	10	15	19	24	29	34	39	44	49	54	60	65	70			
18	—	—	2	6	11	16	21	26	31	37	42	47	53	58	64	70	75	81		
19	—	0	3	7	12	17	22	28	33	39	45	51	57	63	69	74	81	87	93	
20	—	0	3	8	13	18	24	30	36	42	48	54	60	67	73	79	86	92	99	105
21	—	0	3	8	14	19	25	32	38	44	51	58	64	71	78	84	91	98	105	112
22	—	0	4	9	14	21	27	34	40	47	54	61	68	75	82	89	96	104	111	118
23	—	0	4	9	15	22	29	35	43	50	57	64	72	79	87	94	102	109	117	125
24	—	0	4	10	16	23	30	37	45	52	60	68	75	83	91	99	107	115	123	131
25	—	0	5	10	17	24	32	39	47	55	63	71	79	87	96	104	112	121	129	138
26	—	0	5	11	18	25	33	41	49	58	66	74	83	92	100	109	118	127	135	144
27	—	1	5	12	19	27	35	43	52	60	69	78	87	96	105	114	123	132	142	151
28	—	1	5	12	20	28	36	45	54	63	72	81	91	100	109	119	128	138	148	157
29	—	1	6	13	21	29	38	47	56	66	75	85	94	104	114	124	134	144	154	164
30	—	1	6	13	22	30	40	49	58	68	78	88	98	108	119	129	139	150	160	170
31	—	1	6	14	22	32	41	51	61	71	81	92	102	113	123	134	145	155	166	177
32	—	1	7	14	23	33	43	53	63	74	84	95	106	117	128	139	150	161	172	184
33	—	1	7	15	24	34	44	55	65	76	87	98	110	121	132	144	155	167	179	190
34	—	1	7	16	25	35	46	57	68	79	90	102	113	125	137	149	161	173	185	197
35	—	1	8	16	26	37	47	59	70	82	93	105	117	129	142	154	166	179	191	203
36	—	1	8	17	27	38	49	60	72	84	96	109	121	134	146	159	172	184	197	210
37	—	1	8	17	28	39	51	62	75	87	99	112	125	138	151	164	177	190	203	217
38	—	1	9	18	29	40	52	64	77	90	102	116	129	142	155	169	182	196	210	223
39	—	2	9	19	30	41	54	66	79	92	106	119	133	146	160	174	188	202	216	230
40	—	2	9	19	31	43	55	68	81	95	109	122	136	150	165	179	193	208	222	237

(Continued)

TABLE A.11   The Mann-Whitney Test (continued)
Values of $U_{.01}$ $[n_L;n_S]$

n_L	1	2	3	4	5	6	7	8	9	10	11	12	13	14	15	16	17	18	19	20
1	—																			
2	—	—																		
3	—	—	—																	
4	—	—	—	—																
5	—	—	—	0	1															
6	—	—	—	1	2	3														
7	—	—	0	1	3	4	6													
8	—	—	0	2	4	6	7	9												
9	—	—	1	3	5	7	9	11	14											
10	—	—	1	3	6	8	11	13	16	19										
11	—	—	1	4	7	9	12	15	18	22	25									
12	—	—	2	5	8	11	14	17	21	24	28	31								
13	—	0	2	5	9	12	16	20	23	27	31	35	39							
14	—	0	2	6	10	13	17	22	26	30	34	38	43	47						
15	—	0	3	7	11	15	19	24	28	33	37	42	47	51	56					
16	—	0	3	7	12	16	21	26	31	36	41	46	51	56	61	66				
17	—	0	4	8	13	18	23	28	33	38	44	49	55	60	66	71	77			
18	—	0	4	9	14	19	24	30	36	41	47	53	59	65	70	76	82	88		
19	—	1	4	9	15	20	26	32	38	44	50	56	63	69	75	82	88	94	101	
20	—	1	5	10	16	22	28	34	40	47	53	60	67	73	80	87	93	100	107	114
21	—	1	5	11	17	23	30	36	43	50	57	64	71	78	85	92	99	106	113	121
22	—	1	6	11	18	24	31	38	45	53	60	67	75	82	90	97	105	112	120	127
23	—	1	6	12	19	26	33	40	48	55	63	71	79	87	94	102	110	118	126	134
24	—	1	6	13	20	27	35	42	50	58	66	75	83	91	99	108	116	124	133	141
25	—	1	7	13	21	29	36	45	53	61	70	78	87	95	104	113	122	130	139	148
26	—	1	7	14	22	30	38	47	55	64	73	82	91	100	109	118	127	136	146	155
27	—	2	7	15	23	31	40	49	58	67	76	85	95	104	114	123	133	142	152	162
28	—	2	8	16	24	33	42	51	60	70	79	89	99	109	119	129	139	149	159	169
29	—	2	8	16	25	34	43	53	63	73	83	93	103	113	123	134	144	155	165	176
30	—	2	9	17	26	35	45	55	65	76	86	96	107	118	128	139	150	161	172	182
31	—	2	9	18	27	37	47	57	68	78	89	100	111	122	133	144	156	167	178	189
32	—	2	9	18	28	38	49	59	70	81	92	104	115	127	138	150	161	173	185	196
33	—	2	10	19	29	40	50	61	73	84	96	107	119	131	143	155	167	179	191	203
34	—	3	10	20	30	41	52	64	75	87	99	111	123	135	148	160	173	185	198	210
35	—	3	11	20	31	42	54	66	78	90	102	115	127	140	153	165	178	191	204	217
36	—	3	11	21	32	44	56	68	80	93	106	118	131	144	158	171	184	197	211	224
37	—	3	11	22	33	45	57	70	83	96	109	122	135	149	162	176	190	203	217	231
38	—	3	12	22	34	46	59	72	85	99	112	126	139	153	167	181	195	209	224	238
39	—	3	12	23	35	48	61	74	88	101	115	129	144	158	172	187	201	216	230	245
40	—	3	13	24	36	49	63	76	90	104	119	133	148	162	177	192	207	222	237	252

Values of $U_{.025}$ $[n_L;n_S]$

n_L	n_S																			
	1	2	3	4	5	6	7	8	9	10	11	12	13	14	15	16	17	18	19	20
1	—																			
2	—	—																		
3	—	—	—																	
4	—	—	—	0																
5	—	—	0	1	2															
6	—	—	1	2	3	5														
7	—	—	1	3	5	6	8													
8	—	0	2	4	6	8	10	13												
9	—	0	2	4	7	10	12	15	17											
10	—	0	3	5	8	11	14	17	20	23										
11	—	0	3	6	9	13	16	19	23	26	30									
12	—	1	4	7	11	14	18	22	26	29	33	37								
13	—	1	4	8	12	16	20	24	28	33	37	41	45							
14	—	1	5	9	13	17	22	26	31	36	40	45	50	55						
15	—	1	5	10	14	19	24	29	34	39	44	49	54	59	64					
16	—	1	6	11	15	21	26	31	37	42	47	53	59	64	70	75				
17	—	2	6	11	17	22	28	34	39	45	51	57	63	69	75	81	87			
18	—	2	7	12	18	24	30	36	42	48	55	61	67	74	80	86	93	99		
19	—	2	7	13	19	25	32	38	45	52	58	65	72	78	85	92	99	106	113	
20	—	2	8	14	20	27	34	41	48	55	62	69	76	83	90	98	105	112	119	127
21	—	3	8	15	22	29	36	43	50	58	65	73	80	88	96	103	111	119	126	134
22	—	3	9	16	23	30	38	45	53	61	69	77	85	93	101	109	117	125	133	141
23	—	3	9	17	24	32	40	48	56	64	73	81	89	98	106	115	123	132	140	149
24	—	3	10	17	25	33	42	50	59	67	76	85	94	102	111	120	129	138	147	156
25	—	3	10	18	27	35	44	53	62	71	80	89	98	107	117	126	135	145	154	163
26	—	4	11	19	28	37	46	55	64	74	83	93	102	112	122	132	141	151	161	171
27	—	4	11	20	29	38	48	57	67	77	87	97	107	117	127	137	147	158	168	178
28	—	4	12	21	30	40	50	60	70	80	90	101	111	122	132	143	154	164	175	186
29	—	4	13	22	32	42	52	62	73	83	94	105	116	127	138	149	160	171	182	193
30	—	5	13	23	33	43	54	65	76	87	98	109	120	131	143	154	166	177	189	200
31	—	5	14	24	34	45	56	67	78	90	101	113	125	136	148	160	172	184	196	208
32	—	5	14	24	35	46	58	69	81	93	105	117	129	141	153	166	178	190	203	215
33	—	5	15	25	37	48	60	72	84	96	108	121	133	146	159	171	184	197	210	222
34	—	5	15	26	38	50	62	74	87	99	112	125	138	151	164	177	190	203	217	230
35	—	6	16	27	39	51	64	77	89	103	116	129	142	156	169	183	196	210	224	237
36	—	6	16	28	40	53	66	79	92	106	119	133	147	161	174	188	202	216	231	245
37	—	6	17	29	41	55	68	81	95	109	123	137	151	165	180	194	209	223	238	252
38	—	6	17	30	43	56	70	84	98	112	127	141	156	170	185	200	215	230	245	259
39	0	7	18	31	44	58	72	86	101	115	130	145	160	175	190	206	221	236	252	267
40	0	7	18	31	45	59	74	89	103	119	134	149	165	180	196	211	227	243	258	274

(Continued)

TABLE A.11   The Mann-Whitney Test (continued)

Values of $U_{.05}$ $[n_L; n_S]$

n_L	1	2	3	4	5	6	7	8	9	10	11	12	13	14	15	16	17	18	19	20
1	—																			
2	—	—																		
3	—	—	0																	
4	—	—	0	1																
5	—	0	1	2	4															
6	—	0	2	3	5	7														
7	—	0	2	4	6	8	11													
8	—	1	3	5	8	10	13	15												
9	—	1	4	6	9	12	15	18	21											
10	—	1	4	7	11	14	17	20	24	27										
11	—	1	5	8	12	16	19	23	27	31	34									
12	—	2	5	9	13	17	21	26	30	34	38	42								
13	—	2	6	10	15	19	24	28	33	37	42	47	51							
14	—	3	7	11	16	21	26	31	36	41	46	51	56	61						
15	—	3	7	12	18	23	28	33	39	44	50	55	61	66	72					
16	—	3	8	14	19	25	30	36	42	48	54	60	65	71	77	83				
17	—	3	9	15	20	26	33	39	45	51	57	64	70	77	83	89	96			
18	—	4	9	16	22	28	35	41	48	55	61	68	75	82	88	95	102	109		
19	0	4	10	17	23	30	37	44	51	58	65	72	80	87	94	101	109	116	123	
20	0	4	11	18	25	32	39	47	54	62	69	77	84	92	100	107	115	123	130	138
21	0	5	11	19	26	34	41	49	57	65	73	81	89	97	105	113	121	130	138	146
22	0	5	12	20	28	36	44	52	60	68	77	85	94	102	111	119	128	136	145	154
23	0	5	13	21	29	37	46	54	63	72	81	90	98	107	116	125	134	143	152	161
24	0	6	13	22	30	39	48	57	66	75	85	94	103	113	122	131	141	150	160	169
25	0	6	14	23	32	41	50	60	69	79	89	98	108	118	128	137	147	157	167	177
26	0	6	15	24	33	43	53	62	72	82	92	103	113	123	133	143	154	164	174	185
27	0	7	15	25	35	45	55	65	75	86	96	107	117	128	139	149	160	171	182	192
28	0	7	16	26	36	46	57	68	78	89	100	111	122	133	144	156	167	178	189	200
29	0	7	17	27	38	48	59	70	82	93	104	116	127	138	150	162	173	185	196	208
30	0	7	17	28	39	50	61	73	85	96	108	120	132	144	156	168	180	192	204	216
31	0	8	18	29	40	52	64	76	88	100	112	124	136	149	161	174	186	199	211	224
32	0	8	19	30	42	54	66	78	91	103	116	128	141	154	167	180	193	206	218	231
33	0	8	19	31	43	56	68	81	94	107	120	133	146	159	172	186	199	212	226	239
34	0	9	20	32	45	57	70	84	97	110	124	137	151	164	178	192	206	219	233	247
35	0	9	21	33	46	59	73	86	100	114	128	141	156	170	184	198	212	226	241	255
36	0	9	21	34	48	61	75	89	103	117	131	146	160	175	189	204	219	233	248	263
37	0	10	22	35	49	63	77	91	106	121	135	150	165	180	195	210	225	240	255	271
38	0	10	23	36	50	65	79	94	109	124	139	154	170	185	201	216	232	247	263	278
39	1	10	23	38	52	67	82	97	112	128	143	159	175	190	206	222	238	254	270	286
40	1	11	24	39	53	68	84	99	115	131	147	163	179	196	212	228	245	261	278	294

Values of $U_{.10}[n_L; n_S]$

n_L	1	2	3	4	5	6	7	8	9	10	11	12	13	14	15	16	17	18	19	20
1	—																			
2	—	—																		
3	—	0	1																	
4	—	0	1	3																
5	—	1	2	4	5															
6	—	1	3	5	7	9														
7	—	1	4	6	8	11	13													
8	—	2	5	7	10	13	16	19												
9	0	2	5	9	12	15	18	22	25											
10	0	3	6	10	13	17	21	24	28	32										
11	0	3	7	11	15	19	23	27	31	36	40									
12	0	4	8	12	17	21	26	30	35	39	44	49								
13	0	4	9	13	18	23	28	33	38	43	48	53	58							
14	0	5	10	15	20	25	31	36	41	47	52	58	63	69						
15	0	5	10	16	22	27	33	39	45	51	57	63	68	74	80					
16	0	5	11	17	23	29	36	42	48	54	61	67	74	80	86	93				
17	0	6	12	18	25	31	38	45	52	58	65	72	79	85	92	99	106			
18	0	6	13	20	27	34	41	48	55	62	69	77	84	91	98	106	113	120		
19	1	7	14	21	28	36	43	51	58	66	73	81	89	97	104	112	120	128	135	
20	1	7	15	22	30	38	46	54	62	70	78	86	94	102	110	119	127	135	143	151
21	1	8	15	23	32	40	48	56	65	73	82	91	99	108	116	125	134	142	151	160
22	1	8	16	25	33	42	51	59	68	77	86	95	104	113	122	131	141	150	159	168
23	1	9	17	26	35	44	53	62	72	81	90	100	109	119	128	138	147	157	167	176
24	1	9	18	27	36	46	56	65	75	85	95	105	114	124	134	144	154	164	174	184
25	1	9	19	28	38	48	58	68	78	89	99	109	120	130	140	151	161	172	182	193
26	1	10	20	30	40	50	61	71	82	92	103	114	125	136	146	157	168	179	190	201
27	1	10	21	31	41	52	63	74	85	96	107	119	130	141	152	164	175	186	198	209
28	1	11	21	32	43	54	66	77	88	100	112	123	135	147	158	170	182	194	206	217
29	2	11	22	33	45	56	68	80	92	104	116	128	140	152	164	177	189	201	213	226
30	2	12	23	35	46	58	71	83	95	108	120	133	145	158	170	183	196	209	221	234
31	2	12	24	36	48	61	73	86	99	111	124	137	150	163	177	190	203	216	229	242
32	2	13	25	37	50	63	76	89	102	115	129	142	156	169	183	196	210	223	237	251
33	2	13	26	38	51	65	78	92	105	119	133	147	161	175	189	203	217	231	245	259
34	2	13	26	40	53	67	81	95	109	123	137	151	166	180	195	209	224	238	253	267
35	2	14	27	41	55	69	83	98	112	127	141	156	171	186	201	216	230	245	260	275
36	2	14	28	42	56	71	86	100	115	131	146	161	176	191	207	222	237	253	268	284
37	2	15	29	43	58	73	88	103	119	134	150	166	181	197	213	229	244	260	276	292
38	2	15	30	45	60	75	91	106	122	138	154	170	186	203	219	235	251	268	284	301
39	3	16	31	46	61	77	93	109	126	142	158	175	192	208	225	242	258	275	292	309
40	3	16	31	47	63	79	96	112	129	146	163	180	197	214	231	248	265	282	300	317

TABLE A.12 Exact Critical Values for the Kruskal-Wallis Nonparametric Analysis of Variance Test *(when samples are of equal size)*

Number of Samples *m*	Size of Each Sample *n*	α = .01 Reject if *H* Exceeds	Exact P (Type I Error)	α = .05 Reject if *H* Exceeds	Exact P (Type I Error)	α = .10 Reject if *H* Exceeds	Exact P (Type I Error)
3	2	—	—	4.571	.0667	4.571	.0667
						3.714	.2000
	3	7.200	.0036	5.600	.0500	4.622	.1000
		6.489	.0107			4.335	.1321
	4	7.654	.0076	5.692	.0487	4.654	.0966
		7.539	.0107	5.654	.0545	4.500	.1042
	5	8.000	.0095	5.780	.0488	4.560	.0995
		7.980	.0105	5.660	.0509	4.500	.1015
	6	8.222	.0099	5.801	.0490	4.643	.0987
		8.187	.0102	5.719	.0502	4.538	.1010
	7	8.378	.0099	5.818	.0491	4.594	.0993
		8.334	.0101	5.766	.0505	4.549	.1007
	∞	9.210	.0100	5.991	.0500	4.605	.1000
4	2	6.667	.0095	6.167	.0381	5.667	.0762
		6.167	.0381	6.000	.0667	5.500	.1143
	3	8.539	.0084	7.000	.0435	6.026	.0978
		8.436	.0108	6.897	.0502	5.974	.1027
	4	9.287	.0100	7.235	.0492	6.088	.0990
		9.265	.0101	7.213	.0507	6.066	.1003
	∞	11.345	.0100	7.815	.0500	6.251	.1000
5	2	8.291	.0095	7.418	.0487	6.982	.0910
		8.073	.0127	7.309	.0635	6.873	.1016
	3	10.200	.0099	8.333	.0495	7.333	.0992
		10.167	.0104	8.300	.0503	7.300	.1001
	∞	13.277	.0100	9.488	.0500	7.779	.1000

Source: William V. Gehrlein and Erwin M. Saniga, "Some exact critical values for the Kurskal-Wallis statistic," *Journal of Quality Technology,* Vol. 10 (1978), page 74. Reproduced by permission of the American Society for Quality Control.

TABLE A.13 The Spearman-Rank Correlation Test

n	$r_s[n;.001]$	$r_s[n;.005]$	$r_s[n;.01]$	$r_s[n;.025]$	$r_s[n;.05]$	$r_s[n;.10]$
4	—	—	—	—	0.8000	0.8000
5	—	—	0.9000	0.9000	.8000	.7000
6	—	0.9429	0.8857	0.8286	0.7714	0.6000
7	0.9643	.8929	.8571	.7450	.6786	.5357
8	.9286	.8571	.8095	.6905	.5952	.4762
9	.9000	.8167	.7667	.6833	.5833	.4667
10	.8667	.7818	.7333	.6364	.5515	.4424
11	0.8455	0.7545	0.7000	0.6091	0.5273	0.4182
12	.8182	.7273	.6713	.5804	.4965	.3986
13	.7912	.6978	.6429	.5549	.4780	.3791
14	.7670	.6747	.6220	.5341	.4593	.3626
15	.7464	.6536	.6000	.5179	.4429	.3500
16	0.7265	0.6324	0.5824	0.5000	0.4265	0.3382
17	.7083	.6152	.5637	.4853	.4118	.3260
18	.6904	.5975	.5480	.4716	.3994	.3148
19	.6737	.5825	.5333	.4579	.3895	.3070
20	.6586	.5684	.5203	.4451	.3789	.2977
21	0.6455	0.5545	0.5078	0.4351	0.3688	0.2909
22	.6318	.5426	.4963	.4241	.3597	.2829
23	.6186	.5306	.4852	.4150	.3518	.2767
24	.6070	.5200	.4748	.4061	.3435	.2704
25	.5962	.5100	.4654	.3977	.3362	.2646
26	0.5856	0.5002	0.4564	0.3894	0.3299	0.2588
27	.5757	.4915	.4481	.3822	.3236	.2540
28	.5660	.4828	.4401	.3749	.3175	.2490
29	.5567	.4744	.4320	.3685	.3113	.2443
30	.5479	.4665	.4251	.3620	.3059	.2400

Source: Gerald J. Glasser and Robert F. Winter, "Critical values of the coefficient of rank correlation for testing the hypothesis of independence," *Biometrika*, Vol. 48 (1961), page 447. Reproduced by permission of the Biometrika Trustees.

Answers to Selected Odd-Numbered Exercises

CHAPTER 1

1. 1.

Category	Degrees
HRA	90
MFH	108
SFH	144
MH	18

1. 3.

Company	Degrees
F	75
H,U	75
F,C	45
Trans.	30
E	15
S,I	22.5
M	7.5
Tax.	90

1. 5.

Company	Degrees
A	148.50
Clark	40.50
OSLA	31.50
AP	29.25
Com.	27.00
U	20.25

APCO	18.00
O	45.00

1. 7.

Problem	Degrees
I	140.40
U	16.92
E	101.88
S	52.50
C	26.28
O	22.32

1. 9.

Company	Degrees
GM	205.49
F	104.12
C	42.73
AM	7.67

1.11. (a)

Interval	Frequency
4.45–4.95	2
4.95–5.45	16
5.45–5.95	11
5.95–6.45	5
6.45–6.95	0
6.95–7.45	1

1.19. **(a)**

Dollar Value	Number of Stocks
below 10	51
below 20	149
below 30	250
below 40	401
below 50	449
below 60	474
below 70	487
below 80	495
below 90	499
below 100	500

1.21. **(a)**

Score	Number of Units
below 19.5	30
below 39.5	45
below 59.5	50
below 79.5	60
below 100.5	100

1.33.

Airline	Degrees
A	48.00
B	18.67
C	16.00
D	58.67
E	53.33
Na	13.33
No	18.67
P	2.67
T	34.67
U	77.33
W	18.67

1.35.

P/E Ratio	Frequency
0–5	13
6–10	15
11–15	16
16–20	8
21–25	2
26–30	2
31–35	4

1.37

Forecast Range	Frequency
134.0–135.9	3
136.0–137.9	4
138.0–139.9	14
140.0–141.9	18
142.0–143.9	8

CHAPTER 2

2. 1. **(a)** 13
(b) 14
(c) 14
(d) 12

2. 3. **(a)** 444
(b) 460
(c) no mode
(d) 525

2. 5. **(a)** $335,266.67
(b) $244,000
(c) $530,000

2. 7. **(a)** 7
(b) 4
(c) 4
(d) 8.5

2. 9. **(a)** 29.895
(b) 29.995
(c) 34.995
(d) 49.995

2.11. **(a)** 5.335
(b) 5.338
(c) 4.995
(d) 4.995

2.17. **(a)** -7
(b) -42
(c) -42

2.19. 39

2.21. 9.35

2.23. **(a)** 1.87
(b) 1.50
(c) 6
(d) 1.87

2.25. **(a)** 230.07
(b) 230.07
(c) 197.07
(d) 810

2.27. $221,097.10

2.29. 0.49135%

2.31. **(a)** 7.4265
(b) 5.9633

2.33. $3987.18

2.35. **(a)** 32%
(b) 10%
(c) peanuts

2.37. Yes

2.39. 16%

2.41. 3.78%

2.43. 97.22%

2.45. 40 days

2.47. **(a)** $\frac{1}{9} = 11.1\%$
(b) 96%

2.49. 5.54%

2.51. $\frac{1}{36} = 2.78\%$

2.53. 3.55%

2.55.

Applicant	(a)	(b)	(c)	Sums
A	1.35	0.99	−0.31	2.03
B	1.08	0.99	0.00	2.07
C	0.00	−0.74	−1.87	−2.61
D	0.00	0.99	1.25	2.24
E	−1.08	−1.24	0.00	−2.32
F	−1.35	−0.99	0.93	−1.41

(d) Applicant D

2.57.

Rank	Applicant	Standard Score
1	#1	1.00
2	#3	0.60
3	#4	−0.33
4	#2	−0.50

2.59. **(a)** 27.90
(b) 29.50
(c) 30.00
(d) 35.50
(e) 12.038
(f) 8.596
(g) 49.00

(d) 19.5
(e) 7.15
(f) 6
(g) 25

2.63. 2.3%

2.65. 93.75% of the time

2.67. 2.87%

2.69. 18,107 items

2.71. $51\frac{1}{100}$ of 1%

2.61. **(a)** 17
(b) 17
(c) 20

2.73. 8.16%

2.75. Yes

2.77. 80.25%

2.79. 96%

2.81. 6.38%

2.83. 1,248

2.85. 4%

CHAPTER 3

3. 1. **(c)** $W \cup L = \{2,3,7,11,12\}$
$W \cap L = \phi$
$W^c = \{2,3,4,5,6,8,9,10,12\}$
$L^c = \{4,5,6,7,8,9,10,11\}$
$W^c \cup L^c = S$
$W^c \cap L^c = \{4,5,6,8,9,10\}$
$E^c = \{3,5,7,9,11\} = D$
$E^c \cap D = D$
$E^c \cup D = D$

3. 3. **(a)** $\frac{1}{6}$ **(g)** $\frac{1}{8}$
(b) $\frac{1}{18}$ **(h)** $\frac{5}{9}$
(c) $\frac{2}{9}$ **(i)** 0
(d) $\frac{1}{9}$ **(j)** $\frac{1}{3}$
(e) $\frac{1}{2}$ **(k)** 0
(f) $\frac{1}{2}$ **(l)** 1

3. 5. **(a)** $\frac{1}{2}$ **(c)** $\frac{1}{8}$
(b) $\frac{5}{8}$ **(d)** $\frac{3}{8}$

3. 7. $P(T \cap W) = -7/40$

3. 9. **(a)** $\frac{4}{9}$
(b) $\frac{1}{9}$
(c) 1
(d) $\frac{1}{2}$
(3) 0
(f) $\frac{1}{9}$

3.11. $\frac{3}{100}$

3.13 **(b)** 1.85%
(c) 40.32%

3.15. **(b)** 59%
(c) 94.9%
(d) 59%

3.17. **(b)** 71%
(c) 2.1%
(d) 71%

3.19. **(a)** 5.36×10^{20}
     (b) 1.304×10^{39}

3.21. **(a)** 15,504
     (b) one
     (c) $1/15,504 = .0000645$

3.23. 6.295×10^{26}

3.25. **(a)** .2583

     (b)

Sample Size	Acceptance Probability
1	.7368
4	.2583
7	.0681
10	.0108

  (c) Actual No. of

Unacceptable Items	Acceptance Probability
0	1.0000
1	.7895
3	.4696
5	.2583
7	.1277
9	.0542
11	.0181

3.27. **(a)** .9265

  (b) Actual No. of

Unacceptable Items	Acceptance Probability
0	1.0000
2	.9944
4	.9684
6	.9265
8	.8724
10	.8096
12	.7411
14	.6693
16	.5965
18	.5247
20	.4552
22	.3892
24	.3279

26	.2717
28	.2212
30	.1766

3.29. **(a)** true
     (b) true
     (c) false
     (d) false

3.31. **(a)** $3/4$
     (b) 1
     (c) $1/3$
     (d) 0

3.33. **(b)** 35%
     (c) 22.86%

3.35. **(a)** .2858
     (b) .2942

3.37. **(a)** .3505

     (b)

Sample Size	Acceptance Probability
1	.7222
2	.5098
3	.3505
4	.2337
5	.1502
6	.0924
7	.0539
8	.0294
9	.0147
10	.0065

  (c) Actual No. of

Unacceptable Items	Acceptance Probability
0	1.0000
1	.8333
2	.6863
3	.5576
4	.4461
5	.3505
6	.2696
7	.2022
8	.1471
9	.1029
10	.0686

CHAPTER 4

4. 1.

Interval	Probability
4.45–4.95	.0571
4.95–5.45	.4571
5.45–5.95	.3143
5.95–6.45	.1429
6.45–6.95	.0000
6.95–7.45	.0286
	1.0000

4. 3.

Company	Probability
A	.4125
Clark	.1125
OSLA	.0875
AP	.0813
Com	.0750
U	.0563
APCO	.0500
O	.1250
	1.0000

4. 5.

Problem	Probability
I	.3900
U	.0470
E	.2830
S	.1450
C	.0730
O	.0620
	1.0000

4. 7.
- **(a)** $n = 15, \pi = 0.1$
- **(b)** $0,1,2,\ldots,15$
- **(c)** 1.5
- **(d)** 1.162
- **(e)**

k	$P(Y = k)$
0	.2059
1	.3431
2	.2669
3	.1285
4	.0429
5	.0105
6	.0019
7	.0003
8–15	.0000 each

4. 9.
- **(a)** $n = 10, \pi = 0.70$
- **(b)** $0,1,2,\ldots,10$
- **(c)** 7

(d) 1.499

(e)

k	$P(A = k)$
0	.0000
1	.0001
2	.0014
3	.0090
4	.0368
5	.1029
6	.2001
7	.2668
8	.2335
9	.1211
10	.0282

4.11.
- **(a)** $n = 20, \pi = .05$
- **(b)** .3585
- **(c)** .3773
- **(d)** .0003
- **(e)** 1.0000
- **(f)** .0003

4.13.
- **(a)** .2380
- **(b)** .2068
- **(c)** .0099
- **(d)** .4050

4.15. .2686

4.17. .2892

4.19 70.16 minutes

4.21.
- **(a)** .0344
- **(b)** 91.08%

4.23. 170.5

4.25 .4407

4.27. 76.64; 43.36

4.29. no

4.31. 15.87%

4.33. 98.8

4.35. $1.45

4.37. 81.85%

4.39. 93.7%

4.41. .1358

4.43.
- **(a)** .5489
- **(b)** .8531
- **(c)** .9998
- **(d)** 1.0000

4.45. (a) .2485
       (b) .0012
       (c) .0000

4.47. (a) .1847
       (b) .2650

4.49. .0183

4.51. (a) .4628
       (b) .5152
       (c) .3023
       (d) .6604
       (e) .3644

4.53. .7769

4.55. .1293

4.57. (a) .0672
       (b) .0362

4.59. .3128

4.61. .2583

4.63. .9265

4.65. .6404

4.67. .1653

4.69. (a) .1852 mile
       (b) .0672

4.71. (a) 8:31 A.M.
       (b) 3.46 minutes
       (c) $\frac{1}{6}$ = .1667

4.73. (b) 50.4
       (c) 60
       (d) .0080
       (e) .2977

4.75. (a) .6217
       (b) .8962

4.77. .00013

4.79. (a) .1513
       (b) .0084

4.81. (a) 92.8 minutes
       (b) 96.4 minutes
       (c) 103.3 minutes

4.83. (a) .9082
       (b) .5000
       (c) .0918
       (d) 141.6

4.85. 1,767

4.87. (a) 10.08 oz.
       (b) 1.43%

4.89. $258.50

4.91. x_{AB} = 82.8
       x_{BC} = 75.2
       x_{CD} = 64.8
       x_{DF} = 57.2

4.93. (a) .092
       (b) .051
       (c) 15 minutes

4.95. .3505

4.97. 12

4.99. 4.01%

4.101. 1.646

4.103. $1.706

CHAPTER 5

5. 3. (a) 0.52
       (b) 1.04
       (c) 1.44
       (d) 1.88
       (e) 2.17
       (f) 2.41
       (g) 2.88
       (h) 3.09

5. 5. .0793

5. 7. .0286

5. 9. $16,182.69 to $16,266.75

5.11. 132.4 to 136.2

5.13. (a) 1.974 to 2.026
       (b) 1.964 to 2.036
       (c) 1.960 to 2.040

5.15. 7.56 to 11.14

5.17. 5.59 to 6.96

5.19. $3.53 to $4.65

5.21. 14.62 to 19.38

5.23. 20.95 to 23.05

5.25. 8.47 to 13.53

5.27. 11.99 to 14.01 years

5.29. 16.83 to 27.17 thousand

5.31. 9.38 to 17.82

5.33. 3.25 to 5.27

5.35. 160.93 to 287.02

5.37. 1.38 to 3.31

5.39. .39 to .67

5.41. .701 to .829

5.43. .527 to .673

5.45. (a) .370 to .408
(b) .367 to .412
(c) .359 to .419

5.47. .0619 to .1791

5.49. 140

5.51. 82

5.53. 246

5.55. 16,512

5.57. 47

5.59. 317

5.61. 246

5.63. (a) 3.168, 3.132
(b) 10, 11

5.65. (a) .0956, .0244
(b) 5

5.67. (a) 110, 125, 57, 143, 118,
90, 128, 133, 40, 91
(b) $0.973
(c) $0.5918
(d) $1.00

5.71. 952

5.73. (a) 2.29 to 2.71
(b) .243 to .580
(c) 67
(d) 92.98%

5.75. 2.194 to 2.546

5.77. (a) 5.64 to 7.66
(b) 5.35 to 8.25

5.79. .116 to .184

5.81. (a) .0225 to .0375
(b) 27,225

5.83. (a) .549 to .651
(b) 2,401

5.85. (a) 25.85, 8.15
(b) Jan-Apr, 1979, 1981

5.87. .648 to .881

CHAPTER 6

6. 1. (a) low
(b) high

6. 3. (a) high
(b) low

6. 5. $z = -1.8$, yes

6. 7. $z = 2.95$, yes

6. 9. $t = 2.67$, yes

6.11. $t = .583$, yes

6.13. $z = -6$, no

6.15. $t = .1943$, no

6.17. (a) $\beta(2.0) = .0314$
(b) $\beta(2.1) = .1949$
(c) $B(2.2) = .5557$
(d) $\beta(2.3) = .8729$

6.19. (a) $\beta(3.96) = .3446$
(b) $\beta(3.98) = .7815$
(c) $\beta(4.03) = .5753$
(d) $\beta(4.05) = .1611$

6.21. (a) $\beta(21.3) = .0000$
(b) $\beta(21.6) = .0314$
(c) $\beta(21.8) = .3859$
(d) $\beta(21.9) = .6879$

6.23. $z = -1.38$, no

6.25. $t = .862$, no

6.27. $z = 1.074$, no

6.29. $t = 1.28$, no

6.31. $z = 4.136$, yes

6.33. $t = 1.12$, no

6.35. $t = 1.90$, yes

6.37. $t = 2.63$, yes

6.39. $t = 0.62$, no

6.41. $t = 2.82$, yes

6.43. $t = 0.93$, no

6.45. $\chi^2 = 5.83$, yes

6.47. $F = 5$, no

6.49. $F = 1.306$, yes

6.51. $F = 2.32$, yes

6.53. $z = -0.47$, yes

6.55. **(a)** $F = 7$, no
       (b) $F = 7$, no

6.57. $z = 1.85$, yes

6.59. **(a)** $z = .6$, no
       (b) $z = 1.90$, yes

6.61. $z = .459$, no

6.63. $z = .845$, no

6.65. **(a)** $t = -.80$, no
       (b) $\chi^2 = 60$, yes

6.67. **(a)** $z = -7.17$, yes
       (b) $F = 1.44$, no

6.69. $z = 1.37$, no

6.71. $F = 17.19$, no

6.73. $t = 3.54$, yes

6.75. $z = -3.51$, yes

6.77. $z = 3.82$, yes

6.79. **(a)** $z = 1.87$, no
       (b) $z = 1.87$, yes
       (c) $\beta(1.40) = .1660$
              $\beta(1.30) = .4443$
              $\beta(1.20) = .7517$
              $\beta(1.10) = .9345$
       (d) $\beta(1.40) = .0485$
              $\beta(1.30) = .2033$
              $\beta(1.20) = .4960$
              $\beta(1.10) = .7939$

6.81. $t = .0412$, no

6.83. $t = 2.66$, yes

CHAPTER 7

7. 1. **(a)** $8.90 = 7.02 + 1.88$
       (b) $F = 24.83$, there are

7. 3. **(a)** $nVB = 19.25$, $VW = 394$
       (b) $F = 0.39$, no

7. 5. **(a)** $nVB = 8.26$, $VW = 34.10$
       (b) $F = 2.18$, yes

7. 7. $F = 24.83$, they are

7. 9. $F = 0.39$, no

7.11. $F = 2.18$, they were

7.13. $F = 0.46$, no

7.15. **(a)** $F = 23.64$, yes
       (b) $F = 14.77$, yes

7.17. **(a)** $F = 69.93$, yes
       (b) $F = 0.94$, no

7.19. **(a)** $F = 2.41$, no
       (b) $F = 1.03$, no

7.21. **(a)** $nVB = 85.75$,
              $VW = 585.875$
       (b) $F = 1.54$, they do

7.23. **(a)** $F = 2.56$, no
       (b) $F = 5.38$, no

7.25. $F = 0.36$, yes

7.27. **(a)** $F = 115.43$, they do
       (b) $F = 17.37$, they do

CHAPTER 8

8. 1. (a) $y = 1.75 + .25x$
(b) $y = -2 + 1.5x$
(c) $y = 10 - x$

8. 3. (a) $(1,8)$ and $(4,14)$
(b) $(0,8)$ and $(2,2)$
(c) $(2,5)$ and $(5,17)$

8. 5. $y = 1.47 + .10x$
$(5,1.97)$

8. 7. (b) $y = 3.33 + 0.67x$
(c) $(3,5.33)$ and $(6,7.33)$

8. 9. (b) $y = -3.60 + 2.79x$
(c) $(6,13.14)$ and $(11,27.09)$

8.11. (b) $y = 4.2 + 0.96x$
(c) $(2,6.12)$ and $(8,11.88)$

8.13. (b) $y = 37.59 - 7.13x$
(c) $(2,23.33)$ and $(5,1.94)$

8.15. (b) $y = 7.78 + 0.28x$
(c) $(20,13.46)$ and $(60,24.82)$

8.17. 92.31%

8.19. 99.49%

0.21. 88.57%

8.23. 57.45%

8.25. 97.78%

8.27. (b) 0.00
(c) 0.00%

8.29. (b) 10.68%

8.31. $t = 6.93$, yes

8.33. $t = 27.93$, yes

8.35. $t = 3.94$, yes

8.37. $t = 0$, no

8.39. 5.988 to 8.679

8.41. 16.16 to 21.28

8.43. 2.19 to 5.52

8.45. 2.58 to 3.00

8.47. 6.788 to 7.875

8.49. 17.70 to 19.74

8.51. (a) 79.71%
(c) $t = 5.61$, yes

8.53. (a) 69.8%
(c) $t = -3.04$, yes
(d) $y = 28.82 - 1.22x$
(e) 1.55 to 31.61

8.55. (a) 52.63%
(c) $t = 4.47$, yes
(d) $y = .303 + 1.771x$
(e) .634 to 4.506

8.57. (a) 88.81%
(c) $t = 6.90$, yes
(d) $y = 1.063 + 5.722x$
(e) 23.57 to 35.77

8.59. (a) 93.46%
(c) $t = -6.55$, yes
(d) $y = 2148 - 800x$
(e) 282.08 to 493.92

8.61. (a) 95.29%
(c) $t = -7.79$, yes
(d) $y = 112.35 - 5.29x$
(e) 22.40 to 13.40

8.63. (a) 90.34%
(c) $t = 6.118$, yes
(d) $y = -0.65 + 8.70x$
(e) 43.25 to 77.18

8.65. (a) 0.08%
(c) $y = 5.221 + 0.075x$
(d) 5.53%
(e) nothing; see (a)

8.67. (a) 25.44%
(c) $y = 89.41 + 64.78x$
(d) $415,913

8.69. (a) 99.49%
(c) $y = -402.10 + 0.0801x$

8.71. (a) 78.98%
(c) $t = 4.33$, yes
(d) $y = 98.387 + 1.028x$
(e) 110.78 to 157.97

8.73. (a) 97.22%
(b) $t = 10.47$, yes
(d) $y = -7.5 + 0.7x$
(e) $(10,-.5)$ and $(40,20.5)$
(f) 10.47 to 14.43

CHAPTER 9

9. 1. (a) 56.66%
    (b) 79.40%
    (c) 95.01%
    (d) $F = 28.54$, yes

9. 3. (a) 23.53%
    (b) 80.19%
    (c) 87.79%
    (d) $F = 25.17$, yes

9. 5. (a) 0.009%
    (b) 25.44%
    (c) 58.183%

9. 7. (a) $y = 2.51 - 0.95x + 5.60w$
    (b) \$19,239.30

9. 9. (a) $y = 11.35 - 3.27x + 0.81w$
    (b) 35.58

9.11.

source	df	SS	MSS	F
R	2	2703.6	1351.8	15.60
E	7	606.4	86.6	
T	10	3310		

    (a) $F = 15.60$, valid
    (b) $F = 29.27$, yes

9.13.

source	df	SS	MSS	F
R	2	225.64	112.82	44.07
E	3	7.69	2.56	
T	6	233.33		

    (a) $F = 44.07$, valid
    (b) $F = 33.09$, yes

9.15. (a) 94.10% (x), 90.67% (w)
    (b) 94.17%
    (c) $y = 123.7 + 14.1x - 2w$
    (d) 236.3
    (e) 107.7
    (f) 123.7

9.17. (a) 4.23%
    (b) 1.56%
    (c) 4.28%
    (d)

source	df	SS	MSS	F
R	2	3.41	1.71	0.11
E	5	76.23	15.25	
T	8	79.64		

    no
    (e) $F = .0027$, no

CHAPTER 10

10. 1. $247,923

10. 3. 37,620 tons

10. 5. $53,356.21

10. 7. 133.76

10. 9. 131.66

10.11. 131.73

10.13. (b) 82.78%
(c) $y = 321.33 + 13.00x$

10.15. (b) 30.19%
(c) $y = .691 - .0065x$

10.21. (a) $-, -, -, 4.57, 3, 29,$
$2.57, 3.29, 4.71, -,$
$-, -$
(b) $10, 9.2, 7.92, 6.35, 4.61,$
$3.17, 2.30, 1.78, 2.27,$
$3.76, 5.86, 8.31$
(c) It doesn't exist during the ski season.

10.29. (a) 48.47%
(b) $y_k = 31 + 0.7y_{k-1}$
(c) 49.62%
(d) $y_k = 26.39 + .60y_{k-1} + .14y_{k-2}$

10.31. (a) 2.5545 million
(b) 90,978 units
(c) 1,573 hoses
(d) 9817.5 and 10,234 tons
(e) 2.83 million

10.33. (b) 74.23%
(c) $y = 57.27 + 0.82x$
(m) 93.62%
(n) $y_k = 1.7 + .99y_{k-1}$

10.35. (b) 70.23%
(c) $y = 76.19 - 0.11x$

CHAPTER 11

11. 1. 0.64

11. 3. 0.949

11. 5. 0.021

11. 7. (a) 0.762
(b) 0.190
(c) 0.048

11. 9. 0.125

11.11. 0.462

11.13. (a) 73.36%
(b) 32%

11.15. 5.3592 to 5.5432

11.17. (a) 72%
(b) 48.6%

11.19. (a) 12.18%
(b) 21.73%
(c) 4.14%

11.21. (a) 22%
(b) 63.6%

11.23. (a) 8.8%
(b) 68.2% from Foster

11.25. (a) 0.817
(b) 0.1212
(c) 0.0027

CHAPTER 12

12. 1. $z = -.354$, no

12. 3. (a) $\theta = 1, n = 7$, no
(b) $\theta = 5, n = 8$, no

12. 5. $\theta = 7, n = 8$, yes

12. 7. $\theta = 11, n = 15$, no

12. 9. (a) $U = 23$, no
(b) $U = 13$, no

12.11. (a) $U = 29.5$, no
(b) $U = 42$, no

12.13. $U = 59$, yes

12.15. $U = 31$, no

12.17. $H = 4.88$, yes

12.19. $H = 7.06$, no

12.21. **(a)** $r_S = 0.0268$, no

12.23. **(a)** $r_S = .7605$, yes

12.25. $r_S = -0.2143$, no

12.27. $r_S = .0500$, no

12.29. **(a)** $U = 7.5$, it is not
            (b) $U = 7.5$, it is not

(d) $\theta = 2, n = 2$, no
         $(\alpha = .25)$

(e) $\theta = 5, n = 5$, yes

(f) $\alpha = .0312$

12.31. $\theta = 7, n = 8$, no

12.33. $\theta = 3, n = 6$, no

12.35. $H = 2.45$, no

12.37. $H = 1.16$, no

12.39. $r_S = -.2063$, no

CHAPTER 13

13. 1. $\chi^2 = 16.67$, yes

13. 3. $\chi^2 = 49.72$, yes

13. 5. $\chi^2 = 1.78$, yes

13. 7. $\chi^2 = 121.25$, yes

13. 9. $\chi^2 = 49.14$, yes

13.11. By 13.33%

13.13. By 17.65%

13.15. No

13.17. By 12.6%

13.19. By 15%

13.21. $\chi^2 = 3.17$, yes

13.23. $\chi^2 = 19.17$, yes

13.25. $\chi^2 = 6.92$, yes

13.27. $\chi^2 = 41.5$, yes

13.29. $\chi^2 = .301$, yes

13.31. $z = -1.39$, randomly

13.33. $z = -2.15$, yes

13.35. **(a)** $\chi^2 = 1.34$, no
            (b) 0%

13.37. **(a)** $\chi^2 = 190.5$, yes
            (b) 15%

13.39. $z = -3.02$, no

13.41. $\chi^2 = 22.19$, no

13.43. $\chi^2 = 2.26$, yes

13.45. $\chi^2 = 1.30$, yes

Index

X

Z